Ullmann's Pharmaceuticals

Volume 1

Ullmann's Pharmaceuticals

Volume 1

Edited by

Axel Kleemann and Bernhard Kutscher

WILEY-VCH

The Editors

Professor Dr. Axel Kleemann
Amselstrasse 2
63454 Hanau
Germany

Professor Dr. Bernhard Kutscher
Stresemannstraße 9
63477 Maintal
Germany

Cover
Cover Design: Gunther Schulz,
 Fußgönheim, Germany
Cover Image: GettyImages / Jorg Greuel

Library of Congress Card No.:
applied for

British Library Cataloguing-in-Publication Data
A catalogue record for this book is available from the
British Library.

**Bibliographic information published by the Deutsche
Nationalbibliothek**
The Deutsche Nationalbibliothek lists this publication in
the Deutsche Nationalbibliografie; detailed bibliographic
data are available on the Internet at <http://dnb.d-nb.de>.

© 2022 Wiley-VCH GmbH, Boschstraße 12, 69469
Weinheim, Germany

Print ISBN: 978-3-527-34252-5
ePDF ISBN: 978-3-527-80732-1
ePub ISBN: 978-3-527-80733-8

Typesetting Straive, Chennai, India
Printing and Binding CPI Group (UK) Ltd,
Croydon, CR0 4YY

Printed on acid-free paper

C121711_030322

Preface

The field of drug development is rapidly expanding and the pharmaceutical market is still growing worldwide.

How to describe the complex achievements of the last centuries and provide a systematic description of the entire armamentarium of existing drugs in a way that both professionals, but also interested laymen, can understand progress, new mechanism of action, or production methods including references and patents? How to overlook an entirely new and innovative field of therapeutics such as kinase inhibitors with their beneficial role in oncology, but also used as immunosuppressants against inflammatory diseases or as chemoprotectives?

In 1976, the World Health Organization (WHO) established the Anatomical Therapeutic Chemical (ATC) Classification System that classifies the active pharmaceutical ingredients (APIs) of drugs according to the organ or system on which they act and their therapeutic, pharmacological, and chemical properties.

All drugs are grouped in five levels:

- anatomical main group
- therapeutic subgroup
- pharmacological subgroup
- chemical subgroup and
- chemical substance

Alterations in the ATC classification are implemented by the WHO Collaborating Centre for Drug Statistics Methodology in Oslo, Norway. Following current scientific publications and international pharmacopeias, changes are made in assigned new international generic names (INN), in ATC levels, or alterations of defined daily doses (DDD).

To the best of our knowledge, a general comprehensive handbook covering facts about pharmaceutical APIs according to ATC has not been published yet.

The information content was collected from the relevant literature as well as commercial patent data and condensed into a standardized table format.

In the tables of the particular chapters, the individual drug substances are described by their INN, synonyms, brand or trade names, ATC number, structure, CAS registry number, formula, molecular weight, mode of action, medical use, formulation, DDD, originator, country and/or date of approval, production method and references. The references for each drug typically include the basic patents and publications, in which the relevant synthetic or biosynthetic processes for production are described (see example of L01EB07 Dacomitinib below).

By this approach, we achieved to more than 3500 APIs of different structural families and importance as content of Ullmann's Pharmaceuticals. Extracts, nutrients, medical devices as well as homeopathy preparations are not included in this review, focus is on so-called synthetic small molecules and in part biologics.

Each of the 48 chapters starts with a brief introduction into the specific field highlighting the segment's market importance, medical need, new developments, and specific review articles.

The ATC classification system follows a strict hierarchy with 14 main groups (A–V, see the following table).

Code	Contents
A	Alimentary tract and metabolism
B	Blood and blood-forming organs
C	Cardiovascular system
D	Dermatologicals
G	Genito-urinary system and sex hormones
H	Systemic hormonal preparations, excluding sex hormones and insulins
J	Antiinfectives for systemic use
L	Antineoplastic and immunomodulating agents
M	Musculo-skeletal system
N	Nervous system
P	Antiparasitic products, insecticides, and repellents
R	Respiratory system
S	Sensory organs
V	Various

As the first level (Example: **L** Antineoplastic and Immunomodulating Agents), followed by a second level of codes that indicates the therapeutic subgroup with two digits (Example: **L01** Antineoplastic Agents). The third level defines the therapeutic/pharmacological subgroup and consists of one letter (Example: **L01E** Protein Kinase Inhibitors), followed by the fourth level of codes that indicates the chemical/therapeutic/pharmacological subgroup and consists of one letter (Example: **L01EB** Epidermal growth factor receptor (EGFR) tyrosine kinase inhibitors). The fifth level of codes indicates the chemical substance and consists of two digits (Example: **L01EB07** Dacomitinib).

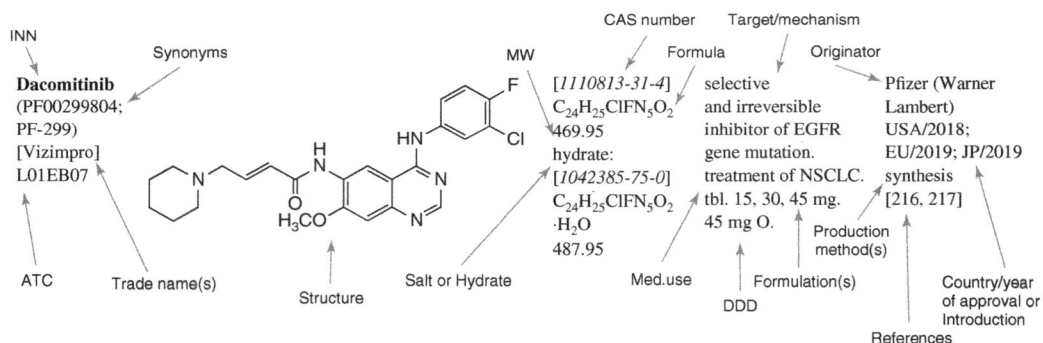

An index helps the reader to rapidly find a specific API via its INN within the book.

Ullmann's Pharmaceuticals is hence the most authoritative and extensive resource for everyone involved in drug-related development, evaluation, marketing, regulatory, teaching, and documentation activities.

This book features selected articles from the online edition of Ullmann's Encyclopedia of Industrial Chemistry (https://onlinelibrary.wiley.com/doi/book/10.1002/14356007) including one article which has not been published before (Vitamins (A11)). Ullmann's Encyclopedia is the world's largest reference in applied chemistry, industrial chemistry, and chemical engineering. First published in 1914, it offers a wealth of comprehensive and well-structured information on all facets of industrial chemistry in more than 1100 articles. True to the tradition of the Ullmann's Encyclopedia, the book Ullmann's Pharmaceuticals covers the topic by providing a perspective from industrial/medicinal chemists for all readers interested in the steadily progressing field of synthetic drugs.

Acknowledgments

We are grateful to the authors Michael C.-D. Fürst, Markus R. Heinrich, Axel Kleemann, Bernhard Kutscher, Anna S. Pirzer, Kurt Ritter, Hagen Trommer, and Heinz Weinberger who skillfully contributed to the 48 chapters. We are also indebted to the publisher for supporting the concept of this new compendium even in challenging times in providing scientific editing, corrections, proofreading and typeset by Melanie Rohn, Jeyasitha Sivalingam, and Mohan Mathiazhagan. Finally, we would like to thank the entire Ullmann's-Team represented by Claudia Ley, Barbara Elvers, Lesley Fenske, Carola Schmidt, and Frank Weinreich for valuable discussions and excellent cooperation during the design and production of this book.

Axel Kleemann and Bernhard Kutscher
Hanau and Maintal, January 2022

Contents

X   Contents

Stomatological Preparations (A01)

AXEL KLEEMANN, Hanau, Germany

1. Introduction

Stomatology is the science dealing with the cavity of the mouth and its diverse diseases, which can affect the skin, the oral mucous membranes, or the teeth. Other terms are oral medicine, dental medicine, and oral and maxillofacial medicine. Many of the active ingredients used in stomatological preparations are classified and described in other Anatomical Therapeutical Chemical (ATC) groups (this applies especially in the case of antiinfectives ($\rightarrow$ Antiinfectives for Systemic Use, 1. Antibacterials; $\rightarrow$ Antiinfectives for Systemic Use, 2. Antimycotics (Antifungals) and Antimycobacterials; $\rightarrow$ Antiinfectives for Systemic Use, 3. Antivirals), corticosteroids ($\rightarrow$ Respiratory Disorders, 2. Nasal Preparations and Decongestants for Topical Use (R01), Section 3 R01AD Corticosteroids), and local anesthetics ($\rightarrow$ Anesthetics (N01), Section 1.2 Local Anesthetics), to which the reader is referred. Preparations for treatment of throat infections are classified in $\rightarrow$ Respiratory Disorders, 3. Throat Preparations (R02). For preparations for mouth hygiene and dental care, see the chapter "Oral Hygiene Products" in Ullmann's Encyclopedia, Online Edition [24].

The ATC Classification of the WHO lists stomatological preparations as A01A according to their indication (A, Alimentary Tract and Metabolism; A01, Stomatological Preparations; and A01A, Stomatological Preparations).

Ullmann's Pharmaceuticals. Edited by Axel Kleemann and Bernhard Kutscher.
© 2022 Wiley-VCH GmbH
Set ISBN: 978-3-527-34252-5/ DOI: 10.1002/14356007.w25_w01

2. A01AA Caries Prophylactic Agents (Table 1)

Table 1. A01AA Caries prophylactic agents

INN (Synonyms) [Brand names] ATC	Structure (Remarks)	[CAS-No.] Formula MW, g/mol	Target (if known) Medical use Formulation DDD	Originator Approval (country/year) Production method [References]
Sodium fluoride A01AA01		[*7681-49-4*] NaF 41.99	added to municipal drinking water and toothpastes in order to maintain dental health by formation of fluoroapatite, a component of tooth enamel. 1.1 mg O (children).	
Sodium monofluorophosphate (MFP; SMFP) A01AA02	NaO, NaO—P⩽O F	[*10163-15-2*] Na$_2$PO$_3$F 143.95	source of fluoride (by hydrolysis) for use in toothpastes.	W. Lange; Ozark; Colgate−Palmolive USA/1967 synthesis [1−3]
Olaflur (C-27-amine fluoride) [Elmex, Dynexaminfluorid] A01AA03	H$_3$C—(17)—N—N—OH · OH OH x 2 HF	[*6818-37-7*] C$_{27}$H$_{60}$F$_2$N$_2$O$_3$ 498.8	amine fluoride, source of fluoride for use in toothpastes, chewing gums, and gels for prevention of dental caries. 1.1 mg O (children).	GABA AG synthesis [4, 5]

3. A01AB Antiinfectives and Antiseptics for Local Oral Treatment (Table 2)

Table 2. A01AB Antiinfectives and antiseptics for local oral treatment

INN (Synonyms) [Brand names] ATC	Structure (Remarks)	[CAS-No.] Formula MW, g/mol	Target (if known) Medical use Formulation DDD	Originator Approval (country/year) Production method [References]
Hydrogen peroxide (dihydrogen dioxide; oxydol; oxygenated water; perhydrol) A01AB02; D08AX01; S02AA06		[7722-84-1] H_2O_2 34.02	strong oxidant, bleaching agent, e.g., in use for teeth and hair; broad spectrum efficacy against bacteria, viruses, and yeasts. disinfectant/antiseptic and for sterilization (e.g., of surgical tools and in vaporized form for room disinfection). use in form of diluted aqueous solutions (3%, 10%, 30%, 50%, and between).	industrial production started in 1908 synthesis (anthraquinone process since 1936; formerly electrolysis) [6, 7]
Chlorhexidine (CHG for digluconate) [Chlorhexamed, Betasept, Hibiscrub, Hibitane, Merfene, HEXtra, Nolvasan, Chlorasept, Unisept, Paroex, Peridex, Periogard] A01AB03; B05CA02; D08AC02; D09AA12; R02AA05 see (→ Respiratory Disorders, 3. Throat Preparations (R02)) S01AX09, S02AA09, S03AA04 **WHO essential medicine**		[55-56-1] $C_{22}H_{30}Cl_2N_{10}$ 505.45 dihydrochloride: [3697-42-5] $C_{22}H_{30}Cl_2N_{10}$ ·2 HCl 578.37 diacetate: [56-95-1] $C_{22}H_{30}Cl_2N_{10}$ ·2 $C_2H_4O_2$ 625.56 D-digluconate: [18472-51-0] $C_{34}H_{54}Cl_2N_{10}O_{14}$ 897.8	bisbiguanide compound with antiseptic and topical antibacterial activity. for reduction of pocket depth in patients with adult periodontitis, used as an adjunct to scaling and root planning procedures. Also for prevention of dental caries, oropharyngeal decontamination in critically ill patients, hand hygiene in health care personnel, general skin cleanser, and catheter site preparation and care. most commonly in use is the digluconate as mouthwashes, toothpastes, oral rinses, gels, and sprays in 0.12–0.20% concentrations. 30 mg O.	ICI synthesis (from 4-chloroaniline, cyanoguanidine and hexa- methylene- diamine) [8]
Amphotericin B A01AB04; J02AA01 see (→ Antiinfectives for Systemic Use, 2. Antimycotics (Antifungals) and Antimycobacterials)			topical (surface) treatment of fungal infections in mouth and gastrointestinal tract without systemic availability.	

(continued)

Table 2. (*Continued*)

INN (Synonyms) [Brand names] ATC	Structure (Remarks)	[CAS-No.] Formula MW, g/mol	Target (if known) Medical use Formulation DDD	Originator Approval (country/year) Production method [References]
8-Hydroxyquinoline (oxine, oxyquinoline; oxychinolin; quinolin-8-ol) [Chinosol] A01AB07; D08AH03; G01AC30; R02AA14 see (→ Respiratory Disorders, 3. Throat Preparations (R02))		[148-24-3] C_9H_7NO 145.16 sulfate: [134-31-6] $(C_9H_7NO)_2 \cdot H_2SO_4$ 388.39	antiseptic and disinfectant (and stabilizer for hydrogen peroxide). sulfate is used in treatment of athlete's foot, vaginitis, as gargle, eyewash, nasal douche, and in hemorrhoidal preparations; constituent of combinations. 80 mg O.	Skraup synthesis (from 2-aminophenol, glycerol and nitrobenzene; "Skraup Synthesis") [9, 10]
Neomycin A01AB08; J01GB05 see (→ Antiinfectives for Systemic Use, 1. Antibacterials) R02AB01 see (→ Respiratory Disorders, 3. Throat Preparations (R02))				
Miconazole A01AB09; J02AB01 see (→ Antiinfectives for Systemic Use, 2. Antimycotics (Antifungals) and Antimycobacterials)				
Natamycin A01AB10; S01AA10;				
Hexetidine [Drossadin, Hexoral, Hextril, Isozid-H, Oraldene, Sterisol, Vagi-Hex] A01AB12; G01AX16; R02AA28 see (→ Respiratory Disorders, 3. Throat Preparations (R02))		[141-94-6] $C_{21}H_{45}N_3$ 339.61	bactericidal and fungicidal cationic antiseptic. mouthwash for local infections and oral hygiene. mouthwash sol. 0.1%; spray 0.2%; vaginal tbl. 10 mg (Vagi-Hex).	Commercial Solvents/Gödecke/Pfizer DE/1966 synthesis [11–13]
Tetracycline A01AB13; J01AA07 see (→ Antiinfectives for Systemic Use, 1. Antibacterials, Section 3.1 J01A Tetracyclines)				
Metronidazole A01AB17; J01XD01 see (→ Antiinfectives for Systemic Use, 1. Antibacterials, Section 3.9 J01X Other Antibacterials)				
Clotrimazole A01AB18; G01AF02				
Chlortetracycline A01AB21; J01AA03 see (→ Antiinfectives for Systemic Use, 1. Antibacterials, Section 3.1 J01A Tetracyclines)				
Doxycycline A01AB22; J01AA02 see (→ Antiinfectives for Systemic Use, 1. Antibacterials, Section 3.1 J01A Tetracyclines)				
Minocycline A01AB23; J01AA08 see (→ Antiinfectives for Systemic Use, 1. Antibacterials, Section 3.1 J01A Tetracyclines)				

Table 2. (*Continued*)

INN (Synonyms) [Brand names] ATC	Structure (Remarks)	[CAS-No.] Formula MW, g/mol	Target (if known) Medical use Formulation DDD	Originator Approval (country/year) Production method [References]
Thymol (thyme camphor; 3-*p*-cymenol; *m*-thymol; 2-isopropyl- 5-methylphenol) [Listerine] A01AB24	CH$_3$ OH H$_3$C CH$_3$	[*89-83-8*] C$_{10}$H$_{14}$O 150.22	component of thyme oil (*Thymus vulgaris* L.) with antiseptic, antibacterial, and antifungal activities. ingredient of toothpastes and some mouth washes, e.g., Listerine.	first isolation by NEUMANN in 1719, and characterized by LALLEMAND in 1853 extraction from thyme and synthesis [14, 15]
Ethacridine lactate (**monohydrate**) [Rivanol, Metifex, Acrinol] A01AB27; B05CA08; D08AA01	NH$_2$ O CH$_3$ H$_2$N N × C$_3$H$_6$O$_3$ × H$_2$O (*R,S*-lactate)	[*6402-23-9*] C$_{15}$H$_{15}$N$_3$O· C$_3$H$_6$O$_3$·H$_2$O 361.40	antiseptic for topical use. solution 0.1%; ointment 0.2%; powder for preparation of solutions.	Hoechst synthesis [16, 17]

Nystatin
A01AB33;
D01AA01
see (→ Dermatologicals (D), 2. Antifungals (D01), Emollients and Protectives (D02), and Antipruritics (D04) for Dermatological Use

4. A01AC Corticosteroids for Local Oral Treatment (Table 3)

Table 3. A01AC Corticosteroids for local oral treatment

INN (Synonyms) [Brand names] ATC	Structure (Remarks)	[CAS-No.] Formula MW, g/mol	Target (if known) Medical use Formulation DDD	Originator Approval (country/year) Production method [References]

Triamcinolone
A01AC01;
H02AB08 see (→ Hormone Therapeutics – Systemic Hormonal Preparations, Excluding Reproductive Hormones and Insulins (H))
R01AD11 see (→ Respiratory Disorders, 2. Nasal Preparations and Decongestants for Topical Use (R01))

Dexamethasone
A01AC02;
H02AB02 see (→ Hormone Therapeutics – Systemic Hormonal Preparations, Excluding Reproductive Hormones and Insulins (H));
R01AD03 see (→ Respiratory Disorders, 2. Nasal Preparations and Decongestants for Topical Use (R01))

Hydrocortisone
A01AC03;
H02AB09 see (→ Hormone Therapeutics – Systemic Hormonal Preparations, Excluding Reproductive Hormones and Insulins (H))

Prednisolone
A01AC04;
H02AB06 see (→ Hormone Therapeutics – Systemic Hormonal Preparations, Excluding Reproductive Hormones and Insulins (H));
R01AD02 see (→ Respiratory Disorders, 2. Nasal Preparations and Decongestants for Topical Use (R01))

Betamethasone
A01AC05;
H02AB01 see (→ Hormone Therapeutics – Systemic Hormonal Preparations, Excluding Reproductive Hormones and Insulins (H));
R01AD06 see (→ Respiratory Disorders, 2. Nasal Preparations and Decongestants for Topical Use (R01))

5. A01AD Other Agents for Local Oral Treatment (Table 4)

Table 4. A01AD Other agents for local oral treatment

INN (Synonyms) [Brand names] ATC	Structure (Remarks)	[CAS-No.] Formula MW, g/mol	Target (if known) Medical use Formulation DDD	Originator Approval (country/year) Production method [References]
Epinephrine A01AD01; C01CA24 see (→ Cardiovascular System (C))				
Benzydamine (benzindamine) [Afloben, Benzyrin, Difflam, Tantum Verde] A01AD02; G02CC03; M01AX07; M02AA05; R02AX03		[642-72-8] $C_{19}H_{23}N_3O$ 309.41 hydrochloride: [132-69-4] $C_{19}H_{23}N_3O \cdot HCl$ 345.87	nonsteroidal anti-inflammatory drug with local anesthetic and analgesic properties. medical use for treatment of inflammatory conditions of mouth and throat, e.g., gingivitis, aphthous ulcers, dental surgery, pharyngitis, and tonsillitis. sol. 1.5 mg/mL; spray 1.5 mg/mL; lozenges 3 mg.	Angelini Francesco synthesis [18, 19]
Acetyl salicylic acid A01AD05 N02BA01 see (→ Analgesics (N02))				
Adrenalone A01AD06 B02BC05 see (→ Antithrombotics (B01) and Antihemorrhagics (B02), Section 3.2.3 B02BC Local Hemostatics)				
Amlexanox (amoxanox; AA-673; CHX-3673) [Aphthasol, Elics, Solfa] A01AD07; R03DX01		[68302-57-8] $C_{16}H_{14}N_2O_4$ 298.30	anti-inflammatory, antiallergic, immunomodulatory, histamine and leukotriene inhibitor. medical use for treatment of recurrent aphthous ulcers (oral cavity) and in Japan bronchial asthma, allergic rhinitis, and conjunctivitis. oral paste 5%.	Takeda; Chemex Pharm. USA/1987 synthesis [20–22]
Choline salicylate [Actasal, Arthropan, Artrobione, Audax, Mundisal] A01AD18		[2016-36-6] $C_{12}H_{19}NO_4$ 241.29	choline salt of salicylic acid with analgesic, antipyretic, and anti-inflammatory properties. medical use for treatment of mouth ulcers, cold sores, denture sore spots, and as main ingredient in teething gels to relieve pain associated with tooth growth in infants. All preparations are combinations with other drugs (e.g., caffeine, dextromethorphan, and diphenhydramine).	Labs. For Pharmaceut. Dev. synthesis [23]

6. A01AE Local Anesthetics for Oral Treatment (Table 5)

Table 5. A01AE Local anesthetics for oral treatment

INN (Synonyms) [Brand names] ATC	Structure	[CAS-No.] Formula MW, g/mol	Target (if known) Medical use Formulation DDD	Originator Approval (country/year) Production method [References]
Lidocaine A01AE01; C01BB01 see (→ Cardiovascular System (C)) N01BB02 see (→ Anesthetics (N01)) R02AD02 see (→ Respiratory Disorders, 3. Throat Preparations (R02))				
Tetracaine A01AE02; N01BA03 see (→ Anesthetics (N01))				

List of Abbreviations

ATC	Anatomical Therapeutic Chemical Classification
DDD	defined daily dose
INN	International Nonproprietary Name
MW	molecular mass
sol.	solution
tbl.	tablets
WHO	World Health Organization

References

1 Ozark, (1949) US 2 481 807, USA-prior.1947.

2 Ozark, (1969) US 3 463 605, USA-prior.1966.

3 Use in Dentifrice Preparations: Colgate-Palmolive, (1966) US 3 227 617, USA-prior.1954.

4 GABA, (1963) US 3 083 143, CH-prior.1957.

5 Toothpaste Preparation: Colgate-Palmolive, (1979) US 4 160 022, USA-prior.1972.

6 Thénard, L.J. (1818) Observations sur des nouvelles combinaisons entre l'oxigène et divers acides, *Annales de chimie et de physique, 2nd series* **8**, 306–312.

7 Goor, G., et al. (1999–2017) Hydrogen Peroxide, in *Ullmann's Encyclopedia of Industrial Chemistry*, 7th edn, J. Wiley & Sons, New York, NY. Online Posting Date: April 15, 2007.

8 ICI, (1954) US 2 684 924, GB-prior.1951.

9 Skraup, Z.H. (1881) DE 14 976.

10 Skraup, Z.H. (1880) *Monatsh. Chem.* **1**, 316; **3**, 536.

11 Senkus, M. (1946) *J. Am. Chem. Soc.* **68**, 1611.

12 Satzinger, G., et al. (1978) *Anal. Profiles Drug Subst.* **7**, 277–295.

13 Commercial Solvents, (1962) US 3 054 797, USA-prior.1961.

14 Neumann, C. (1724) *Philos. Trans. Roy. Soc. Lond.* **33** (389), 321–332.

15 Lallemand, A. (1853) *Compt. Rend.* **37**, 498–500.

16 Hoechst, (1922), DE 360 421; prior.1919.

17 Hoechst, (1923), DE 393 411; prior.1921.

18 Angelini Francesco, (1967), US 3 318 905, IT-prior.1963.

19 Palazzo, G., et al. (1966) *J. Med. Chem.* **9**, 38.

20 Nohara, A., et al. (1985) *J. Med. Chem.* **28**, 559.

21 Takeda, (1979) US 4 143 042, JP-prior.1977.

22 Medical Use: Chemex Pharm., (1994) US 5 362 737, USA-prior.1993.

23 Broh-Kan, R.H. and Sasmor, E.J. (1962) US 3 069 321, USA-prior.1960.

24 Wells, T. (2015) Oral Hygiene Products, *Ullmann's Encyclopedia of Industrial Chemistry*, Wiley-VCH, Weinheim, Electronic Release, February.

Drugs for Acid-Related Disorders (A02)

AXEL KLEEMANN, Hanau, Germany

ULLMANN'S ENCYCLOPEDIA OF INDUSTRIAL CHEMISTRY

1. Introduction

This article deals mainly with gastrointestinal diseases, also called peptic ulcer disease (PUD), comprising the lower esophagus (esophageal reflux disease), stomach (gastric ulcers), and duodenum (duodenal ulcers). For an overview of the pathophysiology. The most common cause for PUDs is infection by *Helicobacter pylori,* responsible for 90% of duodenal ulcers and for about 70% of gastric ulcers. Other causes may be use of nonsteroidal anti-inflammatory drugs (NSAIDs) and to a lesser extent smoking and alcohol. The most common symptom of PUD is burning stomach pain, which is intensified by stomach acid. PUD is a global problem with a lifetime risk of development ranging from 5% to 10%, depending on the socioeconomic status.

Treatment of PUD may include:

1. antibiotic medication in case of *H. pylori* infection (e.g., a triple regimen with two antibiotics (often clarithromycin and metronidazole or amoxicillin) and a proton pump inhibitor (PPI, e.g., pantoprazole or esomeprazole);

2. antacids that neutralize stomach acid for a short period, but without healing effect;

3. antisecretory drugs such as H$_2$-receptor antagonists or PPIs;

4. cytoprotective agents such as sucralfate and misoprostol;

5. surgery in case of acute bleeding or perforation; before the first H$_2$-blocker cimetidine was introduced in 1976, surgery was applied rather often, mostly without real benefit for the patients.

Pharmacology and medical use are well referenced in [1–4]. Treatment with antacids and cytoprotective agents only plays a minor role now.

The antisecretory drugs, especially cimetidine and omeprazole, mark important milestones in target-oriented drug discovery and development. In 1969, the pharmacologist BLACK joined the research laboratory of the company Smith–Kline and French at Welwyn Garden City, England, after he was successful with the development of the β-blocker propranolol at ICI. It was known that histamine stimulates gastric acid production in stomach

Ullmann's Pharmaceuticals. Edited by Axel Kleemann and Bernhard Kutscher.
© 2022 Wiley-VCH GmbH
Set ISBN: 978-3-527-34252-5/ DOI: 10.1002/14356007.a02_321.pub3

cells, but the known antihistamine drugs did not inhibit this stimulation. BLACK had the theory of the existence of a different histamine receptor which could be inhibited. This proved right, and the research team successfully developed receptor-specific antagonists, of which cimetidine was clinically effective, and this first H_2-receptor antagonist was introduced under the trade name Tagamet in UK in 1976 and the following year in USA [5]. It became very successful as the first ever "blockbuster" drug with sales of more than US$ 1 billion a year. It not only was far more effective than the until-then used antacids, but it also was able to prevent the recurrence of ulcers and minimized gastric surgery considerably. BLACK received the Nobel Prize in medicine in 1988 for his work on β-blockers and H_2-blockers [6]. In 1983, Glaxo introduced the second H_2-blocker ranitidine under the brand name Zantac and within a rather short period this drug topped

the sales of Tagamet [7]. Already in 1986, the Zantac sales surpassed US$ 2 billion.

The situation changed when Astra AB introduced the PPI omeprazole in the early 1990s under the trade name Losec (later named Prilosec) [8]. It turned out that the PPIs were more potent than the H_2-inhibitors and they replaced them as drugs of choice. Six PPIs are currently on the market, and they have very similar chemical structures (see Table 7). Both drug groups, H_2-inhibitors and PPIs, are now over-the-counter (OTC) products.

In the following sections, the different drug groups are reviewed and arranged according to their Anatomical Therapeutic Chemical Classification (ATC) of the WHO, in which they are arranged according to their indication and chemical structure. The ATC list of A02 contains quite a number of combination preparations, but here only plain drugs are presented.

2. A02A Antacids

2.1. A02AA Magnesium Compounds (Table 1)

Table 1. A02AA Magnesium compounds

INN (Synonyms) [Brand names] ATC	Structure (Remarks)	[CAS-No.] Formula MW, g/mol	Target (if known) Medical use Formulation DDD	Originator Approval (country/year) Production method [References]
Magnesium carbonate (magnesia alba; magnesite) **Magnesium carbonate hydrate** (magnesium carbonate light) A02AA01; A06AD01	$MgCO_3$ $MgCO_3 \cdot H_2O$ (occurs as natural mineral, magnesite, in hydrated forms, e.g., barringtonite, and in basic forms, e.g., hydromagnesite)	[546-93-0] $CMgO_3$ 84.31 hydrate: [23389-33-5] CH_2MgO_4 102.33	antacid for treatment in duodenal ulcer disease and as laxans; can cause gastric flatulence and acid reflux.	mfg. from dolomite: see Ref. [3], p. 738. preparation of various hydrates [9]
Magnesium oxide A02AA02	MgO (occurs as natural mineral periclase)	[1309-48-4] MgO 40.30	antacid and laxative.	preparation from magnesite [10]
Magnesium hydroxide (magnesium hydrate) [Carmilax, Emgesan, Marinco H] A02AA04	$Mg(OH)_2$ (occurs as natural mineral, e.g., brucite)	[1309-42-8] H_2MgO_2 58.32	antacid.	preparation from magnesium salts with NaOH [11, 12]
Magnesium trisilicate [Magnosil, Petimin, Trisomin] A02AA05	$Mg_2Si_3O_8$	[14987-04-3] $Mg_2Si_3O_8$ 260.86	antacid.	preparation from sodium silicate and magnesium sulfate [13]

2.2. A02AB Aluminum Compounds (Table 2)

Table 2. A02AB Aluminum compounds

INN (Synonyms) [Brand names] ATC	Structure (Remarks)	[CAS-No.] Formula MW, g/mol	Target (if known) Medical use Formulation DDD	Originator Approval (country/year) Production method [References]
Aluminum hydroxide (aluminum hydrate) [Actal, Aldrox, Gaviscon, Liquirit] A02AB01	$Al(OH)_3$ (available also as hydrate, $Al(OH)_3 \cdot n\,H_2O$, algeldrate; occurs also as natural mineral, gibbsite, bayerite, nordstrandite)	[21645-51-2] AlH_3O_3 78.00	antacid.	synthesis [14, 15]
Aluminum phosphate [Phosphalugel] A02AB03	$AlPO_4$ (occurs as natural mineral berlinite)	[7784-30-7] AlO_4P 121.95	antacid.	synthesis (from aluminum sulfate and sodium phosphate)
Dihydroxyaluminum sodium carbonate (carbaldrate) [Kompensan, Minicid] A02AB04	$Al(OH)_2^+\ NaCO_3^-$	[12011-77-7] CH_2AlNaO_5 143.99	antacid. tbl. 300 mg.	synthesis (from aluminum isopropylate und sodium bicarbonate) [16]

2.3. A02AC Calcium Compounds (Table 3)

Table 3. A02AC Calcium compounds

INN (Synonyms) [Brand names] ATC	Structure (Remarks)	[CAS-No.] Formula MW, g/mol	Target (if known) Medical use Formulation DDD	Originator Approval (country/year) Production method [References]
Calcium carbonate [Abanta, Cacit, Calcichew, Calcidia, Caltrate, Citrical, Fixical, Maalox Quick Dissolve, Os-Cal] A02AC01; A12AA04	$CaCO_3$ (occurs as natural mineral, calcite, aragonite, marble)	[471-34-1] $CCaO_3$ 100.09	antacid, calcium supplement, and antidiarrheal agent. tbl. 420, 600, 1000 mg; chewing tbl. 500 mg; powder; suspension.	synthesis (from calcium chloride and sodium carbonate in aqueous sol., gives a microcrystalline powder)

2.4. A02AD Combinations and Complexes of Aluminum, Calcium, and Magnesium Compounds (Table 4)

Table 4. A02AD Combinations and complexes of aluminum, calcium, and magnesium compounds

INN (Synonyms) [Brand names] ATC	Structure (Remarks)	[CAS-No.] Formula MW, g/mol	Target (if known) Medical use Formulation DDD	Originator Approval (country/year) Production method [References]
Magaldrate (AY-5710; monalium hydrate; aluminum magnesium hydroxide sulfate) [Dynese, Riopan, Ripon] A02AD02	$Al_5Mg_{10}(OH)_{31}(SO_4)_2$	[74978-16-8] $Al_5H_{31}Mg_{10}O_{39}S_2$ 1097.3	antacid, treatment of esophagitis, duodenal and gastric ulcers, and gastroesophageal reflux. chewing tbl. 400, 800 mg; suspension 400 mg/5 mL.	Byk–Gulden DE/1983 synthesis [17]
Hydrotalcite (LS-77306) [Ancid, Megalac Hydrotalcit, Nacid, Talcid, Ultacid] A02AD04		[12304-65-3] $Al_2Mg_6(OH)_{16}CO_3$ ·4 H_2O 531.9	antacid. chewing tbl. 500 mg.	synthesis [18, 19]
Almasilate (aluminum magnesium silicate hydrate) [Ervasil, Gelusil, Megalac Almasilat, Simagel, Ultilac] A02AD05	Al_2O_3·MgO·2 SiO_2·x H_2O (found in nature in minerals, e.g., colerainite, zebedassite, leuchtenbergite)	[71205-22-6] anhydrous: [50958-44-6] $Al_2MgO_8Si_2$ 262.4	antacid. suspension 1 g/10 mL.	synthesis

3. A02B Drugs for Peptic Ulcer and Gastroesophageal Reflux Disease (GERD)

3.1. A02BA H$_2$-Receptor Antagonists (Table 5)

Table 5. A02BA H$_2$-Receptor antagonists

INN (Synonyms) [Brand names] ATC	Structure (Remarks)	[CAS-No.] Formula MW, g/mol	Target (if known) Medical use Formulation DDD	Originator Approval (country/year) Production method [References]
Cimetidine (SKF-92334) [Tagamet, Aciloc, Acinil, Cimal, Dyspamet, Stomedine, Ulcedin, Ulcimet] A02BA01		[*51481-61-9*] C$_{10}$H$_{16}$N$_6$S 252.34 hydrochloride: [*70059-30-2*] C$_{10}$H$_{16}$N$_6$S ·HCl 288.80	histamine H$_2$-receptor antagonist, inhibits gastric acid secretion and reduces pepsin output. treatment of heartburn and peptic ulcers. tbl. 200, 400, 800 mg. 0.8 g O, P.	SK&F UK/1976; USA/1977 synthesis [5, 6, 20–23]
Ranitidine (AH-19065) [Azantac, Noctone, Raniben, Ranidil, Raniplex, Sostril, Taural, Terposen, Trigger, Ulcex, Ultidine, Zantac, Zantic] A02BA02 **WHO essential medicine**		[*66357-35-5*] C$_{13}$H$_{22}$N$_4$O$_3$S 314.40 hydrochloride: [*66357-59-3*] C$_{13}$H$_{22}$N$_4$O$_3$S ·HCl 350.86	histamine H$_2$-receptor antagonist, inhibits gastric acid secretion. treatment of peptic ulcers, gastroesophageal reflux disease, and Zollinger–Ellison syndrome. cps. 150, 300 mg. 0.3 g O, P.	Allen & Hanburys (Glaxo) USA/1981 synthesis [7, 24]
Famotidine (YM-11170; MK-208) [Amfamox, Fadul, Famodil, Famosan, Famoxal, Ganor, Gaster, Gastridin, Gastropen, Lecedil, Motiax, Muclox, Pepcid, Pepcidine, Pepdul, Ulfamid] A02BA03		[*76824-35-6*] C$_8$H$_{15}$N$_7$O$_2$S$_3$ 337.44	histamine H$_2$-receptor antagonist, inhibits gastric acid secretion and reduces pepsin output. treatment of peptic ulcer disease, gastroesophageal reflux disease, and Zollinger–Ellison syndrome. amp. 10, 20 mg; f.c. tbl. 10, 20, 40 mg. 40 mg O, P.	Yamanouchi USA/1985 synthesis [25]
Nizatidine (LY-139037; ZE-101; ZL-101) [Axid, Calmaxid, Cronizat, Distaxid, Gastrax, Nizax, Nizaxid, Tazac] A02BA04		[*76963-41-2*] C$_{12}$H$_{21}$N$_5$O$_2$S$_2$ 331.45	histamine H$_2$-receptor antagonist, inhibits gastric acid secretion. treatment of peptic ulcers and gastroesophageal reflux disease. amp. 100, 150, 300 mg; cps. 75, 150, 300 mg; oral susp. 40 mg/5 mL. 0.3 g O, P.	Eli Lilly USA/1987 synthesis [26]

(continued)

Table 5. (*Continued*)

INN (Synonyms) [Brand names] ATC	Structure (Remarks)	[CAS-No.] Formula MW, g/mol	Target (if known) Medical use Formulation DDD	Originator Approval (country/year) Production method [References]
Roxatidine acetate (HOE-760; TZU-0460) [Altat, Gastralgin, Roxit] A02BA06		[78628-28-1] $C_{19}H_{28}N_2O_4$ 348.44 hydrochloride: [93793-83-0] $C_{19}H_{28}N_2O_4$ ·HCl 384.90	histamine H_2-receptor antagonist, inhibits gastric acid secretion. treatment of peptic ulcers and gastroesophageal reflux disease. tbl. 75, 150 mg; cps. 75, 150 mg; amp. 75 mg. 0.15 g O.	Teikoku Hormone synthesis [27]
Ranitidine bismuth citrate (GR-122311X; ranitidine bismutrex) [Pylorid, Tritec] A02BA07		[128345-62-0] $C_{13}H_{22}N_4O_3S$ ·$C_6H_5BiO_7$ 712.48	histamine H_2-receptor antagonist, inhibits gastric acid secretion. treatment of *Helicobacter pylori* and duodenal ulcers. tbl. 400 mg. 0.8 g O.	Glaxo synthesis [28]
Lafutidine (FRG-8813) [Protecadin] A02BA08		[118288-08-7] $C_{22}H_{29}N3O_4S$ 431.55	so-called second generation histamine H_2-receptor antagonist. treatment of gastric and duodenal ulcers and acute and chronic gastritis. tbl. 5 mg. 20 mg O.	Fujirebio JP/2000 synthesis [29]

3.2. A02BB Prostaglandins (Table 6)

Table 6. A02BB Prostaglandins

INN (Synonyms) [Brand names] ATC	Structure (Remarks)	[CAS-No.] Formula MW, g/mol	Target (if known) Medical use Formulation DDD	Originator Approval (country/year) Production method [References]
Misoprostol (SC-29333) [Arthrotec, Cytotec, Gymiso, Misodex] A02BB01; G02AD06 **WHO essential medicine**	(double racemate of (+)- and (−)- enantiomers of the 16-(R)- and 16-(S) forms, the above formula shows only the 16-(S) form)	[59122-46-2] $C_{22}H_{38}O_5$ 382.54	synthetic analogue of prostaglandin E1, inhibits gastric acid and pepsin secretion and enhances mucosal resistance to injury with NSAIDs; oxytocic activity. treatment of gastric ulcers and prevention of gastric ulcers in case of NSAIDs use. Intravaginal use for induction of labor in intrauterine fetal death. tbl. 0.1, 0.2 mg; comb. with diclofenac: s.r. tbl. 50/0.2 mg and 75/0.2 mg. 0.8 mg O.	Searle USA/1985 synthesis [30–32]
Enprostil (RS-84135) [Camleed, Gardrin] A02BB02 no more important		[73121-56-9] $C_{23}H_{28}O_6$ 400.47	synthetic analogue of prostaglandin E2, inhibits gastric acid secretion and has a mucoprotecting effect. treatment of gastroduodenal ulcers.	Syntex synthesis [33]

3.3. A02BC Proton Pump Inhibitors (PPI) (Table 7)

Table 7. A02BC Proton pump inhibitors (PPI)

INN (Synonyms) [Brand names] ATC	Structure (Remarks)	[CAS-No.] Formula MW, g/mol	Target (if known) Medical use Formulation DDD	Originator Approval (country/year) Production method [References]
Omeprazole (H-168/68) [AntraMUPS, Ecomep, Gastrobene, Gastrogard, Losec, Mopral, Omep, Osiren, Parizac, Pepticum, Prilosec, UlcerGard, Zegerid, Zoltum] A02BC01 **WHO essential medicine**		[73590-58-6] $C_{17}H_{19}N_3O_3S$ 345.42 magnesium salt: [95382-33-5] $C_{34}H_{36}MgN_6O_6S_2$ 713.12 sodium salt: [95510-70-6] $C_{17}H_{18}N_3NaO_3S$ 367.41	PPI (irreversible). treatment of gastroesophageal reflux disease, peptic ulcer, and Zollinger−Ellison syndrome; prevention of upper gastrointestinal bleeding in patients at risk. tbl. MUPS 10, 20 mg; cps 20, 40 mg. 20 mg O, P.	AB Hässle (Astra AB) USA/1988 synthesis [8, 34]
Pantoprazole (BY-1023; SKF-96022) [Azidex, Eupantol, Gastroloc, Pantecta, Pantozol, Pantopan, Peptazol, Protium, Protonix, Somac, Zurcal] A02BC02		[102625-70-7] $C_{16}H_{15}F_2N_3O_4S$ 383.37 sodium salt: [138786-67-1] $C_{16}H_{14}F_2N_3NaO_4S$ 405.35	PPI (irreversible). treatment of gastroesophageal reflux disease, peptic ulcer, and Zollinger−Ellison syndrome. tbl. 20, 40 mg. 20 mg O, P.	Byk Gulden DE/1994; USA/2000 synthesis [35, 36]
Lansoprazole (A-65006; AG-1749) [Agopton, Lansox, Lanzor, Limpidex, Ogast, Prevacid, Takepron, Zoton] A02BC03		[103577-45-3] $C_{16}H_{14}F_3N_3O_2S$ 369.36	PPI (irreversible). treatment of gastroesophageal reflux disease, peptic ulcer, and Zollinger−Ellison syndrome. cps. 15, 30 mg. 15 g O.	Takeda JP/1991; USA/1995 synthesis [37]
Rabeprazole (E-3810; pariprazole) [Aciphex, Pariet] A02BC04		[117976-89-3] $C_{18}H_{21}N_3O_3S$ 359.44 sodium salt: [117976-90-6] $C_{18}H_{20}N_3NaO_3S$ 381.43	PPI (partially reversible). treatment of gastroesophageal reflux disease, peptic ulcer, and Zollinger−Ellison syndrome. tbl. 10, 20 mg. 10 mg O, P.	Eisai JP/1997; USA/1999 synthesis [38]

Table 7. (*Continued*)

INN (Synonyms) [Brand names] ATC	Structure (Remarks)	[CAS-No.] Formula MW, g/mol	Target (if known) Medical use Formulation DDD	Originator Approval (country/year) Production method [References]
Esomeprazole (H-199/18; perprazole) [Nexium, Inexium, Axagon, Esopral] A02BC05	(S) form of omeprazole	[*119141-88-7*] $C_{17}H_{19}N_3O_3S$ 345.42 magnesium salt: [*161973-10-0*] $C_{34}H_{36}MgN_6O_6S_2$ 713.14 magnesium salt trihydrate: [*217087-09-7*] $C_{34}H_{36}MgN_6O_6S_2$ ·3 H_2O 767.18	PPI (irreversible). treatment of gastroesophageal reflux disease, peptic ulcer, and Zollinger–Ellison syndrome; prevention of upper gastrointestinal bleeding in patients at risk. amp. 40 mg for inj./infusion; cps. 20, 40 mg; tbl. MUPS 20, 40 mg. 20 mg O, P.	AB Hässle (Astra AB) USA/2000 synthesis [8, 39]
Dexlansoprazole (T-168390; R-lansoprazole; TAK-390) [Dexilant, Kapidex] A02BC06	(R) form of lansoprazole	[*138530-94-6*] $C_{16}H_{14}F_3N_3O_2S$ 369.36	PPI (irreversible). treatment of gastroesophageal reflux disease and erosive esophagitis, peptic ulcer. cps. (delayed release) 30, 60 mg. 30 mg O.	Takeda USA/2009 synthesis [40–42]

3.4. A02BX Other Drugs for Peptic Ulcer and Gastroesophageal Reflux Disease (GERD) (Table 8)

Table 8. A02BX other drugs for peptic ulcer and gastroesophageal reflux disease (GERD)

INN (Synonyms) [Brand names] ATC	Structure (Remarks)	[CAS-No.] Formula MW, g/mol	Target (if known) Medical use Formulation DDD	Originator Approval (country/year) Production method [References]
Carbenoxolone (carbenoxalone; CBX) [Bioplex, Bioral, Duogastrone, Glucophage, Neogel, Pyrogastrone, Sanodin] A02BX01 wfm in almost all countries		[5697-56-3] C$_{34}$H$_{50}$O$_7$ 570.77 disodium salt: [7421-40-1] C$_{34}$H$_{48}$Na$_2$O$_7$ 614.73	inhibitor of 11β-hydroxysteroid dehydrogenase (11β-HSD), which converts cortisone into cortisol. treatment of digestive tract ulcers (esp. peptic ulcer). tbl. 50 mg; granulate 1%.	Biorex semisynthetic (from glycyrrhetic acid and succinic anhydride) [43]
Sucralfate (basic aluminum sucrose sulfate complex) [Antepsin, Carafate, Citogel, Hexagastron, Keal, Succosa, Sucrabest, Sucralfin, Sucrate, Sugast, Sulcrate, Ulcar, Ulcerlmin, Ulcogant] A02BX02	R = SO$_3$[Al$_2$(OH)$_5$]	[54182-58-0] C$_{12}$H$_{54}$Al$_{16}$O$_{75}$S$_8$ 2086.67	locally acting in the stomach as acid buffer for 6–8 h; builds insoluble complexes with albumin and fibrinogen on the surface of ulcers as protective barriers and so prevents further damage from acid and other aggressive substances. treatment of duodenal ulcers. tbl. 1000 mg; powder sachets 1000 mg. 4 g O.	Chugai USA/1981 synthesis [44]
Pirenzepine (LS-519-Cl2) [Duogastral, Durapirenz, Gasteril, Gastrozepin, Leblon, Maghen, Renzepin, Tabe, Ulcoforton, Ulcosan] A02BX03		[28797-61-7] C$_{19}$H$_{21}$N$_5$O$_2$ 351.41 dihydrochloride: [29868-97-1] C$_{19}$H$_{21}$N$_5$O$_2$ ·2 HCl 424.33	selective muscarinic receptor M$_1$ antagonist, parasympatholytic; reduction of gastric acid secretion and muscle spasms. treatment of peptic ulcers. tbl. 50 mg. 0.1 g O, 20 mg P.	Thomae synthesis [45]

Name	Structure	Identifiers	Description	Source/Synthesis
Methylmethionine sulfonium chloride (S-methylmethionine; methiosulfonium chloride; vitamin U; W-106699) [Ardesyl, Merastom, MMSC, Cabagin-U, Vitas-U] A02BX04	H_3C–S^+–CH_3, COOH, NH_2, Cl^-	*[1115-84-0]* $C_6H_{14}ClNO_2S$ 199.69 S-methyl-methionine inner salt: *[4727-40-6]* $C_6H_{13}NO_2S$ 163.24	vitamin U, found in green vegetables (no true vitamin), inhibitory effect on gastric acid secretion. treatment of peptic ulcers, colitis, and gastritis. food additive.	Kowa (?) [46]
Bismuth subcitrate (bismuth tripotassium dicitrate; tripotassium dicitratobismuthate) [Pylera] A02BX05	^-OOC, COO^-, HO, COO^-, Bi^{3+}, 3K^+	*[57644-54-9]* $C_{12}H_{14}BiK_3O_{14}$ 708.51	protection of mucous membrane in the stomach. pylera: fixed dose combination with metronidazole and tetracycline for *Helicobacter pylori* eradication (→ Antiinfectives for Systemic Use, 1. Antibacterials). originally bismuth subcitrate and similar derivatives were synthesized for treatment of syphilis.	Axcan Pharma EU/2006 synthesis combinations [47, 48]
Proglumide (CR-242; W-5219; xylamide) [Milid, Milide, Promid] A02BX06	COOH, CH_3, CH_3, N, O, N, H, O	*[6620-60-6]* $C_{18}H_{26}N_2O_4$ 334.42	cholinergic antagonist and nonselective cholecystokinin (CCK) antagonist. was in use for treatment of stomach ulcers. 1.2 g O.	Rotta synthesis [49]
Gefarnate (DA-688; geranyl farnesylacetate) [Gefanil] A02BX07	H_3C, CH_3, CH_3, CH_3, CH_3, CH_3, O, O	*[51-77-4]* $C_{27}H_{44}O_2$ 400.65	cytoprotectant that has been used in treatment of peptic ulcer and gastritis.	Ist de Angeli semisynthesis (from farnesylacetic acid and geraniol) [50]
Bismuth subsalicylate (bismuth salicylate) [Bismed, Jatrox, Pepto-Bismol] A02BX22	O–Bi–OH, O, O	*[14882-18-9]* $C_7H_5BiO_4$ 362.09	still used in certain countries as antidiarrheal and antacid. chewable tbl. 262 mg. comb. with metronidazole and tetracycline-HCl and an H_2-antagonist for eradication of *Helicobacter pylori* (FDA label) (→ Antiinfectives for Systemic Use, 1. Antibacterials).	Norwich Pharmacal USA/1900 synthesis (from salicylic acid and bismuth hydroxide) [51]

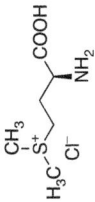

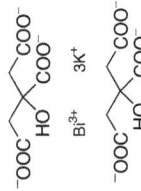

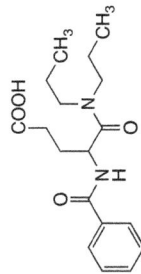

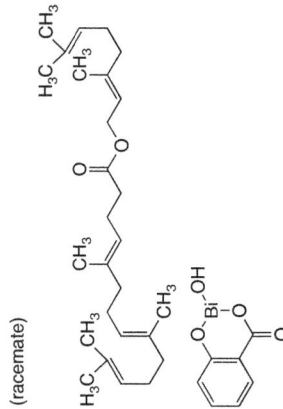

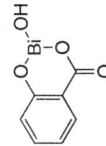

(racemate)

List of Abbreviations

amp.	ampoules
cps.	capsules
GERD	gastroesophageal reflux disease
inj.	injection
mfg.	manufacturing
MUPS	multiple-unit pellet system
NSAID	nonsteroid anti-inflammatory drug
PPI	proton pump inhibitor
PUD	peptic ulcer disease
tbl.	tablets

References

1 Wallace, J.L., Sharkey, K.A., (2011) Chapter 45, Pharmacotherapy of Gastric Acidity, Peptic Ulcers, and Gastroesophageal Reflux Disease, in *Goodman & Gilman's The Pharmacological Basis of Therapeutics*, 12th edn, McGraw-Hill, London, pp. 1309–1322.

2 Sweetman, S. (ed.) (2007) Gastrointestinal Drugs, in *Martindale, The Complete Drug Reference*, 35th edn, Pharmaceutical Press, London, UK/Grayslake, IL, USA, pp. 1526–1602.

3 Swinyard, E.A. (1980) Chapter 40 Gastrointestinal Drugs, in *Remington's Pharmaceutical Sciences*, 16th edn, Mack Publishing Company, Easton, PA, pp. 734–756.

4 Kromer, W., et al. (2012) Antiulcer Drugs, in *Ullmann's Encyclopedia of Industrial Chemistry*, J. Wiley & Sons, New York a02_321.pub2 (20 pp).

5 Ganellin, C.R. (1981) *Chronicles of Drug Discovery*, (eds. J.S. Bindra and D. Lednicer), vol. 1, J. Wiley & Sons, pp. 1–38.

6 Sneader, W. *Drug Discovery – A History*, J. Wiley & Sons, Chichester, pp. 212–215.

7 Bradshaw, J.L. (1993) *Chronicles of Drug Discovery*, (ed. D. Lednicer), vol. 3, J. Wiley & Sons, pp. 45–81.

8 Olbe, L., et al. (2003) A Proton-Pump Inhibitor Expedition: The Case Histories of Omeprazole and Esomeprazole, *Nat. Rev. Drug Discov.* 2, 132–139.

9 Merck & Co., (1965) US 3 169 826, USA-prior.1961.

10 Northwest Magnesite, (1967) US 3 320 029, USA-prior.1964.

11 Dow, (1964) US 3 127 241, USA-prior.1961.

12 FMC, (1966) US 3 232 708, USA-prior.1962.

13 Merck & Co., (1966) US 3 272 594, USA-prior.1961.

14 Dominé-Berges, M. (1950) *Ann. Chim.* 5 (12), 106.

15 Aluminum Comp. of America, (1990) US 4 915 930, USA-prior. 1986.

16 Chattanooga Medicine, (1957) US 2 783 179, USA-prior.1955.

17 Byk-Gulden, (1960) US 2 923 660, DE-prior.1955.

18 Kyowa, (1970) US 3 539 306, JP-prior.1966.

19 Wiyantoko, B., et al. (2015) *Proc. Chem.* 17, 21–26.

20 Black, J.W., et al. (1972) Definition and antagonism of histamine H₂-receptor, *Nature* 236, 385–390.

21 Ganellin, C.R. (1981) *J. Med. Chem.* 24, 913–920.

22 SK&F, (1976) US 3 950 333, GB-prior.1972.

23 SK&F, (1975) US 3 894 151, GB-prior.1972.

24 Allen & Hanburys, (1978) US 4 128 658, GB-prior.1976.

25 Yamanouchi, (1981) US 4 283 408, JP-prior.1979.

26 Eli Lilly, (1983) US 4 375 547, USA-prior.1980.

27 Teikoku Hormone, (1981) US 4 293 557, JP-prior.1979.

28 Glaxo, (1989) GB 2 220 937, GB-prior.1988.

29 Fujirebio, (1990) US 4 912 101, JP-prior.1987.

30 Collins, P.W. and Pappo, R. (1977) *J. Med. Chem.* 20, 1152.

31 Searle, (1975) US 3 965 143, USA-prior.1974.

32 Collins, P.W. (1993) *Chronicles of Drug Discovery*, (ed. D. Lednicer), vol. 3, J. Wiley & Sons, pp. 101–124.

33 Syntex, (1979) US 4 178 457, USA-prior.1978.

34 Hässle, (1981) US 4 255 431, SE-prior.1978.

35 Kohl, B., et al. (1992) *J. Med. Chem.* 35, 1049.

36 Byk Gulden, (1988) US 4 758 579, CH-prior.1984.

37 Takeda, (1987) US 4 689 333, JP-prior.1984.

38 Eisai, (1991) US 5 045 552, JP-prior.1986.

39 Astra, (1999) US 5 948 789, SE-prior.1994.

40 Takeda, (2002) US 6 462 058, JP-prior.1999.

41 Takeda, (2001) US 6 328 994, JP-prior.1998.

42 Takeda, (2008) US 7 431 942, JP-prior.1998.

43 Biorex, (1962) US 3 070 623, GB-prior.1957.

44 Chugai, (1969) US 3 432 489, JP-prior.1965.

45 Thomae, (1972) DE 1 795 183, DE-prior.1968.

46 Atkinson, R.O. and Poppelsdorf, F. (1951) *J. Chem. Soc.* 1368–1378.

47 Borody, T.J. (1993) US 5 196 205, AU-prior.1987.

48 Axcan Pharma, (2002) US 6 350 468, IT-prior.1997.

49 Umetzu, T., et al. (1980) *Eur. J. Pharmacol.* 64, 69.

50 Cardani, C., et al. (1963) *J. Med. Chem.* 6, 457–458.

51 Fischer, B. and Grützner, B. (1893) *Arch. Pharm.* 231, 680.

Drugs for Functional Gastrointestinal Disorders (A03)

Axel Kleemann, Hanau, Germany

ULLMANN'S ENCYCLOPEDIA OF INDUSTRIAL CHEMISTRY

1. Introduction

In the following sections, the different drug groups are reviewed and arranged according to their Anatomical Therapeutic Chemical Classification (ATC) of the WHO, in which they are arranged according to their indication and chemical structure [1]. The article A03 lists antispasmodics for treatment of gastrointestinal spasms and pain and propulsives or prokinetics. The latter ones are agents that stimulate gastrointestinal motility. Most of these are substituted 4-aminobenzamides like the still important metoclopramide, a WHO essential medicine.

Almost all antispasmodics listed under A03A had already been developed and marketed in the 1950s and 1960s, with the exception of serotonin receptor antagonists. They are still in use for management of gastrointestinal spasm and irritable bowel syndrome (IBS). Exact launch dates are not easily available for some of the drugs listed. Many of them have been withdrawn in Western countries but are still available in certain markets in Eastern Europe and Southeast Asia, partly in form of combinations. The main effect of these antispasmodics consists in a relaxant action on smooth muscles, either direct or via an antimuscarinic action on the parasympathetic innervation. Besides the antimuscarinics, there are also drugs with calcium channel blocker activity for relaxation of smooth muscle. In A03B, belladonna alkaloids (tropane alkaloids) and their semisynthetic derivatives are listed, which are still important for treatment of gastrointestinal spasms because of their good parasympatholytic (spasmolytic) effectivity.

Roughly 11% of the population globally are affected by IBS symptoms with a predominance in females (14%) over men (8.9%) with variations in geographic regions ranging from 7% to 21%. Usually people below the age of 50 are affected. There are several different causes for IBS and also several different symptoms (e.g., cramping and pain in abdomen, diarrhea, constipation, bloating, increased gas, and food intolerance) that make treatment difficult.

Ullmann's Pharmaceuticals. Edited by Axel Kleemann and Bernhard Kutscher.
© 2022 Wiley-VCH GmbH
Set ISBN: 978-3-527-34252-5/ DOI: 10.1002/14356007.w09_w02

For review publications on pharmacology and therapeutic application of the drugs of A03, see [2–5]. In the references the reader will find the originator patents wherever available and/or (if existing) the corresponding medicinal chemistry publication. The syntheses routes for most of the pharmaceutical substances (active pharmaceutical ingredients, APIs) will also be found in [6].

The ATC list of A03 contains quite a number of combination preparations, namely A03C — antispasmodics in combination with psycholeptics, A03D — antispasmodics in combination with analgesics, and A03E — antispasmodics and anticholinergics in combination with other drugs, but here only plain drug substances are presented and characterized.

2. A03A Drugs for Functional Gastrointestinal Disorders

2.1. A03AA Synthetic Anticholinergics, Esters with Tertiary Amino Group (Table 1)

Table 1. A03AA Synthetic anticholinergics, esters with tertiary amino group

INN (Synonyms) [Brand names] ATC	Structure (Remarks)	[CAS-No.] Formula MW, g/mol	Target (if known) Medical use Formulation DDD	Originator Approval (country/year) Production method [References]
Oxyphen-cyclimine [Daricon, Spazamin, Setrol] A03AA01		[125-53-1] $C_{20}H_{28}N_2O_3$ 344.46 hydrochloride: [125-52-0] $C_{20}H_{28}N_2O_3$ ·HCl 380.92	antimuscarinic agent with antispasmodic and antisecretory activities. relaxing effect on smooth muscles of the gastrointestinal tract. tbl. 10 mg. 20 mg O.	Pfizer synthesis [7, 8]
Mebeverine [Colofac, Duspatal, Duspatalin] A03AA04		[3625-06-7] $C_{25}H_{35}NO_5$ 429.56 hydrochloride: [2753-45-9] $C_{25}H_{35}NO_5$ ·HCl 466.02	spasmolytic for muscles in and around the gut. used for treatment of the symptoms of irritable bowel syndrome. cps. 100 mg; s.r. cps. 200 mg. 0.4 g O.	Philips N.V. USA/1965 synthesis [9]
Trimebutine (TM-906) [Cerekinon, Debridat, Digerent, Modulon, Polibutin, Spabucol, Transacalm] A03AA05		[39133-31-8] $C_{22}H_{29}NO_5$ 387.48 maleate: [34140-59-5] $C_{22}H_{29}NO_5$ ·$C_4H_4O_4$ 503.55	antimuscarinic. for use as spasmolytic in the treatment of irritable bowel syndrome. f.c. tbl. 100 mg 0.6 g O.	Jouveinal synthesis [10]

Table 1. (*Continued*)

INN (Synonyms) [Brand names] ATC	Structure (Remarks)	[CAS-No.] Formula MW, g/mol	Target (if known) Medical use Formulation DDD	Originator Approval (country/year) Production method [References]
Rociverine (LG-30158) [Rilaten] A03AA06		[53716-44-2] $C_{20}H_{37}NO_3$ 339.52	antimuscarinic. for use as spasmolytic in the GI tract. tbl. 10 mg; 25 mg; vials for inf. 20 mg.	Guidotti synthesis [11]
Dicycloverine (dicyclomine) [Bentyl, Bentylol, Merbentyl, Wyovin] A03AA07		[77-19-0] $C_{19}H_{35}NO_2$ 309.49 hydrochloride: [67-92-5] $C_{19}H_{35}NO_2$ ·HCl 345.95	relaxes smooth muscles in the GI tract; treatment of irritable bowel syndrome. tbl. 10, 20 mg; cps. 10 mg. 80 mg O, P.	Merrell USA/1950 synthesis [12, 13]
Piperidolate (JB-305) [Crapinon, Dactil] A03AA30 wfm in many countries		[82-98-4] $C_{21}H_{25}NO_2$ 323.44 hydrochloride: [129-77-1] $C_{21}H_{25}NO_2$ ·HCl 359.89	antimuscarinic for use as spasmolytic in the GI tract. tbl. 50 mg.	Merrell synthesis [14, 15]

2.2. A03AB Synthetic Anticholinergics, Quaternary Ammonium Compounds (Table 2)

Table 2. A03AB Synthetic anticholinergics, quaternary ammonium compounds

INN (Synonyms) [Brand names] ATC	Structure (Remarks)	[CAS-No.] Formula MW, g/mol	Target (if known) Medical use Formulation DDD	Originator Approval (country/year) Production method [References]
Glycopyrronium bromide (AHR-504; glycopyrrolate) [Nodapton, Robinul, Tarodyn] A03AB02; D11AA01; R03BB06		[596-51-0] $C_{19}H_{28}BrNO_3$ 398.34	antimuscarinic agent. decrease of gastric acid secretion; used for treatment of stomach ulcers, normally in combination with other agents. important for treatment of COPD by inhalation. amp. 0.2, 0.5 mg/mL. 3 mg O; 0.3 mg P.	A.H. Robins synthesis [16]
Oxyphenonium bromide (Ba-5473; C-5473) [Antrenyl, Spasmophen] A03AB03 wfm		[50-10-2] $C_{21}H_{34}BrNO_3$ 428.41	antimuscarinic agent. med. use as antispasmodic as adjunct in the treatment of gastric and duodenal ulcer and of visceral spasms; mydriatic. tbl. 5 mg. 25 mg O.	Ciba synthesis [17]
Propantheline bromide [Corrigast, Ercotina, Pro-Banthine, Neo-Metantyl, Pantheline] A03AB05		[50-34-0] $C_{23}H_{30}BrNO_3$ 448.40	antimuscarinic agent. relieves cramps or spasms of stomach, gut, and bladder; can be used to treat symptoms of irritable bowel syndrome; suppresses excessive sweating (hyperhidrosis). tbl. 7.5, 15 mg. 60 mg O, P.	Searle synthesis [18, 19]
Otilonium bromide (SP-63) [Doralin, Menoctyl, Pasminox, Spasen, Spasmoctyl, Spasmomen] A03AB06		[26095-59-0] $C_{29}H_{43}BrN_2O_4$ 563.58	antimuscarinic agent; treatment of intestinal spasms and irritable bowel syndrome. tbl. 40 mg.	Menarini synthesis [20, 21]

Table 2. (*Continued*)

INN (Synonyms) [Brand names] ATC	Structure (Remarks)	[CAS-No.] Formula MW, g/mol	Target (if known) Medical use Formulation DDD	Originator Approval (country/year) Production method [References]
Methantheline bromide (MTB-51; SC-2910) [Banthine bromide, Avagal, Uldumont, Vagantin, Metaxan, Xanteline, Gastron, Gastrosedan, Doladene] A03AB07		[53-46-3] $C_{21}H_{26}BrNO_3$ 420.35	antimuscarinic agent. used for treatment of peptic ulcer, urinary incontinence, and hyperhidrosis. tbl. 50 mg. 150 mg O.	Searle synthesis [18, 19]
Tridihexethyl chloride (921 C) [Pathilon, Pathibamate] A03AB08 wfm		[4310-35-4] $C_{21}H_{36}ClNO$ 353.98	antimuscarinic agent. was in use as adjunct to antacid therapy in peptic ulcer and treatment of irritable colon, mucous colitis, and acute enterocolitis. tbl. 25 mg; s.r. cps. 75 mg; sol. for inj. 10 mg/mL.	American Cyanamid (Lederle) synthesis [22]
Isopropamide iodide (R-79) [Darbid, Priamide, Tyrimide, Ornatos, Stelabid] A03AB09 wfm in many countries		[71-81-8] $C_{23}H_{33}IN_2O$ 480.43	antimuscarinic agent. was used in the treatment of peptic ulcers. tbl. 5 mg; inj. sol. 3 mg/2 mL.	Janssen synthesis [23]
Mepenzolate bromide [Cantil, Trancolon] A03AB12		[76-90-4] $C_{21}H_{26}BrNO_3$ 420.35	antimuscarinic agent, decreases gastric acid and pepsin secretion. adjunct in the treatment of peptic ulcer. tbl. 25 mg. 0.1 g O.	Lakeside Labs. synthesis [24]
Pipenzolate bromide (JB-323) [Asilon, Espasal, Spasmotal] A03AB14 controlled substance (CSA) wfm in many countries		[125-51-9] $C_{22}H_{28}BrNO_3$ 434.37	antimuscarinic agent. was developed for treatment of gastrointestinal spasms. hallucinogen!	Lakeside Labs synthesis [25]
Diphemanil methylsulfate [Prantal] A03AB15		[62-97-5] $C_{21}H_{27}NO_4S$ 389.51	antimuscarinic agent. decreases gastric acid, saliva, and sweat; med. use for treatment of peptic ulcer and hyperhidrosis. tbl. 100 mg.	Schering Corp. synthesis [26]

(*continued*)

Table 2. (*Continued*)

INN (Synonyms) [Brand names] ATC	Structure (Remarks)	[CAS-No.] Formula MW, g/mol	Target (if known) Medical use Formulation DDD	Originator Approval (country/year) Production method [References]
Tiemonium iodide [Visceralgine, Tiemozyl] A03AB17 **tiemonium methylsulfate** [Visceralgine; Vasodal; Colchimax]		[*144-12-7*] $C_{18}H_{24}INO_2S$ 445.36 tiemonium methylsulfate: [*6504-57-0*] $C_{19}H_{27}NO_6S_2$ 429.6	antimuscarinic agent. treatment of visceral spasms/gastrointestinal disorders. tbl. 50 mg (methylsulfate); amp. 5 mg/2 mL (methylsulfate).	Mauvernay synthesis [27]
Timepidium bromide (SA-504) [Mepidium, Sesden, Timepin, Zesun] A03AB19		[*35035-05-3*] $C_{17}H_{22}BrNOS_2$ 400.39	antimuscarinic agent, decreases gastric acid and pepsin secretion. treatment of visceral spasms; adjunct in the treatment of peptic ulcer. cps. 30 mg; tbl. 30 mg; amp. 7.5 mg/1 mL.	Tanabe synthesis [28, 29]
Trospium chloride [Regurin, Relaspium, Sanctura, Spasmex, Spasmolyt, Spasmo-Urgenin, Trospi, Trospium Pfleger, Urivesc] A03AB20; G04BD09		[*10405-02-4*] $C_{25}H_{30}ClNO_3$ 427.97	antimuscarinic agent. treatment of the symptoms of overactive bladder. amp. 1.2 and 2.0 mg/2 mL sol. for i.v. inj.; f.c. tbl. 5, 10, 15, 20, 30, 45 mg; s.r. cps. 60 mg.	Pfleger; Madaus DE/1999; USA/2004 synthesis [30]
Fenpiverinium bromide (fenpipramide methylbromide) [Baralgin] A03AB20		[*125-60-0*] $C_{22}H_{29}BrN_2O$ 417.39	antimuscarinic agent. in some countries a comb. with metamizol sodium or diclofenac and pitofenone hydrochloride is in use for muscle spasms and pain, e.g., under the names Baralgin, Combispas, and Novaspas.	Hoechst synthesis [31, 32]

2.3. A03AC Synthetic Antispasmotics, Amides with Tertiary Amines (Table 3)

Table 3. A03AC Synthetic antispasmotics, amides with tertiary amines

INN (Synonyms) [Brand names] ATC	Structure (Remarks)	[CAS-No.] Formula MW, g/mol	Target (if known) Medical use Formulation DDD	Originator Approval (country/year) Production method [References]
Tiropramide (CR-605) [Alfospas, Maiorad, Piromide, Spastro, Tiram, Tiromac, Tiromin] A03AC05 (mainly marketed in Asian countries)	(racemate)	[*55837-29-1*] $C_{28}H_{41}N_3O_3$ 467.65 hydrochloride: [*53567-47-8*] $C_{28}H_{41}N_3O_3$ ·HCl 504.11	antispasmodic. treatment of functional bowel disorders. tbl. 100, 200 mg.	Rotta synthesis [33]

2.4. A03AD Papaverine and Derivatives (Table 4)

Table 4. A03AD Papaverine and derivatives

INN (Synonyms) [Brand names] ATC	Structure (Remarks)	[CAS-No.] Formula MW, g/mol	Target (if known) Medical use Formulation DDD	Originator Approval (country/year) Production method [References]
Papaverine [numerous brand names which were withdrawn from the markets; mainly generics] A03AD01; C04AX38; G04BE02 (→ Cardiovascular System (C))	H_3CO, H_3CO structure with OCH_3, OCH_3	[58-74-2] $C_{20}H_{21}NO_4$ 339.39 hydrochloride: [61-25-6] $C_{20}H_{21}NO_4$ ·HCl 375.85	inhibitor of cAMP phosphodiesterase. medical use to treat spasms of the gastrointestinal tract, bile ducts, and ureter and as cerebral and coronary vasodilator in subarachnoid hemorrhage and treatment of erectile dysfunction. vials 30 mg/mL. 0.1 g O, P.	discovered by Georg Merck in 1848 as about 1% constituent of opium. synthesis
Moxaverine [Eupaverin, Kollateral] A03AD30; C04AX42; G04BE02 (→ Cardiovascular System (C))	H_3CO, H_3CO structure with CH_3	[10539-19-2] $C_{20}H_{21}NO_2$ 307.39 hydrochloride: [1163-37-7] $C_{20}H_{21}NO_2$ ·HCl 343.85	inhibitor of cAMP phosphodiesterase, spasmolytic and peripheral vasodilator. medical use a. o. for treatment of spasms in the gastrointestinal tract. drg. 150 mg.	Orgamol synthesis [34]
Drotaverine [No-Spa] A03AD02	H_3C, O, H_3C, O structure with O—CH_3, O—CH_3, NH	[14009-24-6] $C_{24}H_{31}NO_4$ 397.52 hydrochloride: [985-12-6] $C_{24}H_{31}NO_4$ ·HCl 434.	selective inhibitor of phosphodiesterase-4 without anticholinergic effects. medical use as antispasmodic, mainly in Asian and Central and East European countries; treatment of functional bowel disorders. tbl. 40 mg, 80 mg. 0.1 g O, P.	Chinoin synthesis [35]

2.5. A03AE Serotonin Receptor Antagonists (Table 5)

Table 5. A03AE Serotonin receptor antagonists

INN (Synonyms) [Brand names] ATC	Structure (Remarks)	[CAS-No.] Formula MW, g/mol	Target (if known) Medical use Formulation DDD	Originator Approval (country/year) Production method [References]
Alosetron (GR-68755) [Lotronex] A03AE01	structure with HN, O, N, H_3C, N—CH_3	[122852-42-0] $C_{17}H_{18}N_4O$ 294.36 hydrochloride: [122852-69-1] $C_{17}H_{18}N_4O$·HCl 330.82	selective inhibitor of 5-HT$_3$-receptor agonist. treatment of irritable bowel syndrome. tbl. 0.5 mg, 1 mg. 1 mg O.	Glaxo USA/2000 synthesis [36]

2.6. A03AX Other Drugs for Functional Gastrointestinal Disorders (Table 6)

Table 6. A03AX Other drugs for functional gastrointestinal disorders

INN (Synonyms) [Brand names] ATC	Structure (Remarks)	[CAS-No.] Formula MW, g/mol	Target (if known) Medical use Formulation DDD	Originator Approval (country/year) Production method [References]
Pinaverium bromide (LAT-1717) [Dicetel, Eldicet, Disten, Nulite] A03AX04		[53251-94-8] $C_{26}H_{41}Br_2NO_4$ 591.43	calcium channel blocker with antispasmodic activity. treatment of symptoms of irritable bowel syndrome. f.c. tbl. 50 mg, 100 mg. 0.15 g O.	Solvay synthesis [37, 38]
Fenoverine [Fexadin, Spasmopriv, Syncrospas] A03AX05		[37561-27-6] $C_{26}H_{25}N_3O_3S$ 459.56	calcium channel blocker with antispasmodic activity. treatment of muscle aches and spasms and irritable bowel disease. cps. 100 mg.	Bouchara Vaillant-Defresne synthesis [39]
Alverine (dipropyline, phenpropamine) [Spasmaverine, Alvercol, Spasmonal] A03AX08		[150-59-4] $C_{20}H_{27}N$ 281.44 citrate: [5560-59-8] $C_{20}H_{27}N$ $\cdot C_6H_8O_7$ 473.57	smooth muscle relaxant. prevents muscle spasms in case of functional gastrointestinal disorders such as irritable bowel syndrome. cps. 60 mg, 120 mg; tbl. 40 mg. combin. with simeticone is available in some countries.	Norgine synthesis [40, 41]
Trepibutone (AA-149) [Supacal] A03AX09 wfm		[41826-92-0] $C_{16}H_{22}O_6$ 310.35	promoter of bile and pancreatic juice, antispasmodic. treatment of gastrointestinal symptoms associated with chronic pancreatitis.	Takeda synthesis [42]
Caroverine [Spasmium, Tinnex, Tinnitin] A03AX11		[23465-76-1] $C_{22}H_{27}N_3O_2$ 365.48 hydrochloride: [55750-05-5] $C_{22}H_{27}N_3O_2$ $\cdot HCl$ 401.93	unspecific calcium channel blocker and antagonist of non-NMDA and NMDA glutamate receptors, smooth muscle relaxant, and antispasmodic. treatment of spasms in smooth muscles incl. of intestine and also of tinnitus. amp. 160 mg; cps. 20 mg.	Phafag AG (Austria) synthesis [43]

Table 6. (*Continued*)

INN (Synonyms) [Brand names] ATC	Structure (Remarks)	[CAS-No.] Formula MW, g/mol	Target (if known) Medical use Formulation DDD	Originator Approval (country/year) Production method [References]
Phloroglucinol (1,3,5-benzenetriol) [Spasfon-Lyoc, Flospan, Spassirex] A03AX12		[*108-73-6*] $C_6H_6O_3$ 126.11 dihydrate: [*6099-90-7*] $C_6H_6O_3 \cdot 2H_2O$ 162.14	antispasmodic and antioxidant. treatment of spasmodic bowel, biliary tract, bladder, and uterine pain. tbl. 80 mg, 160 mg.	synthesis [44]
Valethamate bromide [Epidosin, Epimol, Resitan, Shinmetane, Valosin] A03AX14		[*90-22-2*] $C_{19}H_{32}BrNO_2$ 386.37	blocker of acetylcholine. antispasmodic for treatment of symptoms such as gastrointestinal spasms and pain during menstruation, and urinary tract and bile stone colic. tbl. 10 mg; amp. 8 mg.	Kali-Chemie synthesis [45, 46]

3. A03B Belladonna and Derivatives, Plain

3.1. A03BA Belladonna Alkaloids, Tertiary Amines (Table 7)

Table 7. A03BA Belladonna alkaloids, tertiary amines

INN (Synonyms) [Brand names] ATC	Structure (Remarks)	[CAS-No.] Formula MW, g/mol	Target (if known) Medical use Formulation DDD	Originator Approval (country/year) Production method [References]
Atropine (D,L-hyoscyamine; D,L-tropyl tropate; tropine tropate) [predominantly generics; AtroPen, Atropisol, Atropt] A03BA01; S01FA01 **WHO essential medicine**		[51-55-8] $C_{17}H_{23}NO_3$ 289.38 sulfate monohydrate: [5908-99-6] $(C_{17}H_{23}NO_3)_2$ $\cdot H_2SO_4 \cdot H_2O$ 694.84	competitive antagonist of muscarinic acetylcholine receptors M_1-M_5; parasympatholytic (spasmolytic), inhibits salivary and mucus glands. treatment of gastrointestinal spasms and spasms/pain of urinary and bile tracts. tbl. 0.5 mg; amp. 0.25, 0.5, 1 mg. 1.5 mg O, P.	belladonna (tropane) alkaloid, formed by racemization of hyoscyamine during alkaline work-up of the extract from *Hyoscyamus niger* or *Atropa belladonna*. first isolation: [47] extraction process: [48]
Hyoscyamine (L-tropyl tropate; (S)-tropine tropate; (S)-atropine; duboisine) [Anaspaz, Daturine, Cystospaz, Ed-Spaz, Egacene, HyoMax, Hyosyne, Levbid, Levsin, Oscimin, Peptard, Symax] A03BA03		[101-31-5] $C_{17}H_{23}NO_3$ 289.38 sulfate dihydrate: [6835-16-1] $(C_{17}H_{23}NO_3)_2$ $\cdot H_2SO_4 \cdot 2 H_2O$ 712.85	competitive antagonist of muscarinic acetylcholine receptors M_1-M_5; parasympatholytic (spasmolytic), inhibits salivary and mucus glands. treatment of gastrointestinal spasms and spasms/pain of urinary and bile tracts. elixir, oral, 0.125 mg/5 mL; sol. oral, 0.125 mg/mL; sol. for inj. 0.5 mg/mL; tbl. 0.125 mg; tbl. ER (12 h) 0.375 mg; tbl. sublingual 0.125 mg. 1.2 mg O.	belladonna (tropan) alkaloid from *Hyoscyamus niger,* *Atropa belladonna* and some other plants by acidic extract or by resolution of atropine. isolation: [49] resolution of atropine: [50]

3.2. A03BB Belladonna Alkaloids, Semisynthetic, Quaternary Ammonium Compounds (Table 8)

Table 8. A03BB Belladonna alkaloids, semisynthetic, quaternary ammonium compounds

INN (Synonyms) [Brand names] ATC	Structure (Remarks)	[CAS-No.] Formula MW. g/mol	Target (if known) Medical use Formulation DDD	Originator Approval (country/year) Production method [References]
Butylscopol-ammonium bromide (hyoscine butylbromide; butylscopolamine bromide; scopolamine butylbromide) [Buscapina, Buscolysin, Buscopan, Scoburen, Spasman] A03BB01		[149-64-4] $C_{21}H_{30}BrNO_4$ 440.38	blocker of muscarinic receptors located on postganglionic parasympathetic nerve endings and on smooth muscle cells and so blocking the activity of acetylcholine. This leads to spasmolytic (antispasmodic) effects in the gastrointestinal, urinary, uterine, and biliary tracts. amp. 20 mg; tbl. 10 mg, 20 mg; drg. 10 mg; suppos. 7.5 mg, 10 mg. 60 mg O. P. R.	Boehringer Ing. USA/1951; DE/1951 semisynthetic (from scopolamine and butylbromide) [51]
WHO essential medicine				
Methscopol-amine bromide (hyoscine methyl bromide; methylscopolamine bromide; scopolamine methobromide) [Extendryl, Holopon, Rescon, Pamine] A03BB03		[155-41-9] $C_{18}H_{24}BrNO_4$ 398.30	blocker of muscarinic receptors located on postganglionic parasympathetic nerve endings and on smooth muscle cells and so blocking the activity of acetylcholine. This leads to spasmolytic (antispasmodic) effects in the gastrointestinal, urinary, uterine, and biliary tracts. tbl. 2.5 mg, 5 mg. 12 mg O. P.	E. Merck USA/1947 semisynthetic (from scopolamine and methylbromide) [52, 53]
Fentonium bromide (phentonium bromide; FA-402; Z-326) [Ulcesium] A03BB04 wfm in many countries		[5868-06-4] $C_{31}H_{34}BrNO_4$ 564.52	blocker of muscarinic receptors located on postganglionic parasympathetic nerve endings and on smooth muscle cells and so blocking the activity of acetylcholine. This leads to spasmolytic (antispasmodic) effects in the gastrointestinal tract. tbl. 20 mg. 60 mg O.	Zambon; Sanofi-Aventis synthesis [54]
Cimetropium bromide (DA-3177) [Alginor] A03BB05 wfm in many countries		[51598-60.8] $C_{21}H_{28}BrNO_4$ 438.36	blocker of muscarinic receptors located on postganglionic parasympathetic nerve endings and on smooth muscle cells and so blocking the activity of acetylcholine. This leads to spasmolytic (antispasmodic) effects in the gastrointestinal tract. med. use for treatment of irritable bowel syndrome. amp. 5 mg/mL; tbl. 50 mg; suppos. 50 mg.	De Angeli synthesis [55]

4. A03F Propulsives

4.1. A03FA Propulsives (Table 9)

Table 9. A03FA Propulsives

INN (Synonyms) [Brand names] ATC	Structure (Remarks)	[CAS-No.] Formula MW, g/mol	Target (if known) Medical use Formulation DDD	Originator Approval (country/year) Production method [References]
Metoclopramide (MCP; DEL-1267; AHR-3070-C) [Cerukal, Emetid, Gastromax, Gastronerton, Gastrosil, Maxolon, Migpriv, Pamid, Paspertin, Primperan, Reglan] A03FA01 **WHO essential medicine**		[364-62-5] $C_{14}H_{22}ClN_3O_2$ 299.80 monohydrochloride monohydrate: [54143-57-6] $C_{14}H_{22}ClN_3O_2$ $\cdot HCl \cdot H_2O$ 354.27	dopamin D_2 receptor and 5-HT$_3$ receptor antagonist. treatment of nausea and vomiting, migraine headache, and gastroparesis. amp. 10 mg/2 mL; cps. 10 mg, 30 mg; s.r. cps. 30 mg; sol. 1 mg/mL; tbl. 5 mg, 10 mg; suppos. 10 mg, 20 mg. 30 mg O, P, R.	Soc. d'Etudes Sci. Ind. de l'Ile-de-France/ Delagrange FR/1964; DE/1965; USA/1979 synthesis [56]
Cisapride (R-51619) [Acenalin, Alimix, Prepulsid, Risamol] A03FA02 wfm in many countries because of generating arrhythmias	(3*RS*, 4*SR*)-structure; racemic mixture of both enantiomers.	[81098-60-4] $C_{23}H_{29}ClFN_3O_4$ 465.95 monohydrate: [260779-88-2] $C_{23}H_{29}ClFN_3O_4$ $\cdot H_2O$ 483.97	serotonin 5-HT$_4$ receptor agonist and indirect parasympathomimetic; increases gastrointestinal motility. was used for treatment of gastroesophageal reflux disease (GERD). tbl. 2.5, 5, 10, 20, 25, 50, 100 mg. 30 mg O, R.	Janssen USA/1988 synthesis [57]
Domperidone (R-33812) [Biperidys, Gastronorm, Motilium, Peridon, Motylio, Nauzelin, Peridys] A03FA03		[57808-66-9] $C_{22}H_{24}ClN_3O_2$ 425.92	dopamin D_2 receptor blocker. med. use as antiemetic and gastric prokinetic agent. tbl. 5 mg, 10 mg; suppos. 10, 30, 60 mg; susp. 10 mg/mL. 30 mg O, P; 0.12 g R.	Janssen DE/1979 synthesis [58]

Table 9. (*Continued*)

INN (Synonyms) [Brand names] ATC	Structure (Remarks)	[CAS-No.] Formula MW, g/mol	Target (if known) Medical use Formulation DDD	Originator Approval (country/year) Production method [References]
Bromopride [Cascapride, Emepride, Emoril, Opridan, Plesium, Praiden, Valopride, Viaben, Viadil] A03FA04		[4093-35-0] $C_{14}H_{22}BrN_3O_2$ 344.25 hydrochloride: [52423-56-0] $C_{14}H_{22}BrN_3O_2$ ·HCl 380.71	dopamine D_2 receptor antagonist with prokinetic properties, closely related to metoclopramide. treatment of nausea and vomiting, and GERD. amp. 10 mg; cps. 10 mg; drops 13.3 mg. 20 mg O, P.	Soc. d'Etudes Sci Ind. de l'Ile-de-France synthesis [56]
Alizapride (MS-5080) [Nausilen, Plitican, Vergentan] A03FA05		[59338-93-1] $C_{16}H_{21}N_5O_2$ 315.38 hydrochloride: [59338-87-3] $C_{16}H_{21}N_5O_2$ ·HCl 351.84	dopamine D_2 receptor antagonist with antiemetic and prokinetic properties. amp. 50 mg/2 mL; tbl. 50 mg. 0.15 g O, P.	Delagrange synthesis [59]
Clebopride [Amicos, Clanzol, Clast, Cleboril, Cleprid, Motilex] A03FA06		[55905-53-8] $C_{20}H_{24}ClN_3O_2$ 373.88 malate: [57645-91-7] $C_{20}H_{24}ClN_3O_2$ ·$C_4H_6O_5$ 507.97	dopamine D_2 receptor antagonist with antiemetic and prokinetic properties. tbl. 0.5 mg; syrup 0.5 mg/5 mL; amp. 1 mg/2 mL.	Anphar; Asahi synthesis [60]
Itopride (HSR-803; HC-803) [Dagla, Itax, Ganaton, Itogard, Itomed, Zirid] A03FA07		[122898-67-3] $C_{20}H_{26}N_2O_4$ 358.44 hydrochloride: [122892-31-3] $C_{20}H_{26}N_2O_4$ ·HCl 394.90	dopamine D_2 receptor antagonist with antiemetic and prokinetic properties and inhibitor of acetylcholinesterase. tbl. 50 mg.	Hokuriku Pharm. synthesis [61]
Cinitapride (LAS-17177) [Cidine] A03FA08		[66564-14-5] $C_{21}H_{30}N_4O_4$ 402.50 tartrate: [96623-56-2] $C_{21}H_{30}N_4O_4$ ·$C_4H_6O_6$	cinitapride is a gastroprokinetic agent and antiulcer benzamide with agonist activity at $5\text{-}HT_1$ and $5\text{-}HT_4$ receptors and antagonist activity at $5\text{-}HT_2$ receptors. tbl. 100, 200, 300, 400, 600, 800 mg; f.c. tbl. 200, 300, 400, 800 mg; sol. for inj. 150 mg/mL.	Anphar; Almirall E/1990 synthesis [62]

List of Abbreviations

amp.	ampoule(s)
API	active pharmaceutical ingredients
ATC	Anatomical Therapeutic Chemical Classification
cAMP	cyclic adenosine monophosphate
COPD	chronic obstructive pulmonary disease
cps.	capsules
CSA	United States Controlled Substances Act
DDD	defined daily dose
drg.	dragees
ER	extended release
f.c. tbl	film-coated tablets
GERD	gastroesophageal reflux disease
GI	gastrointestinal
IBS	irritable bowel syndrome
inj.	injection
INN	international nonproprietary name
MW	molecular mass
NMDA	N-methyl D-aspartic acid
sol.	solution
s.r. cps.	slow-release capsules
suppos.	suppositories
wfm	withdrawn from market
WHO	World Health Organization

References

1 WHO Collaborating Centre for Drug Statistics Methodology (2020) *Guidelines for ATC Classification and DDD Assignment 2019*, WHO Collaborating Centre for Drug Statistics Methodology, Oslo, Norway.

2 Wallace, J.L. and Sharkey, K.A. (2011) Chapter 46, Treatment of Disorders of Bowel Motility and Water Flux; Antiemetics; Agents Used in Biliary and Pancreatic Disease, in *Goodman & Gilman's The Pharmacological Basis of Therapeutics*, (eds. L.L. Brunton, et al.), 12th edn, McGraw-Hill, London, pp. 1323–1349.

3 Sweetman, S.C. (2007) Gastrointestinal Drugs, in *Martindale, The Complete Drug Reference*, 35th edn, Pharmaceutical Press, London, UK and Grayslake, IL, pp. 1526–1602.

4 Osol, A., Chase, G.D., Gennaro, A.R., et al. eds. (1980) Chapter 40, Gastrointestinal Drugs, in *Remington's Pharmaceutical Sciences*, 16th edn, Mack Publishing Company, Easton, PA, pp. 734–756.

5 Bungardt, E. and Mutschler, E., Spasmolytics. DOI: 10.1002/14356007.a24_515.

6 Kleemann, A., Engel, J., Kutscher, B. and Reichert, D. eds. *Pharmaceutical Substances Online*, Thieme Chemistry, Stuttgart.

7 Faust, J.A., et al. (1959) *J. Am. Chem. Soc.* **81**, 2214.
8 Pfizer, (1956) DE 1 058 515 (USA-prior. 1955).
9 Philips, (1966) US 3 265 577 (NL-prior. 1957).
10 Jouveinal, (1963) FR 1 344 455 (FR-prior. 1962).
11 Guidotti, (1972) US 3 700 675 (USA-prior. 1967).
12 Tilford, C.H., et al. (1947) *J. Am. Chem. Soc.* **69**, 2903.
13 Merrell, (1949) US 2 474 796 (USA-prior. 1946).
14 Biel, J.H., et al. (1952) *J. Am. Chem. Soc.* **74**, 1485.
15 Lakeside Labs., (1959) US 2 918 407 (USA-prior. 1950).
16 Robins, A.H. (1960) US 2 956 062 (USA-prior. 1959).
17 Ciba, (1949) CH 259 948 (CH-prior. 1944).
18 Cusic, J.W. and Robinson, R.A. (1951) *J. Org. Chem.* **16**, 1921.
19 Searle, (1953) US 2 659 732 (USA-prior. 1950).
20 Ghelardoni, M., et al. (1973) *J. Med. Chem.* **16**, 1063.
21 Menarini, (1970) US 3 536 723 (IT-prior. 1966).
22 American Cyanamid, (1959) US 2 913 494 (USA-prior. 1957).
23 Janssen, (1955) GB 772 921 (NL-prior. 1954).
24 Lakeside, (1959) US 2 918 408 (USA-prior. 1950).
25 Lakeside, (1959) US 2 918 406 (USA-prior. 1950).
26 Schering Corp., (1956) US 2 739 968 (USA-prior. 1951).
27 Mauvernay, (1961) GB 953 386 (FR-prior. 1960).
28 Kawazu, M., et al. (1972) *J. Med. Chem.* **15**, 914.
29 Tanabe, (1973) US 3 764 607 (JP-prior. 1970).
30 Kawazu, M., et al. (1972) *J. Med. Chem.* **15**, 914.
31 Hoechst, (1943) DE 731 560 (DE-prior. 1941).
32 Hoechst, (1952) DE 858 552 (DE-prior. 1952).
33 Rotta, (1977) US 4 004 008 (IT-prior. 1974).
34 Orgamol (1963) GB 1 030 022 (CH-prior. 1962).
35 Chinoin, (1962) BE 621 917 (HU-prior. 1961).
36 Glaxo, (1994) US 5 360 800 (GB-prior. 1987).
37 Baronnet, R., et al. (1974) *Eur. J. Med. Chem.* **9**, 182.
38 Soc. Berri-Balzac, (1974) US 3 845 048 (USA-prior. 1971).
39 Buzas, A. and Pierre, R. (1972) FR 2 092 639 (FR-prior. 1970).
40 Külz, F., et al. (1939) *Ber. Dtsch. Chem. Ges.* **72**, 2161.
41 Stühmer, W. and Elbrächter, E.-A. (1954) *Arch. Pharm. Ber. Dtsch. Pharm. Ges.* **287**, 139.
42 Takeda, (1976) US 3 943 169 (JP-prior. 1971).
43 Donau-Pharm, (1962) US 3 028 384 (AT-prior. 1959).
44 Hwu, J.R. and Tsay, S.-C. (1990) *J. Org. Chem.* **55**, 5987.
45 Kali-Chemie, (1958) DE 969 245 (DE-prior. 1953).
46 Kali-Chemie, (1958) DE 971 136 (DE-prior. 1953).
47 Geiger, P.L. (1833) *Annalen* **5**, 43.
48 Chemnitius, F. (1927) *J. Prakt. Chem.* **116**, 276. ("Über die Technik der Atropindarstellung").
49 Ladenburg, A. (1881) *Ann. Chem.* **206**, 274.
50 Werner, G. and Miltenberger, K. (1960) *Ann. Chem.* **631**, 163.
51 Boehringer Ing., (1950) DE 856 890 (DE-prior. 1950).
52 E. Merck, (1902) DE 145 996.
53 Upjohn, (1956) US 2 753 288 (USA-prior. 1952).
54 Whitefin Holding S. A., (1967) US 3 356 682 (USA-prior. 1964).
55 De Angeli, (1974) US 3 853 886 (GB-prior. 1972).
56 Soc. d'Etudes Sci Ind. de l'Ile-de-France, (1965) US 3 177 252 (FR-prior. 1961).
57 Janssen, (1990) US 4 962 115 (USA-prior. 1981).
58 Janssen, (1978) US 4 066 772 (USA-prior. 1975).
59 Delagrange, (1977) US 4 039 672 (DE-prior. 1975).
60 Anphar, (1979) US 4 138 492 (GB-prior. 1974).
61 Hokuriku, (1991) US 4 983 633 (JP-prior. 1987).
62 Walton, (1991) US 5 026 858 (GB-prior. 1976).

Antiemetics and Antinauseants (A04)

Axel Kleemann, Hanau, Germany

ULLMANN'S ENCYCLOPEDIA OF INDUSTRIAL CHEMISTRY

1. Introduction — A04A Antiemetics and Antinauseants

In the following chapters, the different drug groups are reviewed and arranged according to their Anatomical Therapeutic Chemical Classification (ATC) of the WHO, in which they are arranged according to their indication and chemical structure [1].

Chapter A04 deals with drugs for treatment of nausea and vomiting, two symptoms which are best explained in [2]: "The act of emesis and the sensation of nausea that accompanies it generally are viewed as protective reflexes that serve to rid the stomach and intestine of toxic substances and prevent their further ingestion." The underlying physiological processes and mechanisms are also explained in [2] and may not be treated here.

The general classification of antiemetics according to [2] follows the predominant receptor on which they act:

- 5-HT$_3$ receptor antagonists (e.g., ondansetron)
- centrally acting dopamine receptor antagonists, e.g., metoclopramide ($\rightarrow$ Drugs for Functional Gastrointestinal Disorders (A03))
- histamine H$_1$ receptor antagonists, e.g., promethazine ($\rightarrow$ Hypnotics and Sedatives (N05C))
- muscarinic receptor antagonists, e.g., hyoscine = scopolamine ($\rightarrow$ Hypnotics and Sedatives (N05C))
- neurokinin receptor antagonists (e.g., aprepitant)
- cannabinoid receptor agonists (e.g., dronabinol)

Unfortunately, the classification in Chapter A04 differs somewhat with this pharmacological classification. In addition, it has to be kept in mind that antihistamines, which are often used as antiemetics, are classified in R06 ($\rightarrow$ Respiratory Disorders, 7. Antihistamines for Systemic Use (R06)) and dopamine antagonists such as metoclopramide under propulsives (A03FA) in $\rightarrow$ Drugs for Functional Gastrointestinal Disorders (A03) and $\rightarrow$ Drugs for Acid-Related Disorders (A02).

The mixed chapter "A04AD Other Antiemetics" lists scopolamine, the old phenothiazine serotonin antagonists metopimazine and triflupromazine, the cannabinoids dronabinol and nabilone, and finally the newer NK$_1$ antagonists aprepitant, rolapitant, and netupitant which are especially useful in prevention and treatment of chemotherapy-induced nausea and vomiting.

In the references for the single-drug substances, the reader will find the originator patents wherever available and/or, if existing, the corresponding medicinal chemistry publication. The syntheses routes for most of the pharmaceutical substances (APIs, active pharmaceutical ingredients) will also be found in [3].

Ullmann's Pharmaceuticals. Edited by Axel Kleemann and Bernhard Kutscher.
© 2022 Wiley-VCH GmbH
Set ISBN: 978-3-527-34252-5/ DOI: 10.1002/14356007.a03_009.pub3

2. A04AA Serotonin (5HT$_3$) Antagonists (Table 1)

Table 1. A04AA Serotonin (5HT$_3$) antagonists

INN (Synonyms) [Brand names] ATC	Structure (Remarks)	[CAS-No.] Formula MW, g/mol	Target (if known) Medical use Formulation DDD	Originator Approval (country/year) Production method [References]
Ondansetron (GR-38032F; GR-C507/75; SN-307) [Axisetron, Cellondan, Zofran, Zophren, Zotrix] A04AA01 **WHO essential medicine**	(racemate)	[99614-02-5] C$_{18}$H$_{19}$N$_3$O 293.37 hydrochloride dihydrate: [99614-01-4] C$_{18}$H$_{19}$N$_3$O ·HCl·2 H$_2$O 365.86	5-HT$_3$ receptor antagonist. medical use for prevention of nausea and vomiting caused by cancer chemotherapy, radiation therapy, or postsurgery and treatment of gastroenteritis. amp. 4 mg/2 mL, 8 mg/4 mL; f.c. tbl. 4 mg, 8 mg. 16 mg O, P, R.	Glaxo USA/1990 synthesis [4, 5]
Granisetron (BRL-43694A) [Axigran, Granisol, Kevatril, Kytril, Sancuso, Sustol] A04AA02		[109889-09-0] C$_{18}$H$_{24}$N$_4$O 312.42 hydrochloride: [107007-99-8] C$_{18}$H$_{24}$N$_4$O ·HCl 348.88	5-HT$_3$ receptor antagonist. medical use for prevention of nausea and vomiting caused by cancer chemotherapy, radiation therapy, or postsurgery and treatment of gastroenteritis. amp. 1 mg/1 mL, 3 mg/3 mL; transdermal (TD) patch ER 34.3 mg/52 cm^2; tbl. 1 mg, 2 mg. 2 mg O, 3 mg P, 3.1 mg TD patch.	Beecham GB/1991; USA/1994 synthesis [6, 7]
Tropisetron (ICS-205-930) [Navoban, Novaban] A04AA03		[89565-68-4] C$_{17}$H$_{20}$N$_2$O$_2$ 284.36 hydrochloride: [105826-92-4] C$_{17}$H$_{20}$N$_2$O$_2$ ·HCl 320.82	5-HT$_3$ receptor antagonist. medical use for prevention of nausea and vomiting caused by cancer chemotherapy, radiation therapy, or postsurgery and treatment of gastroenteritis. cps. 5 mg; amp. 5 mg/5 mL. 5 mg O, P.	Sandoz USA/1992 synthesis [8]
Dolasetron (MDL-73147; MDL-73147EF) [Anemet, Anzemet] A04AA04 wfm (2011)		[115956-12-2] C$_{19}$H$_{20}$N$_2$O$_3$ 324.38 methanesulfonate: [115956-13-3] C$_{19}$H$_{20}$N$_2$O$_3$ ·CH$_3$SO$_3$H 420.48	5-HT$_3$ receptor antagonist. medical use for prevention of nausea and vomiting caused by cancer chemotherapy, radiation therapy, or postsurgery and treatment of gastroenteritis. amp. 12.5 mg/0.625 mL, 100 mg/5 mL; f.c. tbl. 50 mg, 200 mg. 0.2 g O, 0.1 g P.	Merrell-Dow Pharm. DE/1997 synthesis [9]
Palonosetron (RS-25259-197) [Aloxi, Onicit; Akynzeo] A04AA05		[135729-61-2] C$_{19}$H$_{24}$N$_2$O 296.41 hydrochloride: [135729-62-3] C$_{19}$H$_{24}$N$_2$O ·HCl 332.87	5-HT$_3$ receptor antagonist. medical use for prevention and treatment of nausea and vomiting caused by cancer chemotherapy. Also comb. with NK$_1$ receptor antagonist netupitant (see Table 3) for same indication (Akynzeo). cps. 0.5 mg; amp. 0.25 mg/5 mL. 0.5 mg O, 0.25 mg P.	Syntex USA/2003; DE/2005 synthesis [10, 11]

3. A04AB Antihistamines (Table 2) (→ Respiratory Disorders, 7. Antihistamines for Systemic Use (R06))

Table 2. A04AB Antihistamines

INN (Synonyms) [Brand names] ATC	Structure (Remarks)	[CAS-No.] Formula MW, g/mol	Target (if known) Medical use Formulation DDD	Originator Approval (country/year) Production method [References]
Dimenhydrinate (diphenhydramine teoclate) [Amosyt, Antemin, Arlevert, Dramamine, Dramina, Gravol, Nausicalm, Superpep, Travel-gum, Viabom, Vomex A, Xamamina] A04AB02; R06AA52		[523-87-5] $C_{17}H_{21}NO$ $\cdot C_7H_7ClN_4O_2$ 469.97	1:1 adduct (salt) of the histamine H1 antagonist diphenhydramine and the mild stimulant 8-chlorotheophylline which counteracts drowsiness. treatment of motion sickness and nausea. amp. 62 mg/10 mL; syrup 33 mg/10 mL; tbl. 50 mg; drg. 50 mg; cps. 120 mg, 150 mg; suppos. 40, 70, 80, 150 mg.	Searle USA/1949 synthesis [12, 13]
Thiethylperazine (GS-95; NSC-130044) [Torecan, Toresten, Norzine] A04AB03; R06AD03 wfm 2014		[1420-55-9] $C_{22}H_{29}N_3S_2$ 399.62 dimaleate: [1179-69-7] $C_{22}H_{29}N_3S_2$ $\cdot 2\,C_4H_4O_4$ 631.76	dopamine D_2 receptor antagonist. medical use as antiemetic. f.c. tbl. 6.5 mg; suppos. 6.5 mg.	Sandoz CH/1960 synthesis [14]
Meclozine (meclizine; histamethizine; UCB-5062) [Agyrax, Ancolan, Antivert, Bonamine, Bonine, Postafen] A04AB04; R06AE05	(racemate)	[569-65-3] $C_{25}H_{27}ClN_2$ 390.96 dihydrochloride monohydrate: [31884-77-2] $C_{25}H_{27}ClN_2$ $\cdot 2\,HCl\cdot H_2O$ 481.89	H_1 antihistamine for treatment of motion sickness. tbl. 12.5, 25, 50 mg; suppos. 50 mg. 37.5 mg O, 50 mg R.	UCB GB/1953 synthesis [15]
Diphenhydramine (benzhydramine; DPH) [Benadryl, Benocten, Betadorm, Detensor, Histaderm, Lunadon, Nytol, Sedopretten, Sominex, Somnium, Sunsan, Unisom Sleepgels, Zantall] A04AB05; D04AA32; N05CM20; R06AA02 (→ Dermatologicals (D), 2. Antifungals (D01), Emollients and Protectives (D02), and Antipruritics (D04) for Dermatological Use); (→ Hypnotics and Sedatives (N05C))		[58-73-1] $C_{17}H_{21}NO$ 255.36 hydrochloride: [147-24-0] $C_{17}H_{21}NO\cdot HCl$ 291.82	inverse agonist of histamine H_1 receptor and a competitive antagonist of muscarinic acetylcholine receptors and sodium channel blocker (local anesthetic). medical use as antiallergic, sedative, antiemetic, in motion sickness, insomnia, and as antiparkinson drug. tbl. 25 mg, 50 mg; s.r. cps. 30 mg; drops 12.5 mg; suppos. 10, 20, 50 mg.	Parke Davis USA/1946 synthesis [16, 17]

4. A04AD Other Antiemetics (Table 3)

Table 3. A04AD Other antiemetics

INN (Synonyms) [Brand names] ATC	Structure (Remarks)	[CAS-No.] Formula MW, g/mol	Target (if known) Medical use Formulation DDD	Originator Approval (country/year) Production method [References]
Scopolamine (hyoscine; 6,7-epoxytropine) [Scopoderm TTS, Transcop, Transderm Scop] A04AD01; N05CM05; S01FA02 (→ Hypnotics and Sedatives (N05C)) **WHO essential medicine**		[51-34-3] $C_{17}H_{21}NO_4$ 303.36 hydrochloride: [55-16-3] $C_{17}H_{21}NO_4$ ·HCl hydrobromide: [114-49-8] $C_{17}H_{21}NO_4$ ·HBr 384.27 hydrobromide trihydrate: $C_{17}H_{21}NO_4$ ·HBr·3 H_2O 438.32	competitive blocker/antagonist of muscarinic acetylcholine receptors. medical use for postoperative nausea and vomiting, eye drops, and transdermal patch against motion sickness and sea sickness. amp. 0.3, 0.5, 1 mg/mL. eye drops 0.25%; patch 1.5 mg/2.5 cm². 0.9 mg O, P.	A. Ladenburg (isolation, 1881) extraction (from *Datura metel* or *Duboisia*) [19]; CNS effects in humans [18]
Metopimazine (EXP-999; NG-101; RP-9965) [Vogalene] A04AD05		[14008-44-7] $C_{22}H_{27}N_3O_3S_2$ 445.60	antagonist of dopamine D_2 and D_3 receptors. gastroprokinetic/ antiemetic. cps. 7.5 mg, 15 mg; tbl. 2.5 mg; sol. 5 mg/5 mL; suppos. 5 mg. 15 mg O, P.	Rhone-Poulenc synthesis [20]
Triflupromazine (RP-7746) [Psyquil, Siquil, Vesprin] A04AD06; N05AA05 (→ Antipsychotics (Neuroleptics, N05A)) wfm in many countries, withdrawn from German market in 2003		[146-54-3] $C_{18}H_{19}F_3N_2S$ 352.42 hydrochloride: [1098-60-8] 388.88	D_1 and D_2 receptor antagonists. was used as neuroleptic (antipsychotic) and antiemetic drug. tbl. 10, 25, 50 mg.	Squibb; Smith Kline & French (SK&F) synthesis (from 2-trifluoromethylpheno-thiazine) [21–23]
Dronabinol ((−)-*trans*-Δ⁹- Tetrahydrocannabinol; Δ⁹-THC; QCD-84924; principal active constituent of cannabis) [Dronabinol, Marinol, Syndros] A04AD10 controlled substance: WHO schedule IV; CSA schedule III		[1972-08-3] $C_{21}H_{30}O_2$ 314.47	targets and binds to cannabinoid receptors (CBR) in the CNS and acts directly on the appetite and vomiting control centers in the brain to stimulate appetite and prevent emesis. It also induces analgesia. cps. 2.5, 5, 10 mg. (Marinol contains synthetic product)	Solvay USA/1986 extraction from medicinal cannabis, e.g., *C. sativa* L. [24]; synthesis starting from limonene and olivetol [25–28]

Table 3. (*Continued*)

INN (Synonyms) [Brand names] ATC	Structure (Remarks)	[CAS-No.] Formula MW, g/mol	Target (if known) Medical use Formulation DDD	Originator Approval (country/year) Production method [References]
Nabilone (LY-109514) [Cesamet, Canemes] A04AD11		[51022-71-0] $C_{24}H_{36}O_3$ 372.55	synthetic cannabinoid that mimics the natural THC. medical use as antiemetic (especially for chemotherapy-induced nausea and vomiting) and as an adjunct analgesic for neuropathic pain and for chronic pain management. cps. 1 mg. 3 mg O.	Eli Lilly USA/1985 synthesis [29, 30]
Aprepitant (MK-0869) [Emend] A04AD12 **WHO essential medicine**		[170729-80-3] $C_{23}H_{21}F_7N_4O_3$ 534.43	neurokinin-1 (NK_1) receptor for substance P antagonist. medical use for prevention of chemotherapy-induced as well as postoperative nausea and vomiting. cps. 80 mg, 125 mg. 95 mg O.	Merck & Co. USA/2003; EU/2003 synthesis [31, 32]
Fosaprepitant dimeglumine (MK-0517; water-soluble injectable prodrug of aprepitant) [Ivemend] A04AD12		[265121-04-8] $C_{23}H_{22}F_7N_4O_6P$ $\cdot 2\ C_7H_{17}NO_5$ 1004.84	prodrug of aprepitant. vials 150 mg for preparation of an inj. sol. 150 mg P.	Merck & Co. USA/2008; EU/2008 synthesis [33, 34]
Rolapitant (SCH-619734) [Varubi, Varuby] A04AD14		[552292-08-7] $C_{25}H_{26}F_6N_2O_2$ 500.48 hydrochloride monohydrate: [914462-92-3] $C_{25}H_{26}F_6N_2O_2$ $\cdot HCl \cdot H_2O$ 554.9	selective and long-acting NK_1 receptor antagonist. treatment and prevention of chemotherapy-induced nausea and vomiting. f.c. tbl. 90 mg. 0.18 g O.	Schering Corp. USA/2015; EU/2017 synthesis [35, 36]
Netupitant (RO 67-3189) [Akynzeo (comb. with palonosetron)] A04AA55		[290297-26-6] $C_{30}H_{32}F_6N_4O$ 578.6 monohydro-chloride: [290296-54-7] $C_{30}H_{32}F_6N_4O$ $\cdot HCl$ 615.1	selective and long-acting NK_1 receptor antagonist. treatment and prevention of chemotherapy-induced nausea and vomiting (including the highly emetogenic effect of cisplatin), in the form of a fixed-dose comb. with palonosetron (see Table 1). cps. 300 mg netupitant and 0.5 mg palonosetron.	Roche; Helsinn Health Care USA/2014; EU/2015 synthesis [37, 38] compositions [39]

List of Abbreviations

amp.	ampoule(s)
API	active pharmaceutical ingredient
ATC	Anatomical Therapeutic Chemical Classification
CBR	cannabinoid receptor
cps.	capsule(s)
CSA	United States Controlled Substances Act
DDD	defined daily dose
drg.	dragees
ER	extended release
f.c. tbl.	film-coated tablets
inj.	injection
INN	international nonproprietary name
MW	molecular mass
NK	neurokinin
sol.	solution
suppos.	suppositories
THC	tetrahydrocannabinol
wfm	withdrawn from market
WHO	World Health Organization

References

1 WHO Collaborating Centre for Drug Statistics Methodology (2020) *Guidelines for ATC Classification and DDD Assignment 2019*, WHO Collaborating Centre for Drug Statistics Methodology, Oslo, Norway.

2 Wallace, J.L. and Sharkey, K.A. (2011) Chapter 46, Treatment of Disorders of Bowel Motility and Water Flux; Antiemetics; Agents Used in Biliary and Pancreatic Disease, in *Goodman & Gilman's The Pharmacological Basis of Therapeutics*, (eds. L.L. Brunton, et al.), 12th edn, McGraw-Hill, London, pp. 1323–1349.

3 Kleeman, A., Engel, J., Kutscher, B. and Reichert, D. eds. *Pharmaceutical Substances Online*, Thieme Chemistry, Stuttgart.

4 Glaxo, (1987) US 4 695 578 (GB-prior. 1984).

5 Glaxo, (1988) US 4 753 789 (GB-prior. 1985).

6 Bermudrez, J. (1994) *Bioorg. Med. Chem. Lett.* **4**, 2376.

7 Beecham, (1989) US 4 886 808 (GB-prior. 1985).

8 Sandoz, (1988) US 4 789 673 (CH-prior. 1982).

9 Merrell Dow, (1990) US 4 906 755 (USA-prior. 1986).

10 Clark, R.D., et al. (1993) *J. Med. Chem.* **36**, 2645.

11 Syntex, (1993) US 5 202 333 (USA-prior. 1989).

12 Searle, (1950) US 2 499 058 (USA-prior. 1949).

13 Searle, (1950) US 2 534 813 (USA-prior. 1950).

14 Bourquin, J.-P., et al. (1958) *Helv. Chim. Acta* **41**, 1072.

15 UCB, (1955) US 2 709 169 (BE-prior. 1951).

16 Parke Davis, (1947) US 2 421 714 (USA-prior. 1944).

17 Parke Davis, (1947) US 2 427 878 (USA-prior. 1947).

18 Safer, D.J., et al. (1971) *Biol. Psychiatry* **3**, 347.

19 Chemnitius, F. (1928) *J. Prakt. Chem.* **120**, 221.

20 Rhone-Poulenc, (1962) DE 1 092 476 (FR-prior. 1958).

21 Craig, P.N., et al. (1959) *J. Org. Chem.* **22**, 709.

22 Squibb, (1969) DE 1 095 836 (USA-prior. 1955, 1956).

23 SK&F, (1960) US 2 921 069 (USA-prior. 1956).

24 McKinney, L.O. (1981) US 4 279 824 (USA-prior. 1979).

25 Sheehan Inst. for Research, (1978) US 4 116 979 (USA-prior. 1975).

26 Aerojet, (1983) US 4 381 399 (USA-prior. 1981).

27 Mechoulam, R., et al. (1976) *Chem. Rev.* **76**, 75.

28 WHO, (2018) Critical Review, Delta-9-Tetrahydro-Cannabinol.

29 Archer, R.A., et al. (1977) *J. Org. Chem.* **42**, 2277.

30 Eli Lilly, (1976) US 3 968 125 (USA-prior).

31 Hale, J.J., et al. (1998) *J. Med. Chem.* **41**, 4607.

32 Merck & Co., (1998) US 5 719 147 (USA-prior. 1992).

33 Hale, J.J., et al. (2000) *J. Med. Chem.* **43**, 1234.

34 Merck & Co., (1997) US 5 691 336 (USA-prior. 1994).

35 Schering Corp., (2006) US 7 049 320 (USA-prior. 2001).

36 OPKO Health, (2014) US 8 796 299 (USA-prior. 2001).

37 Harrington, P.J., et al. (2006) *Org. Process Res. Dev.* **10**, 1157.

38 Roche, (2001) US 6 297 375 (EP-prior. 1999).

39 Helsinn, (2012) US 8 623 826 (USA-prior. 2009).

Bile and Liver Therapy (A05)

AXEL KLEEMANN, Hanau, Germany

1. Introduction

In this article, the bile and liver therapeutics are arranged according to the Anatomical Therapeutical Chemical (ATC) Classification System of the WHO. The individual active pharmaceutical ingredients (APIs) are described by their INN (international nonproprietary name), synonyms, brand or trade names, ATC number, structure, CAS registry number, formula, molecular mass, mode of action, medical use, formulation, DDD (defined daily dose), originator, country and/or date of approval, production method, and references. The references for each drug typically include the basic patents and publications in which the relevant synthetic or biosynthetic processes for production are described. As the bile acids are natural substances (except the semisynthetic obeticholic and dehydrocholic acids), it is difficult to find out an originator, and for this reason the first isolations are referenced.

Bovine bile (fel tauri, also called "ox bile") is listed as A05AA05. Besides its main use as dietary supplement without clear medical indication, it still finds some use in homeopathic preparations, but as it is no chemical entity, it is not included here. The "other drugs for bile therapy" in Section 2.3 had been used in the 1950s–1970s and are now obsolete because of their limited clinical efficiency, and only two of them are still on the market in very few countries.

A special situation can be seen with the so-called lipotropics classified in A05BA, which are based on the proteinogenic amino acids arginine and ornithine and the Krebs-cycle metabolite α-ketoglutaric acid. They are still used in some countries for treatment of liver diseases, although their clinical efficacy in reduction of hyperammonemia to alleviate hepatic encephalopathy is quite weak. These lipotropics are even no longer mentioned in the relevant textbooks of pharmacology. The other substances of this group are no longer in use; glycyrrhizic acid has some use as sweetener.

Ullmann's Pharmaceuticals. Edited by Axel Kleemann and Bernhard Kutscher.
© 2022 Wiley-VCH GmbH
Set ISBN: 978-3-527-34252-5/ DOI: 10.1002/14356007.a12_143.pub2

2. A05A Bile Therapy

2.1. A05AA Bile Acids and Derivatives (Table 1)

Table 1. A05AA Bile acids and derivatives

INN (Synonyms) [Brand names] ATC	Structure (Remarks)	[CAS-No.] Formula MW, g/mol	Target (if known) Medical use Formulation DDD	Originator Approval (country/year) Production method [References]
Chenodeoxycholic acid (CDC; chenodiol; chenocholic acid; 3α,7α-dihydroxy-5β-cholan-24-oic acid) [Chendol, Chenocol, Chenofalk, Chenossil, Cholanorm, Fluibil, Henohol, Xenbilox] A05AA01		[474-25-9] $C_{24}H_{40}O_4$ 392.58	besides cholic acid main natural constituent of bile acid, occurring as N-glycine and N-taurine conjugate. Forms mixed micelles with lecithin in bile, which solubilize cholesterol and thus facilitates its excretion. medical use for dissolving gallstones. tbl. 250 mg; cps. 100, 250 mg. 1 g O.	Dr. Falk Pharma/et al. USA/1983 semisynthetic (from cholic acid) [1–6]
Ursodeoxycholic acid (ursodiol; UDCA) [Actigall, Cholit Ursan, Delursan, Desol, Deursil, Litursol, Urdes, Ursacol, Urso, Ursochol, Ursofalk, Ursolvan) A05AA02	(epimer (7β) of chenodeoxycholic acid)	[128-13-2] $C_{24}H_{40}O_4$ 392.58	mammalian bile acid found first in the bile of bears (Ursidae) as conjugate with taurine. used to dissolve cholesterol gallstones (anticholelithogenic) and to treat cholestatic forms of liver diseases including primary biliary cirrhosis; several times more potent than chenodiol. f.c. tbl. 250, 400, 500 mg; cps. 250, 300 mg; susp. 50 mg/mL. 0.75 g O.	(bear bile as cholagogue was for long time a Chinese and Japanese folk medicine) Alfa Farmaceutici synthesis (from cholic or chenodeoxycholic acid) [1, 7–10]
Cholic acid (3α,5β,7α,12α)-3,7,12-trihydroxy-24-oic acid) [Cholbam, Kolbam, Orphacol] A05AA03		[81-25-4] $C_{24}H_{40}O_5$ 408.58	cholic acid is a major primary bile acid besides chenodeoxycholic acid produced in the liver and is usually conjugated with glycine or taurine. treatment of inborn errors in primary bile acid synthesis due to 3β-hydroxy-Δ5-C27-steroid oxidoreductase deficiency or Δ43-oxosteroid-5β-reductase deficiency in infants, children and adolescents aged 1 month to 18 years and adults. cps. 50, 250 mg. 0.25 g O children.	isolation: Wieland, H. (1939) extraction from beef bile [11]

Table 1. (*Continued*)

INN (Synonyms) [Brand names] ATC	Structure (Remarks)	[CAS-No.] Formula MW, g/mol	Target (if known) Medical use Formulation DDD	Originator Approval (country/year) Production method [References]
Obeticholic acid (OCA; 6α-ethyl-chenodeoxycholic acid; INT-747) [Ocaliva] A05AA04		[459789-99-2] $C_{26}H_{44}O_4$ 420.6	potent agonist of the farnesoid X nuclear receptor (FXR). treatment of liver diseases including primary biliary cholangitis. tbl. 5, 10 mg. 10 mg O.	intercept Pharmaceut./Dainippon Sumitomo Pharma USA/2016 (orphan drug) EU/2016 semisynthesis [12–16]
Dehydrocholic acid ((5β)-3,7,12-trioxo-cholan-24-oic acid) [Bilidren, Cholagon, Decholin, Dehychol, Ketochol] A05AA11		[81-23-2] $C_{24}H_{34}O_5$ 402.53 sodium salt: [145-41-5] $C_{24}H_{33}NaO_5$ 424.51	choleretic (stimulating effect on bile secretion by the liver). as a drug it was discontinued in most countries, and it is in use only in very few combination drugs in Asian markets and as a dietary supplement.	semisynthesis (oxidation of cholic acid) [17, 18]

2.2. A05AB Preparations for Biliary Tract Therapy

There is only one substance mentioned, *N*-hydroxymethylnicotinamide, which can no longer be identified as an ingredient of marketed drugs.

2.3. A05AX Other Drugs for Bile Therapy

This ATC group mainly lists rather old choleretics/cholagogues, of which only hymecromone and fenipentol have a certain interest in a few markets. The other three APIs (piprozolin, cyclobutyrol, and febuprol) are no longer in medical use (Table 2).

Table 2. A05AX Other drugs for bile therapy

INN (Synonyms) [Brand names] ATC	Structure (Remarks)	[CAS-No.] Formula MW, g/mol	Target (if known) Medical use Formulation DDD	Originator Approval (country/year) Production method [References]
Hymecromone (imecromone; LM-94; 4-methyl-umbelliferon; 4-MU; 7-hydroxy-4-methylcoumarin) [Cantabiline, Cholestil, Cholonerton, Cholspasmin, Tridemon] A05AX02		[90-33-5] $C_{10}H_8O_3$ 176.17	choleretic and antispasmotic drug for bile therapy. amp. 200 mg; tbl. 400 mg; cps. 200, 400 mg. 1 g O.	Lipha synthesis (from resorcinol and ethyl acetoacetate) [19–21]
Fenipentol (PC 1; phenylbutylcarbinol) [Euralan, Febichol, Pancoral, Pentabil] A05AX07	(racemate)	[583-03-9] $C_{11}H_{16}O$ 164.25	choleretic for bile therapy. cps. 100 mg.	Thomae synthesis [22]

3. A05B Liver Therapy, Lipotropics

3.1. A05BA Liver Therapy (Table 3)

Table 3. A05BA Liver therapy

INN (Synonyms) [Brand names] ATC	Structure (Remarks)	[CAS-No.] Formula MW, g/mol	Target (if known) Medical use Formulation DDD	Originator Approval (country/year) Production method [References]
Arginine glutamate (L-arginine-L-glutamate) [Dynamisan forte, Modumate] A05BA01		[4320-30-3] $C_{11}H_{23}N_5O_6$ 321.33	the semiessential amino acid L-arginine is a donor of nitric oxide (NO) which is a relaxing factor of blood vessels. ammonium detoxicant in case of hepatic failure. mainly used as tonic (5 g/day), treatment of fatigue (and erectile dysfunction). drinking amp. 5 g/10 mL; sachet 5 g.	mixture of the two natural amino acids, both produced by fermentation [23, 24]
Citiolone (AHCTL; N-acetyl-homocysteine thiolactone; BO-714) [Citiolase, Reducdyn] A05BA04	(racemate)	[1195-16-0] $C_6H_9NO_2S$ 159.20	radical scavenger. used for treatment of hepatic disorders and constituent in some amino acid infusion solutions. tbl. 50 mg.	Nordmark synthesis (from methionine) [25, 26]
Ornithine oxoglurate (L-ornithine α-ketoglutarate; ornithine oxoglutarate) [Cetornan, Ornicetil] A05BA06		[34414-83-0] $C_{10}H_{18}N_2O_7$ 278.26	used for building muscle protein after surgery or stroke and treatment of brain disorders caused by liver disease and as dietary supplement. sachets 5 g powder for preparing oral solutions.	AT/1973 [27, 28]
Tidiacic arginine (thiazolidine-2,4-dicarboxylic acid compound with L-arginine; arginine tidiacicate) [Tiadilon] A05BA07		[30986-62-0] $C_{11}H_{21}N_5O_6S$ 351.38	hepatoprotective drug. cps. 100 mg.	Lab. CRINEX FR/1985 synthesis [29]
Ornithine aspartate (L-ornithine-L-aspartate; LOLA) [Hepa-Merz, Hepa-Vibolex, Heparut] A05BA07		[3230-94-2] $C_9H_{19}N_3O_6$ 265.26	amino acids. treatment of reduced brain function in patients with advanced liver disease (hepatic encephalopathy) in order to reduce toxic ammonia in the blood; constituent of amino acid infusion solutions. granulate 3 g, 6 g; amp. 5 g/10 mL for infusion.	Merz DE/2004 fermentation (both amino acids) and cocryst. reviews: [30–32]

List of Abbreviations

amp.	ampoules
API	active pharmaceutical ingredient
ATC	Anatomical Therapeutical Chemical Classification
cps.	capsules
DDD	defined daily dose
f.c. tbl.	film-coated tablets
FXR	farnesoid X nuclear receptor
infus.	infusion
INN	international nonproprietary name
susp.	suspension
WHO	World Health Organization

References

1 Fieser, L.F. and Rajagopalan, S. (1950) *J. Am. Chem. Soc.* **72**, 5530.

2 Hauser, E., et al. (1960) *Helv. Chim. Acta* **43**, 1595.

3 Hofmann, A.F. (1963) *Acta Chem. Scand.* **17**, 173.

4 Sato, Y. and Ikekawa, N. (1959) *J. Org. Chem.* **24**, 1367.

5 Saltzman, W.A. and Intell. Property Dev. Corp., (1974) US 3 833 620, USA-prior. 1972.

6 Purification: Diamalt, (1979) US 4 163 017, DE-prior. 1976.

7 First isolation: Shoda, M. (1927) *J. Biochem. (Tokyo)* **7**, 505.

8 Erregierre SpA, (1983) US 4 379 093, IT-prior. 1981.

9 Zambon, (1984) US 4 486 352, IT-prior. 1982.

10 Critical review syntheses: Tonin, F. and Arends, I.W.C.E. (2018) *Beilstein J. Org. Chem.* **14**, 470–483.

11 Wieland, H. and Siebert, Z. (1939) *Physiol. Chem.* **262** (1), 3569–3572.

12 Pellicciari, R., et al. (2002) *J. Med. Chem.* **45**, 3569.

13 Feng, W.-D., et al. (2019) *Org. Process Res. Dev.* **23**, 1979–1989.

14 Intercept, (2006) US 7 138 390, USA-prior. 2001.

15 Intercept, (2011) US 8 058 267, USA-prior. 2001.

16 Intercept, (2013) US 8 377 916, USA-prior. 2001.

17 Sterling Drug, (1951) US 2 576 728, USA-prior. 1950.

18 Merck & Co., (1960) US 2 966 499, USA-prior. 1958.

19 Pechmann, H. von and Duisberg, C. (1883) *Ber.* **16**, 2119.

20 Woods, L.L. and Sapp, J. (1962) *J. Org. Chem.* **27**, 3703.

21 Lipha, (1961) FR-M 1 430.

22 Thomae, (1963) US 3 084 100, USA-prior. 1961.

23 General Mills, (1958) US 2 851 482, USA-prior. 1957.

24 Zieve, L., et al. (1989) *Metab. Brain Dis.* **4**, 113.

25 Degussa, (1963) DE 1 134 683, DE-prior. 1961.

26 Kirnberger, E.J., et al. (1958) *Arzneim.-Forsch.* **8**, 72.

27 Michel, H., et al. (1971) *Presse Med.* **79**, 867–868.

28 Szam, I., et al. (1974) *Wien Med. Wochenschr.* **124** (20), 319–325.

29 Rizzo, S. (1986) *Int. J. Clin. Pharmacol. Res.* **6**, 225.

30 Dhiman, R.K. and Chawla, Y.K. (2008) *Trop. Gastroenterol.* **29**, 6–12.

31 Rücker, D. (1997) *Pharmazeut. Zeitung* (40), 29.09.1997.

32 Degussa, (1995) US 5 405 761, DE-prior. 1990.

Drugs for Constipation (A06)

AXEL KLEEMANN, Hanau, Germany

ULLMANN'S ENCYCLOPEDIA OF INDUSTRIAL CHEMISTRY

1. A06A Drugs for Constipation — Introduction

The gastrointestinal (GI) tract as the digestive system is very complex [1]. It is a hollow muscular tube starting from the oral cavity with food entering the mouth and ending at the rectum and anus where digested food residues are expelled. Between the start and end are the pharynx, esophagus (oesophagus), stomach, small, and large intestine (gut). The small intestine is subdivided into the duodenum, jejunum, and ileum. Accessory organs such as salivary glands, liver, pancreas, and gall bladder secrete enzymes, which are necessary for the digestive processes. In addition, special cells in the tract release the so-called digestive hormones (gastrin, secretin, cholecystokinin, and ghrelin). Important as regulator of the the physiological gut function is the enteric nervous system besides the autonomic nervous system. The GI tract contains trillions of microbes (gut microbiota and gut flora; part of human microbiome), which also contribute to metabolism and immune status.

Without going deep into the details, it is necessary to be aware of some facts: digestion takes place mainly in the stomach and small intestine, where food is broken down into its basic nutritive building blocks, which are absorbed via the epithelium of the small intestine and enter the circulation. In the large intestine excess water is reabsorbed, and undigested and secreted waste material (the stool) is excreted via defecation. The defecation process needs well-functioning smooth muscles to effect the harmonized peristaltic movements in the large intestine. If any of the complex systems or processes is not working well, then GI disease symptoms such as nausea, vomiting, diarrhea, malabsorption, or constipation will occur. As there are many different functions, many causes and targets for therapeutical interventions have to be considered. An important GI disorder is constipation, where many different kinds of drugs are in medical use, as can be seen by the following chapters and tables. It is estimated that 25% of the population of the USA, more commonly women and elderly people, complain of constipation [1]. The main causes for constipation are low-fiber diet, low liquid intake, certain diseases (e.g., irritable bowel syndrome, celiac disease, and colon cancer), and side effects of many drugs (e.g., anticholinergics and calcium antagonists), whereby intake of opioids leads to constipation in 90% of the patients.

Ullmann's Pharmaceuticals. Edited by Axel Kleemann and Bernhard Kutscher.
© 2022 Wiley-VCH GmbH
Set ISBN: 978-3-527-34252-5/ DOI: 10.1002/14356007.a15_183.pub3

It has to be mentioned that the terms cathartics, laxatives, and purgatives often are used interchangeably. In the following chapters, the different drug groups are reviewed and arranged according to their Anatomical Therapeutic Chemical Classification (ATC) of the WHO in which they are arranged according to their indication and chemical structure [2]. Publications on pharmacology and therapeutic application of the drugs of A06 are thoroughly reviewed in [1]. The synthesis routes for most of the active pharmaceutical ingredients (APIs) will be found in [3].

The group A06AC lists bulk-forming laxatives of natural origin such as linseed, psylla seed products, and methyl cellulose but no defined molecules, and so this group is omitted here. The group A06AG lists "oil" under A06AG06, but this is a term for many different products, and therefore it is not possible to include this here.

2. A06AA Softeners, Emollients (Table 1)

Table 1. A06AA Softeners, emollients

INN (Synonyms) [Brand names] ATC	Structure (Remarks)	[CAS-No.] Formula MW, g/mol	Target (if known) Medical use Formulation DDD	Originator Approval (country/year) Production method [References]
Docusate sodium (DSS; sodium dioctyl sulfosuccinate) [Colace, Comfolax, Coprola, Dioctylal, Dioctyl, Diotilan, Disonate, Doxinate, Doxol, Dulcivac, Jamylene, Molatoc, Molcer, Nevax, Norgalax, Otitex, Otowax, Regutol, Soliwax, Velmol, Waxsol, Yal] A06AA02 **WHO essential medicine**		[577-11-7] $C_{20}H_{37}NaO_7S$ 444.56	anionic surfactant; stool softener, management of constipation; softening wax in the ear as ear drops 0.5% or 5%. cps. 50, 100 mg; suppos. 10 mg; drinking amp. 50, 100 mg. 0.15 g O.	American Cyanamid synthesis [4, 5]

3. A06AB Contact Laxatives (Table 2)

Oxyphenisatine is still listed under A06AB01. It was on the market but was withdrawn worldwide in the 1980s because of liver toxicity. So it is not included here.

Table 2. A06AB Contact laxatives

INN (Synonyms) [Brand names] ATC	Structure (Remarks)	[CAS-No.] Formula MW, g/mol	Target (if known) Medical use Formulation DDD	Originator Approval (country/year) Production method [References]
Bisacodyl [Bicol, Correctol, Drix, Dulcolax, Durolax, Florisan, Mediolax, Muxol, Protolax, Tavolax] A06AB02; A06AG02		[603-50-9] $C_{22}H_{19}NO_4$ 361.40	acts with a parasympathetic effect on mucosal sensory nerves, thereby increasing peristaltic contractions. also used as a laxative in pre- and postoperative treatment. tbl. 5 mg; suppos. 5, 10 mg; enema 10 mg/37 mL. 10 mg O, R.	Thomae DE/1953 synthesis [6]
Dantron (danthrone; chrysazin; 1,8-dihydroxy-anthraquinone) [Altan, Antrapurol, Codalax, Diaquone, Fenogar, Istizin, Modane, Modaton, Normax] A06AB03; A06AG03 wfm in many countries		[117-10-2] $C_{14}H_8O_4$ 240.21	used as a laxative in many countries, but now only in a few countries in limited/restricted use. It is considered as a carcinogen. tbl. 37.5, 75 mg; tbl. 5 mg; suppos. 10 mg; cps. 200 mg poloxamer + 25 mg dantron (= Codalax). 50 mg O.	(Generic) DE/1913 synthesis [7]
Phenolphthalein (3,3-bis(p-hydroxy-phenyl)phthalide) A06AB04 wfm		[77-09-8] $C_{20}H_{14}O_4$ 318.33	similar to dantron, the substance is an effective laxative and was about 100 years in medical use before it was suspected to be a carcinogen and has been withdrawn as a drug in all countries.	DE/1902 synthesis [8–11]

(continued)

Table 2. (*Continued*)

INN (Synonyms) [Brand names] ATC	Structure (Remarks)	[CAS-No.] Formula MW, g/mol	Target (if known) Medical use Formulation DDD	Originator Approval (country/year) Production method [References]
Sennoside (sennosides; mixture of sennosides A and B; senna glycosides) [Bekunis, Colonorm, Darmol, Dragees 19 Senna, Glysennid, Nytilax, Pursennid, Sennocol, Sennokot, Senolax; many comb. preparations ww] A06AB06 **WHO essential medicine**	(diastereoisomeric mixture of A and B; different in 9,9′-position, i.e., sennoside A: 9R,9′R; sennoside B: 9R,9′S)	sennoside A: [81-27-6] $C_{42}H_{38}O_{20}$ 862.75 sennoside B: [128-57-4] $C_{42}H_{38}O_{20}$ 862.75	anthraquinone glycoside; stimulant laxative for treatment of constipation and for bowel evacuation before diagnostic procedures or surgery. available in different oral formulations (tbl., cps., sol., and granules) and rectal formulations, e.g., tbl. 8.6, 15, 20, 25 mg; sol. 8.8 mg/5 mL.	the plant senna belongs to the legume family and there are more than 300 species known. The type species for the extracts is *Senna alexandrina*. The extracts are known and used since the early medieval times. [12–14]
Picosulfate sodium (sodium picosulfate) [Agiolax Pico, Darmol Pico, Dulcolax, Guttalax, Laxoberal, Laxoberon, Neopax, Picolax, Prepopik] A06AB08		[10040-45-6] $C_{18}H_{13}NNa_2O_8S_2$ 481.40	contact stimulant laxative/cathartic for treatment of constipation or to prepare the large bowel before colonoscopy or surgery. cps. 2.5, 5 mg; drops 7.5 mg/mL. 5 mg O.	Istituto de Angeli; Thomae synthesis [15–17]

4. A06AD Osmotically Acting Laxatives (Table 3)

Table 3. A06AD Osmotically acting laxatives

INN (Synonyms) [Brand names] ATC	Structure (Remarks)	[CAS-No.] Formula MW, g/mol	Target (if known) Medical use Formulation DDD	Originator Approval (country/year) Production method [References]
Magnesium carbonate [Magnofit, Magnosolv] A06AD01; A02AA01 (→ Drugs for Acid-Related Disorders (A02))	$MgCO_3$	anhydrous $MgCO_3$ (magnesite): [546-93-0] $CMgO_3$ 84.31 hydrate: [23389-33-5] monohydrate: [13717-00-5] dihydrate: [5145-48-2] trihydrate: [14457-83-1] pentahydrate: [61042-72-6]	magnesite: used as an antacid, usually in combination with aluminum hydroxide. hydrate: used as a laxative; food additive/supplement many different formulations and doses. 7 g O.	commercially introduced in England as antacid and purgative at the beginning of the 18th century. synthesis (from $MgCl_2$ and Na_2CO_3) [18]
Magnesium oxide (calcined magnesia; magnesia usta) [Magcal, Maglite, Magox] A06AD02; A12CC10; A02AA02 (→ Drugs for Acid-Related Disorders (A02))	MgO	[1309-48-4] MgO 40.30	antacid and mild laxative, similar to magnesium hydroxide; food additive/supplement many different formulations and doses. 7 g O.	used as an antacid and a purgative since mid of 18th century. semisynthesis (from magnesite ores) [19]
Magnesium sulfate (heptahydrate: bitter salt; bittersalz; epsomite) [Epsom salt] A06AD04; A12CC02; B05XA05; D11AX05; V04CC02 **WHO essential medicine**	$MgSO_4$	[7487-88-9] MgO_4S 120.36 heptahydrate: [10034-99-8] $MgO_4S \cdot 7\,H_2O$ 246.47	a. o. calcium channel blocker. medical use for treatment of magnesium deficiency, antiarrhythmic, bron-chodilator/management of severe asthma attacks (inhal.), eclampsia (as anticonvulsant), and as laxative/cathartic. many different formulations and doses. 7 g O.	isolated by N. Grew from the medicinal waters at Epsom/South of England in 1697. [20]
Lactulose (4-β-D-galactosido-D-fructose) [Bifinorm, Bifiteral, Cholac, Constilac, Duphalac, Eugalac, Gatinar, Generlac, Lactuflor, Laevilac, Legendal, Normase, Tulotract] A06AD11 **WHO essential medicine**		[4618-18-2] $C_{12}H_{22}O_{11}$ 342.30	synthetic disaccharide for treatment of constipation and hepatic encephalopathy. sol. 10 g/15 mL. 12.5 g O.	F. Petuely DE/1964 semisynthesis (from lactose under alkaline conditions) [21, 22]

(continued)

Table 3. (*Continued*)

INN (Synonyms) [Brand names] ATC	Structure (Remarks)	[CAS-No.] Formula MW, g/mol	Target (if known) Medical use Formulation DDD	Originator Approval (country/year) Production method [References]
Lactitol (4-*O*-β-D-galactopyranosyl-D-glucitol; β-galactoside sorbitol; lactite; lactobiosit; lactositol) [Importal, Pizensy, Portolac] A06AD12		[585-86-4] $C_{12}H_{24}O_{11}$ 344.31 monohydrate: [81025-04-9] $C_{12}H_{24}O_{11} \cdot H_2O$ 362.33 dihydrate: [81025-03-8] $C_{12}H_{24}O_{11}$ $\cdot 2\,H_2O$ 380.34	medical use as laxative, cathartic, and excipient; food additive (low caloric sweetener). powder for sol. preparation; syrup 66.67 g/100 mL 10 g O.	USA/2020 semisynthesis (from lactose by hydrogenation in presence of Raney nickel as catalyst) [23]
Sodium sulfate (decahydrate: Glauber's salt) A06AD13	Na_2SO_4	[7757-82-6] Na_2O_4S 142.04 decahydrate: [7757-72-3] $Na_2O_4S \cdot 10\,H_2O$ 322.19	osmotic laxative, daily dose 5–10 g. In combination with high molecular mass macrogol in dilute solution it is used for prompt bowel evacuation before investigational procedures or surgery.	Johann Glauber AT/1624 occurs in mineral waters [24]
Macrogol (polyethylene glycol; PEG) [Carbowax, Forlax, Idrolax, MiraLax, Movicol, ClearLax, Golytely, Pluracol E, Poly-G] A06AD15	$n \geq 4$ (average value) MW (ranges): PEG 200: $n = 4$, MW 190–210; liquid; PEG 400: $n = 8.2–9.1$, MW 380–420; liq.; PEG 600: $n = 12.5–13.9$, MW 570–630; liq.; PEG 1500: $n = 29–36$, MW 1300–1600; white powder; PEG 3350/PEG 4000: $n = 68–84$, MW 3000–3700; white powder or creamy flakes; PEG 6000: $n = 158–204$, MW 7000–9000; white powder or creamy flakes	[25322-68-3]	osmotic laxative for treatment of short- and long-term constipation of various causes; also used for bowel preparation before surgery or colonoscopy; also used as an excipient in different pharmaceutical products, e.g., the low molecular mass variants as solvents in oral liquids and soft capsules, and the solid variants as bases for ointments, tablet binders, lubricants, and film coatings. macrogol 3350 powder for preparation of solution/laxative. 10 g O.	Norgine DE/1995; USA/1999 synthesis (from ethylene oxide) [25]
Mannitol (D-mannitol; mannit; manna sugar) [Manicol, Mannidex, Osmitrol, Osmosal, Resectisol] A06AD16; B05BC01; B05CX04; R05CB16; V04CX04 **WHO essential medicine**		[69-65-8] $C_6H_{14}O_6$ 182.17	sugar alcohol, which is found in many natural products, so in many plants. Used as a sweetener in diabetic food; medical use as mild laxative and mainly as osmotic diuretic, administered by injection/infusion. In addition, mannitol is in use as an excipient.	discovery by J. L. Proust in 1806 semisynthesis (by hydrogenation of fructose or mannose) [26–29]

Table 3. (*Continued*)

INN (Synonyms) [Brand names] ATC	Structure (Remarks)	[CAS-No.] Formula MW, g/mol	Target (if known) Medical use Formulation DDD	Originator Approval (country/year) Production method [References]
Sorbitol (D-glucitol; D-sorbitol) [Resulax, Sorbilax, Sorbostyl, Sorbilande, and many more] A06AD18; A06AG07; B05CX02; V04CC0		[50-70-4] $C_6H_{14}O_6$ 182.17	sugar alcohol, occurs in many fruits and berries; low-caloric sweetener for use in diabetic food; medical use as osmotic diuretic, also in mixture with mannitol, and as mild laxative and in enemas.	semisynthesis (catalytic hydrogenation of D-glucose) [29]
Magnesium citrate (trimagnesium dicitrate; magnesium citrate tribasic) A06AD19; A12CC04; B05CB03		[3344-18-1] $C_{12}H_{10}Mg_3O_{14}$ 451.11	cathartic agent in use for bowel preparation prior to a colonoscopy and remedy of occasional constipation and mainly as dietary (mineral) supplement.	synthesis [30, 31]

5. A06AG Enemas (Table 4)

Table 4. A06AG Enemas

INN (Synonyms) [Brand names] ATC	Structure (Remarks)	[CAS-No.] Formula MW, g/mol	Target (if known) Medical use Formulation DDD	Originator Approval (country/year) Production method [References]
Sodium phosphate (monobasic + dibasic sodium phosphate) [OsmoPrep, Visicol] A06AG01	$NaH_2PO_4 + Na_2HPO_4$	monobasic: [7558-80-7] H_2NaO_4P 119.98 dibasic: [7558-79-4] HNa_2O_4P 141.96	cathartic (purgative; saline laxative) agent in use for bowel preparation prior to a colonoscopy. tbl. 1.5 g containing 1.102 g monobasic and 0.398 g dibasic sodium phosphate; 118 mL delivered dose with 19 g monobasic and 7 g dibasic sodium phosphate (laxative).	Salix Pharmaceuticals USA/2005 synthesis [32, 33]
Bisacodyl A06AG02	(see A06AB02; Chap. 3)			
Dantron A06AG03	(see A06AB03; Chap. 3)			
Glycerol (glycerine) A06AG04; A06AX01; QA16QA03	OH HO⁀⁀OH	[56-81-5] $C_3H_8O_3$ 92.09	fats are fatty acid esters of glycerol. by rectal application as suppository or enema glycerol acts as hyperosmotic laxative and is used for colon evacuation prior to colonoscopy. glycerol is used as a lubricant, solvent, and humectant in many pharmaceutical and personal care preparations. It can also be used as a sweetener.	discovered 1779 by Scheele (hydrolysis of olive oil) semisynthetic (by saponification of fats, byproduct of soap production) or synthetic (from allyl alcohol or allylchloride) [34]
Sorbitol A06AG07	(see A06AD18; Chap. 4)			
Docusate sodium A06AG10	(see A06AA02; Chap. 2)			
Sodium lauryl sulfoacetate (SLSA; dodecyl sodium sulfoacetate) [Lathanol; LAL] A06AG11	H_3C⁀⁀ structure with Na^+	[1847-58-1] $C_{14}H_{27}NaO_5S$ 330.42	surfactant for use in enema-type laxative products, e.g., combination with glycerol and sodium citrate (brand names Microlax, Micolette, Micro-enema for treatment of constipation); used also in cosmetics.	Pharmacia SE/1960; USA/1962 synthesis [35]

6. A06AH Peripheral Opioid Receptor Antagonists (Table 5)

Table 5. A06AH Peripheral opioid receptor antagonists

INN (Synonyms) [Brand names] ATC	Structure (Remarks)	[CAS-No.] Formula MW, g/mol	Target (if known) Medical use Formulation DDD	Originator Approval (country/year) Production method [References]
Methylnaltrexone bromide (methylnaltre-xonium; naltrexone methobromide) [Relistor] A06AH01		[73232-52-7] $C_{21}H_{26}BrNO_4$ 436.35	antagonist of peripheral opioid receptors in the GI tract, the bladder, and the skin and decreases opioid-related constipation, urinary retention, and pruritis. It does not affect the central analgesic effects of opioids, as it cannot cross the blood–brain barrier. tbl. 150 mg; prefilled syringes for single-dose 8 mg/0.4 mL and 12 mg/0.6 mL. 6 mg P.	Progenics/Wyeth EU/2008; USA/2008 synthesis (from naltrexone and methyl bromide [36, 37]
Alvimopan (ADL-8-2698; LY-246736) [Entereg] A06AH02		[156053-89-3] $C_{25}H_{32}N_2O_4$ 424.54	peripheral selective µ-opioid receptor antagonist in the GI tract. Avoidance and/or treatment of postoperative ileus after partial large or small bowel resection with primary anastomosis. cps. 12 mg.	Eli Lilly/Adolor Corp. USA/2008 synthesis [38–41]
Naloxegol (AZ-13337019/ oxalate; NKTR-118/oxalate; PEGylated naloxol) [Movantik, Moventig] A06AH03		[854601-70-0] $C_{34}H_{53}NO_{11}$ 651.8 oxalate: [1354744-91-4] $C_{34}H_{53}NO_{11} \cdot C_2H_2O_4$ 741.8	peripherally acting µ-opioid receptor antagonist for treatment of opioid-induced constipation. tbl. 12.5, 25 mg. 25 mg O.	AstraZeneca/Nektar Therapeutics USA/2014; EU/2015 synthesis (from naloxone) [42]

Table 5. (*Continued*)

INN (Synonyms) [Brand names] ATC	Structure (Remarks)	[CAS-No.] Formula MW, g/mol	Target (if known) Medical use Formulation DDD	Originator Approval (country/year) Production method [References]
Naloxone (EN-15304) [Evzio, Nalone, Narcan, Narcanti, Suboxone] A06AH04		[*465-65-6*] $C_{19}H_{21}NO_4$ 327.38 hydrochloride: [*357-08-4*] $C_{19}H_{21}NO_4 \cdot HCl$ 363.84	μ-opioid receptor antagonist in the CNS; reverses the effects of opioid analgesics and inhibits their actions like analgesia, euphoria, sedation, respiratory depression, miosis, bradycardia, and physical dependence. In emergency situations it is used as an i.v. application to reverse respiratory depression following overdosing of heroin, morphine, or other opioids. amp. 0.04, 0.2, 0.4 mg/mL cps. 4 mg in comb. with tilidine; tbl.0.5 mg in comb. with pentazocine; vials 0.4 mg/mg, 1 mg/mL, 10 mg/mL; tbl.0.5 mg in comb. with 2 mg buprenorphine (suboxone), see (→ Analgesics (N02)).	Lewenstein, M.J., Fishman, J. /Sankyo USA/1971 synthesis (starting, e.g., with oxycodone) [43]
Naldemedine (S-297,995) [Symproic, Rizmoic] A06AH05		[*916072-89-4*] $C_{32}H_{34}N_4O_6$ 570.64 tosylate: [*1345728-04-2*] $C_{39}H_{42}N_4O_9S$ 742.84	peripherally selective μ-opioid receptor antagonist. treatment of opioid-induced constipation. tbl. 0.2 mg. 0.2 mg O.	Shionogi JP/2017; USA/2017; EU/2019 synthesis (starting, e.g., with naltrexone) [44–46]

7. A06AX Other Drugs for Constipation (Table 6)

Table 6. A06AX Other drugs for constipation

INN (Synonyms) [Brand names] ATC	Structure (Remarks)	Target (if known) Medical use Formulation DDD	[CAS-No.] Formula MW, g/mol	Originator Approval (country/year) Production method [References]
Glycerol A06AX01	(see A06AG04; Chap. 5)			
Lubiprostone (RU-0211; SPI-0211) [Amitiza] A06AX03		derivative of prostaglandin E_1, which acts as chloride channel (ClC-2) activator in the membrane of GI epithelium and releases an efflux of chloride ions and thereby drawing water into the GI lumen, resulting in softening the stool and so in laxative activity. cps. 8 μg, 24 μg.	[333963-40-9] $C_{20}H_{32}F_2O_3$ 390.47 monocyclic diketo tautomer: [136790-76-6]	Sucampo USA/2006; UK/2014 synthesis [47, 48]
Linaclotide (ASP-0456; MD-1100; MM-416775) [Linzess, Constella] A06AX04	H-Cys1-Cys2-Glu3-Tyr4-Cys5-Cys6-Asn7-Pro8-Ala9-Cys10-Thr11-Gly12-Cys13-Tyr14OH (all amino acids in L-form)	agonist of intestinal guanylate cyclase type C (GC-C) with secretagogue, analgesic, and laxative activities. treatment of chronic constipation and irritable bowel syndrome. caps. 72, 145, 290 μg. 0.29 mg O.	[851199-59-2] $C_{59}H_{79}N_{15}O_{21}S_6$ 1526.73 acetate: [851199-60-5] $C_{59}H_{79}N_{15}O_{21}S_6 \cdot C_2H_4O_2$ 1586.78	Microbia (since 2010 named Ironwood) USA/2012; EU/2012 synthesis [49–51]
Prucalopride (R-93877; R-108512) [Motegrity, Prudac, Resolor] A06AX05		selective 5-HT$_4$ receptor agonist with potent prokinetic activity. therapy of chronic idiopathic constipation; laxative. tbl. 1 mg, 2 mg. 2 mg O.	[179474-81-8] $C_{18}H_{26}ClN_3O_3$ 367.87 succinate: [179474-85-2] $C_{18}H_{26}ClN_3O_3 \cdot C_4H_6O_4$ 486	Janssen; Shire EU/2009; USA/2018 synthesis [52, 53]

(continued)

Table 6. (*Continued*)

INN (Synonyms) [Brand names] ATC	Structure (Remarks)	[CAS-No.] Formula MW, g/mol	Target (if known) Medical use Formulation DDD	Originator Approval (country/year) Production method [References]
Tegaserod (HTF-919; SDZ-HTF-919) [Zelmac, Zelnorm] A06AX06		[145158-71-0] $C_{16}H_{23}N_5O$ 301.39 maleate: [189188-57-6] $C_{16}H_{23}N_5O \cdot C_4H_4O_4$ 417.47	5-HT$_4$ receptor agonist. treatment of irritable bowel syndrome with constipation. tbl. 2 mg, 6 mg. 12 mg O.	Sandoz USA/2002 (withdrawn in 2007, reintroduced in 2019) synthesis [54, 55]
Plecanatide (SP-304) [Trulance] A06AX07	H-Asn^1-Asp^2-Glu^3-Cys^4-Glu^5-Leu^6-Cys^7-Val^8-Asn^9-Val^{10}-Ala^{11}-Cys^{12}-Thr^{13}-Gly^{14}-Cys^{15}-Leu^{16}-OH	[467426-54-6] $C_{65}H_{104}N_{18}O_{26}S_4$ 1681.9	agonist of intestinal guanylate cyclase type C (GC-C). treatment of chronic idiopathic constipation and irritable bowel syndrome. tbl. 3 mg. 3 mg O.	Synergy Pharm. USA/2017 synthesis [56, 57]

List of Abbreviations

amp.   ampoule(s)
API   active pharmaceutical ingredient
ATC   Anatomical Therapeutic Chemical
                (Classification)
CNS   central nervous system
cps.   capsules
DDD   defined daily dose
GI   gastrointestinal
inj.   injection
INN   international nonproprietary name
MW   molecular mass
PEG   polyethylene glycol
sol.   solution
suppos.   suppositories
tbl.   tablet(s)
wfm   withdrawn from market
WHO   World Health Organization

References

1 Goodman & Gilman's (2011) Chapter 45, Pharmacotherapy of Gastric Acidity, Peptic Ulcers, and Gastroesophageal Reflux Disease, in *The Pharmacological Basis of Therapeutics*, 12th edn, McGraw-Hill, London, pp. 1323–1349.

2 WHO Collaborating Centre for Drug Statistics Methodology, Guidelines for ATC Classification and DDD Assignment 2019, Oslo, Norway.

3 Kleemann, A., Engel, J., Kutscher, B. and Reichert, D. *Pharmaceutical Substances Online*, Thieme Chemistry, Stuttgart.

4 American Cyanamid, (1936) US 2 028 091, USA-prior. 1933.

5 American Cyanamid, (1939) US 2 176 423, USA-prior. 1936.

6 Thomae, (1956), US 2 764 590, DE-prior. 1952.

7 Bayer (1904), DE 197 607.

8 Vamossy, Z.V. (1902) *Therap. D. Gegenwart* 201.

9 Baeyer, A.V. (1871) *Ber.* **4**, 658.

10 Baeyer, A.V. (1880) *Ann.* **202**, 68.

11 Ex Lax, (1940) US 2 192 485, USA-prior. 1938.

12 Stoll, A., et al. (1949) *Helv. Chim. Acta* **32**, 1892.

13 Byk-Gulden, (1958) GB 804 232, DE-prior. 1956.

14 Nattermann, (1970) US 3 517 269, DE-prior. 1966.

15 Istituto de Angeli, (1970) US 3 528 986, USA-prior. 1966.

16 Thomae, (1970) DE 1 904 322, DE-prior. 1969.

17 Prepopik: Ferring, (2017) US 9 827 231, KR-prior. 19.03.2014.

18 Merck & Co., (1965), US 3 169 826, USA-prior. 1961.

19 Northwest Magnesite, (1967) US 3 320 029, USA-prior. 1964.

20 Grew, N. (1697) *A Treatise on the Nature and Use of the Bitter Purging Salt*, Joseph Bridges, London.

21 Montgomery, E.M. and Hudson, C.S. (1930) *J. Am. Chem. Soc.* **52**, 2101.

22 Petuely, F. and North Amer. Philips Comp., (1966), US 3 272 705, USA-prior. 1953.

23 Wolfrom, M.L., et al. (1938) *J. Am. Chem. Soc.* (60), 571.

24 Sneader, W. (2005) *Drug Discovery – A History*, J. Wiley & Sons, Chichester, p. 64.

25 Fordyce, R., et al. (1939) *J. Am. Chem. Soc.* **61**, 1905; 1910.

26 Crystallization of Mannitol: Atlas Chemical Ind., (1972) US 3 632 656, USA-prior. 1967.

27 Towa Kasei Kogyo Co., (1978) US 4 083 881, JP-prior. 1975.

28 Roquette Freres, (1995) US 5 466 795, FR-prior. 1992.

29 Review: Ullmann's article on "Sugar Alcohols" a25_413pub3(1)pdf.

30 Pfizer, (1941) US 2 260 004, USA-prior. 1939.

31 Cumberland Swan, (2003) US 6 514 537, USA-prior. 2000.

32 Aronchick, C.A., (1997) US 5 616 346, USA-prior. 1993.

33 Salix, (2010) US 7 687 075, USA-prior. 19.11.2003.

34 *Ullmann's Encyclopedia of Industrial Chemistry online*, a12_477.pub2.

35 Pharmacia, (1965) US 3 211 614, USA-prior. 1960.

36 Boehringer Ing., (1979) US 4 176 186, USA-prior. 1978.

37 Wyeth, (2012) US 8 247 425, USA-prior. 2008.

38 Eli Lilly, (1993) US 5 250 542, USA-prior. 1991.

39 Eli Lilly, (1995) US 5 434 171, USA-prior. 1993.

40 Adolor, (2002) US 6 469 030, USA-prior. 2001.

41 Zimmerman, D.M., et al. (1994) *J. Med. Chem.* **37**, 2262.

42 Nektar, (2010) US 7 786 133, USA-prior. 16.12.2003.

43 Lewenstein, M.J., Fishman, J. (1966) US 3 254 088, USA-prior. 1961.

44 Shionogi, (2011) US 8 084 460, JP-prior. 25.05.2005.

45 Shionogi, (2017) US RE 46375, JP-prior. 25.05.2005.

46 Shionogi, (2016) US 9 464 094, JP-prior. 12.11.2010.

47 Sucampo, (2002) US 6 414 016, USA-prior. 05.09.2000.

48 Sucampo, (2006) US 7 064 148, USA-prior. 31.08.2001.

49 Thayer, A. (2013) *Chem. Engng. News*, March 4 16–17.

50 Microbia, (2007) US 7 304 036, USA-prior. 28.01.2003.

51 Ironwood Pharm., (2012) US 8 110 553, USA-prior. 28.01.2003.

52 Janssen, (1998) US 5 854 260, PCT-prior. 1995.

53 Janssen, (2001) US 6 310 077, EP-prior. 1994.

54 Buchheit, K.-H., et al. (1995) *J. Med. Chem.* **38**, 2331.

55 Sandoz, (1996) US 5 510 353, GB-prior. 1991.

56 Synergy, (2010) US 7 799 897, USA-prior. 29.03.2001.

57 Synergy, (2018) US 10 011 637, USA-prior. 05.06.2013.

Antidiarrheals, Intestinal Antiinflammatory and Antiinfective Agents (A07)

Axel Kleemann, Hanau, Germany

1. Introduction

In the following chapters, the different drug groups are reviewed and arranged according to their Anatomical Therapeutic Chemical Classification (ATC) of the WHO in which they are arranged according to their indication and chemical structure [1]. It has to be mentioned that only drug substances (APIs, active pharmaceutical ingredients) with clearly defined structures are referenced here. This is the reason why chapters A07BA, A07BC, A07BP, A07CA, A07EF, A07FA, A07XA, A07XH, and A07XP, comprising adsorbents, combinations, or microorganisms, have been omitted.

According to Ref. [2], diarrhea results from disorders of intestinal water and electrolyte transport. There are quite a number of different treatment possibilities, which is being reflected in the classification groups of the Chapter A07. Treatment of severe cases, in which dehydration and electrolyte imbalances are a dominating risk particularly in infants, consists in oral rehydration therapy with a balanced combination of potassium chloride, sodium chloride, sodium citrate, and D-glucose, listed in A07CA (but as a combination not referenced here).

Among the antiinfectives listed in A07A are well-proven antibiotics (A07AA), sulfonamides (A07AB), imidazole derivatives (A07AC), and other antiinfectives (A07AX) for treatment of diarrhea caused by bacteria or protozoa. With the exception of paromomycin and fidaxomicin, the antibiotics of A07AA are described under ATC codes D01 and J01 to which they are referred in Table 1.

Mild cases of acute diarrhea can be treated by antipropulsives, which act by decreasing intestinal motility through their effects on μ- and/or δ-opioid receptors on enteric nerves. They are opioids and listed in A07D, the most prominent and described drug substance being loperamide. It looks like all these drugs were developed by the Belgian company Janssen, a member of the Johnson & Johnson group.

Ulcerative colitis and Crohn's disease are both inflammatory bowel diseases (IBD; chronic idiopathic inflammatory disorders of the GI tract) and often accompanied by severe diarrhea besides other symptoms. Crohn's disease was and still is treated preferably by glucocorticoids, although progress has been seen with immunosuppressive agents and TNFα-blockers such as infliximab,

Ullmann's Pharmaceuticals. Edited by Axel Kleemann and Bernhard Kutscher.
© 2022 Wiley-VCH GmbH
Set ISBN: 978-3-527-34252-5/ DOI: 10.1002/14356007.w02_w04

adalimumab, and natalizumab, all three being monoclonal antibodies. The glucocorticoids for this purpose are listed in A07EA, see Section 4.1. They are well described in the groups H02AB, D07AC, and R01AD; the corresponding references are given. Ulcerative colitis may also be treated with cromoglicic acid, see A07EB01 in Table 5. Both diseases, especially mild to moderate cases, can be treated successfully with 5-aminosalicylic acid (5-ASA, mesalazine) which is together with its prodrugs listed in A07EC, see Table 6. Finally, it is noteworthy that the peripheral enkephalinase inhibitor racecadotril is in use for treatment of acute diarrhea, see A07XA and Table 7. The syntheses of almost all described APIs can be found in [3]. In the references the original patents (US patents whenever available) are cited, where possible, and the relevant publications on medicinal chemistry, where existing.

2. A07A Intestinal Antiinfectives

This group comprises locally acting anti-infectives. Antiinfectives for systemic use are listed in ($\rightarrow$ Antiinfectives for Systemic Use, 1. Antibacterials; $\rightarrow$ Antiinfectives for Systemic Use, 2. Antimycotics (Antifungals) and Antimycobacterials; $\rightarrow$ Antiinfectives for Systemic Use, 3. Antivirals).

2.1. A07AA Antibiotics (Table 1)

Table 1. A07AA Antibiotics

INN (Synonyms) [Brand names] ATC	Structure (Remarks)	[CAS-No.] Formula MW, g/mol	Target (if known) Medical use Formulation DDD	Originator Approval (country/year) Production method [References]
Neomycin A07AA01: see **J01GB05** (→ Antiinfectives for Systemic Use, 1. Antibacterials); R02AB01; G01AA14; A01AB08; B05CA09; R01AX08; S01AA03; S03AA01				
Nystatin A07AA02: see **D01AA01** (→ Dermatologicals (D). 2. Antifungals (D01), Emollients and Protectives (D02), and Antipruritics (D04) for Dermatological Use); A01AB33; G01AA01				
Natamycin A07AA03: see **D01AA02** (→ Dermatologicals (D). 2. Antifungals (D01), Emollients and Protectives (D02), and Antipruritics (D04) for Dermatological Use); A01AB10; G01AA02; S01AA10				
Streptomycin A07AA04: see **J01GA01** (→ Antiinfectives for Systemic Use, 1. Antibacterials)				
Polymyxin B A07AA05: see **J01XB02** (→ Antiinfectives for Systemic Use, 1. Antibacterials)				
Paromomycin (aminosidin; catenulin; Fl-5853; R-400; SF-767B) [Aminoxidin, Aminosidine, Catenulin, Crestomycin, Estomycin, Hydroxymycin, Farmiglucin, Farminosidin, Gabbromicina, Gabroral, Humagel, Humatin, Monomycin A, Neomycin E, Paucimycin, Pargonyl, Paramicina, Paricina, Sinosid] A07AA06 **WHO essential medicine**		[7542-37-2] $C_{23}H_{45}N_5O_{14}$ 615.63 sulfate: [1263-89-4] $C_{23}H_{45}N_5O_{14}$ $\cdot H_2SO_4$ 713.71	aminoglycoside antibiotic that inhibits protein synthesis and is used for treatment of intestinal parasitic infections as amoebiasis, cryptosporidiosis, prophylaxis and therapy of portosystemic encephalopathy, and in some countries for leishmaniasis (also in vet. application). cps. 250 mg; powder 1000 mg for prep. of oral sol. 3 g O.	Parke Davis; Pfizer USA/1960 fermentation of *Streptomyces rimosus forma paromomycinus* [4, 5]

(continued)

Table 1. (*Continued*)

INN (Synonyms) [Brand names] ATC	Structure (Remarks)	[CAS-No.] Formula MW, g/mol	Target (if known) Medical use Formulation DDD	Originator Approval (country/year) Production method [References]
Amphotericin B A07AA07; see **J02AA01** (→ Antiinfectives for Systemic Use, 2. Antimycotics (Antifungals) and Antimycobacterials)				
Kanamycin A07AA08; see **J01GB04** (→ Antiinfectives for Systemic Use, 1. Antibacterials)				
Vancomycin A07AA09; see **J01XA01** (→ Antiinfectives for Systemic Use, 1. Antibacterials)				
Colistin A07AA10; see **J01XB01** (→ Antiinfectives for Systemic Use, 1. Antibacterials)				
Rifaximin A07AA11; see **D06AX11** (→ Dermatologicals (D), 3. Antipsoriatics (D05), Antibiotics and Chemotherapeutics (D06), and Corticosteroids (D07) for Dermatological Use)				
Fidaxomicin (OPT-80; PAR-101: lipiarmycin; tiacumicin B) [Dificid, Dificlir] A07AA12		[873857-62-6] $C_{52}H_{74}Cl_2O_{18}$ 1058.05	macrolide antibiotic, inhibitor of bacterial RNA polymerase; has activity against pathogenic *Clostridium difficile*. treatment of *Clostridium difficile*-associated diarrhea. tbl. 200 mg; oral suspension 40 mg/mL. 0.4 g O.	Optimer Pharmaceuticals USA/2011; EU/2011 fermentation of *Dactylosporangium aurantiacum hamdenensis* [6, 7]

2.2. A07AB Sulfonamides (Table 2)

Table 2. A07AB Sulfonamides

INN (Synonyms) [Brand names] ATC	Structure (Remarks)	[CAS-No.] Formula MW, g/mol	Target (if known) Medical use Formulation DDD	Originator Approval (country/year) Production method [References]
Phthalylsulfathiazole [Neothalidine, Sulfathalidine, Talidine, Thalazole] A07AB02 wfm in many countries		[85-73-4] $C_{17}H_{13}N_3O_5S_2$ 403.43	broad-spectrum antimicrobial agent for treatment of dysentery, colitis, and gastroenteritis. tbl. 500 mg. 9 g O.	Sharp & Dohme USA/1945 synthesis (from sulfathiazole) [8, 9]
Sulfaguanidine (RP-2275) [Diacta, Guanidan, Guanicil, Shigatox] A07AB03 wfm in many countries		[57-67-0] $C_7H_{10}N_4O_2S$ 214.24	antimicrobial agent for treatment of enteric infections. Current use mainly in veterinary medicine. tbl. 0.5 g. 4 g O.	American Cyanamid synthesis [10]
Succinylsulfathiazole [Thiacyl, Cremomycin, Cremostrep, Cremosuxidine, Streptoguanidin] A07AB04 wfm in many countries		[116-43-8] $C_{13}H_{13}N_3O_5S_2$ 355.38	antimicrobial agent for treatment of enteric infections. tbl. 0.5 g.	Sharp & Dohme synthesis (from sulfathiazole) [11, 12]
Sulfaguanole [Enterocura] A07AB06 wfm		[27031-08-9] $C_{12}H_{15}N_5O_3S$ 309.34	antimicrobial agent for treatment of enteric infections.	Nordmark DE/1973 synthesis [13]

2.3. A07AC Imidazole Derivatives

In this subgroup only one drug substance, the locally active antifungal miconazole for treatment of gastrointestinal mycosis, is classified under A07AC01. See its description under **J02AB01** (→ Antiinfectives for Systemic Use, 2. Antimycotics (Antifungals) and Antimycobacterials). Additionally, it is listed under A01AB09, D01AC02, G01AF04, and S02AA13.

2.4. A07AX Other Intestinal Antiinfectives (Table 3)

Table 3. A07AX Other intestinal antiinfectives

INN (Synonyms) [Brand names] ATC	Structure (Remarks)	Target (if known) Medical use Formulation DDD	[CAS-No.] Formula MW. g/mol	Originator Approval (country/year) Production method [References]
Broxyquinoline (5,7-dibromo-8-quinoline) [Colepur, Intensopan, Intestopan] A07AX01; G01AC06; P01AA01 (→ Genitourinary System and Sex Hormones (G); → Antiparasitics (PC))		intestinal antiseptic and disinfectant, antiprotozoal agent.	[521-74-4] $C_9H_5Br_2NO$ 302.95	synthesis (by bromination of 8-quinolinol) [14]
Acetarsol (acetarsone) [Stovarsol, Spirozid, Polygynax] A07AX02; G01AB01; P01CD02 (→ Genitourinary System and Sex Hormones (G); → Antiparasitics (PC))		antiprotozoal agent that was used for treatment of congenital syphilis.	[97-44-9] $C_8H_{10}AsNO_5$ 275.09	Ernest Fourneau 1925/FR synthesis [15, 16]
Nifuroxazide (RC-27109) [Ambatrol, Antinal, Bacifurane, Diarlidan, Diax, Dicoferin, Enterovid, Ercefuryl, Panfurex, Pentofuryl] A07AX03		intestinal antiseptic for treatment of colitis and diarrhea caused by bacteria. cps. 200 mg; syrup 100 mg. 0.6 g O.	[965-52-6] $C_{12}H_9N_3O_5$ 275.22	Robert et Carriere synthesis [17]
Nifurzide (LS-153043) [Ricridène] A07AX04		intestinal antiseptic for treatment of diarrhea caused by bacteria. cps. 150 mg. susp. 40 mg/mL.	[39978-42-2] $C_{12}H_8N_4O_6S$ 336.28	Lipha synthesis [18, 19]

Name	Structure	Registry / Formula	Uses	Synthesis
Chloroxine (5,7-dichloro-8-quinolinol) [Capitol, Endiaron, Triaderm, Valpeda] A07AX05	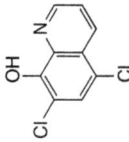	[773-76-2] $C_9H_5Cl_2NO$ 214.05	antibacterial for treatment of diarrhea and dandruff and seborrheic dermatitis (as shampoo or dermal cream).	synthesis (chlorination of 8-quinolinol) [20]
Furazolidone (NF-180) [Furoxone, Dependal-M, Giarlam, Medaron, Nifulidone, Ortazol, Roptazol, Tikofuran, Topazone] A07AX06; G01AX06 (→ Genitourinary System and Sex Hormones (G))	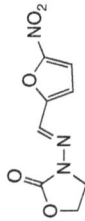	[67-45-8] $C_8H_7N_3O_5$ 225.16	antibacterial with broad spectrum and it is also a monoamine oxidase inhibitor. in use for treatment of diarrhea and enteritis caused by bacteria or protozoal infections. tbl. 100 mg; liquid 50 mg/15 mL.	Norwich Pharm. synthesis [21–23]
Ethacridine (acrinol, aethacridin; H-779) [Acrinol, Acrolactine, Ethodin, Metifex, Rimaon, Rivanol, Vucine] A07AX07; B05CA08; D08AA01 (→ Dermatologicals (D), 4. Antiseptics and Disinfectants (D08), Anti-Acne Preparations (D10), and Other Dermatological Preparations (D11))	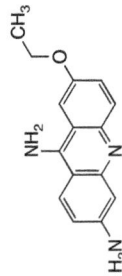	[442-16-0] $C_{15}H_{15}N_3O$ 253.31 lactate (R,S)-mono-hydrate: [6402-23-9] $C_{15}H_{15}N_3O$ ·$C_3H_6O_3$·H_2O 361.40 (→ Stomatological Preparations (A01))	topical antiseptic, effective against most gram-positive bacteria; used as 0.1% solution. drg. 200 mg; eye drops 1 mg/g; sol. 0.1%, 0.2%, 0.5%; tbl. 100 mg.	Hoechst DE/1921 synthesis [24, 25]

3. A07D Antipropulsives

3.1. A07DA Antipropulsives (Table 4)

Table 4. A07DA Antipropulsives

INN (Synonyms) [Brand names] ATC	Structure (Remarks)	[CAS-No.] Formula MW, g/mol	Target (if known) Medical use Formulation DDD	Originator Approval (country/year) Production method [References]
Diphenoxylate (R-1132) [All drugs are comb. with atropine: Atridol, Atrolate, Co-phenotrope, Dhamotil, Diarsed, Intard, Lomotil, Lonox, Reasec, Tropergen] A07DA01 controlled substance (schedule II, CSA)		[915-30-0] $C_{30}H_{32}N_2O_2$ 452.60 hydrochloride: [3810-80-8] $C_{30}H_{32}N_2O_2$ ·HCl 489.06	centrally active synthetic opioid substance; medical use in comb. with atropine sulfate for treatment of diarrhea (by slowing gastrointestinal motility). Active metabolite is difenoxin (A07DA04). tbl. 2.5 mg diphe-noxylate/0.025 mg atropine sulfate. 15 mg O.	Janssen DE/1960 synthesis [26]
Loperamide (R-18553) [Arestal, Binaldan, Dissenten, Enterobene, Imodium, Imosec, Lopedium, Lopemin, Lopimed, Normakut, Suprasec, Tebloc] A07DA03 **WHO essential medicine**		[53179-11-6] $C_{29}H_{33}ClN_2O_2$ 477.05 hydrochloride: [34552-83-5] $C_{29}H_{33}ClN_2O_2$ ·HCl 513.50	μ-opioid receptor agonist in the myenteric plexus of the large intestine; does not cross the blood–brain barrier. med. use for treatment of diarrhea. tbl. 1 mg, 2 mg; cps. 1 mg, 2 mg; syrup 1 mg/5 mL. 10 mg O.	Janssen DE/1976; USA/1976 synthesis [27, 28]
Difenoxin (McN-JR-15403-11; R-15403] [Lyspafen, Motofen] A07DA04 controlled substance (schedule I), in comb. with atropine schedule IV (CSA)		[28782-42-5] $C_{28}H_{28}N_2O_2$ 424.54 hydrochloride: [35607-36-4] $C_{28}H_{28}N_2O_2$ ·HCl 461.01	active metabolite of diphenoxylate (A07DA01), marketed in comb. with atropine to reduce the potential for abuse. med. use for treatment of diarrhea. tbl. 0.5 mg.	Janssen USA/1978; DE/1980 synthesis [29]

Table 4. (*Continued*)

INN (Synonyms) [Brand names] ATC	Structure (Remarks)	[CAS-No.] Formula MW, g/mol	Target (if known) Medical use Formulation DDD	Originator Approval (country/year) Production method [References]
Eluxadoline (JnJ-27018966) [Truberzi, Viberzi] A07DA06		[*864821-90-9*] $C_{32}H_{35}N_5O_5$ 569.65	μ- and κ-opioid receptor agonist and δ-opioid receptor antagonist with local action in the enteric nervous system. med. use for treatment of diarrhea and abdominal pain in patients with diarrhea-predominant irritable bowel syndrome. tbl. 75 mg, 100 mg. 0.2 g O.	Janssen; Actavis USA/2015 synthesis [30]

4. A07E Intestinal Antiinflammatory Agents

4.1. A07EA Corticosteroids Acting Locally

This group comprises glucocorticoids, which have been well described in the groups H02AB (→ Hormone Therapeutics – Systemic Hormonal Preparations, Excluding Reproductive Hormones and Insulins (H)), D07AC (→ Dermatologicals (D), 3. Antipsoriatics (D05), Antibiotics and Chemotherapeutics (D06), and Corticosteroids (D07) for Dermatological Use), and R01AD (→ Respiratory Disorders, 2. Nasal Preparations and Decongestants for Topical Use (R01)).

In the following, substances of group A07EA are listed and the corresponding interlinks are given:

- Prednisolone A07EA01, see D07AA01 (→ Dermatologicals (D), 3. Antipsoriatics (D05), Antibiotics and Chemotherapeutics (D06), and Corticosteroids (D07) for Dermatological Use); D07AA03; H02AB06; R01AD02; S01BA04; S02BA03; S03BA02.
- Hydrocortisone A07EA02, see H02AB09 (→ Hormone Therapeutics – Systemic

Hormonal Preparations, Excluding Reproductive Hormones and Insulins (H)); D07AA02; S01BA02; S02BA01.
- Prednisone A07EA03, see H02AB07 (→ Hormone Therapeutics – Systemic Hormonal Preparations, Excluding Reproductive Hormones and Insulins (H)).
- Betamethasone A07EA04, see H02AB01 (→ Hormone Therapeutics – Systemic Hormonal Preparations, Excluding Reproductive Hormones and Insulins (H)); D07AC01; R01AD06; R03BA04; S01BA06; S02BA07; S03BA03.
- Tixocortol A07EA05, see R01AD07 (→ Respiratory Disorders, 2. Nasal Preparations and Decongestants for Topical Use (R01)).
- Budesonide A07EA06, see D07AC09 (→ Dermatologicals (D), 3. Antipsoriatics (D05), Antibiotics and Chemotherapeutics (D06), and Corticosteroids (D07) for Dermatological Use).
- Beclometasone A07EA07, see D07AC15 (→ Dermatologicals (D), 3. Antipsoriatics (D05), Antibiotics and Chemotherapeutics (D06), and Corticosteroids (D07) for Dermatological Use); R01AD01.

4.2. A07EB Antiallergic Agents, Excluding Corticosteroids (Table 5)

Table 5. A07EB Antiallergic agents, excluding corticosteroids

INN (Synonyms) [Brand names] ATC	Structure (Remarks)	[CAS-No.] Formula MW, g/mol	Target (if known) Medical use Formulation DDD	Originator Approval (country/year) Production method [References]
Cromoglicic acid (cromoglycic acid; cromolyn sodium; DSCG; FPL-670) [Aarane, Intal, Lomudal, Nalcrom, Nasalcrom, Opticrom, Rynacrom, Vicrom] A07EB01; D11AH03; R01AC01; R03BC01; S01GX01. (→ Dermatologicals (D), 4. Antiseptics and Disinfectants (D08), Anti-Acne Preparations (D10), and Other Dermatological Preparations (D11)) (→ Respiratory Disorders, 2. Nasal Preparations and Decongestants for Topical Use (R01))		[16110-51-3] $C_{23}H_{16}O_{11}$ 468.37 disodium salt: [15826-37-6] $C_{23}H_{14}Na_2O_{11}$ 512.33	mast cell stabilizer with antiinflammatory activity; inhibits release of histamine and leukotrienes. treatment of ulcerative colitis (besides allergic rhinitis, asthma, allergic conjunctivitis, and mastocytosis). aerosol 1 mg/0.05 mL, 20 mg/mL; cps. 20 mg, 100 mg; eye drops 20 mg/mL; gran. 10%, 100 mg, 200 mg; nasal drops 20 mg/mL; nasal spray 2.8 mg/0.14 mL, 20 mg/mL; ophthalmic drops 10 mg/0.5 mL, 20 mg/mL (as disodium salt); sol. 20 mg/mL. 0.8 g O; 40 mg N.	Fisons DE/1969 synthesis [31]

4.3. A07EC Aminosalicylic Acid and Similar Agents (Table 6)

Table 6. A07EC Aminosalicylic acid and similar agents

INN (Synonyms) [Brand names] ATC	Structure (Remarks)	[CAS-No.] Formula MW, g/mol	Target (if known) Medical use Formulation DDD	Originator Approval (country/year) Production method [References]
Sulfasalazine (salazosulfapyridine; sulphasalazine) [Azulfidine, Colo-Pleon, Salazopyrin] A07EC01; M01CX02 (→ Drugs for the Musculo-Skeletal System (M)) **WHO essential medicine**		[599-79-1] $C_{18}H_{14}N_4O_5S$ 398.39	antiinflammatory agent for treatment of inflammatory bowel disease incl. ulcerative colitis and Crohn's disease and rheumatoid arthritis. It is metabolized in the gut into sulfapyridine and 5-aminosalicylic acid (mesalazine). f.c. tbl. 500 mg. 2 g O. R.	Pharmacia USA/1950 synthesis [32]
Mesalazine (mesalamine; 5-aminosalicylic acid; 5-ASA) [Asacol, Asazine, Canasa, Claversal, Lixacol, Mesagran, Mesasal, Pentasa, Rowasa, Salofalk] A07EC02		[89-57-6] $C_7H_7NO_3$ 153.14	antiinflammatory agent for treatment of mild to moderate cases of inflammatory bowel disease incl. ulcerative colitis and Crohn's disease. tbl. 250, 400, 500, 800, 1000, 1600 mg; s.r. tbl. 1200 mg; gran. 1000, 1500, 2000 mg; s.r. gran. (Granu-Stix) 500, 1000, 1500, 3000 mg; cps. 250 mg, 400 mg; suppos. 250, 500, 1000 mg; enema 2 g/30 mL, 4 g/60 mL. 1.5 g O. R.	USA/1987 synthesis (from salicylic acid by nitration and reduction of 5-nitrosalicylic acid) [2, 33, 34]

(continued)

Table 6. (*Continued*)

INN (Synonyms) [Brand names] ATC	Structure (Remarks)	[CAS-No.] Formula MW, g/mol	Target (if known) Medical use Formulation DDD	Originator Approval (country/year) Production method [References]
Olsalazine (azodisal sodium; disodium azodisalicylate; C.I.14130; CI-91B) [Dipentum] A07EC03	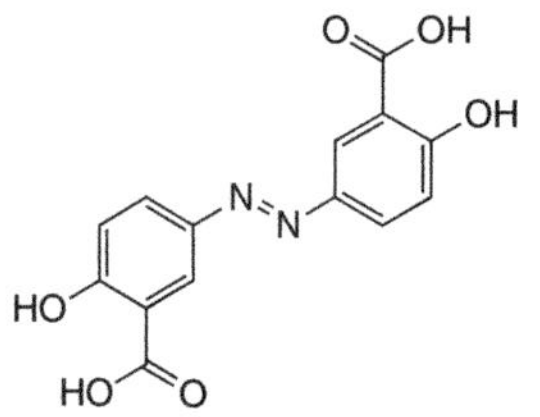	[15722-48-2] $C_{14}H_{10}N_2O_6$ 302.24 disodium salt: [6054-98-4] $C_{14}H_8N_2Na_2O_6$ 346.21	prodrug of mesalazine; treatment of inflammatory bowel disease/ulcerative colitis; releases mesalazine in the large intestine. tbl. 500 mg; cps. 250 mg. 1 g O.	Pharmacia USA/1990 synthesis [35, 36]
Balsalazide (BX-661A) [Colazal, Colazide, Giazo] A07EC04	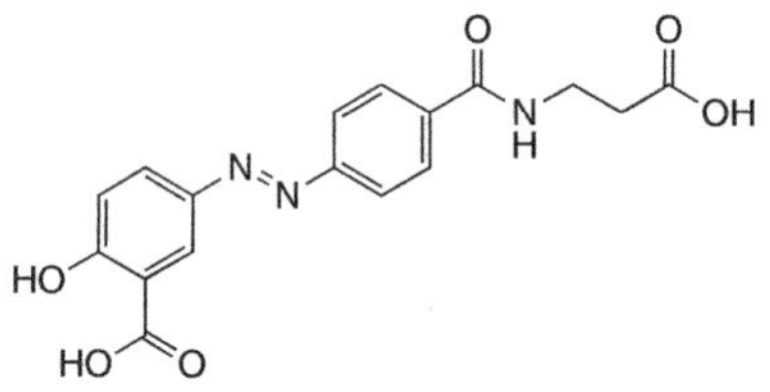	[80573-04-2] $C_{17}H_{15}N_3O_6$ 357.32 disodium salt dihydrate: [150399-21-6] $C_{17}H_{13}N_3Na_2O_6$ ·2H$_2$O 437.32	prodrug of mesalazine; treatment of inflammatory bowel disease/ulcerative colitis; releases mesalazine in the large intestine. cps. 750 mg. 6.75 g O.	Biorex USA/1997 synthesis [37]

5. A07X Other Antidiarrheals

5.1. A07XA Other Antidiarrheals (Table 7)

Table 7. A07XA Other antidiarrheals

INN (Synonyms) [Brand names] ATC	Structure (Remarks)	[CAS-No.] Formula MW, g/mol	Target (if known) Medical use Formulation DDD	Originator Approval (country/year) Production method [References]
Racecadotril (acetorphan) [Hidrasec, Tiorfan] A07XA04	(racemate)	[*81110-73-8*] $C_{21}H_{23}NO_4S$ 385.48	peripheral enkephalinase inhibitor with an antisecretory effect, it reduces the secretion of water and electrolytes into the intestine. med. use for treatment of acute diarrhea in children and adults. cps. 100 mg; gran. 10 mg, 30 mg. 0.3 g O; 0.1 g O children.	Bioprojet FR/1990 synthesis [38]

List of Abbreviations

API	active pharmaceutical ingredient
ATC	Anatomical Therapeutic Chemical Classification
comb.	combination
cps.	capsule(s)
CSA	United States Controlled Substances Act
drg.	dragees
f.c. tbl.	film-coated tablets
GI	gastrointestinal
gran.	granulate
IBD	chronic idiopathic inflammatory disorders
prep.	preparation
sol.	solution
s.r.	sustained release
suppos.	suppositories
susp.	suspension
tbl.	tablet(s)
vet.	veterinary
WHO	World Health Organization

References

1 WHO Collaborating Centre for Drug Statistics Methodology (2019) *Guidelines for ATC Classification and DDD Assignment*, Oslo, Norway.

2 Chapter 47, Pharmacotherapy of Inflammatory Bowel Disease (2011) in *Goodman & Gilman's The Pharmacological Basis of Therapeutics*, 12th edn, McGraw-Hill, London, pp. 1351–1362.

3 Kleemann, A., Engel, J., Kutscher, B. and Reichert, D. *Pharmaceutical Substances Online*, Thieme Chemistry, Stuttgart.

4 Parke Davis, (1959) US 2 916 485, USA-prior.1959.

5 Pfizer, (1959) US 2 895 876, USA-prior.1954.

6 Abbott Labs, (1990) US 4 918 174, USA-prior.1986.

7 Optimer, (2009) US 7 507 564, USA-prior.29.07.2002.

8 Sharp & Dohme, (1943) US 2 324 013, USA-prior.1941.

9 Sharp & Dohme, (1943) US 2 324 015, USA-prior.1941.

10 American Cyanamid, (1940) US 2 218 490, USA-prior.1940.

11 Sharp & Dohme, (1943) US 2 324 013, USA-prior.1941.

12 Sharp & Dohme, (1943) US 2 324 014, USA-prior.1941.

13 Nordmark, (1971) US 3 562 258, USA-prior.1969.

14 Bedall, K., et al. (1881) *Ber. Dtsch. Chem. Ges.* **14**, 1367.

15 Raiziss, G.W. and Fisher, B.C. (1926) *J. Am. Chem. Soc.* **48**, 1323.

16 Hoechst, (1913) US 1 077 462, USA-prior.1912.

17 Robert et Carriere, (1966) US 3 290 213, FR-prior.1961.

18 Szarvasi, E., et al. (1973) *J. Med. Chem.* **16**, 281.

19 Lipha, (1974) US 3 847 911, FR-prior.1971.

20 Hebebrand, A. (1888) *Ber. Dtsch. Chem. Ges.* **21**, 2977.

21 Gever, G., et al. (1955) *J. Am. Chem. Soc.* **77**, 2277.

22 Norwich, (1956) US 2 759 931, USA-prior.1953.

23 Norwich, (1960) US 2 927 110, USA-prior.1958.

24 Hoechst, (1922) DE 360 421, DE-prior.1919.

25 Hoechst, (1924) DE 393 411, DE-prior.1921.

26 Janssen, (1959) US 2 898 340, NL-prior.1957.

27 Stokbroekx, R.A., et al. (1973) *J. Med. Chem.* **16**, 782.

28 Janssen, (1973) US 3 714 159, USA-prior.1971.

29 Janssen, (1972) US 3 646 207, USA-prior.1968.

30 Janssen, (2010) US 7 741 356, USA-prior.14.03.2004.

31 Fisons, (1968) US 3 419 578, GB-prior.1965.

32 Pharmacia, (1946) US 2 396 145, SE-prior.1940.

33 Weil, H. (1922) *Ber. Dtsch. Chem. Ges.* **55**, 2664.

34 Prantera, C., et al. (1999) Mesalamine in the treatment of mild to moderate active Crohn's ileitis, *Gastroenterology* **116**, 521–526.

35 Pharmacia, (1985) US 4 528 367, SE-prior.1980.

36 Pharmacia, (1985) US 4 559 330, SE-prior.1980.

37 Biorex, (1983) US 4 412 992, GB-prior.1980.

38 Bioprojet, (1985) US 4 513 009, FR-prior.1980.

Anti-Obesity Drugs (A08)

Kurt Ritter, Frankfurt, Germany

ULLMANN'S ENCYCLOPEDIA OF INDUSTRIAL CHEMISTRY

1. Introduction

The World Health Organization (WHO) defines obesity as "a condition in which percentage body fat is increased to an extent in which health and well-being are impaired," and, due to the alarming prevalence increase, declared it as a "global epidemic". According to the fact sheet of the WHO, worldwide obesity has tripled since 1975, in numbers, 650 million people were obese in 2016 [1, 2]. Estimations indicate a future increase in industrialized nations and even more in developing countries such as China, India, and Brazil. Of special concern is the high rate of overweight or obesity in children.

Obesity is the consequence of an imbalance between calories consumed and calories expended over a certain time period. A synergistic interplay of one's genetic susceptibility and the access to high-energy food as well as the adaption of a "Western" lifestyle (sedentary behavior) contributes to the increased risk of becoming obese. In adults the body mass index (BMI), the ratio between weight to the squared height (kg/m^2) of a subject, is used as an easy way to estimate body fat percentage [3]. A BMI between 25 and 30 kg/m^2 is considered overweight, a BMI over 30 kg/m^2 defines obesity. In the clinic, severe or morbid obesity is indicated by values of BMI in the Class III (BMI ≥ 40 kg/m^2) and Class II (35 $kg/m^2 \leq$ BMI ≥ 39.9 kg/m^2 in the presence of comorbidities). However, the simple calculation of the BMI does not take into account the tissue distribution of fat and is often overestimated as predictor. Thus, there is an ongoing controversy on defining metabolically healthy and unhealthy obesity [3]. Another method, measurement of waist circumference, provides an additional risk predictor for the increase in abdominal fat, but this is limited for patients with a BMI of 25 to 35 kg/m^2.

For a long time obesity was not considered as a disease. However, obesity is a predisposing factor for many diseases and it is associated with a higher risk for cardiovascular diseases (hypertension, myocardial infarction, stroke), type 2 diabetes, nonalcoholic fatty liver disease, nonalcoholic steatohepatitis, obstructive sleep apnea, musculoskeletal disorders such as osteoarthritis, and certain types of cancer such as liver or colon cancer.

Treatment options for obesity are often reduced to the term "weight loss" and include:

- Long-lasting lifestyle adjustment (dietary changes, reduction of sedentary behavior and increase in physical activity)
- Pharmacotherapy [4–8]
- Bariatric surgery (for severe or morbid obesity) [3]

Unfortunately, despite many different strategies in lifestyle adjustment, the percentage of obese people losing a considerable amount of weight and maintaining the new level of weight remains quite low. Rebound effects are common with diet changes. Numerous food supplements claim to burn fat without actual proof.

Ullmann's Pharmaceuticals. Edited by Axel Kleemann and Bernhard Kutscher.
© 2022 Wiley-VCH GmbH
Set ISBN: 978-3-527-34252-5/ DOI: 10.1002/14356007.a02_313.pub2

Bariatric surgery sealing off most of the stomach to reduce the meal size of consumable food is currently applied only in severe obese people [3]. Its efficacy in achieving long-term weight loss carries along the risk of peri-operative mortality, postsurgical complications and the need for supplementation of certain minerals and vitamins. On the other hand, especially the surgical procedure of gastric bypass (Rouxen-Y) enables immediate partial or complete reduction of antidiabetic medication of obese type 2 diabetic patients and improves or eliminates medical conditions such as hypertension, sleep apnea or acid reflux.

In the attempt to achieve long-lasting weight loss, lifestyle modifications might benefit from an integrated pharmacological intervention. The chronic use of such drugs for this indication demands a high level of drug safety. The main approaches to reduce weight include 1) suppression of appetite, 2) induction of satiety (centrally acting drugs), 3) decrease of intestinal food absorption (peripherally acting drugs), or 4) increase of metabolic rate (enhancing energy expenditure) [4–8]. Placebo-adjusted weight reductions of 3–10% are usually achieved, however there are huge differences in response to the pharmacological treatment. Another common issue is lack of compliance, especially when weight loss reaches a plateau after some time on drug treatment. Discontinuation of the drug therapy usually causes a rebound in weight. Dependency development or abuse problems are a threat for drugs on the market, especially for the centrally acting medications.

The history of anti-obesity pharmacotherapy started with the short success story (1933–1938) of the then very popular diet pills containing 2,4-dinitrophenol. The uncoupling mechanism of 2,4-dinitrophenol in mitochondria turns the metabolism of fat towards heat generation instead of production of ATP, thus raising body temperature and leading to weight loss. However, severe side effects, including hyperthermia-related deaths, risk of cataracts, renal failure and neuropathy lead to a general ban of pills containing this compound. Since that, the rise of highly praised anti-obesity drugs and their fall, either restricted use or withdrawal, due to unacceptable side effects has been a common place in this field (see Table 1) [9, 10].

Glucagon-like peptide-1 (GLP-1) receptor agonists, originally developed as injectables for type 2 diabetes (Hormone Therapeutics – Systemic Hormonal Preparations, Excluding Reproductive Hormones and Insulins (H), Table 16), or dual GLP-1/glucagon receptor agonists provide new treatment options for obese patients. In 2014, the peptide liraglutide (A10BJ Glucagon-like peptide-1 (GLP-1) analogues, A10BJ02) became the first GLP-1 receptor agonist approved by the Food and Drug Administration for use in long-term weight management [8, 9].

2. A08AA Centrally Acting Anti-Obesity Products (Table 1)

Energy homeostasis is controlled by the hypothalamus combining signals from the periphery of the body and initiating anabolic or catabolic pathways. Neuronal circuits are involved in the desire for food, the reward for food intake and mood changes thereafter. Thus, centrally acting anti-obesity drugs interfere with the release and processing of neurotransmitters or interaction with their receptors, e.g., in the dopamine, opioid, or endocannabinoid system [5, 8]. Monoamine neurotransmitters such as serotonin, dopamine and noradrenalin induce satiety, thus anti-obesity drugs either mimic their action on their CNS receptors (agonists), increase their release (stimulants) or prevent their clearance from the synapses (reuptake inhibitors).

Amphetamine and amphetamine-like analogs induce weight loss by suppressing appetite and increasing locomotor activity via stimulating of the sympathetic nervous system (sympathomimetic). Widely prescribed in the 1950s to 1970s for obesity, treatment was abandoned due to cardiovascular side effects, hallucinations, addiction, and abuse problems. Phentermine, a structural analogue to amphetamine, but with lower abuse potential, is approved as monotherapy only for short-term use (less than twelve weeks). The combination of phentermine, inhibiting the clearance of serotonin, with fenfluramine, increasing the release of serotonin, (fen-phen) was withdrawn from market in 1999 due to the occurrence of heart

Table 1. A08AA Centrally acting anti-obesity products

INN (Synonyms) [Brand names] ATC	Structure (Remarks)	[CAS-No.] Formula MW (g/mol)	Target (if known) Medical use Formulation DDD	Originator Approval (country/year) Production method [References]
Phentermine [Adipex-P, Adipex Retard, Duromine Elvenir, Fastin, Ifa Acxion, Ionamin, Panbesy, Redusa, Redusa Forte, Razin, Sentis, Suprenza, Terfamex] A08AA01		[122-09-8] $C_{10}H_{15}N$ 149.23	stimulation of noradrenaline release approved for short term treatment (a few weeks) of obesity adjunct in a regimen of weight reduction based on caloric restriction. tbl. 30 mg, O, once daily	Marion Merrell Dow (now Sanofi) USA/1959 Australia/1991 withdrawn in EU 1999 Synthesis: [15]
Phentermine and Topiramate [Qsymia, Onexa]		Hydrochloride [1197-21-3] $C_{10}H_{15}N \cdot HCl$ 185.7 [97240-79-4] $C_{12}H_{21}NO_8S$ 339.37	extended release combination of phentermine and topiramate. Topimarate is a GABA receptor agonist and kainite/AMPA receptor antagonist. tbl. 3.75 mg phentermine/23 mg topiramate; 7.5 mg/46 mg; 11.25 mg/69 mg; 15 mg/92 mg, O, once daily	Combination Osymia Vivus USA/2012 Synthesis of topiramate (as anticonvulsant drug, McNeil Pharmaceuticals, part of Johnson & Johnson): [16, 17]
Fenfluramine [Dimafen, Pesos, Ponderal, Ponderax, Pondimin] A08AA02		[458-24-2] $C_{12}H_{16}F_3N$ 231.26	selective serotonin reuptake inhibitor, modulator of vesicular monoamine transporter 2 tbl. 120 mg, O	withdrawn from market 1997 USA/1973
		Hydrochloride [404-82-0] $C_{12}H_{16}F_3N \cdot HCl$ 267.72		
Amfepramone [Anorex, Linea Vales, Prefamone, Regenon, Tenuate, Tenuate retard, Tenuate Dospan, Tepanil] A08AA03		[90-84-6] $C_{13}H_{19}NO$ 205.30	noradrenalin-releasing drug, approved for short term treatment (a few weeks) of obesity adjunct in a regimen of weight reduction based on caloric restriction tbl. 21.2 mg (Regenon), three times daily, O 25 mg immediate release (Tenuate), three times daily, O 75 mg controlled release (Tenuate Dospan), once daily, O	Temmler Werke USA /1959 Germany/1957 (reintroduction 2003) withdrawn from many markets (Turkey, Sweden, Oman, United Arab Emirates, France, UK, USA, Norway, Venezuela, Brazil) [18, 19]
		Hydrochloride [134-80-5] $C_{13}H_{19}NO \cdot HCl$ 241.76		

(continued)

Table 1. (*Continued*)

INN (Synonyms) [Brand names] ATC	Structure (Remarks)	[CAS-No.] Formula MW (g/mol)	Target (if known) Medical use Formulation DDD	Originator Approval (country/year) Production method [References]
Dexfenfluramine [Adifax, Glypolix, Isomeride, Redux] A08AA04		[*3239-44-9*] $C_{12}H_{16}F_3N$ 231.26	selective serotonin reuptake inhibitor, modulator of vesicular monoamine transporter 2, D-enantiomer of fenfluramine 30 mg, O	Science Union USA/1996 withdrawn from market 1997 Synthesis: [20]
		Hydrochloride [*3239-45-0*] $C_{12}H_{16}F_3N \cdot HCl$ 267.72		
Mazindol [Sanorex, Teronac, Mazildene, Mazanor] A08AA05		[*22232-71-9*] $C_{16}H_{13}ClN_2O$ 284.74	sympathomimetic amine (stimulating central nervous system) for short-term treatment of obesity tbl., 1 mg, O	Sandoz Pharmaceuticals (now Novartis AG) 1970 USA/1973 Japan/1992 withdrawn form market in Germany, Italy, UK, USA, Brazil [21, 22]
Etilamfetamine [Apetinil, Adiparthrol] **A08AA06**		[*457-87-4*] $C_{11}H_{17}N$ 163.26	sympathomimetic amine (stimulating central nervous system) for short-term treatment of obesity	withdrawn from market [23]
Cathine ((+)-norpseudo-ephedrine) [Antidipositum ×112S, Mirapront N, Fasupond] A08AA07		[*492-39-7*] $C_9H_{13}NO$ 151.21		USA/ 1947 withdrawn from market in Germany, Brazil, Malaysia, Singapore, USA, Oman, Canada, Cuba, India [24]
Clobenzorex [Asenlix, Dinintel, Finedal, Rexigen] A08AA08		[*13364-32-4*] $C_{16}H_{18}ClN$ 259.78	sympathomimetic amine (stimulating central nervous system) for short-term treatment of obesity tbl. 30 mg (hydrochloride), O	Mexico (Aventis) withdrawn from market in USA, Mauritius, Oman 2000 [25]
		Hydrochloride [*5843-53-8*] $C_{16}H_{18}ClN \cdot HCl$ 296.24		
Mefenorex [Rondimen, Pondinil, Anexate] A08AA09		[*17243-57-1*] $C_{12}H_{18}ClN$ 211.73	sympathomimetic amine (stimulating central nervous system) for short-term treatment of obesity tbl. 40 mg (hydrochloride), O	Hoffman-La Roche withdrawn from market in many countries, 1999 Europe, USA, Oman Synthesis: [26]

Table 1. (*Continued*)

INN (Synonyms) [Brand names] ATC	Structure (Remarks)	[CAS-No.] Formula MW (g/mol)	Target (if known) Medical use Formulation DDD	Originator Approval (country/ year) Production method [References]
		Hydrochloride [5586-87-8] $C_{12}H_{18}ClN \cdot HCl$ 248.20		
Sibutramine [Meridia, Reductil, Obescare, Raductil, Reductase, Reduxade, Ectiva, Sibutral, Zelium] A08AA10		[106650-56-0] $C_{17}H_{26}N$ 279.86	selective noradrenalin/ serotonin reuptake inhibitor tbl. 5 mg, 10 mg, 15 mg, O	Boots Ph. (now part of AbbVie) USA/1997 withdrawn from market 2010 EU/2001 withdrawn from market 2010 Maleate in South Korea 2011 Mesilate (Slimmer) South Korea 2007 countries, Australia, Canada, Mexico, New Zealand [12, 27]
		X = HCl · H_2O (Hydrochloride) [1274688-22-0] $C_{17}H_{26}N \cdot HCl \cdot H_2O$ 334.33 X = $HO_2CCH_2CO_2H$ (Maleate) [897657-71-5] $C_{17}H_{26}N \cdot C_3H_4O_4$ 383.92 X = $CH_3SO_3H \cdot 1/2 H_2O$ Mesilate hemihydrate [676598-09-7] $C_{17}H_{26}N \cdot C_3H_4SO_3 \cdot 1/2 H_2O$ 383.97		
Lorcaserin **Lorcaserin hydrochloride** [Belviq, Belviq-XR, Venespri] A08AA11		[616202-92-7] $C_{11}H_{14}ClN$ 195.69	selective serotonin 2C receptor agonist in adjunct to a diet and increased physical activity for chronic weight management in obese adult patients, or overweight (BMI >27) with comorbidities tbl. 10 mg. O, twice daily extended release tbl. 10 mg, O, twice daily, 20 mg, O, once daily	Arena/Eisai USA/2012 Mexico, Brasil /2016 Israel/2017 [13, 28–30]
		Hydrochloride: [856681-05-5] $C_{11}H_{15}Cl_2N \cdot 0.5 H_2O$, 241.16		

(continued)

Table 1. (*Continued*)

INN (Synonyms) [Brand names] ATC	Structure (Remarks)	[CAS-No.] Formula MW (g/mol)	Target (if known) Medical use Formulation DDD	Originator Approval (country/year) Production method [References]
Ephedrine, combinations A08AA56		[299-42-3] $C_{10}H_{15}NO$ 165.24 [50906-05-3] 0.5 H_2O [50-98-6] HCl [134-72-5] H_2SO_4	sympathomimetic amine (stimulating central nervous system) for short-term treatment of obesity isolated from ephedra, dietary supplements containing ephedrine	prohibited in USA/2004
Bupropion and naltrexone [Contrave, Mysimba] A08AA62		Bupropion [34911-55-2] $C_{13}H_{18}ClNO$ 239.74	combination of bupropion, noradrenalin-dopamine reuptake inhibitor, and nicotinic acetylcholine receptor antagonist, and naltrexone, opioid receptor antagonist, augmenting bupropion's activation of proopiomelanocortin film-coated, extended-release tablet tbl. 8 mg of naltrexone hydrochloride/90 mg of bupropion hydrochloride, O, twice daily	Orexigen Therapeutics USA/2014 EU/2015 South Korea/2016 [31] Synthesis naltrexone (trade names naltrexone: REVIA, DEPAD) [32]
		Hydrochloride [31677-93-7] $C_{13}H_{18}ClNO \cdot HCl$ 276.2		
		Naltrexone [16590-41-3] $C_{20}H_{23}NO_4$ 341.41		
		Hydrochloride [16676-29-2] $C_{20}H_{23}NO_4 \cdot HCl$ 377.86		

valve damage and pulmonary arterial hypertension. Dexfenfluramine, the D-enantiomer of fenfluramine, was also withdrawn [9, 10]. Recently fixed-dose combinations of phentermine with topiramate, a GABA-receptor agonist and kainite/AMPA-receptor antagonist, have gained approval for long-term use [6–8].

Another fixed-dose combination, contrave, containing bupropion, a noradrenalin-dopamine reuptake inhibitor and nicotinic acetylcholine receptor antagonist, and naltrexone, an opioid receptor antagonist, was introduced to the market in an extended release form. Combinations of drugs might have the advantage of increased efficacy and reduction of side effects as compared to the administration of each component alone.

An increase in cardiovascular adverse events was the reason for withdrawal of sibutramine, a noradrenalin and serotonin reuptake inhibitor, from the market in USA and Europe in 2010 [9, 10, 12].

One year treatment with lorcaserin, a highly potent, selective serotonin 2C receptor (5-HT$_{2C}$) agonist, leads to placebo-adjusted weight loss of 3–3.8% in obese patients [6–13]. Its selectivity against the two other serotonin receptor subtypes, the 5-HT$_{2A}$ and the 5-HT$_{2B}$ receptor, results in an improved safety profile. Agonism of the 5-HT$_{2B}$ receptor is strongly implicated in the drug-induced cardiac valve disease (valvulopathy) seen with the fen-phen combination. Adverse effects for lorcaserin include headache, dizziness, or nausea.

3. A08AB Peripherally Acting Anti-Obesity Products (Table 2)

Orlistat, a peripherally acting drug, is a covalent inhibitor of gastrointestinal lipases, thus reducing the absorption of dietary fats [6–8]. Gastrointestinal side effects include diarrhea, flatulence, bloating, abdominal pain and oily stool leading often to discontinuation of therapy. Furthermore supplementation of fat-soluble vitamins is necessary during drug treatment. The lower dose tablet is available as over-the-counter drug.

4. A08AX and A10BJ Other Anti-Obesity Drugs (Table 3)

The anorectic drug rimonabant, approved in Europe in 2006, induces appetite suppression by inverse agonism of the cannabinoid receptor 1 [14]. However, this mode of action interferes also with the receptor's other multiple functions. Indeed, rimonabant was withdrawn from the European market in 2008 due to an elevated risk of psychiatric adverse events such as increase in depression or suicidal thoughts [9, 10].

Table 2. A08AB Peripherally acting anti-obesity products

INN (Synonyms) [Brand names] ATC	Structure (Remarks)	[CAS-No.] Formula MW (g/mol)	Target (if known) Medical use Formulation DDD	Originator Approval (country/year) Production method [References]
Orlistat [Xenical, Alli] A08AB01		[96829-58-2] C$_{29}$H$_{53}$NO$_5$ 495.74	covalent inhibitor of gastrointestinal lipase inhibitor for obesity management by inhibiting the absorption of dietary fats. tbl. 60 mg (Alli) 120 mg (Xenical), O, three times a day	Roche USA/1999 EU/1998 [33–36]

Table 3. A08AX Other anti-obesity drugs

INN (Synonyms) [Brand names] ATC	Structure (Remarks)	[CAS-No.] Formula MW (g/mol)	Target (if known) Medical use Formulation DDD	Originator Approval (country/year) Production method [References]
Rimonabant [Acomplia, Zimulti] A08AX01		[168273-06-1] $C_{22}H_{21}Cl_3N_4O$ 463.80 [158681-13-1] $C_{22}H_{21}Cl_3N_4O \cdot HCl$ 500.25	cannabinoid receptor 1 (CB_1) antagonist tbl. 20 mg, O, once daily	Sanofi EU/2006 withdrawn from market 2009 in EU and India [14, 37–39]
Liraglutide [Saxenda] A10BJ02	H-His-Ala-Glu-Gly-Thr-Phe-Thr-Ser-Asp-Val-Ser-Ser-Tyr-Leu-Glu-Gly-Gln-Ala-Ala-(N-ε-(γ-(N-α-hexadecanoyl)-Glu)Lys-Glu-Phe-Ile-Ala-Trp-Leu-Val-Arg-Gly-Arg-Gly-OH	[204656-20-2] $C_{172}H_{265}N_{43}O_{51}$ 3751.26	glucagon-like peptide-1 receptor agonist subcutaneous injection treatment for adults, who are obese or overweight with at least one weight-related comorbid condition. 3 mg, P, once daily	NovoNordisk USA/2015 EU/2015 Synthesis: [9, 40–44]

Abbreviations

AMPA	α-amino-3-hydroxy-5-methyl-4-isoxazolepropionic acid
BMI	body mass index
DDD	daily drug dose
GABA	γ-aminobutyric acid
GLP-1	glucagon-like peptide-1
5-HT	serotonin
INN	international nonproprietary name
MW	molecular weight
O	oral
P	parenteral
tbl.	tablet
WHO	World Health Organization

References

1 Report of a WHO consultation (2000) Obesity: preventing and managing the global epidemic. *World Health Organ Tech. Rep. Ser.* **894**, 1–253.

2 World Health Organisation (WHO) (2018) Obesity and overweight, http://www.who.int/mediacentre/factsheets/fs311/en (accessed 20th February 2018).

3 De Lorenzo, A., Soldati, L., Sarlo, F., Calvani, M., Di Lorenzo, N. and Di Renzo, L. (2016) New obesity classification criteria as a tool for bariatric surgery indication. *World J Gastroenterol.* **22**, 681–703.

4 Valentino, M.A., Lin, J.E. and Waldman, S.A. (2010) Central and peripheral molecular targets for anti-obesity pharmacotherapy. *Clin. Pharmacol. Ther.* **87**, 652–662.

5 Adam, R.A.H. (2013) Mechanisms underlying current and future anti-obesity drugs. *Trends in Neurosc.* **36**, 133–140.

6 Gadde, K.M. (2014) Current pharmacotherapy for obesity: extrapolation of clinical trials data to practice. *Expert. Opin. Pharmacother* **15**, 809–822.

7 Kakkar, A.K. and Dahiya, N. (2015) Drug treatment of obesity: current status and future prospects. *Eur. J. Int. Med.* **26**, 89–94.

8 Narayanaswami, V. and Dwoskin, L.P. (2017) Obesity: current and potential pharmacotherapeutics and targets. *Pharmacology & Therapeutics* **170**, 116–147.

9 Knudsen, L.B. et al. (2000) Potent derivatives of glucagon-like peptide-1 with pharmacokinetic properties suitable for once daily administration. *J. Med. Chem.* **43**, 1664–1669.

10 Kang, J.G. and Park, C.Y. (2012) Anti-obesity drugs: a review about their effects and safety. *Diab. Metab. J.* **36**, 13–25.

11 Onakpoya, I.J., Heneghan, C.J., and Aronson, J.K. (2016) Post-marketing withdrawal of anti-obesity medicinal products because of adverse drug reactions. *BMC Medicine* **14** 191–201.

12 Jeffery, J.E., Kerrigan, F., Miller, T.K., Smith, G.J. and Tometzki, G.B. (1996) Synthesis of sibutramine, a novel cyclobutylalkylamine useful in the treatment of obesity, and its major human metabolites. *J. Chem. Soc., Perk. Trans. 1*, **21**, 2583–2589.

13 Smith, B.M. et al. (2008) Discovery and structure–activity relationship of (1R)-8-chloro-2,3,4,5-tetrahydro-1-methyl-1H-3-benzazepine (lorcaserin), a selective serotonin 5-HT2C receptor agonist for the treatment of obesity. *J. Med. Chem.* **51**, 305–313.

14 Rinaldi-Carmona, M. et al. (1994) SR141716A, a potent and selective antagonist of the brain cannabinoid receptor. *FEBS letters* **350**, 240–244.

15 Merrell Pharmac. (1942) US 2 408 345, US-prior. 13.4.1942.

16 Maryanoff, B.E., Nortey, S.O., Gardocki, J.F., Shank, R.P and Dodgson, S.P. (1987) Anticonvulsant O-alkyl sulfamates. 2,3:4,5-Bis-O-(1-methylethylidene)-β-D-fructopyranose sulfamate and related compounds, *J. Med. Chem.* **30**, 880–887.

17 McNeil Lab. (1983) US 4 513 006, US-prior. 26.9.1983.

18 Temmler Werke (1958) US 3 001 910, US-prior. 16.4.1958.

19 Hyde, J.F., Browning, E. and Adams, R. (1928) Synthetic homologs of D,L-ephedrine, *J. Am. Chem. Soc.* **50**, 2287–2292.

20 Sience Union (1964) GB 1 078 186, priority date 27.6.1964, US.

21 Aeberli, P., Eden, P., Goger, J.H., Houlihan, W.J. and Penberthy, C. (1975) 5-Aryl-2,3-dihydro-5*H*-imidazo[2,1-a]isoindol-5-ols. Novel class of anorectic agents. *J. Med. Chem.* **18**, 177–182.

22 Sandoz (1968) US 3 597 445, US-prior. 19.6.1968.

23 Junet, R. (1956) Ethylamphetamine in the treatment of obesity. *Praxis* **45**, 986–988.

24 Pfanz, H., Wieduwilt, H. (1956) DD 13 785, DDR-prior. 8.2.1956.

25 Societe Industrielle F.A. (1964) FR 142306, FR-prior. 23.11.1964.

26 Hoffmann-La Roche (1959) DE 1 210 873, DE-prior. 18.3.1959.

27 Boots Comp. BLC (1985) GB 2 184 122, UK-prior. 17.12.1985.

28 Arena Pharm. (2002) WO 200 3086 306, US-prior. 12.2.2002.

29 Arena Pharm. (2003) WO 2005 003 096, US-prior. 17.6.2003.

30 Arena Pharm. (2006) WO 2007 120 517, US-prior. 3.4.2006.

31 Orexigen Therapeutics (2012) WO 2012 075 453, US-prior. 3.12.2012.

32 Endo Lab. (1996) US 3 332 950, US-prior. 6.12.1996.

33 Roche (1984) EP 189 577, CH-prior. 21.12.1984.

34 Roche (1996) EP 803 576, EP-prior. 26.4.1996.

35 Barbier, P. and Schneider, F. (1987) Syntheses of tetrahydrolipstatin and absolute configuration of tetrahydrolipstatin and lipstatin. *Helv. Chim. Acta* **70**, 196–202.

36 Barbier, P. and Schneider, F. (1988) Synthesis of tetrahydrolipstatin and tetrahydroesterastin, compounds with a β-lactone moiety. Stereoselective hydrogenation of a β-keto δ-lactone and conversion of the δ-lactone into a β-lactone. *J. Org. Chem.* **53**, 1218–1221.

37 Sanofi-Synthelabo (1995) FR 2 831 883, EP-prior.

38 Sanofi (1992) US 5 624 941, FR-prior. 23.6.1992, 2.12.1993, 20.6.1994.

39 Kotagiri, V.K. et al. (2007) An improved synthesis of rimonabant: anti-obesity drug. *Org. Proc.Res. Dev.* **11** 910–912.

40 Novo Nordisk (1993) WO 9 517 510, DK-prior. 23.12.1993.

41 Novo Nordisk (1996) WO 9 808 871, DK-prior. 30.8.1996.

42 Novo Nordisk (1996) US 6 458 924, DK-prior. 30.8.1996.

43 Novo Nordisk (1996) US 7 235 627, DK-prior. 30.8.1996.

44 Novo Nordisk (1998) WO 9 943 707, DK-prior. 27.2.1998.

Antidiabetic Drugs (A10)

KURT RITTER, Frankfurt, Germany

ULLMANN'S ENCYCLOPEDIA OF INDUSTRIAL CHEMISTRY

1. Introduction

Diabetes mellitus is a disease which makes the human body unable to maintain basic glucose homeostasis, the continuous regulation of the blood glucose level after digestion of food and during fasting periods. Glucose is the sole energy source for brain cells, however, elevated blood glucose levels over a longer period can be detrimental to blood vessels and peripheral tissues. Main regulation of the blood glucose level occurs through secretion of two counteracting hormones. Insulin from the β-cells within the islets of the pancreas is secreted in response to an elevated blood glucose level, especially after digestion of food, whereas glucagon from the α-cells of the islets is released in situations of low blood glucose levels.

Two main forms of diabetes have been described: Type 1 diabetes with a prevalence below 10% and type 2 diabetes with a prevalence above 90%.

Type 1 diabetes (T1D) is an autoimmune disease in which the body's own immune cells attack and destroy the insulin-producing β-cells of the pancreas. The cause is still unknown as it is for other autoimmune diseases. The rapid onset of the disease occurs usually in childhood, thus T1D is often referred to as "juvenile diabetes". The T1D patients have to rely on lifelong treatment with exogenous insulin or insulin analogues.

Type 2 diabetes (T2D), a systemic, progressive disease, is multifactorial with a slow onset, often in later stages of life. Typical risk factors for developing type 2 diabetes are being overweight or obese, lack of physical activity, and poor diet ("high-fat, high-carb"), but also genetic predisposition and epigenetic disturbances during pregnancy can be contributing factors. Most patients are overweight or obese at diagnosis and often have additional metabolic disorders, especially of lipid metabolism. The observed worldwide increase in obesity is followed by a growing

Ullmann's Pharmaceuticals. Edited by Axel Kleemann and Bernhard Kutscher.
© 2022 Wiley-VCH GmbH
Set ISBN: 978-3-527-34252-5/ DOI: 10.1002/14356007.a03_001.pub3

global epidemic of T2D. The International Diabetes Federation (IDF) predicts a further raise of T2D patients from 415 million adult (2013) to 640 million by 2040 [1]. A huge impact on the healthcare system is expected to cover the costs for the treatment of diabetes and its complications.

Several metabolic irregularities are established in T2D patients. Secretion of insulin from the islets in the pancreas is altered. It is elevated in the prediabetic stage or early stage of the disease, but reduced or absent in the late stages of the disease due to substantial loss of β-cells. Resistance against the action of insulin (insulin resistance) is observed in different cells and tissues, leading to reduced uptake of glucose in liver and muscle cells, and hepatic overproduction of endogenous glucose. Furthermore, changes in fat metabolism and fat storage in adipocytes due to insulin resistance results in increased plasma levels of free fatty acids.

Chronic elevated glucose levels (hyperglycemia) can lead to micro- and macrovascular complications, long-term damage, dysfunction and even organ failure. For example, damage to the kidney might lead to renal failure and dialysis, damage to the retina in the eyes can cause blindness, and damage to nerves and blood vessels can result in foot wounds, ulcers, and ultimately to amputations. Regarding macrovascular complications, T2D patients have an increased risk for cardiovascular diseases (two to four times higher) and for stroke (two times higher) when compared to nondiabetics.

Diabetes may be diagnosed based on one of these plasma glucose criteria: a) Elevated fasting plasma glucose level (FPG), above 126 mg/dL vs 70–99 mg/dL in a healthy adult, b) elevated two-hour plasma glucose levels, above 200 mg/dL vs below 140 mg/dL after a 75-g oral glucose administration in a tolerance test (OGTT). Another criterion is the increased percentage of glycated hemoglobin (HbA_{1c}) in red blood cells, above or equal to 6.5% as compared to 4.0–5.5% in healthy adults. The HbA_{1c} value reflects a three months average plasma glucose concentration without the influence of day-to-day variation of the plasma glucose measurements. The reduction of the HbA_{1c} value correlates with the downregulation of plasma glucose by an effective antidiabetic therapy over a long term period.

Treatment options for T2D patients depend on the stage of the disease [2–4]. Changes in lifestyle such as an increase in physical exercise, and diet for reduction of body weight are first recommended and will be accompanied by administration of an oral drug, usually metformin. If treatment goals are not reached, combinations of two or three drugs with complementary mechanism of action are the standard of care. Addition of an injectable drug such as a glucagon-like peptide-1 (GLP-1) analog or insulin might be of advantage at this stage. At a later stage of the disease, an intensified treatment with insulin or insulin analogues might be necessary. The frequent need for escalating therapy reflects the fact that blood glucose control by antidiabetic drugs slows the further loss of the islets β-cell function, but cannot prevent it.

Exogenous insulin or insulin analogues are a direct supply for the insufficient insulin production of T2D patients. The other currently marketed antidiabetic drugs normalize the blood glucose level by enhancing the body's sensitivity to insulin (insulin sensitizers, i.e., metformin) or increasing the insulin secretion from the β-cells, (insulin secretagogues, i.e., sulfonylureas, glinides, DPP-4 inhibitors, GLP-1 analogues). The latest class of antidiabetics, the sodium-glucose cotransporter 2 (SGLT2) inhibitors, prevents the reabsorption of glucose in the kidneys.

An important challenge in the treatment of T2D patients is the prevention of cardiovascular diseases. Intensive glycemic control by current therapies delays the onset and lowers the progression of microvascular complications. However, a similar benefit has not been observed in the clinic with regard to the number of cardiovascular events such as strokes or heart attacks, or in reduction of mortality, due to macrovascular complications. Some studies indicate that new antidiabetic drugs, such as the SGLT2 inhibitors empagliflozin and canagliflozin, as well as the GLP1-analogues liraglutide and semaglutide showed statistically significant reductions in cardiovascular morbidity and mortality in high-risk T2D patients for the first time.

2. A10A Insulins and Analogues (Table 1)

In 1921, FREDERICK BANTING and CHARLES BEST, Toronto, Canada, discovered the hormone insulin and demonstrated its key role in glucose homeostasis by injecting it into dogs after having made them diabetic. In 1922, the first T1D patient was treated with insulin isolated from fetal calf pancreas with the help of JAMES COLLIP. Refined pure insulin could be obtained by support of Eli Lilly. Until the mid 1980s, treatment of diabetic patients (T1D, T2D) with animal-derived insulins, either of porcine (one amino acid (AA) different to human insulin) or bovine origin (three AA different to human insulin) was the standard of care. More and more human insulin or human insulin analogues produced by recombinant technologies have replaced these insulins. Since 2006 they have completely conquered the market [5].

Human insulin is a peptide hormone of 51 amino acids. As a heterodimer, it consists of two strands, A-strand (21 AA) and B-strand (30 AA), which are linked together by two disulfide bonds (Cys-7 (A-strand)–Cys-7 (B-strand) and Cys-19 (B-strand)–Cys-20 (A-strand)). An intrachain linkage between Cys-6 and Cys-11 is found in the A-strand. Human insulin is produced and released by the β-cells in the pancreas. It has many physiological effects in different tissues. Foremost, it regulates blood glucose levels by stimulating its uptake into liver and muscle cells and downregulates hepatic glucose production. Furthermore, insulin suppresses lipolysis and promotes fat storage in adipocytes regulating also the levels of free fatty acids. The physiological storage form of insulin in β-cells is a zinc-coordinated hexamer. After secretion into blood, a rapid equilibrium between hexa-, di-, and monomeric forms occurs. Monomeric insulin binds to the insulin receptor on the cell membrane.

The insulin analogues are distinguished by their onset of action and their half-life as compared with wildtype human insulin [6, 7]. Rapid-acting or fast-acting insulins have a faster onset (10–15 min) than natural insulin (30 min) and a duration of action between 2 and 3 h. They are mostly used in connection with meals to lower postprandial glucose levels. Changes in the amino acid sequence decrease the stability of the different oligomers and thus increase the rapid availability of the active monomeric form.

Long-acting insulins have a slower onset (1–3 h), but show a prolonged continuous delivery of insulin. They regulate the blood glucose during periods of fasting, especially during sleep, thus, they are often referred to as basal insulins. The prolonged duration of action (20–40 h) is achieved by increasing half-life by reducing metabolism and clearance. One way to modify insulins is to attach long fatty acid moieties to the side chain of lysine-29 to achieve binding to serum albumin in the blood. The equilibrium with albumin and the stability of the hexamers itself defines the duration action. Examples are insulin degludec and detemir.

The long-acting insulin glargine works through a different mechanism. Specific modification in its sequence led to good solubility and stability at acidic pH 4 and decreased solubility at physiological pH of 7.4 due to the increase of the isoelectric point to 6.9. After subcutaneous injection of the slightly acidic solution, precipitation of this insulin analog due to the pH change, generates a depot form in the subcutaneous tissue. Slow redissolution from this depot results in a steady release into the blood stream. Fast metabolism produces two main metabolites rapidly dissociating into monomeric active forms.

Type 1 diabetic patients have to add a rapid acting insulin to the basal insulin during meal time, whereas type 2 diabetics can combine their basal insulin with other oral or injectable drugs. A major risk of the combinations with insulin therapy is hypoglycemia (i.e., after extensive exercise, due to reduced food intake or extended stress). Hypoglycemia can be life-threatening, particularly the nocturnal hypoglycemia in type 1 diabetics. Another side effect of insulin treatment is weight gain in T2D patients, especially of concern in already obese patients.

Short-acting insulins can be administered via oral inhalation. Two inhalable human insulins, Exubera, and Afrezza, were approved but later taken from the market due to patient to patient variability, reimbursement issues and low market penetration.

Table 1. A10A Insulins and analogues

INN (Synonyms) [Brand names] ATC	Structure (amino acid or side chain modification in bold blue: difference to human insulin; Cys in bold: forming sulfide bridge)	[CAS-No.] Formula MW (g/mol)	Target (if known) Medical use Formulation DDD	Originator Approval (country/year) Production method [References]
Insulin (human) [Humulin, Humulin 70/30, Humulin R U-500,] A10AB01, A10AC01, A10AD01 A10AE01 **WHO essential medicine**	A-strand: H-Gly-Ile-Val-Glu-Gln-**Cys**-**Cys**-Thr-Ser-Ile-**Cys**-Ser-Leu-Tyr-Gln-Leu-Glu-Asn-Tyr-**Cys**-Gln B-strand: H-Phe-Val-Asn-Gln-His-Leu-**Cys**-Gly-Ser-His-Leu-Val-Glu-Ala-Leu-Tyr-Leu-Val-**Cys**-Gly-Glu-Arg-Gly-Phe-Phe-Tyr-Thr-Pro-Lys-Thr Location — Description Cys7(A)–Cys7(B) — disulfide bridge Cys19(B)–Cys20(A) — disulfide bridge Cys6(A)–Cys11(A) — disulfide bridge	[*11061-68-0*] $C_{257}H_{383}N_{65}O_{77}S_6$ 5808.63	human insulin for the treatment of type 1 and type 2 diabetes. subcutaneous injection available as short-acting (Humulin R), intermediate-acting (Humulin N), high concentrated (Humulin R U-500), and premixed formulations (Humulin 70/30). DDD 40 units.	Eli Lilly and Genentech Humulin USA/1982 EU/1982 Japan/1996 China/1997 Humulin70/30 USA/1989 Humulin R U-500 USA/1997 produced by recombinant DNA technology [8–12]
Insulin (bovine) A10AB02	A-strand: H-Gly-Ile-Val-Glu-Gln-**Cys**-**Cys**-**Ala**-Ser-**Val**-**Cys**-Ser-Leu-Tyr-Gln-Leu-Glu-Asn-Tyr-**Cys**-Gln B-strand: H-Phe-Val-Asn-Gln-His-Leu-**Cys**-Gly-Ser-His-Leu-Val-Glu-Ala-Leu-Tyr-Leu-Val-**Cys**-Gly-Glu-Arg-Gly-Phe-Phe-Tyr-Thr-Pro-Lys-**Ala** Location — Description Cys7(A)–Cys7(B) — disulfide bridge Cys19(B)–Cys20(A) — disulfide bridge Cys6(A)–Cys11(A) — disulfide bridge	[*11070-73-8*] $C_{254}H_{377}N_{65}O_{75}S_6$ 5733.55	DDD 40 units.	withdrawn from market
Insulin (pork) A10AB03	A-strand: H-Gly-Ile-Val-Glu-Gln-**Cys**-**Cys**-Thr-Ser-Ile-**Cys**-Ser-Leu-Tyr-Gln-Leu-Glu-Asn-Tyr-**Cys**-Gln B-strand: H-Phe-Val-Asn-Gln-His-Leu-**Cys**-Gly-Ser-His-Leu-Val-Glu-Ala-Leu-Tyr-Leu-Val-**Cys**-Gly-Glu-Arg-Gly-Phe-Phe-Tyr-Thr-Pro-Lys-**Ala** Location — Description Cys7(A)–Cys7(B) — disulfide bridge Cys19(B)–Cys20(A) — disulfide bridge Cys6(A)–Cys11(A) — disulfide bridge	[*12584-58-6*] $C_{257}H_{387}N_{65}O_{76}S_6$ 5795.6	DDD 40 units.	withdrawn from market
Insulin lispro [Humalog, Humalog Mix50/50, Lispro, Eglucent, Liprolog, Bio-Lysprol] A10AB04; A10AC04; A10AD04	28^B-L-lysine-29^B-L-proline-insulin (human) A-strand: H-Gly-Ile-Val-Glu-Gln-**Cys**-**Cys**-Thr-Ser-Ile-**Cys**-Ser-Leu-Tyr-Gln-Leu-Glu-Asn-Tyr-**Cys**-Gln B-strand: H-Phe-Val-Asn-Gln-His-Leu-**Cys**-Gly-Ser-His-Leu-Val-Glu-Ala-Leu-Tyr-Leu-Val-**Cys**-Gly-Glu-Arg-Gly-Phe-Phe-Tyr-Thr-**Lys**-**Pro**-Thr Location — Description Cys7(A)–Cys7(B) — disulfide bridge Cys19(B)–Cys20(A) — disulfide bridge Cys6(A)–Cys11(A) — disulfide bridge	[*133107-64-9*] $C_{257}H_{383}N_{65}O_{77}S_6$ 5808.63	rapid-acting human insulin analog (changes: Lys(B28), Pro (B29)) for the treatment of type 1 and type 2 diabetes. subcutaneous injection Humalog U-100 insulin lispro 100 units/mL (in 10 mL vials, as 3 mL Humalog KwikPen (prefilled), as 3 mL Humalog Junior KwikPen (prefilled), as 3 mL cartridges), U-200 insulin lispro 200 units/mL (as 3 mL Humalog KwikPen (prefilled)). Humalog Mix50/50 50%: insulin lispro protamine suspension and 50% insulin lispro injection 100 units/mL. DDD 40 units.	Eli Lilly Humalog USA/1996 EU/1996 Japan/2001 Humalog Mix US/1999 Japan/2004 produced by recombinant DNA technology utilizing a nonpathogenic laboratory strain of *Escherichia coli* [13, 14] formulation: [15–19]

Table 1. (*Continued*)

INN (Synonyms) [Brand names] ATC	Structure (amino acid or side chain modification in bold blue: difference to human insulin; Cys in bold: forming sulfide bridge)	[CAS-No.] Formula MW (g/mol)	Target (if known) Medical use Formulation DDD	Originator Approval (country/year) Production method [References]
Insulin aspart [NovoLog, NovoRapid, NovoRapid Mix, NovoMix30, NovoLog Mix, NovoTwist, NovoFine] A10AB05; A10AD05	28^B-L-aspartic acid-insulin (human) A-strand: H-Gly-Ile-Val-Glu-Gln-**Cys**-**Cys**-Thr-Ser-Ile-**Cys**-Ser-Leu-Tyr-Gln-Leu-Glu-Asn-Tyr-**Cys**-Gln B-strand: H-Phe-Val-Asn-Gln-His-Leu-**Cys**-Gly-Ser-His-Leu-Val-Glu-Ala-Leu-Tyr-Leu-Val-**Cys**-Gly-Glu-Arg-Gly-Phe-Phe-Tyr-Thr-**Asp**-Lys-Thr Location — Description Cys7(A)–Cys7(B) — disulfide bridge Cys19(B)–Cys20(A) — disulfide bridge Cys6(A)–Cys11(A) — disulfide bridge	[*116094-23-6*] $C_{256}H_{381}N_{65}O_{79}S_6$ 5825.8	rapid acting human insulin analog (changes: Asp(B28)) for the treatment of type 1 and type 2 diabetes. subcutaneous injection NovoLog insulin aspart 100 units/mL: in 10 mL vials, as 3 mL PenFill cartridges for the 3 mL PenFill cartridge delivery device with NovoFine® disposable needles, as 3 mL Novolog FlexPen, as 3 mL Novolog FlexTouch. NovoLog Mix 50/50: human insulin analog suspension (50% insulin aspart protamine crystals and 50% soluble insulin aspart). 100 units/mL. NovoLog Mix 70/30: human insulin analog (70% insulin aspart protamine crystals and 30% soluble insulin aspart). 100 units/mL. DDD 40 units.	Novo Nordisk NovoLog US/2000 EU/1999 Japan/2001 produced by recombinant DNA technology utilizing *Saccharomyces cerevisiae* formulation: [22] NovoLog Mix EU/2005 US/2006 [23]
Insulin glulisine [Apidra, Shorant] A10AB06	3^B-L-lysine-29^B-L-glutamic acid-insulin (human) A-strand: H-Gly-Ile-Val-Glu-Gln-**Cys**-**Cys**-Thr-Ser-Ile-**Cys**-Ser-Leu-Tyr-Gln-Leu-Glu-Asn-Tyr-**Cys**-Gln B-strand: H-Phe-Val-**Lys**-Gln-His-Leu-**Cys**-Gly-Ser-His-Leu-Val-Glu-Ala-Leu-Tyr-Leu-Val-**Cys**-Gly-Glu-Arg-Gly-Phe-Phe-Tyr-Thr-Pro-**Glu**-Thr Location — Description Cys7(A)–Cys7(B) — disulfide bridge Cys19(B)–Cys20(A) — disulfide bridge Cys6(A)–Cys11(A) — disulfide bridge	[*207748-29-6*] $C_{258}H_{384}N_{64}O_{78}S_6$ 5823	rapid-acting human insulin analog (changes: Lys (B3), Glu(B29)) for the treatment of type 1 and type 2 diabetes. subcutaneous injection Apidra 100 units/mL. in 10 mL vials or 3 mL SoloStar prefilled pen. DDD 40 units.	Hoechst Marion Roussel (then Aventis (now part of Sanofi) EU/2004 USA/2004 Japan/2009 produced by recombinant DNA technology utilizing a non-pathogenic laboratory strain of *Escherichia coli* (K12) [24, 25] formulation: [26]
Insulin glargine [Lantus; Toujeo] A10AE04	21^A-glycine-30^B-L-arginine-31^B-L-arginine-insulin (human) A-strand: H-Gly-Ile-Val-Glu-Gln-**Cys**-**Cys**-Thr-Ser-Ile-**Cys**-Ser-Leu-Tyr-Gln-Leu-Glu-Asn-Tyr-**Cys**-**Gly** B-strand: H-Phe-Val-Asn-Gln-His-Leu-**Cys**-Gly-Ser-His-Leu-Val-Glu-Ala-Leu-Tyr-Leu-Val-**Cys**-Gly-Glu-Arg-Gly-Phe-Phe-Tyr-Thr-Pro-Lys-**Arg**-**Arg** Location — Description Cys7(A)–Cys7(B) — disulfide bridge Cys19(B)–Cys20(A) — disulfide bridge Cys6(A)–Cys11(A) — disulfide bridge	[*160337-95-1*] $C_{267}H_{404}N_{72}O_{78}S_6$ 6063	long-acting basal human insulin analog (changes: Gly (A21), added Arg(B30), Arg (31)) for the treatment of type 1 and type 2 diabetes. subcutaneous injection Lantus 100 units/mL in 10 mL vial, as 3 mL SoloStar prefilled pen. Toujeo 300 units/mL as 1.5 mL Toujeo SoloStar disposable prefilled pen (450 units/1.5 mL). DDD 40 units.	Sanofi Lantus EU/2000 USA/2001 Japan/2003 Toujeo EU/2015 USA/2015 Japan/2015 produced by recombinant DNA technology utilizing a nonpathogenic laboratory strain of *Escherichia coli* (K12) [27–29] formulation: [30–32] Toujeo: [33]

(continued)

Table 1. (*Continued*)

INN (Synonyms) [Brand names] ATC	Structure (amino acid or side chain modification in bold blue: difference to human insulin; Cys in bold: forming sulfide bridge)	[CAS-No.] Formula MW (g/mol)	Target (if known) Medical use Formulation DDD	Originator Approval (country/year) Production method [References]
Insulin detemir [Levemir, Revemir] A10AE05	29^B-$[N^6$-$[N$-(14-carboxy-1-triadecyl)-L-lysine](des-30^B-L-threonine-insulin (human) A-strand: H-Gly-Ile-Val-Glu-Gln-**Cys**-**Cys**-Thr-Ser-Ile-**Cys**-Ser-Leu-Tyr-Gln-Leu-Glu-Asn-Tyr-**Cys**-Gln B-strand: H-Phe-Val-Asn-Gln-His-Leu-**Cys**-Gly-Ser-His-Leu-Val-Glu-Ala-Leu-Tyr-Leu-Val-**Cys**-Gly-Glu-Arg-Gly-Phe-Phe-Tyr-Thr-Pro-Lys(**N^6-tetradecanoyl**) Location — Description Cys7(A)–Cys7(B) — disulfide bridge Cys19(B)–Cys20(A) — disulfide bridge Cys6(A)–Cys11(A) — disulfide bridge	[*169148-63-4*] $C_{267}H_{402}N_{64}O_{76}S_6$ 5916.9	long-acting basal human insulin analog (changes: Lys (N^6- tetradecanoyl) (B29), des Thr(B30)), for the treatment of type 1 and type 2 diabetes. subcutaneous injection. Levemir 100 units/mL in 10 mL vial, as 3 mL Levemir® FlexTouch®. DDD 40 units.	Novo Nordisk EU/2004 USA/2005 Japan/2007 China/2009 expression of recombinant DNA in *Saccharomyces cerevisiae* followed by chemical modification [34–37] formulation: [38]
Insulin degludec [Tresiba, Degludec] A10AE06	29^B-$[N^6$-$[N$-(15-carboxy-1-oxopentadecyl)-L-γ-glutamyl]-L-lysine]-(des-30^B-L-threonine-insulin insulin (human) A-strand: H-Gly-Ile-Val-Glu-Gln-**Cys**-**Cys**-Thr-Ser-Ile-**Cys**-Ser-Leu-Tyr-Gln-Leu-Glu-Asn-Tyr-**Cys**-Gln B-strand: H-Phe-Val-Asn-Gln-His-Leu-**Cys**-Gly-Ser-His-Leu-Val-Glu-Ala-Leu-Tyr-Leu-Val-**Cys**-Gly-Glu-Arg-Gly-Phe-Phe-Tyr-Thr-Pro-Lys[**N^6-[N-(15-carboxy-1-oxopentadecyl)-L-γ-glutamyl]**] Location — Description Cys7(A)–Cys7(B) — disulfide bridge Cys19(B)–Cys20(A) — disulfide bridge Cys6(A)–Cys11(A) — disulfide bridge	[*844439-96-9*] $C_{274}H_{411}N_{65}O_{81}S_6$ 6103.97	long-acting basal human insulin analog for the treatment of type 1 and type 2 diabetes. subcutaneous injection. 100 units/mL, 200 units/mL (both as 3 mL FlexTouch disposable prefilled pen). DDD 40 units.	Novo Nordisk EU/2013 USA/2015 Japan/2016 expression of recombinant DNA in *Saccharomyces cerevisiae* followed by chemical modification [39]
Insulin degludec and insulin aspart [Ryzodeg; IDegAsp; DegludecPlus] A10AD06		insulin degludec [*844439-96-9*] $C_{274}H_{411}N_{65}O_{81}S_6$ 6103.97 insulin aspart [*116094-23-6*] $C_{256}H_{381}N_{65}O_{79}S_6$ 5825.8	combination of a long-acting (insulin degludec) and rapid-acting insulin (insulin aspart) for the treatment of type 1 and type 2 diabetes. Ryzodeg 70/30 solution containing 70% insulin degludec and 30% insulin aspart for subcutaneous injection once or twice daily with any main meal: 100 units/mL (U-100) in a 3 mL FlexTouch disposable prefilled pen. DDD 40 units.	Novo Nordisk EU/2013 USA/2015 produced by recombinant DNA technology utilizing *Saccharomyces cerevisiae*
Insulin degludec and liraglutide [DegLira; Xultophy] A10AE56		insulin degludec [*844439-96-9*] $C_{274}H_{411}N_{65}O_{81}S_6$ 6103.97 liraglutide [*204656-20-2*] $C_{172}H_{265}N_{43}O_{51}$ 3751.20	combination of a long-acting basal human insulin analog and a GLP-1 analog for the treatment of type 2 diabetes. Xultophy 100/3.6 injection: 100 units/mL insulin degludec 3.6 mg/mL liraglutide in 3 mL pre-filled, disposable, single-patient-use pen injector. DDD 40 units.	Novo Nordisk EU/2014 USA/2016 Japan/2012

3. A10B Blood Glucose-Lowering Drugs, Excl. Insulins

3.1. A10BA Biguanides (Table 2)

French Lilac or "goat's rue" has been used in folk medicine since the Middle Ages to treat symptoms of diabetes such as frequent urination or thirst without knowing its underlying mode of action. The isolation of the active ingredients, galegine, with a guanidine core, encouraged clinical research with similar guanidine-like molecules leading to the use of the antihyperglycemic biguanidines such as phenformin and metformin since the mid 1950s. Phenformin as well as a third available biguanide, buformin, were withdrawn

from the market due to the high risk for lactic acidosis. Metformin has an improved safety profile, causing less hypoglycemia or lactate acidosis, it is, therefore, used as a first-line treatment in T2D patients and also in combinations with many other antidiabetic drugs [40, 41]. Its mechanism of action is not completely understood. One main action is the downregulation of the formation of new glucose (gluconeogenesis) in the liver and the increase of glucose uptake in tissues, both via activation of AMP kinases. Two types of metformin are on the market, an intermediate release and a slow release form, the later having less gastrointestinal side effects. Due to the high dose used and its unchanged excretion via the kidneys, metformin is not recommended for patients with renal insufficiency.

Table 2. A10BA Biguanides

INN (Synonyms) [Brand names] ATC	Structure	[CAS-No.] Formula MW (g/mol)	Target (if known) Medical use Formulation DDD	Originator Approval (country/year) Production method [References]
Phenformin A10BA01		[114-86-3] $C_{10}H_{15}N_5$ 205.26	oral antidiabetic. DDD 0.1 g.	withdrawn from market
Metformin [Glucophage, Diabetosan, Diabex, Fluamine, Flumamine, Gliguanid, Metgluco, Glucofit, Metfogamma, Siofor, Diaformin, Glifage] Metformine -extended release [Diabex XR, Fortamet, Glucophage XR Glumetza] A10BA02 **WHO essential medicine**		Metformin [657-24-9] $C_4H_{11}N_5$ 129.16 Metformin hydrochloride [1115-70-4] $C_4H_{11}N_5 \cdot HCl$ 165.63	oral antidiabetic. tbl. (once or twice daily) 500, 850, 1000 mg, O. Metformin IR (immediate release). tbl. (once or twice daily) 500, 750, 1000 mg, O. Metformin SR (slow release) or XR (extended release). DDD 2 g. many fixed dose-combinations with other oral antidiabetic drugs (see A10BD).	Merck Sante (now Merck KG) France/1957 UK/1958 Canada/1972 USA/1994 [42]
Buformin [Silubin] A10BA03		[692-13-7] $C_6H_{15}N_5$ 157.22	oral antidiabetic. DDD 0.2 g.	Grünenthal withdrawn from many markets available in Romania (Silubin Retard), Hungary, Taiwan, and Japan synthesis: [43]

3.2. A10BB Sulfonylureas (Table 3)

Sulfonylureas have been used to treat T2D patients since the 1950s. Their name stems from a common pharmacophore, an acidic arylsulfonylurea moiety [44]. These orally available drugs block ATP-sensitive potassium (K_{ATP}) channels on the cell membrane of pancreatic β-cells, which leads to the depolarization of the cell membrane followed by opening of voltage-gated calcium channels. The instream of calcium increases the secretion of insulin from vesicles within the cell membrane consequently lowering blood glucose levels. The "first generation" sulfonylureas (i.e., chlorpropamide, tolbutamide, tolazamide, carbutamide, acetohexamide) were replaced by "second generation" (i.e., glibenclamide, glibornuride, glipizide, gliquidone, gliclazide, metahexamide, glisoxepide) and "third generation" sulfonylureas (glimepiride) with increased potency and reduced side effects (toxicity, drug-drug interactions). Sulfonylureas (SUs) can induce severe hypoglycemia by overstimulating insulin release due to their non-dependency on actual blood glucose levels. Loss of efficacy during long-term treatment is also observed. Furthermore, they induce weight gain and are usually not recommended for patients with renal impairment. They are still prescribed as addition or in fixed combination with other oral antidiabetic drugs due to their low price or for patients with intolerance to other drugs.

Table 3. A10BB Sulfonylureas

INN (Synonyms) [Brand names] ATC	Structure	[CAS-No.] Formula MW (g/mol)	Target (if known) Medical use Formulation DDD	Originator Approval (country/year) Production method [References]
Glibenclamide [Daonil, DiaBeta, Euglucan, Euglucon, Gliben, Glibenbeta, Gliben-CT, Glibenclamid dura, Glibendoc, Glibenhexal, GlibenLich, Glib-ratiophar, Glucovance, Glukovital, Glyburide, Maninil, Miglucan] A10BB01		[*10238-21-8*] $C_{23}H_{28}ClN_3O_5S$ 494.01	oral antidiabetic binding to inhibitory regulatory subunit SU receptor 1 of the K_{ATP} channel in pancreatic beta cell (second generation of SUs). tbl. (once daily) 1.25, 1.75, 2.5, 3.5, 5 mg, O; tbl. (once daily) 1.5, 3, 6 mg, O (micronized version). DDD 10 mg.	Hoechst AG (now part of Sanofi)/ Boehringer Mannheim (now part of Roche) Europe/1969 USA/1992 synthesis: [45–47]
Chlorpropamide [Abemide, Diabemide, Diabinese] A10BB02		[*94-20-2*] $C_{10}H_{13}ClN_2O_3S$ 276.74	oral antidiabetic binding to inhibitory regulatory subunit SU receptor 1 of the K_{ATP} channel in pancreatic beta cell (first generation of SUs). tbl. (once daily) 100, 250 mg, O. DDD 375 mg.	Pfizer USA/1958 synthesis: [48–50]

Table 3. (*Continued*)

INN (Synonyms) [Brand names] ATC	Structure	[CAS-No.] Formula MW (g/mol)	Target (if known) Medical use Formulation DDD	Originator Approval (country/year) Production method [References]
Tolbutamide [Arcosal, Diamol, Diatol, Orinase, Rastinon] A10BB03		[*64-77-7*] $C_{12}H_{18}N_2O_3S$ 270.35	oral antidiabetic binding to inhibitory regulatory subunit SU receptor 1 of the K_{ATP} channel in pancreatic beta cell (first generation of SUs). tbl., 500 mg O. DDD 1.5 g.	withdrawn from major markets Hoechst AG and Upjohn Company synthesis: [51, 52]
Glibornuride [Glutril, Gluborid] A10BB04		[*26944-48-9*] $C_{18}H_{26}N_2O_4S$ 366.48	oral antidiabetic binding to inhibitory regulatory subunit SU receptor 1 of the K_{ATP} channel in pancreatic beta cell (second generation of SUs). tbl. (once daily) 25 mg, O. DDD 38 mg.	Hoffmann-LaRoche Switzerland/ 1971 Germany/1972 France/1973 synthesis: [53–55]
Tolazamide [Tolinase, Norglycin] A10BB05		[*1156-19-0*] $C_{14}H_{21}N_3O_3S$ 311.40	oral antidiabetic binding to inhibitory regulatory subunit SU receptor 1 of the K_{ATP} channel in pancreatic beta cell (first generation of SUs). tbl. (once daily) 100, 250, 500 mg, O. DDD 500 mg.	withdrawn from market Upjohn Company USA/1966 synthesis: [56, 57]
Carbutamide [Glucidoral, Insoral, Invenol, Nadisan] A10BB06		[*339-43-5*] $C_{11}H_{17}N_3O_3S$ 271.34	oral antidiabetic binding to inhibitory regulatory subunit SU receptor 1 of the K_{ATP} channel in pancreatic beta cell (first generation of SUs). tbl. (once daily) 500 mg, O. DDD 750 mg.	withdrawn from market Boehringer Mannheim [58]

(*continued*)

Table 3. (*Continued*)

INN (Synonyms) [Brand names] ATC	Structure	[CAS-No.] Formula MW (g/mol)	Target (if known) Medical use Formulation DDD	Originator Approval (country/year) Production method [References]
Glipizide [Glibenese, Glucotrol, Glucotrol XL, Ozidia] A10BB07		[29094-61-9] $C_{21}H_{27}N_5O_4S$ 445.54	oral antidiabetic binding to inhibitory regulatory subunit SU receptor 1 of the K_{ATP} channel in pancreatic beta cell (second generation of SUs). tbl. (once daily) 5, 10 mg, O. tbl. (once daily) 2.5, 5, 10 mg O (extended release Glucotrol XL). DDD 10 mg.	withdrawn from market in Germany and UK Carlo Erba (now part of Pfizer) USA/1984 synthesis: [59, 60]
Gliquidone [Glurenorm, Glurenor] A10BB08		[33342-05-1] $C_{27}H_{33}N_3O_6S$ 527.63	oral antidiabetic binding to inhibitory regulatory subunit SU receptor 1 of the K_{ATP} channel in pancreatic beta cell (second generation of sulfonylureas). tbl. (once daily) 30 mg, O. DDD 60 mg.	Boehringer Ingelheim Germany/1975 Italy/1975 China/1990 synthesis: [61]
Gliclazide [Diamicron, Diamicron MR, Diaprel, Gliclada, Glimicron] A10BB09 **WHO essential medicine**		[21187-98-4] $C_{15}H_{21}N_3O_3S$ 323.41	oral antidiabetic binding to inhibitory regulatory subunit SU receptor 1 of the K_{ATP} channel in pancreatic beta cell (third generation of SUs). tbl. (once daily) 30, 40, 60, 80 mg, O (extended release version). DDD 60 mg	Science union/Servier approved in many countries (WHO Model List of Essential Medicines) synthesis: [62, 63]
Metahexamide [Isodiane] A10BB10		[565-33-3] $C_{14}H_{21}N_3O_3S$ 311.40	oral antidiabetic binding to inhibitory regulatory subunit SU receptor 1 of the K_{ATP} channel in pancreatic beta cell (first generation of SUs). tbl. (once daily) 100 mg, O.	withdrawn from market Servier

Table 3. (*Continued*)

INN (Synonyms) [Brand names] ATC	Structure	[CAS-No.] Formula MW (g/mol)	Target (if known) Medical use Formulation DDD	Originator Approval (country/year) Production method [References]
Glisoxepide [Glucoben, Glysepin, Pro-Diaban] A10BB11		[25046-79-1] $C_{20}H_{27}N_5O_5S$ 449.52	oral antidiabetic binding to inhibitory regulatory subunit SU receptor 1 of the K_{ATP} channel in pancreatic beta cell (second generation of SUs). tbl. (once daily) 4 mg, O.	withdrawn from market Bayer AG synthesis: [64, 65]
Glimepiride [Amarel, Amaryl, Avaglim, Avandaryl, Duetact, Solosa] A10BB12		[93479-97-1] $C_{24}H_{34}N_4O_5S$ 490.62	oral antidiabetic binding to inhibitory regulatory subunit SU receptor 1 of the K_{ATP} channel in pancreatic beta cell (second generation of SUs). tbl. (once daily) 1, 2, 4 mg, O. DDD 2 mg.	Hoechst AG (now Sanofi) Sweden/1995 Europe/1996 USA/1995 Japan/2000 synthesis: [66, 67] controlled release: [68–70]
Acetohexamide [Dimelin, Dimelor, Dymelor] A10BB31		[968-81-0] $C_{15}H_{20}N_2O_4S$ 324.40	oral antidiabetic binding to inhibitory regulatory subunit SU receptor 1 of the K_{ATP} channel in pancreatic beta cell (first generation of SUs). tbl. (once daily) 250, 500 mg. DDD 500 mg.	withdrawn from market Eli Lilly USA/1964 [71, 72]

3.3. A10BC Sulfonamides (Heterocyclic) (Table 4)

The sulfonamide glymidine is an orally available antidiabetic drug with some structural similiarities to the sulfonyl urea class, however, it was only launched in few countries in Europe in the 1960s.

Table 4. A10BC Sulfonamides (heterocyclic)

INN (Synonyms) [Brand names] ATC	Structure	[CAS-No.] Formula MW (g/mol)	Target (if known) Medical use Formulation DDD	Originator Approval (country/year) Production method [References]
Glymidine [Redul, Gondafon, Glycanol] A10BC01		[339-44-6] $C_{13}H_{15}N_3O_4S$ 309.35	oral antidiabetic. tbl. (once daily) 500 mg, O. DDD 1 g.	withdrawn from market Schering AG (now part of Bayer AG) Germany/1964 UK/1966 synthesis: [73, 74]

3.4. A10BD Combinations of Oral Blood Glucose-Lowering Drugs (Table 5)

In order to reach long-term glycemic control in T2D patients, studies with combination of two or more drugs have led to the development of fixed-dose combination regimens, mostly with orally available drugs [75, 76]. Fixed-dose combinations of metformin with all other orally available drug classes (sulfonylureas, DPP-4 inhibitors, SGLT2 inhibitors, thiazolidinediones) have been launched. The latest combinations consist of DPP-4 inhibitors with SGLT2 inhibitors. A fixed-dose combination achieves efficacy at lower dosages of each component as compared with the monotherapies, thus minimizing risk of side effects. In a study with the commercially most successful fixed-dose combination of metformin and sitagliptin (DPP-4 inhibitor), twice as many patients reached the desired goal of HbA1c below 7% as compared with patients on metformin alone. Other factors like cost effectiveness and increased patient's compliance favor the use of fixed-dose combinations. In addition, simplifying treatment regimens or reducing the often high pill burden due to other comorbidities has been beneficial to the patients.

Table 5. A10BD Combinations of oral blood glucose-lowering drugs

INN (Synonyms) [Brand names] ATC	Structure	[CAS-No.] Formula MW (g/mol)	Target (if known) Medical use Formulation DDD	Originator Approval (country/year) Production method [References]
Phenformin and sulfonylureas A10BD01				withdrawn from market
Metformin and sulfonylureas A10BD02		Metformin [657-24-9] $C_4H_{11}N_5$ 129.16		
		Metformin hydrochloride [1115-70-4] $C_4H_{11}N_5 \cdot HCl$ 165.63		
Metformin and rosiglitazone [Avandamet] A10BD03		Metformin hydrochloride [1115-70-4] $C_4H_{11}N_5 \cdot HCl$ 165.63	oral fixed-dose combination of the PPAR-gamma agonist rosiglitazone maleate and metformin hydrochloride for treatment of T2D. tbl. (film-coated, twice daily) 2 mg/500 mg, 4 mg/500 mg, 2 mg/1000 mg, 4 mg rosiglitazone / 1000 mg metformin hydrochloride.	GlaxoSmithKline USA/2002 EU/2003 Korea/2007 China/2011 [77]
		Rosiglitazone maleate [155141-29-0] $C_{18}H_{19}N_3O_3$ $S \cdot C_4H_4O_4$ 473.52		

Table 5. (*Continued*)

INN (Synonyms) [Brand names] ATC	Structure	[CAS-No.] Formula MW (g/mol)	Target (if known) Medical use Formulation DDD	Originator Approval (country/year) Production method [References]
Glimepiride and rosiglitazone [Avandaryl, Avaglim, SB-797620] A10BD04	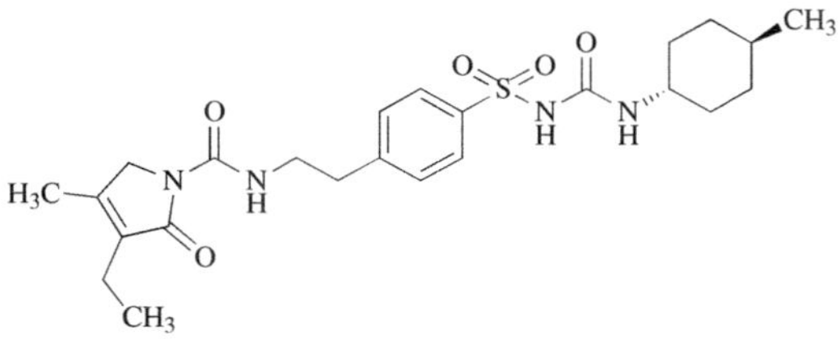	Glimepiride [*93479-97-1*] $C_{24}H_{34}N_4O_5S$ 490.62	oral fixed-dose tablet combination of the PPAR-gamma agonist rosiglitazone maleate and the sulfonylurea blood glucose lowering compound glimepiride, for the treatment of T2D. tbl. (once daily) 4 mg/1 mg, 4 mg/2 mg, 4 mg/4 mg, 8 mg/2 mg, and 8 mg rosiglitazone maleate /4 mg glimepiride, O.	GlaxoSmithKline USA/2005 EU/2006 Korea/2007 [78]
	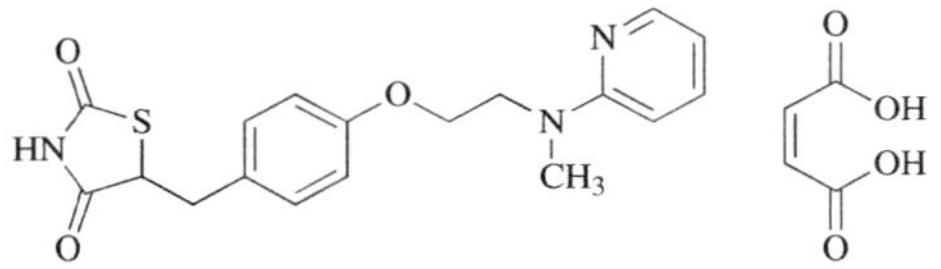	Rosiglitazone maleate [*155141-29-0*] $C_{18}H_{19}N_3O_3$ $S \cdot C_4H_4O_4$ 473.52		
Metformin and pioglitazone [Actoplus Met, Competact, Actosmet] A10BD05	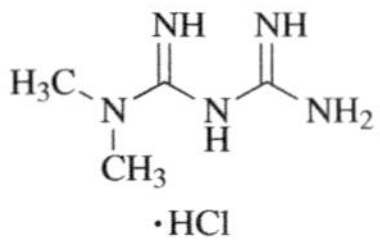	Metformin hydrochloride [*1115-70-4*] $C_4H_{11}N_5 \cdot HCl$ 165.63	oral fixed-dose tablet combination of the PPAR-gamma agonist pioglitazone hydrochloride and an immediate release form of metformin hydrochloride, for the treatment of T2D. tbl. 15 mg/500 mg (twice daily), 15 mg pioglitazone/850 mg metformin hydrochloride (once daily), O. extended-release formulation of Actoplus Met (Actoplus Met XR).	Takeda Pharmaceutical USA/2005 EU/2006 Japan/2010 [79, 80]
	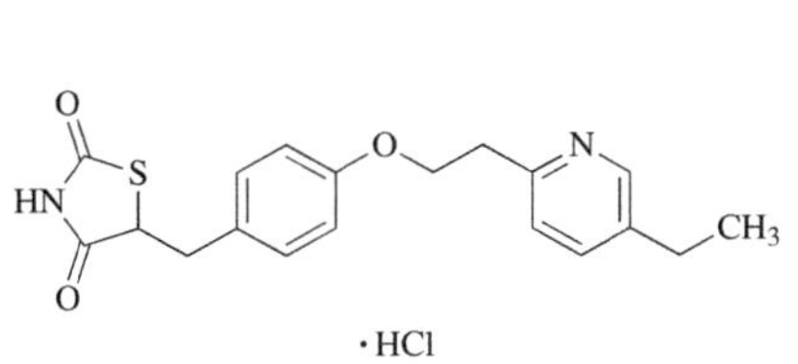	Pioglitazone hydrochloride [*112529-15-4*] $C_{19}H_{20}N_2O_3$ $S \cdot HCl$ 392.90		

(continued)

Table 5. (*Continued*)

INN (Synonyms) [Brand names] ATC	Structure	[CAS-No.] Formula MW (g/mol)	Target (if known) Medical use Formulation DDD	Originator Approval (country/year) Production method [References]
Glimepiride and pioglitazone [Duetact, Tandemact, Sonias LD, Sonias HD, Actosryl] A10BD06	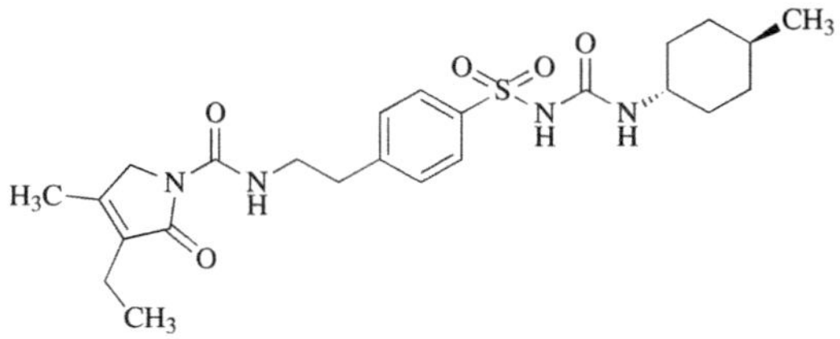	Glimepiride [*93479-97-1*] $C_{24}H_{34}N_4O_5S$ 490.62	oral fixed-dose tablet combination of the PPAR-gamma agonist pioglitazone hydrochloride and the sulfonylurea blood glucose lowering compound glimepiride, for the treatment of T2D. tbl. (once daily) 30 mg/2 mg, 30 mg pioglitazone (as base)/4 mg glimepiride), O.	Takeda Pharmaceutical USA/2006 EU/2007 Japan/2011 [79, 81]
	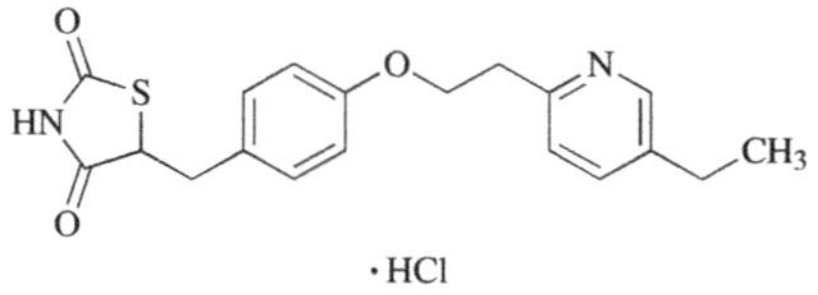	Pioglitazone hydrochloride [*112529-15-4*] $C_{19}H_{20}N_2O_3S \cdot HCl$ 392.90		
Metformin and sitagliptin [Janumet, Velmetia, Efficib, Ristfor, Janumet XR] A10BD07	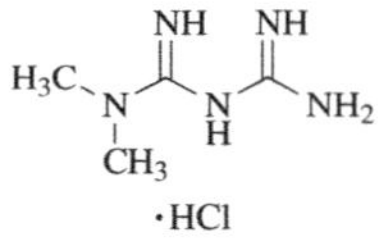	Metformin hydrochloride [*1115-70-4*] $C_4H_{11}N_5 \cdot HCl$ 165.63	oral fixed-dose combination of the DPP-4 inhibitor sitagliptin phosphate monohydrate and metformin hydrochloride for treatment of T2D. Janumet tbl. (twice daily, film-coated) 50 mg/500 mg, 50 mg sitagliptin/1000 mg metformin hydrochloride, O.	Merck Sharp & Dohme Ltd. Janumet USA/2007 EU/2008 Canada/2009 Australia/2009 Korea/2009 China/2012 [82]
	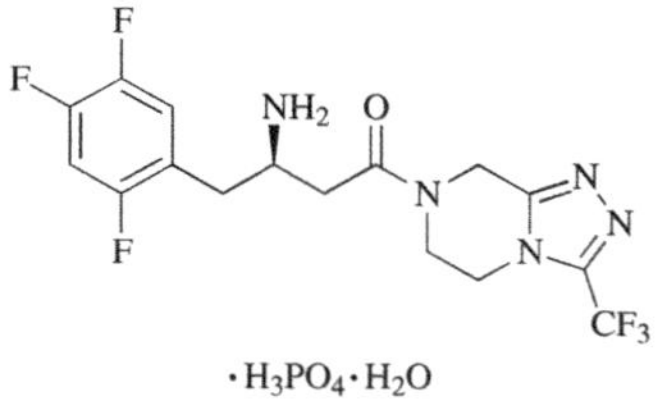	Sitagliptin phosphate monohydrate [*654671-77-9*] $C_{16}H_{15}F_6N_5O \cdot H_3O_4P \cdot H_2O$ 523.32	Janumet XR: sitagliptin phosphate monohydrate and metformin HCl extended-release, tbl. (once daily, film-coated) 50 mg/500 mg, 100 mg/500 mg, 100 mg sitagliptin/1,000 mg metformin extended-release, O.	Janumet XR USA/2012 Korea/2013 [83]

Structures shown in column:

Glimepiride structure (·HCl noted under pioglitazone structure with H_3C, CH_3, S, O, N groups).

Pioglitazone: thiazolidinedione (HN, S, O) linked to phenyl-O-ethyl-pyridine with CH_3; · HCl

Metformin: H_3C–N(CH_3)–C(=NH)–NH–C(=NH)–NH_2 · HCl

Sitagliptin: trifluoromethyl-phenyl with F, F, F; NH_2, O, triazolopyrazine ring, CF_3; · $H_3PO_4 \cdot H_2O$

Table 5. (*Continued*)

INN (Synonyms) [Brand names] ATC	Structure	[CAS-No.] Formula MW (g/mol)	Target (if known) Medical use Formulation DDD	Originator Approval (country/year) Production method [References]
Metformin and vildagliptin [Eucreas, Zomarist, Icandra, Galvusmet, EquMet, Equmet LD, Equmet HD, Jalra-M, Galvumet, Sobrea] A10BD08	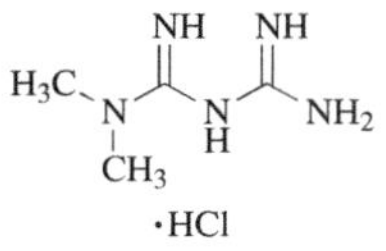	Metformin hydrochloride [*1115-70-4*] $C_4H_{11}N_5 \cdot HCl$ 165.63	oral fixed-dose tablet combination of the DPP-4 inhibitor vildagliptin and metformin hydrochloride for the treatment of T2D. tbl. (twice daily, film-coated) 50 mg/850 mg, 50 mg vildagliptin/1000 mg metformin hydrochloride, O.	Novartis AG EU/2007 Korea/2008 China/2014 Japan/2015 [84] formulation: [85]
	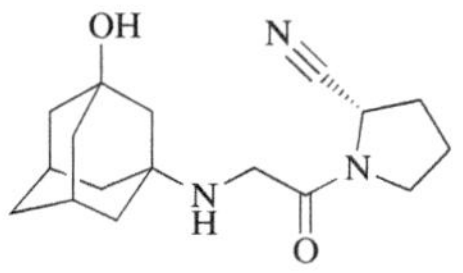	Vildagliptin [*274901-16-5*] $C_{17}H_{25}N_3O_2$ 303.40		
Pioglitazone and alogliptin [Liovel, Oseni, Incresync, Nesina Act] A10BD09	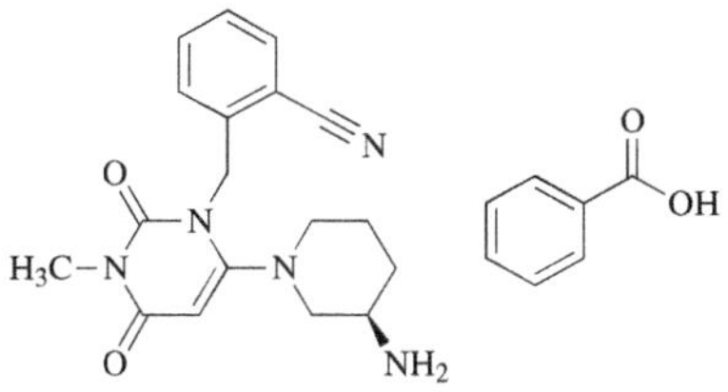	Alogliptin benzoate [*850649-62-6*] $C_{18}H_{21}N_5O_2 \cdot C_7H_6O_2$ 461.51	oral fixed-dose tablet combination of the PPAR-gamma agonist pioglitazone hydrochloride and the DPP-4 inhibitor alogliptin benzoate for treatment of T2D. tbl. (once daily) 25 mg/15 mg, 25 mg/30 mg, 25 mg/45 mg, 12.5 mg/15 mg, 12.5 mg/30 mg, and 12.5 mg/45 mg tablets, O.	Takeda Pharmaceutical USA/2013 EU/2013 [86]
	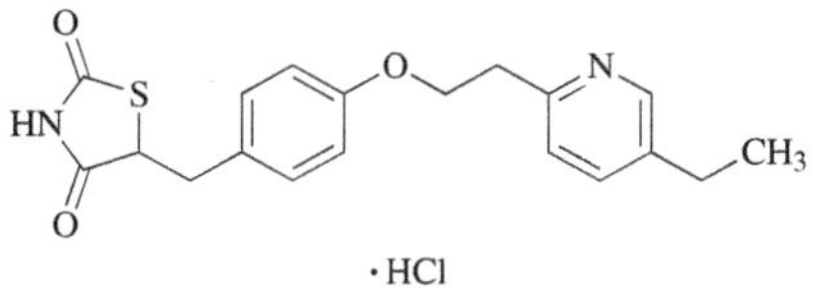	Pioglitazone hydrochloride [*112529-15-4*] $C_{19}H_{20}N_2O_3S \cdot HCl$ 392.90		

(*continued*)

Table 5. (*Continued*)

INN (Synonyms) [Brand names] ATC	Structure	[CAS-No.] Formula MW (g/mol)	Target (if known) Medical use Formulation DDD	Originator Approval (country/year) Production method [References]
Metformin and saxagliptin [Kombglyze, Komboglyze FDC, Combiglyza] A10BD10	·HCl	Metformin hydrochloride [*1115-70-4*] $C_4H_{11}N_5 \cdot HCl$ 165.63	oral fixed-dose tablet combination of DPP-4 inhibitor saxagliptin monohydrate and metformin hydrochloride for treatment of T2D. two forms: Komboglyze: an immediate-release fixed-dose combination of saxagliptin (2.5 mg) and metformin hydrochloride (1000 mg). Kombiglyze XR (Riax-M), an extended-release fixed-dose combination of saxagliptin and metformin hydrochloride.	Bristol-Myers Squibb (Astra Zeneca acquired diabetes business from BMS, 2014) Komboglyze: EU/2011 Canada/2013 Australia/2013 [87] formulation: [88] Kombiglyze XR: USA/2011
	· H₂O	Saxagliptin monohydrate [*945667-22-1*] $C_{18}H_{25}N_3O_2 \cdot H_2O$ 333.43		
Metformin and linagliptin [Jentadueto, Trajenta Duo, Trajentamet, Ondero Me] A10BD11	·HCl	Metformin hydrochloride [*1115-70-4*] $C_4H_{11}N_5 \cdot HCl$ 165.63	oral fixed-dose tablet combination of the DPP-4 inhibitor linagliptin (BI-1356, Ondero) and immediate-release (IR) metformin, a biguanide, for treatment of T2D. tbl. (twice daily, film-coated), 2.5 mg/850 mg, 2.5 mg linagliptin/1000 mg metformin, O.	Boehringer Ingelheim and Eli Lilly USA/2012 EU/2012 [89, 90]
		Linagliptin [*668270-12-0*] $C_{25}H_{28}N_8O_2$ 472.54		

Table 5. (*Continued*)

INN (Synonyms) [Brand names] ATC	Structure	[CAS-No.] Formula MW (g/mol)	Target (if known) Medical use Formulation DDD	Originator Approval (country/year) Production method [References]
Sitagliptin and pioglitazone [Janacti] A10BD12	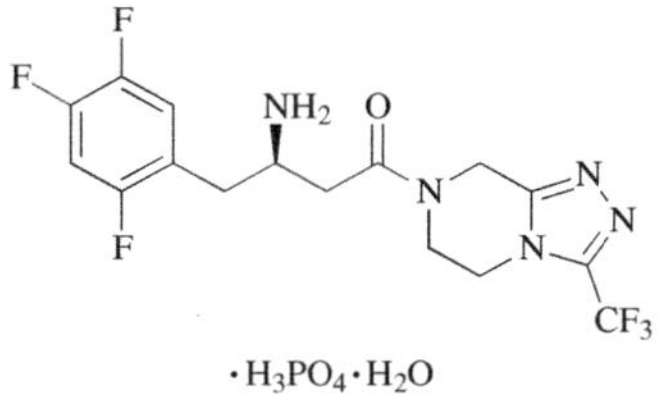	Sitagliptin phosphate monohydrate [*654671-77-9*] $C_{16}H_{15}F_6N_5O \cdot H_3O_4P \cdot H_2O$ 523.32	oral fixed-dose tablet combination, DPP-4 inhibitor sitagliptin and PPAR-gamma agonist pioglitazone for treatment of T2D. tbl. (once daily) 100 mg sitagliptin and 30 mg pioglitazone; 100 mg sitagliptin and 45 mg pioglitazone, O.	withdrawn from market Merck Sharp & Dohme Ltd. USA [91, 92]
	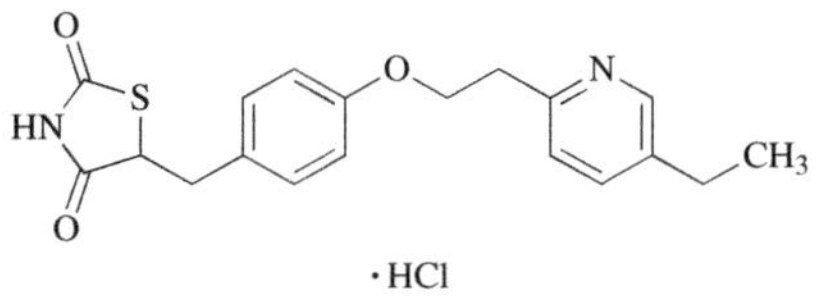	Pioglitazone hydrochloride [*112529-15-4*] $C_{19}H_{20}N_2O_3S \cdot HCl$ 392.90		
Metformin and alogliptin [Kazano, Nesina Met, Vipdomet, Inisync] A10BD13	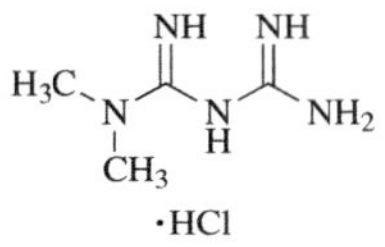	Metformin hydrochloride [*1115-70-4*] $C_4H_{11}N_5 \cdot HCl$ 165.63	oral fixed-dose tablet combination of the DPP-4 inhibitor alogliptin benzoate with metformin hydrochloride for treatment of T2D hydrochloride: tbl. (twice daily) 12.5 mg/500 mg, 12.5 mg alogliptin/1000 mg metformin, O.	Takeda Pharmaceutical USA/2013 EU/2013 Japan/2016 [93, 94]
	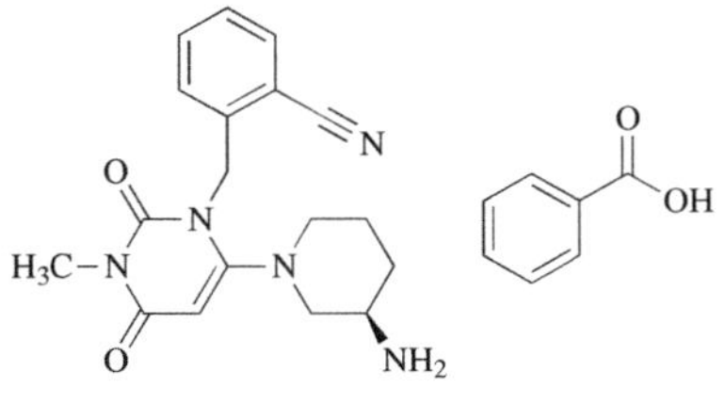	Alogliptin benzoate [*850649-62-6*] $C_{18}H_{21}N_5O_2 \cdot C_7H_6O_2$ 461.51		
Metformin and repaglinide [PrandiMet, NovoGard, NovoMet, MetPrandin] A10BD14	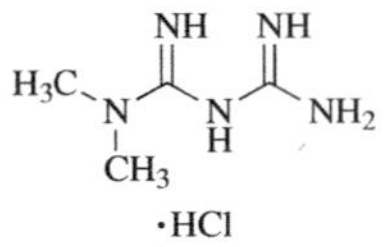	Metformin hydrochloride [*1115-70-4*] $C_4H_{11}N_5 \cdot HCl$ 165.63	oral fixed-dose tablet combination tablet of metformin hydrochloride plus the insulin secretagogue repaglinide for treatment of T2D. tbl. (two to three times daily) 1 mg/500 mg, 2 mg repaglinide/500 mg metformin, O.	Novo Nordisk/Shionogi Pharma USA/2008 [95]
	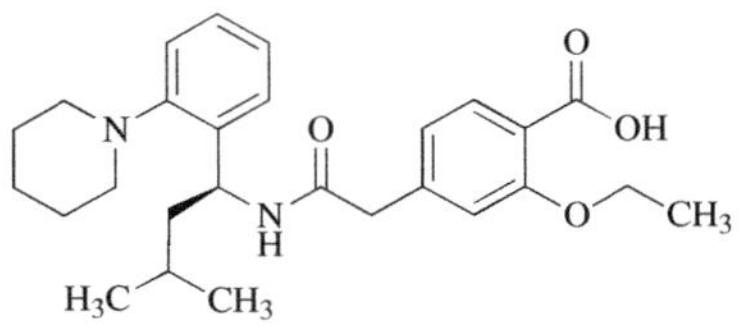	Repaglinide [*135062-02-1*] $C_{27}H_{36}N_2O_4$ 452.6		

(continued)

Table 5. (*Continued*)

INN (Synonyms) [Brand names] ATC	Structure	[CAS-No.] Formula MW (g/mol)	Target (if known) Medical use Formulation DDD	Originator Approval (country/year) Production method [References]
Metformin and dapagliflozin [Xigduo, Ebymect, Xigduo XR, Oxramet] A10BD15		Metformin hydrochloride [*1115-70-4*] $C_4H_{11}N_5 \cdot HCl$ 165.63	oral fixed-dose tablet combination of the SGLT2 inhibitor dapagliflozin propanediol and metformin hydrochloride for treatment of T2D. tbl. (twice daily, film-coated) 5 mg/850 mg, 5 mg dapagliflozin/1000 mg metformin, O.	Bristol-Myers Squibb (Astra Zeneca acquired diabetes business from BMS, 2014) USA/2014 EU/2014 [96–100]
		Dapagliflozin propanediol monohydrate [*1971128-01-4*] $C_{21}H_{25}ClO_6 \cdot C_3H_8O_2 \cdot H_2O$ 484.98		
Metformin and canagliflozin [Vokanamet, Invokamet] A10BD16		Metformin hydrochloride [*1115-70-4*] $C_4H_{11}N_5 \cdot HCl$ 165.63	oral fixed-dose tablet combination of the SGLT2 inhibitor canagliflozin hemihydrate and an immediate-release formulation of metformin hydrochloride for treatment of T2D. tbl. (once daily, film-coated) 50 mg/850 mg, 50 mg/1000 mg, 150 mg/850 mg, 150 mg canagliflozin/1000 mg metformin, O.	Johnson & Johnson, under license from Mitsubishi Tanabe Pharma UK/2014 EU/2015 USA/2015 Canada/2016 [101]
		Canagliflozin hemihydrate [*928672-86-0*] $C_{24}H_{25}FO_5S \cdot 1/2\,H_2O$ 453.53		
Metformin and acarbose [Glucobay-M] A10BD17		Metformin hydrochloride [*1115-70-4*] $C_4H_{11}N_5 \cdot HCl$ 165.63	oral fixed-dose tablet combination of the alpha-glucosidase inhibitor, acarbose (50 mg) and metformin (500 mg) for treatment of T2D. (two to three times daily).	Bayer AG India/2012

Table 5. (*Continued*)

INN (Synonyms) [Brand names] ATC	Structure	[CAS-No.] Formula MW (g/mol)	Target (if known) Medical use Formulation DDD	Originator Approval (country/year) Production method [References]
	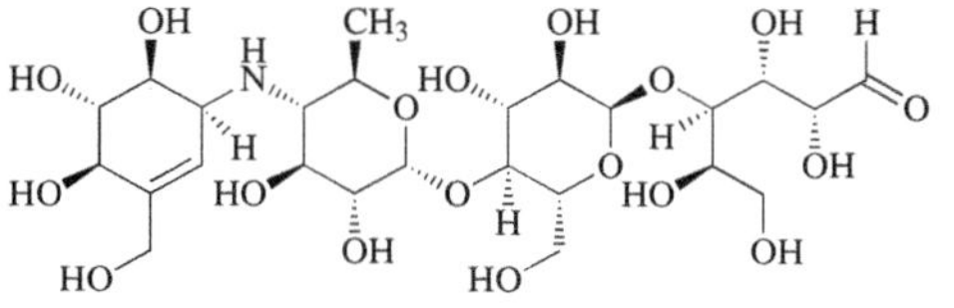	Acarbose [56180-94-0] $C_{25}H_{43}NO_{18}$ 645.61		
Metformin and gemigliptin [ZemiMet] A10BD18	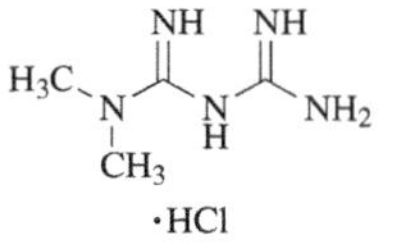	Metformin hydrochloride [1115-70-4] $C_4H_{11}N_5 \cdot HCl$ 165.63	oral fixed-dose, sustained release tablet combination of the DPP-4 inhibitor gemigliptin, and metformin hydrochloride, for the treatment of T2D. tbl. (once daily) 25 mg gemigliptin/500 mg metformin, 50 mg/1000 mg, O.	LG Life Sciences Korea/2013 [102]
	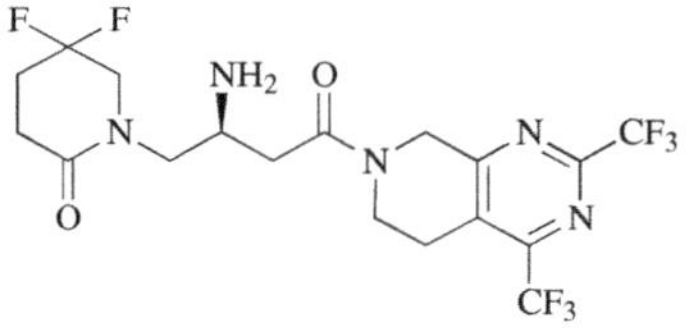	Gemigliptin [911637-19-9] $C_{18}H_{19}F_8N_5O_2$ 489.36		
Linagliptin and empagliflozin [Glyxambi] A10BD19	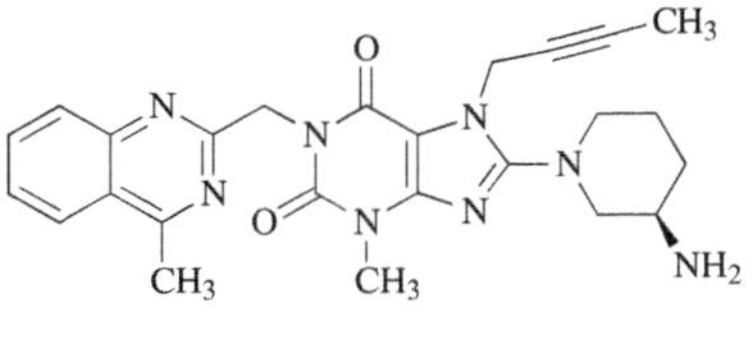	Linagliptin [668270-12-0] $C_{25}H_{28}N_8O_2$ 472.54	oral fixed-dose tablet combination of the DPP-4 inhibitor linagliptin and the SGLT2 inhibitor empagliflozin for treatment of T2D. tbl. (once daily) 5 mg/10 mg, 5 mg linagliptin/25 mg empagliflozin.	Boehringer Ingelheim and Eli Lilly USA/2015 EU/2016 [103–105]
	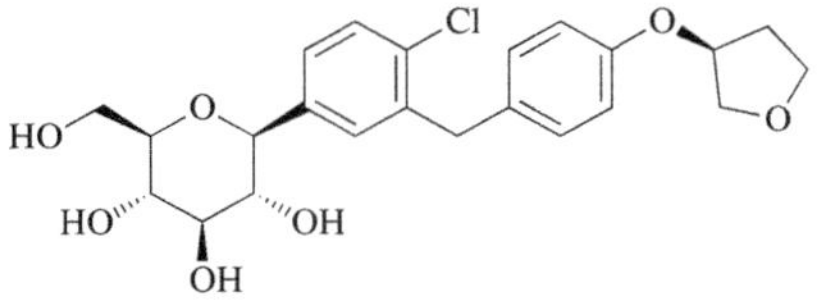	Empagliflozin [864070-44-0] $C_{23}H_{27}ClO_7$ 450.91		

(continued)

Table 5. (*Continued*)

INN (Synonyms) [Brand names] ATC	Structure	[CAS-No.] Formula MW (g/mol)	Target (if known) Medical use Formulation DDD	Originator Approval (country/year) Production method [References]
Metformin and empagliflozin [Synjardy, Synjardy XR] A10BD20		Metformin hydrochloride [1115-70-4] $C_4H_{11}N_5 \cdot HCl$ 165.63	oral fixed-dose tablet combination of the sodium glucose transporter-2 (SGLT-2) inhibitor empagliflozin (and immediate-release (IR) metformin hydrochloride for treatment of T2D. Synjardy: tbl. (twice daily) 5 mg/850 mg, 5 mg/1000 mg, 12.5 mg/850 mg, 12.5 mg empagliflozin /1000 mg metformin hydrochloride, O. Synjardy XR with extended-release formulation of metformin hydrochloride. tbl. (once daily) 5 mg/1000 mg, 10 mg/1000 mg, 12.5 mg/1000 mg, 25 mg empagliflozin/1000 mg metformin hydrochloride, O.	Boehringer Ingelheim and Eli Lilly Synjardy: EU/2015 USA/2015 [106–108] Synjardy XR: USA/2017
		Empagliflozin [864070-44-0] $C_{23}H_{27}ClO_7$ 450.91		
Saxagliptin and dapagliflozin [SaxaDapa, Qtern] A10BD21		Saxagliptin monohydrate [945667-22-1] $C_{18}H_{25}N_3O_2 \cdot H_2O$ 333.43	oral fixed dose tablet combination of the DPP-4 inhibitor saxagliptin monohydrate and the SGLT2 inhibitor dapagliflozin propanediol for treatment of T2D. tbl. (once daily) 5 mg saxagliptin monohydrate/10 mg dapagliflozin propanediol, O.	Bristol-Myers Squibb (Astra Zeneca acquired diabetes business from BMS, 2014) USA/2017 Europe/2016 Korea/2017 [109–113] formulation: [114]
		Dapagliflozin propanediol monohydrate [1971128-01-4] $C_{21}H_{25}ClO_6 \cdot C_3H_8O_2 \cdot H_2O$ 484.98		

Table 5. (*Continued*)

INN (Synonyms) [Brand names] ATC	Structure	[CAS-No.] Formula MW (g/mol)	Target (if known) Medical use Formulation DDD	Originator Approval (country/year) Production method [References]
Metformin and evogliptin [Sugamet] A10BD22	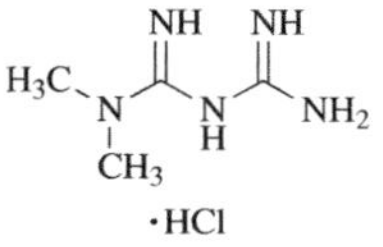	Metformin hydrochloride [*1115-70-4*] $C_4H_{11}N_5 \cdot HCl$ 165.63	oral fixed dose tablet combination of extended release form of metformin hydrochloride and the SGLT2 inhibitor evogliptin for treatment of T2D. tbl. (once daily) 5 mg evogliptin/1000 mg metformin hydrochloride XR, O.	Dong-A ST Korea/2016
	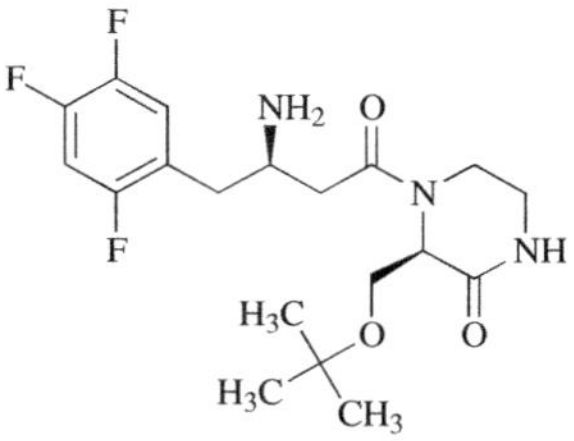	Evogliptin [*1222102-29-5*] $C_{19}H_{26}F_3N_3O_3$ 401.42		
Metformin hydrochloride and ertugliflozin [Segluromet] A10BD23	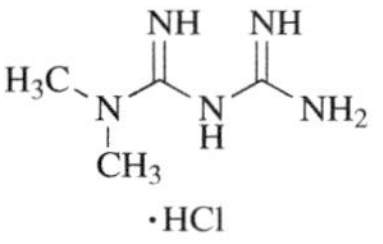	Metformin hydrochloride [*1115-70-4*] $C_4H_{11}N_5 \cdot HCl$ 165.63	oral fixed dose tablet combination of metformin hydrochloride (500 or 1000 mg) and the SGLT2 inhibitor ertugliflozin (2.5 or 7.5 mg) for treatment of T2D. tbl. (once daily) 2.5 mg/500 mg, 7.5 mg/500 mg, 2.5 mg/1000 mg, 7.5 mg ertugliflozin/1000 mg metformin hydrochloride, O.	Merck Sharpe Dohme Ltd. and Pfizer USA/2017 [115–117]
	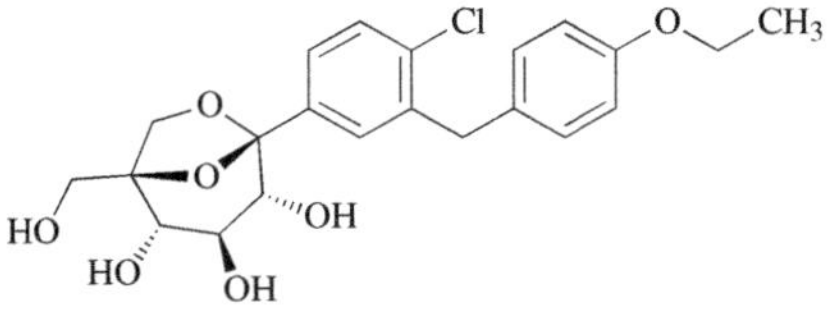	Ertugliflozin [*1210344-57-2*] $C_{22}H_{25}ClO_7$ 436.89		

(continued)

Table 5. (*Continued*)

INN (Synonyms) [Brand names] ATC	Structure	[CAS-No.] Formula MW (g/mol)	Target (if known) Medical use Formulation DDD	Originator Approval (country/year) Production method [References]
Sitagliptin and ertugliflozin [Steglujan] A10BD24		Sitagliptin [486460-32-6] $C_{16}H_{15}F_6N_5O$ 407.31 Ertugliflozin [1210344-57-2] $C_{22}H_{25}ClO_7$ 436.89	oral fixed dose tablet combination of the DPP-4 inhibitor sitagliptin and the SGLT2 inhibitor ertugliflozin for treatment of T2D. tbl. (once daily) 5 mg/100 mg; 15 mg ertugliflozin/100 mg sitagliptin, O.	Merck Sharpe Dohme Ltd. and Pfizer USA/2017 [118, 119]

3.5. A10BF Alpha-Glucosidase Inhibitors (Table 6)

Food contains glucose or other monosaccharides mainly bound in form of complex carbohydrates (polysaccharides such as starch) or simpler disaccharides. Special enzymes along the digestive system convert them into monosaccharides which are absorbed via special transporters in the gut. Orally administered inhibitors of enzymes such as alpha-glucosidase prevent or delay the glucose absorption and thus have a blood glucose-lowering effect. These compounds are either oligosaccharides such as acarbose [125, 126] or amino-/imino sugars such as miglitol and voglibose [131]. They have to be taken before or at the start of a meal for a maximal effect. Due to their mechanism of action, gastrointestinal side effects, such as abdominal pain, diarrhea, and flatulence are common.

Table 6. A10BF Alpha-glucosidase inhibitors

INN (Synonyms) [Brand names] ATC	Structure	[CAS-No.] Formula MW (g/mol)	Target (if known) Medical use Formulation DDD	Originator Approval (country/year) Production method [References]
Acarbose [Glicobase, Glucobay, Glucor Prandase, Precose] A10BF01		[56180-94-0] $C_{25}H_{43}NO_{18}$ 645.61	oral alpha-glucosidase inhibitor for treatment of T2D. tbl. (three times daily) 25, 50, 100 mg, O. DDD 300 mg.	Bayer AG EU/1988 Japan/1993 USA/1995 Canada/1999 China/2010 [120] purification process: [121–126]

Table 6. (*Continued*)

INN (Synonyms) [Brand names] ATC	Structure	[CAS-No.] Formula MW (g/mol)	Target (if known) Medical use Formulation DDD	Originator Approval (country/year) Production method [References]
Miglitol [Diastobol, Glyset, Seibule] A10BF02		[72432-03-2] C$_8$H$_{17}$NO$_5$ 207.22	oral alpha-glucosidase inhibitor for treatment of T2D. tbl. (three times daily) 25, 50, 100 mg (coated), O. DDD 300 mg.	Bayer AG EU, USA/1996 Japan/2005 China/2002 synthesis: [127–130]
Voglibose [Basen, Glustat, Volix] A10BF03		[83480-29-9] C$_{10}$H$_{21}$NO$_7$ 267.28	oral alpha-glucosidase inhibitor for treatment of T2D. tbl. (three times daily) 0.2, 0.3 mg, O. DDD 0.6 mg.	Takeda Pharmaceuticals Japan/1994 South Korea/2001 China/2002 Synthesis: [131, 132]

3.6. A10BG Thiazolidinediones (Table 7)

Thiazolidinediones, also known as glitazones, are orally available antidiabetic drugs. They are agonists of the peroxisome proliferator-activated receptors (PPARs), a group of nuclear hormone receptors regulating certain genes in metabolism. Especially PPARγ is a key regulator of fatty acid storage and glucose utilization. PPAR agonists induce changes in adipose tissue (i.e., increase of insulin-sensitive fat cells), lead to reduction of hepatic glucose production and increase glucose and free fatty acid uptake and their utilization in peripheral tissue (fat, muscle). However, the two main representatives of this drug class, pioglitazone [145] and rosiglitazone [141, 142], have been (partially) withdrawn from the market. Pioglitazone was withdrawn in France and Germany due to a study indicating an increased risk of bladder cancer or studies showing higher incidences of bone fractures in female patients. The selling restriction for rosiglitazone based on studies anticipating a higher risk of cardiovascular events was lifted in 2013 in the USA, but the drug is still suspended in European markets.

Table 7. A10BG Thiazolidinediones

INN (Synonyms) [Brand names] ATC	Structure	[CAS-No.] Formula MW (g/mol)	Target (if known) Medical use Formulation DDD	Originator Approval (country/year) Production method [References]
Troglitazone [Rezulin, Resulin, Romozin, Noscal] A10BG01		[97322-87-7] C$_{24}$H$_{27}$NO$_5$S 441.54	oral available agonist of PPAR alpha and gamma for the treatment of T2D. tbl. 100, 200, 300, 400 mg. DDD 400 mg.	withdrawn from market Daichi Sankyo UK/1997 USA/1997 Japan/1997 [133–135]

(continued)

Table 7. (*Continued*)

INN (Synonyms) [Brand names] ATC	Structure	[CAS-No.] Formula MW (g/mol)	Target (if known) Medical use Formulation DDD	Originator Approval (country/year) Production method [References]
Rosiglitazone maleate [Avandia] A10BG02		Rosiglitazone [122320-73-4] $C_{18}H_{19}N_3O_3S$ 357.43	oral available agonist of PPAR gamma for the treatment of T2D. Rosiglitazone maleate tbl. (once daily) 2 mg, 4 mg, 8 mg (as free base) film-coated, O. DDD 6 mg.	GlaxoSmithKline USA/1999 (restrictions on the drug imposed by the FDA from 2010 to 2015) EU/2000 (withdrawn in the EU from September 2010 due to the adverse cardiovascular data) China/2002 [136–138] synthesis maleate salt and other derivatives: [139–142]
		Rosiglitazone maleate [155141-29-0] $C_{18}H_{19}N_3O_3$ $S \cdot C_4H_4O_4$ 473.52		
Pioglitazone hydrochloride [Actos] A10BG03		Pioglitazone [111025-46-8] $C_{19}H_{20}N_2O_3S$ 356.44	oral available agonist of PPAR-gamma for the treatment of T2D. Pioglitazone hydrochloride tbl. (once daily), 15, 30, 45 mg (as free base), O. DDD 30 mg.	Takeda Chem. Ind. US/1999 Japan/1999 EU/2003 China/2004 (withdrawn in France 2011) synthesis: [143–146]
	· HCl	Pioglitazone hydrochloride [112529-15-4] $C_{19}H_{20}N_2O_3S \cdot HCl$ 392.90		

3.7. A10BH Dipeptidyl Peptidase 4 (DPP-4) Inhibitors (Table 8)

The incretin effect describes the occurrence of increased insulin secretion elicited by oral versus intravenous administration of glucose. Incretins, a class of peptidic gut hormones, are secreted from gut cells into circulation in response to digestion of glucose-containing food. The two main incretins, glucagon-like peptide-1 (GLP-1, 30 AA) and gastric inhibitory peptide (or glucose-dependent insulinotropic polypeptide, GIP, 42 AA) increase the secretion of insulin from pancreatic β-cells, inhibit the release of glucagon, the opponent hormone to insulin, from the pancreatic α-cells and interfere with food intake by reducing gastric emptying. In T2D patients the incretin effect is diminished due to reduced secretion of GLP-1 and diminished GIP activity. The

serine protease, dipeptidyl peptidase 4, rapidly inactivates both GLP-1 and GIP. Dipeptidyl peptidase 4 (DPP-4) inhibitors, also known as gliptins, enhance the half-life of endogenous incretin peptides and thus lower blood glucose levels in a strictly blood glucose-dependent fashion [147–149]. Thus, these orally available drugs for T2D patients have a low risk of hypoglycemia, are weight neutral and in general well tolerated. Two classes of DPP-4 inhibitors can be distinguished, covalent reversible inhibitors (substrate-like) such as vildagliptin [157] or saxagliptin [160] and nonsubstrate-like inhibitors such as sitagliptin [153], linagliptin [171], or alogliptin [165]. Their use in the treatment of T2D patients has increased over the years since the introduction of the first gliptin in 2006, especially in combination with metformin. Contraindication is depending on the specific DPP-4 inhibitor to be administered.

Table 8. A10BH Dipeptidyl peptidase 4 (DPP-4) inhibitors

INN (Synonyms) [Brand names] ATC	Structure	[CAS-No.] Formula MW (g/mol)	Target (if known) Medical use Formulation DDD	Originator Approval (country/year) Production method [References]
Sitagliptin **Sitagliptin monophosphate** [Januvia, Tesavel, Xelevia] A10BH01		Sitagliptin [486460-32-6] $C_{16}H_{15}F_6N_5O$ 407.31	orally available selective DPP-4 inhibitor for treatment of T2D. tbl. (once daily) 25 mg, 50 mg, 100 mg (as free base) film-coated, O. DDD 100 mg.	Merck Sharp & Dohme Ltd. USA/2006 EU, Japan, China/2009 synthesis: [150–154]
	·H_3PO_4·H_2O	Sitagliptin phosphate monohydrate [654671-77-9] $C_{16}H_{15}F_6N_5O$·H_3PO_4·H_2O 523.32		
Vildagliptin [Galvus] A10BH02		[274901-16-5] $C_{17}H_{25}N_3O_2$ 303.40	orally available selective DPP-4 inhibitor for treatment of T2D. tbl. (once daily) 50 mg, O. DDD 100 mg.	Novartis AG EU/2007 Japan/2010 China/2012 synthesis: [155–157]

(continued)

Table 8. (*Continued*)

INN (Synonyms) [Brand names] ATC	Structure	[CAS-No.] Formula MW (g/mol)	Target (if known) Medical use Formulation DDD	Originator Approval (country/year) Production method [References]
Saxagliptin [Onglyza] A10BH03		Saxagliptin [*361442-04-8*] $C_{18}H_{25}N_3O_2$ 315.41	orally available selective DPP-4 inhibitor for treatment of T2D. tbl. (once daily) 5 mg, O. DDD 5 mg.	Bristol-Myers Squibb (2014, AstraZeneca acquired BMS diabetes business) USA, EU/2009 China/2011 Japan/2013 synthesis: [158–161]
	·H₂O	Saxagliptin monohydrate [*945667-22-1*] $C_{18}H_{25}N_3O_2 \cdot H_2O$ 333.43		
Alogliptin benzoate [Nesina, Vipidia] A10BH04		Alogliptin [*850649-61-5*] $C_{18}H_{21}N_5O_2$ 339.40	orally available selective DPP-4 inhibitor for treatment of T2D. tbl. (once daily) 8.5, 17, 34 mg film-coated, alogliptin benzoate, O. DDD 25 mg.	Syrrx (now part of Takeda Pharmaceutical) Japan/2010 USA/2013 EU/2013 synthesis: [162–165]
		Alogliptin benzoate [*850649-62-6*] $C_{18}H_{21}N_5O_2 \cdot C_7H_6O_2$ 461.51		
Linagliptin [Tradjenta, Trajenta] A10BH05		[*668270-12-0*] $C_{25}H_{28}N_8O_2$ 472.54	orally available selective DPP-4 inhibitor for treatment of T2D. tbl. (once daily) 5 mg film-coated, O. DDD 5 mg.	Boehringer Ingelheim Corp./Eli Lilly USA, EU, Japan/2011 synthesis: [166–171]
Gemigliptin [Zemiglo] A10BH06		Gemigliptin [*911637-19-9*] $C_{18}H_{19}F_8N_5O_2$ 489.36	orally available selective DPP-4 inhibitor for treatment of T2D. tbl. (once daily) 50 mg film-coated, O. DDD 50 mg.	LG Life Sciences Korea/2012 synthesis: [172, 173]

Table 8. (*Continued*)

INN (Synonyms) [Brand names] ATC	Structure	[CAS-No.] Formula MW (g/mol)	Target (if known) Medical use Formulation DDD	Originator Approval (country/year) Production method [References]
	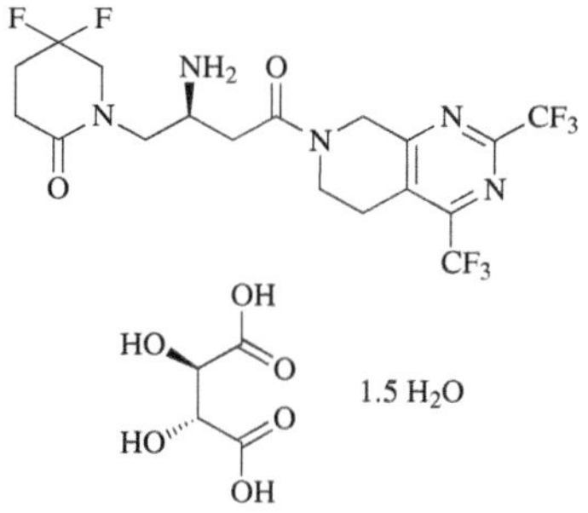	Gemigliptin tartrate hydrate (2:2:3) [*1375415-82-9*] $C_{18}H_{19}F_8N_5O_2 \cdot C_4H_6O_6 \cdot$ 1.5 H_2O 666.47		
Evogliptin [Suganon] A10BH07	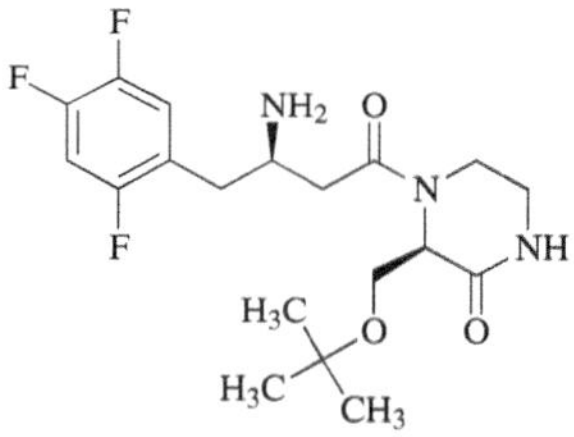	[*1222102-29-5*] $C_{19}H_{26}F_3N_3O_3$ 401.42	orally available selective DPP-4 inhibitor for treatment of T2D. tbl. (once daily) 5 mg, O. DDD 5 mg.	Dong-A Pharmaceutical Korea/2015 synthesis: [174–177]
Sitagliptin monophosphate and Simvastatin [Juvisync] A10BH51	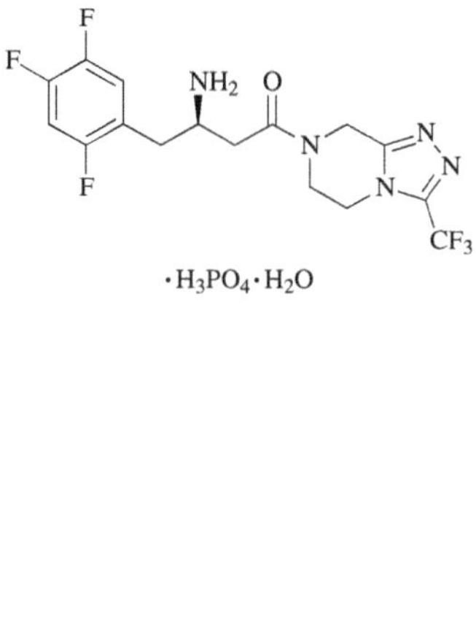	Sitagliptin phosphate monohydrate [*654671-78-0*] $C_{16}H_{15}F_6N_5O \cdot H_3PO_4 \cdot H_2O$ 523.32	orally available selective DPP-4 inhibitor in combination with lipid-lowering HMG-CoA reductase inhibitor simvastatin for treatment of T2D. tbl. 10 mg sitagliptin/100 mg simvastatin, 20 mg/100 mg, 40 mg/100 mg, 10 mg sitagliptin/50 mg simvastatin, 20 mg/50 mg, 40 mg/50 mg.	withdrawn from market Merck Sharp & Dohme Ltd. USA/2011
	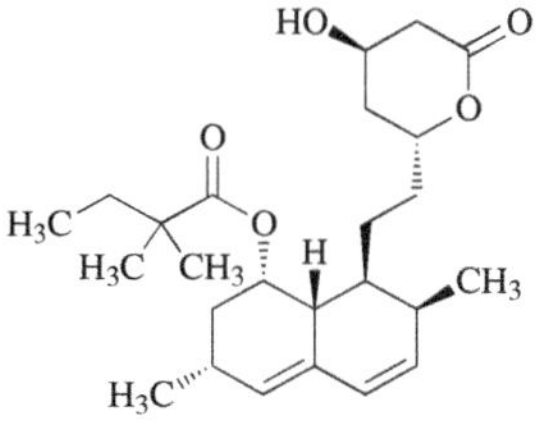	Simvastatin [*79902-63-9*] $C_{25}H_{38}O_5$ 418.57		

3.8. A10BJ Glucagon-Like Peptide-1 (GLP-1) Analogues (Table 9)

The era of glucagon-like peptide-1 (GLP-1) analogues ("incretin mimetics") started in 1992 with the isolation of exendin-4, a 39 AA long peptide in the saliva of the Gila monster, which showed GLP-1-like activity and had increased stability against DPP-4. Its synthetic version, exenatide has been marketed since 2005 as an antidiabetic administered by injection twice a day. GLP-1 analogues on the market have distinct pharmacokinetics and pharmacodynamics [178–182]. Short-acting GLP-1 analogues with half-lifes of 2–4 h such as exenatide [188] and lixisenatide lower postprandial glucose levels and delay gastric emptying. Long-acting

GLP-1 analogues with half-life in human from 10 h up to 14 days such as liraglutide (once daily) [194], exenatide-LAR, dulaglutide [200] and semaglutide [202] (once weekly) lower the fasting blood glucose and show a stronger stimulation of insulin secretion. Strategies for prolonging the duration of action include exchange of amino acids to increase stability against DPP-4 degradation, attachment of fatty acid derivatives for plasma albumin binding (liraglutide) or fusion to the Fc region of immunoglobulin B (dulaglutide). In contrast to insulins, both classes of GLP-1 analogues induce body weight loss and do not cause hypoglycemia. Common side effects are of gastrointestinal nature including diarrhea, nausea, and vomiting.

Table 9. A10BJ Glucagon-like peptide-1 (GLP-1) analogues

INN (Synonyms) [Brand names] ATC	Structure	[CAS-No.] Formula MW (g/mol)	Target (if known) Medical use Formulation DDD	Originator Approval (country/year) Production method [References]
Exenatide [Byetta] **Exenatide-LAR** [Bydureon] A10BJ01	sequence: H-His-Gly-Glu-Gly-Thr-Phe-Thr-Ser-Asp-Leu-Ser-Lys-Gln-Met-Glu-Glu-Glu-Ala-Val-Arg-Leu-Phe-Ile-Glu-Trp-Leu-Lys-Asn-Gly-Gly-Pro-Ser-Ser-Gly-Ala-Pro-Pro-Pro-Ser-NH$_2$	[141758-74-9] $C_{184}H_{282}N_{50}O_{60}S$ 4186.64	glucagon-like peptide-1 receptor agonist. subcutaneous injection Exenatide [Byetta]: twice daily 0.05 mg per dose P., 60 doses, 1.2 mL and 0.010 mg per dose, 60 doses, 2.4 mL prefilled pen. Exenatide-LAR [Bydureon]: once weekly 0.286 mg P. depot inj. = 15 µg. DDD 0.286 mg, P. (depot inj.), 15 µg, P.	Amylin (now a subsidiary of AstraZeneca) Exenatide [Byetta]: USA/2005 EU/2006 China/2009 Japan/2010 Exenatide-LAR [Bydureon] EU/2011 USA, Japan/2012 synthesis: [183–188]
Liraglutide [Victoza] A10BJ02 Liraglutide is also listed in Section 8	sequence: H-His-Ala-Glu-Gly-Thr-Phe-Thr-Ser-Asp-Val-Ser-Ser-Tyr-Leu-Glu-Gly-Gln-Ala-Ala-(N-ε-(γ-(N-α-hexadecanoyl)-Glu)Lys-Glu-Phe-Ile-Ala-Trp-Leu-Val-Arg-Gly-Arg-Gly-OH	[204656-20-2] $C_{172}H_{265}N_{43}O_{51}$ 3751.26	glucagon-like peptide-1 receptor agonist. subcutaneous injection. once daily, 6 mg/mL solution in a prefilled pen for doses of 0.6 mg, 1.2 mg, or 1.8 mg P. DDD 1.2 mg.	Novo Nordisk EU/2009 USA, Japan/2010 China/2011 synthesis: [189–194]
Lixisenatide [Lyxumia, Adlyxin] A10BJ03	sequence: H-His-Gly-Glu-Gly-Thr-Phe-Thr-Ser-Asp-Leu-Ser-Lys-Gln-Met-Glu-Glu-Glu-Ala-Val-Arg-Leu-Phe-Ile-Glu-Trp-Leu-Lys-Asn-Gly-Gly-Pro-Ser-Ser-Gly-Ala-Pro-Pro-Ser-Lys-Lys-Lys-Lsy-Lys-Lys-NH$_2$	[320367-13-3] $C_{215}H_{347}N_{61}O_{65}S$ 4858.49	glucagon-like peptide-1 receptor agonist. subcutaneous injection once daily. P. 0.050 mg/mL in 3 mL solution (15 µg) 0.100 mg/mL in 3 mL solution (30 µg). DDD 20 µg.	Zealand Pharma and Sanofi EU/2013 USA, Japan/2016 synthesis: [195, 196]

Table 9. (*Continued*)

INN (Synonyms) [Brand names] ATC	Structure	[CAS-No.] Formula MW (g/mol)	Target (if known) Medical use Formulation DDD	Originator Approval (country/year) Production method [References]
Albiglutide [Eperzan, Tanzeum] A10BJ04	fusion protein of DPP-IV resistant, modified GLP-1 dimer (H-His-Ala-Glu-Gly-Thr-Phe-Thr-Ser-Asp-Val-Ser-Ser-Tyr-Leu-Glu-Gly-Gln-Ala-Ala-Lys-Glu-Phe-Ile-Ala-Trp-Leu-Val-Lys-Gly-Arg-His-Ala-Glu-Gly-Thr-Phe-Thr-Ser-Asp-Val-Ser-Ser-Tyr-Leu-Glu-Gly-Gln-Ala-Ala-Lys-Glu-Phe-Ile-Ala-Trp-Leu-Val-Lys-Gly-Arg-) to human albumin	[782500-75-8] $C_{3232}H_{5032}N_{864}$ $O_{979}S_{41}$ 72970	glucagon-like peptide-1 receptor agonist. subcutaneous injection once weekly, P. 30 mg or 50 mg albiglutide per 0.5 mL solution following reconstitution. DDD 5.7 mg.	withdrawn from market (July 2017) Human Genome Sciences Inc /GSK EU/USA, 2014 Korea/2015 produced in *Saccharomyces cerevisiae* cells by recombinant DNA technology. lyophilized formulations: [197–199]
Dulaglutide [Trulicity] A10BJ05	fusion protein (homodimer) of modified human (7-37)-GLP-1) [Gly^8, Glu^{22}, Gly^{36}] with peptide linker (16 AA=Gly-Gly-Gly-Gly-Ser-Gly-Gly-Gly-Gly-Ser-Gyl-Gly-Gly-Gly-Ser-Ala-) and immunoglobulin G4 (synthetic human Fc fragment); H-His-Gly-Glu-Gly-Thr-Phe-Thr-Ser-Asp-Val-Ser-Ser-Tyr-Leu-Glu-Glu-Gln-Ala-Ala-Lys-Glu-Phe-Ile-Ala-Trp-Leu-Val-Lys-Gly-Gly-Gly-Gly-Gly-Gly-Ser-Gly-Gly-Gly-Gly-Ser-Gyl-Gly-Gly-Gly-Ser-Ala-desLys^{229}-[Pro^{10},Ala^{16},Ala^{17}] human immunoglobulin heavy constant γ4 chain H-CH2-CH3 fragment dimer with (55-55′:58-58′)-bisdisulfide bridges	[923950-08-7] $C_{2646}H_{4044}N_{704}$ $O_{836}S_{18}$ 59669.81	glucagon-like peptide-1 receptor agonist. subcutaneous injection once weekly, P. 0.75 mg/0.5 mL and 1.5 mg/0.5 mL solution. DDD 0.16 mg.	Eli Lilly EU/USA, 2014 Japan/2015 synthesis: [200] produced in mammalian cell culture by recombinant DNA technology.
Semaglutide [Ozempic] A10BJ06	H-His-Aib-Glu-Gly-Thr-Phe-Thr-Ser-Asp-Val-Ser-Ser-Tyr-Leu-Glu-Gly-Gln-Ala-Ala-(N-(17-carboxy-1-oxoheptadecyl)-L-γ-glutamyl-2-[2-(2-aminoethoxy)ethoxy]-acetyl-2-[2-(2-aminoethoxy)ethoxy]-acetyl]Lys-Glu-Phe-Ile-Ala-Trp-Leu-Val-Arg-Gly-Arg-Gly-OH	[910463-68-2] $C_{187}H_{291}N_{45}O_{59}$ 4113.64	glucagon-like peptide-1 receptor agonist. subcutaneous injection once weekly, P. 0.5 mg and 1 mg.	Novo Nordisk USA, 2017 synthesis: [201, 202]

3.9. A10BK Sodium–Glucose Cotransporter 2 (SGLT2) Inhibitors (Table 10)

The sodium–glucose cotransporter 2 (SGLT2) is highly expressed in the proximal tubule of the kidneys and is a low affinity, but high capacity glucose transporter. Over 90% of the filtered glucose in the kidney is reabsorbed by SGLT2. Highly potent, selective, orally available SGLT2 inhibitors named gliflozins lower blood glucose levels in T2D patients by increasing urinary excretion of glucose [203–205]. This unique mechanism has a negative caloric balance reducing body weight by 2–3 kg. Furthermore, diuretic and natiuretic effects due to their mode of action of SGLT2 inhibitors result in a reduction of blood pressure. First cardiovascular studies with SGLT2 inhibitors (empagliflozin [224] and canagliflozin [219]) have shown benefits for high-risk patients in terms of reduced CV mortality and hospitalization. Risk of genital or urinary tract infections, usually easily treated with antibiotics, is increased with the use of SGLT2 inhibitors as well as the risk for ketoacidosis. Other side effects are often compound-specific.

Table 10. A10BK Sodium–glucose cotransporter 2 (SGLT2) inhibitors

INN (Synonyms) [Brand names] ATC	Structure	[CAS-No.] Formula MW (g/mol)	Target (if known) Medical use Formulation DDD	Originator Approval (country/year) Production method [References]
Dapagliflozin **Dapagliflozin** **propanediol** **monohydrate** [Forxiga, Farxiga] A10BK01		Dapagliflozin [461432-26-8] $C_{21}H_{25}ClO_6$ 408.87	oral available selective SGLT2 inhibitor for treatment of T2D. tbl. (once daily) 5, 10 mg film-coated, O. DDD 10 mg.	Bristol-Myers Squibb (AstraZeneca acquired BMS diabetes business, 2014) EU/2012 USA, Japan/2014 China/2017 synthesis: [206–208] formulation (Dapagliflozin propanediol monohydrate): [209–213]
		Dapagliflozin propanediol monohydrate [1971128-01-4] $C_{21}H_{25}ClO_6 \cdot C_3H_8O_2 \cdot H_2O$ 484.98		
Canagliflozin [Invokana] A10BK02		Canagliflozin [842133-18-0] $C_{24}H_{25}FO_5S$ 444.52	oral available selective SGLT2 inhibitor for treatment of T2D. tbl. (once daily) 100, 300 mg, O. DDD 200 mg.	Mitsubishi Tanabe Pharma and Janssen Pharmaceutica USA, EU/2013 Japan/2014 synthesis: [214–219]
		Canagliflozin hemihydrate [928672-86-0] $C_{24}H_{25}FO_5S \cdot 1/2 H_2O$ 453.53		
Empagliflozin [Jardiance] A10BK03		[864070-44-0] $C_{23}H_{27}ClO_7$ 450.91	oral available selective SGLT2 inhibitor for treatment of T2D. tbl. (once daily) 10, 25 mg (film-coated) O. DDD 17.5 mg.	Boehringer Ingelheim and Eli Lilly EU, USA, CH/2014 synthesis: [220–224]

Table 10. *(Continued)*

INN (Synonyms) [Brand names] ATC	Structure	[CAS-No.] Formula MW (g/mol)	Target (if known) Medical use Formulation DDD	Originator Approval (country/year) Production method [References]
Ertugliflozin [Steglatro] A10BK04		[1210344-57-2] $C_{22}H_{25}ClO_7$ 436.89	oral available selective SGLT2 inhibitor for treatment of T2D. tbl. (once daily) 5, 15 mg, O.	Pfizer and Merck Sharp Dohme Ltd. USA/2017 [225–227] synthesis: [228, 229]
Ipragliflozin [Suglat] A10BK05		Ipragliflozin [761423-87-4] $C_{21}H_{21}FO_5S$ 404.46	oral available selective SGLT2 inhibitor for treatment of T2D. tbl. (once daily) 25, 50 mg, O.	Kotobuki Pharmaceutical and Astellas Pharma Japan/2014 Korea/2014 synthesis: [230–234]
		Ipragliflozin L-proline [951382-34-6] $C_{21}H_{21}FO_5S \cdot C_5H_9NO_2$ 515.59		

3.10. A10BX Other Blood Glucose-Lowering Drugs, Excl. Insulins (Table 11)

Glinides such as repaglinide, nateglinide or mitiglinide, have similar antidiabetic effects as sulfonylureas due to their binding to the ATP-sensitive potassium channel (K_{ATP}) on beta-cells [235, 238]. They have a faster onset and shorter duration of action than sulfonylureas and are usually taken just before a meal to induce insulin secretion. Mild hypoglycemia and weight gain are observed with their use. Pramlintide, an injectable analogue of the peptide hormone amylin, is approved for T1D (amylin deficient) and T2D patients (reduced levels of amylin). As the endogenous hormone, usually cosecreted with insulin, it delays gastric emptying, induces satiety and suppresses the secretion of glucagon.

4. A10X Other Drugs Used in Diabetes (Table 12)

The enzyme aldose reductase converts glucose to sorbitol. Due to hyperglycemia in diabetic patients, increase of sorbitol by this pathway in tissues such as the lenses of the eye or the glomeruli in the kidney might attribute to the damage of these tissues. The aldose reductase inhibitor tolrestat [254], approved for treatment of such diabetic complications, was withdrawn from market in 1997 because of the risk of severe liver toxicity.

Table 11. A10BX Other blood glucose-lowering drugs, excl. insulins

INN (Synonyms) [Brand names] ATC	Structure	[CAS-No.] Formula MW (g/mol)	Target (if known) Medical use Formulation DDD	Originator Approval (country/year) Production method [References]
Guar gum A10BX01				
Repaglinide [NovoNorm, Prandin, GlucoNorm, Surepost] A10BX02		[135062-02-1] $C_{27}H_{36}N_2O_4$ 452.59	oral antidiabetic blocking the K_{ATP} channel in pancreatic beta cell. tbl. (three to four times daily) 0.5, 1, 2 mg, O. DDD 4 mg.	Boehringer Ingelheim (licensed to Novo Nordisk) EU/1998 USA/1997 Japan/2011 synthesis: [236–238]
Nateglinide [Fastic; Starlix; Starsis] A10BX03		[105816-04-4] $C_{19}H_{27}NO_3$ 317.42	oral antidiabetic blocking the K_{ATP} channel in pancreatic beta-cell. tbl. (three times daily) 60, 120, 180 mg, film-coated, O. DDD 360 mg.	Ajinomoto and Novartis AG Japan/1999 EU/2001 USA/2000 China/2003 Synthesis: [239, 240]
Pramlintide acetate [Symlin] A10BX05	H-Lys-**Cys**-Asn-Thr-Ala-Thr-**Cys**-Ala-Thr-Gln-Arg-Leu-Ala-Asn-Phe-Leu-Val-His-Ser-Ser-Asn-Asn-Phe-Gly-Pro-Ile-Leu-Pro-Pro-Thr-Asn-Val-Gly-Ser-Asn-Thr-Tyr-NH$_2$ · CH$_3$CO$_2$H (Disulfide bridge between Cys2 and Cys8)	[151126-32-8] $C_{171}H_{267}N_{51}O_{53}$ $S_2·C_2H_4O_2$ 3949.4	synthetic analog of human neuroendocrine hormone amylin for treatment of diabetes patients (Type 1 and 2) receiving insulin. 1.5 mL or 2.7 mL disposable multidose pen containing 1 mg/mL pramlintide (as acetate).	Amylin Pharmaceuticals (now part of Astra-Zeneca) USA/2001 [241]
Benfluorex [Mediator, Mediaxal] A10BX06		[23602-78-0] $C_{19}H_{20}F_3NO_2$ 351.36	appetite suppressant. tbl. 150 mg, O. DDD 450 mg.	withdrawn from market due to the risk of heart valve disease (fenfluramine-like cardiovascular side-effects) Science Union (Servier) France/1974 [242–244]
Mitiglinide [Glufast; Glinsuna; Kuai ru tuo;] A10BX08		[145375-43-5] $C_{19}H_{25}NO_3$ 315.41	oral antidiabetic blocking the K_{ATP} channel in pancreatic beta-cell. tbl. (three times daily) 5, 10 mg, O. DDD 30 mg.	Kissei Pharmaceuticals Japan/2004 China/2009 Synthesis: [245–247]

Table 12. A10X Other drugs used in diabetes

INN (Synonyms) [Brand names] ATC	Structure	[CAS-No.] Formula MW (g/mol)	Target (if known) Medical use Formulation DDD	Originator Approval (country/year) Production method [References]
Tolrestat [Alredase, Lorestat] A10XA01		[82964-04-3] $C_{16}H_{14}F_3NO_3S$ 357.35	aldose reductase inhibitor tbl. (once daily) 200 mg, O.	withdrawn from market Wyeth-Ayerst Italy/1995 [248–254]

Abbreviations

AA	Amino acid
ATP	Adenosine triphosphate
CV	Cardiovascular
DDD	Daily drug dose
DPP-4	Dipeptidyl peptidase-4
FFA	Free fatty acids
FPG	Fasting plasma glucose level
GLP-1	Glucagon-like peptide-1
GIP	Glucose-dependent insulino-tropic polypeptide
HbA_{1c}	Glycated hemoglobin
IDF	International Diabetes Federation
INN	International nonproprietary name
K_{ATP} channel	ATP-sensitive potassium channel
MW	Molecular weight
NICE	National Institute for Health and Care Management
O.	Oral
OGTT	Oral glucose tolerance test
P.	Parenteral
PPAR	Peroxisome proliferator activated receptor
SGTL1	Sodium-glucose cotransporter 1
SGLT2	Sodium-glucose cotransporter 2
SU	sulfonylurea
tbl.	Tablet
T1D	Type 1 diabetes mellitus
T2D	Type 2 diabetes mellitus
wfm	withdrawn from market
WHO	World Health Organization

References

1 Guidelines: NICE (National Institute for Health and Care Management) (2015) https://www.nice.org.uk/guidance/ng28/chapter/1-Recommendations (accessed 6 March 2018)

2 Inzucchi, S.E. et al. (2015) Management of hyperglycemia in type 2 diabetes. *Diabetes Care* **38**, 140–149.

3 Pfeiffer, A.F.H., Klein H.H. (2014) The treatment of type 2 diabetes. *Dtsch. Ärztebl. Int.* **111**, 69–82.

4 Edmondson, S.D., Weber A.E., Elliott, J., Roth, B.D., Wexler, R.R. (2015) Cardiovascular and metabolic diseases: 50 years of progress. *Ann. Rep. Med. Chem.* **50**, 83–117.

5 Johnson, I.S. (1983) Human insulin from recombinant DNA technology. *Science* **219**, 632–637.

6 Gilroy, C.A., Luginbuhl K.M., Chilkoti, A. (2016) Controlled release of biologics for the treatment of type 2 diabetes. *J. Control. Release* **240**, 151–164.

7 Hilgenfeld, R., Seipke, G., Berchtold, H., Owens, D.R. (2014) The evolution of insulin glargine and its continuing contribution to diabetes care. *Drugs* **74**, 911–927.

8 Genentech Inc. (1979) US 4 356 270, US-prior. 5.11.1979.

9 Genentech Inc. (1979) US 4 366 246, US-prior. 5.11.1979.

10 Genentech Inc. (1982) US 4 431 739, US-prior. 30.7.1982.

11 Eli Lilly (1982) US 4 559 302, US-prior. 1.11.1982.

12 Goeddel, D.V. et al. (1979) Expression in Escherichia coli of chemically synthesized genes for human insulin. *Proceed. Natl. Acad. Soc.* **76**, 106–10.

13 Eli Lilly (1989) EP 383 472, US-prior. 9.2.1989, 4.8.1989.

14 Eli Lilly (1994) EP 692 489, EP-prior. 16.6.1994.

15 Eli Lilly (1994) US 5 461 031, US-prior. 16.6.1994.

16 Eli Lilly (1994) US 5 474 978, US-prior. 16.6.1994.

17 Eli Lilly (1995) US 735 048, EP-prior. 31.3.1995.

18 Eli Lilly (1997) WO 9 856 406, US-prior. 13.6.1997.

19 Howey, D.C., Bowsher, R.R., Brunelle, R.L., Woodworth J.R. (1994) [Lys(B28), Pro(B29)]-human insulin. A rapidly absorbed analogue of human insulin. *Diabetes* **43**, 396–402.

20 NovoNordisk (1985) US 5 618 913, DK-prior. 30.8.1985, 14.10.1985.

21 NovoNordisk (1985) EP 214 826, EP-prior. 30.8.1985, 14.10.1985.

22 NovoNordisk (1996) US 5 866 538, DK-prior. 20.6.1996.

23 NovoNordisk (1993) WO 9 500 550, DK-prior. 21.6.1993, 28.9.1993 US.

24 Hoechst Marion Roussel (1997) EP 885 961, EP-prior. 20.6.1997.

25 Aventis (1997) US 06 221 633, DE-prior. 20.6.1997.

26 Aventis (2001) WO 02 076 495, DE-prior. 23.3.2001.

27 Hoechst AG (1988) EP 368 187, EP-prior. 8.11.1988.

28 Hoechst AG (1988) US 6 100 376, DE-prior. 8.11.1988.

29 Hoechst Marion Roussel (1998) WO 9 964 598, DE-prior. 6.6.1998.

30 Aventis (2002) WO 03 105 888, DE-prior. 18.6.2002.

31 Aventis (2002) US 7 476 652, DE-prior. 18.6.2002.

32 Aventis (2002) US 7 713 930, DE-prior. 18.6.2002.

33 Sanofi (2010) WO 2011 144 673, EP-prior. 19.5.2010, 13.7.2010, 10.2.2011.

34 NovoNordisk (1993) US 5 750 497, DK-prior. 17.9.1993.

35 NovoNordisk (1993) US 6 011 007, DK-prior. 17.9.1993.

36 NovoNordisk (1993) US 06 869 930, DK-prior. 17.9.1993.

37 NovoNordisk (2003) WO 2 005 047 508, DK-prior. 14.11.2003.

38 NovoNordisk (1996) US 5 866 538, DK-prior. 20.6.1996.

39 NovoNordisk (2003) WO 2 00 5012 347, DK-prior. 5.8.2003, 14.8.2003,US.

40 Witters, L.A. (2001) The blooming of the French lilac. *J. Clin. Invest.* **108**, 1105–1107.

41 Ferrannini, E. (2014) The target of metformin in type 2 diabetes. *New Engl. J. Med.* **371**, 1547–1548.

42 Aron S.A.R.L. (1975) FR 2 322 860, FR-prior. 5.9.1975.

43 Shapiro, S.L., Parrino, V.A., Freedman, L. (1959) Hypoglycemic Agents. III. N1-Alkyl- and Aralkylbiguanides. *J. Am. Chem. Soc.* **81**, 3728–3736.

44 Deacon, C.F., Lebovitz, H.E. (2016) Comparative review of dipeptidyl peptidase-4 inhibitors and sulphonylureas. *Diabetes Obes. Metab.* **18**, 333–347.

45 Hoechst AG (1966) DE 128 837, CA-prior. 21.7.1966.

46 Hoechst AG (1965) US 3 454 635, DE-prior. 2.12.1965.

47 Aumüller, W. et al. (1966) Ein neues hochwirksames orales Antidiabeticum. *Arzneim.-Forsch.* **16**, 1640–1641.

48 Pfizer (1958) US 3 013 072, US-prior. 1958.

49 Pfizer (1957) US 3 349 124, US-prior. 20.5.1957.

50 Ruschig, H., Korger, G., Aumüller, W., Hagner, H. Weyer, R. (1958) Neue peroral wirksame blutzuckersenkende Substanzen. *Arzneim.-Forsch.* **8**, 448–454.

51 Upjohn Co. (1955) Hoechst; US 2 968 158, DE-prior. 8.8.1955.

52 Hoechst AG (1955) DE 974 062, DE-prior. 9.8.1955.

53 Hoffmann-La Roche (1966) DE 1 695 201, CH-prior. 28.10.1966.

54 Hoffmann-La Roche (1968) US 3 654 357, CH-prior. 26.4.1968.

55 Bretschneider, H., Hohenlohe-Oehringen, K., Graßmayr, K. (1969) Arylsulfonylureido- und Arylsulfonylamidoacyl-derivate von Oxy- und Oxo-cycloalkanen als potentielle Antidiabetica. *Monatshefte für Chemie* **100**, 2133–2135.

56 Upjohn Co. (1959) US 3 063 903, US-prior. 9.6.1959.

57 Wright, J.B., Willette, R.E. (1962) Antidiabetic agents. N4-arylsulfonylsemicarbazides. *J. Med. Pharm. Chem.* **5**, 815–822.

58 Boehringer Mannh. (1953) US 2 907 692, DE-prior. 11.2.1953.

59 Carlo Erba (1969) DE 2 012 138, I-prior. 26.3.1969.

60 Carlo Erba (1969) US 3 669 966, I-prior. 26.3.1969.

61 Boehringer Ingelheim (1970) US 3 708 486, DE-prior. 5.1.1970, 17.4.1969.

62 Science Union (1966) FR 1 510 714, UK-prior. 10.2.1966.

63 Science Union (1966) US 3 501 495, UK-prior. 10.2.1966.

64 Bayer AG (1967) DE 1 670 952, DE-prior. 25.11.1967.

65 Bayer AG (1967) US 3 668 215, DE-prior. 25.11.1967.

66 Hoechst AG (1979) US 4 378 785, DE-prior. 19.12.1979.

67 Hoechst AG (1979) DE 2 951 135, DE-prior. 19.12.1979.

68 Hoechst AG (1993) DE 4 336 159, DE-prior. 22.1.1993.

69 Weyer, R., Hitzel V. (1988) Acylureidoalkylphenylsulfony-lureas with blood glucose lowering efficacy. *Arzneim.-Forsch.* **38**, 1079–1080.

70 Tanwar, D.K., Surendrabhai, V.R., Gill, M.S. (2017) An efficient and practical process for the synthesis of glimepiride. *Synlett* **18**, 2495–2498 (and refs. therein).

71 Eli Lilly (1960) US 3 320 312, US-prior. 28.4.1960.

72 Marshall, F.J., Sigal, M.V., Sullivan, H.R., Cesnik, C., Root, M.A. (1963) Further studies on N-arylsulfonyl-N'-alkylureas. *J. Med. Chem.* **6**, 60–63.

73 Schering AG (1961) DE 1 445 142, DE-prior. 23.2.1961.

74 Schering AG (1961) US 3 275 635, DE-prior. 18.10.1960, 22.2.1961, 23.2.1961.

75 Abdulsalim, S., Vayalil, M.P., Miraj, S.S. (2016) New fixed dose chemical combinations: the way forward for better diabetes type II management? *Expert Opin. Pharmacother.* **17**, 2207–2214.

76 Harris, S.B. (2016) The power of two: an update on fixed-dose combinations for type 2 diabetes. *Exp. Rev. Clin. Pharmacology* **9**, 1453–1462.

77 SmithKline Beecham (1998) WO 9 857 634, UK-prior. 18.6.1997, 25.3.1998.

78 SmithKline Beecham (1998) WO 9 857 649, UK-prior. 18.6.1997, 27.3.1998.

79 Takeda Chem. (1995) EP 861 666, JP-prior. 20.6.1995.

80 Takeda Chem. (2004) WO 2 005 099 760, JP-prior. 14.4.2004.

81 Takeda Chem. (2003) WO 2 005 041 962, JP-prior. 31.10.2003.

82 Merck Sharp & Dohme Ltd (2008) WO 2 009 099 734, US-prior. 5.2.2008.

83 Merck Sharp & Dohme Ltd (2008) WO 2 009 111 200, US-prior. 4.3.2008.

84 Novartis AG (2004) WO 2 005 117 861, US-prior. 4.6.2004, 28.8.2004.

85 Novartis AG (2005) WO 2 007 041 053, US-prior. 25.9.2005.

86 Takeda Pharm. (2007) WO 2 008 093 882, JP-prior. 1.2.2007.

87 Bristol-Myers Squibb (2000) WO 200 168 603, US-prior. 10.3.2000.

88 Bristol-Myers Squibb (2004) US 8 628 799, US-prior. 25.5.2004.

89 Boehringer Ingelheim (2011) WO 2 012 120 040, EP-prior. 7.3.2011, 15.3.2011.

90 Boehringer Ingelheim (2017) WO 2 017 211 979, EP-prior. 10.6.2016, 1.2.2017.

91 Merck Sharp & Dohme Ltd. (2009) WO 2 010 147 768, US-prior. 15.6.2009, 6.11.2009.

92 Merck Sharp & Dohme Ltd. (2009) WO 2 011 049 773, US-prior. 23.10.2009.

93 Takeda Pharm. (2006) WO 2 007 033 266, US-prior. 14.9.2005, 15.5.2006.

94 Takeda Pharm. (2007) WO 2 009 011 451, JP-prior. 19.7.2007.

95 Novo Nordisk (2006) WO 2 008 037 807, EP-prior. 29.6.2006.

96 Bristol-Myers Squibb (2002) WO 03 099 836, US-prior. 20.5.2002.

97 Bristol-Myers Squibb (2001) US 6 936 590, US-prior. 13.3.2001.

98 Bristol-Myers Squibb, Astra-Zeneca (2009) WO 2 010 138 535, US-prior. 27.5.2009.

99 Bristol-Myers Squibb, Astra-Zeneca (2009) WO 2 011 060 290, US-prior. 13.11.2009.

100 Bristol-Myers Squibb, Astra-Zeneca (2009) WO 2 011 060 256, US-prior. 13.11.2009.

101 Janssen Pharm. (2010) WO 2 011 005 811, US-prior. 8.7.2009, 6.7.2010.

102 LG Life Sciences (2012) WO 2 014 058 188, Korea-prior. 8.10.2012.

103 Boehringer Ingelheim (1997) WO 2 008 055 940, US-prior. 8.11.1997.

104 Boehringer Ingelheim (2009) WO 2 010 092 125, US-prior. 13.2.2009.

105 Boehringer Ingelheim (2009) WO 2 011 039 337, US-prior. 2.10.2009.

106 Boehringer Ingelheim (1997) WO 2 008 055 940, US-prior. 8.11.1997.

107 Boehringer Ingelheim (2009) WO 2 010 092 125, US-prior. 13.2.2009.

108 Boehringer Ingelheim (2012) WO 2 010 092 125, US-prior. 7.3.2012.

109 Bristol-Myers Squibb (2000) WO 200 127 128, US-prior. 12.10.1999, 5.4.2000.

110 Bristol-Myers Squibb (2000) WO 200 168 603, US-prior. 10.3.2000.

111 Bristol-Myers Squibb (2002) WO 03 099 836, US-prior. 20.5.2002.

112 Bristol-Myers Squibb (2003) US 6 936 590, US-prior. 19.6.2003.

113 Bristol-Myers Squibb (2004) WO 2 005 117 841, US-prior. 28.5.2004.

114 Bristol-Myers Squibb (2007) WO 2 008 116 179, US-prior. 22.3.2007.

115 Pfizer (2008) US 8 080 580, US-prior. 24.8.2008.

116 Pfizer (2009) WO 2 010 023 594, US-prior. 28.8.2008, 21.7.2009.

117 Merck Sharp Dohme Ltd (2013) WO 2 014 159 151, US-prior. 14.4.2013, 7.10.2013, 5.12.2013.

118 Merck Sharpe Dome Ltd (2001) WO 2 003 004 498, US-prior. 6.7.2001.

119 Pfizer (2009) WO 2 010 023 594, US-prior. 28.8.2008, 21.7.2009.

120 Bayer AG (1973) US 4 062 950, DE-prior. 22.9.1973.

121 Bayer AG (1985) US 4 904 769, DE-prior. 13.12.1985.

122 Bayer AG (1985) DE 3 543 999, DE-prior. 13.12.1985.

123 Bayer AG (1984) DE 3 439 008, DE-prior. 25.10.1984.

124 Boedeker, B.G.D. (1996) Bioprocess technologies depending on the molecular-structure of pharmaceutical products. *Chimia* **50**, 412–413.

125 Schmidt, D.D. et al. (1977) Alpha-glucosidase inhibitors. New complex oligosaccharides of microbial origin. *Naturwissenschaften* **64**, 535–536.

126 Ogawa, S., Shibata, Y. (1988) Total synthesis of acarbose and adiposin-2. *J. Chem. Soc., Chem. Comm.* **9**, 605–606.

127 Bayer AG (1986) DE 3 611 841, DE-prior. 9.4.1986.

128 Bayer AG (1977) US 4 639 436, DE-prior. 27.8.1977, 24.12.1977.

129 Bayer AG (1979) DE 2 758 025 DE-prior. 24.12.1977.

130 Bayer AG (1981) EP 49 858, DE-prior. 15.10.1981.

131 Takeda Pharm. (1981) EP 56 194, JP-prior. 5.1.1981, 2.5.1981, 6.10.2981.

132 Horii, S. et al. (1986) Synthesis and alpha-D-glucosidase inhibitory activity of *N*-substituted valiolamine derivatives as potential oral antidiabetic agents. *J. Med. Chem.* **29**, 1038–46.

133 Sankyo Co. (1983) EP 139 421, JP-prior. 30.8.1983.

134 Sankyo Co. (1983) US 4 572 912, JP-prior. 30.8.1983.

135 Sankyo Co. (1985) EP 207 581, JP-prior. 26.2.1985.

136 Beecham Group (1988) US 5 002 953, UK-prior. 4.9.1987, 30.11.1987, 4.2.1988.

137 Beecham Group (1987) EP 306 228, UK-prior. 4.9.1987.

138 Beecham Group (1991) WO 9 310 254, UK-prior. 19.11.1991.

139 Beecham Group (1988) US 5 646 169, UK-prior. 4.9.1987, 30.11.1987, 4.2.1988.

140 Beecham Group (1988) EP 306 228, UK-prior. 4.9.1987, 30.11.1987, 4.2.1988.

141 Cantello, B.C.C. et al. (1994) The synthesis of BRL-49653 – a novel potent antihyperglycemic agent. *Bioorg. & Med. Chem. Lett.* **4**, 1181–1184.

142 Cantello, B.C.C. et al. (1994). *w*-Heterocyclylamino)alkoxy] benzyl]-2,4-thiazolidinediones as potent antihyperglycemic agents. *J. Med. Chem.* **37**, 3977–3978.

143 Takeda Chem. Ind. (1985) US 4 687 777, JP-prior. 19.1.1985.

144 Takeda Chem. Ind. (1985) EP 193 256, JP-prior. 19.1.1985.

145 Sohda, T. et al. (1990) Studies on antidiabetic agents. Synthesis and hypoglycemic activity of 5-[4-(pyridylalkoxy)benzyl]-2,4-thiazolidinediones. *Arzneim.-Forsch.* **40**, 37–42.

146 Madivada, L.R. et al. (2009) An improved process for pioglitazone and its pharmaceutically acceptable salt. *Org. Proc. Res. & Dev.* **13**, 1190–1194 (and refs. therein).

147 Scheen, A.J. (2012) Review of gliptins in 2011. *Exp. Op. Pharmacotherapy* **13**, 81–99.

148 Deacon, C.F., Holst, J.J. (2013) DPP-4 inhibitors for the treatment of T2D: comparison, efficacy and safety. *Expert Opin. Pharmacother.* **14**, 2047–2058.

149 Deacon, C.F., Lebovitz, H.E. (2016) Comparative review of dipeptidyl peptidase-4 inhibitors and sulphonylureas. *Diabetes Obes. Metab.* **18**, 333–347.

150 Merck Sharp & Dohme Ltd. (2003) WO 2 004 085 661, US-prior. 24.3.2003.

151 Merck Sharp & Dohme Ltd. (2003) WO 2 005 003 135, US-prior. 24.6.2003.

152 Merck Sharp & Dohme Ltd. (2004) WO 2 006 065 826, US-prior. 15.12.2004.

153 Kim, D. et al. (2005) (2*R*)-4-oxo-4-[3-(trifluoromethyl)-5,6-dihydro[1,2,4]triazolo[4,3-a]pyrazin-7(8*H*)-yl]-1-(2,4,5-trifluorophenyl)butan-2-amine: A potent, orally active dipeptidyl peptidase IV inhibitor for the treatment of type 2 diabetes. *J. Med. Chem.* **48**, 141–151.

154 Hansen, K.B. et al. (2005) First generation process for the preparation of the DPP-IV inhibitor sitagliptin. *Org. Proc. Res. & Dev.* **9**, 634–639.

155 Novartis AG (1998) WO 2 000 034 241, US-prior. 10.12.1998.

156 Novartis AG (1998) US 6 166 063, US-prior. 10.12.1998.

157 Villhauer, E.B. et al. (2003) 1-[[(3-Hydroxy-1-adamantyl) amino]acetyl]-2-cyano-(S)-pyrrolidine: a potent, selective, and orally bioavailable dipeptidyl peptidase IV inhibitor with antihyperglycemic properties. *J. Med. Chem.* **46**, 2774–2789.

158 Bristol-Myers Squibb (2000) US 6 395 767, US-prior. 10.3.2000.

159 Bristol-Myers Squibb (2005) US 6 995 183, US-prior. 17.2.2005.

160 Augeri, D.J. et al. (2005) Discovery and preclinical profile of saxagliptin (BMS-477118): a highly potent, long-acting, orally active dipeptidyl peptidase IV inhibitor for the treatment of type 2 diabetes. *J. Med. Chem.* **48**, 5025–5037.

161 Savage, S.A. et al. (2009) Preparation of saxagliptin, a novel DPP-IV inhibitor. *Org. Proc. Res. & Dev.* **13**, 1169–1176.

162 Syrrx (2004) WO 2 005 095 381, US-prior. 15.3.2004.

163 Takeda Pharm. Comp. (2005) WO 2 007 033 266, US-prior. 14.9.2005.

164 Takeda Pharm. Comp. (2005) WO 2 007 035 629, US-prior. 16.9.2005.

165 Feng, J. et al. (2007) Discovery of alogliptin: a potent, selective, bioavailable, and efficacious inhibitor of dipeptidyl peptidase IV. *J. Med. Chem.* **50**, 2297–2300.

166 Boehringer Ingelheim (2002) WO 2 004 018 468, DE-prior. 21.8.2002.

167 Boehringer Ingelheim (2002) DE 10 238 243, DE-prior. 21.8.2002.
168 Boehringer Ingelheim (2002) US 7 407 955, DE-prior. 21.8.2002.
169 Boehringer Ingelheim (2002) US 8 119 648, DE-prior. 21.8.2002.
170 Boehringer Ingelheim (2002) US 8 178 541, DE-prior. 21.8.2002.
171 Eckhardt, M. et al. (2007) 8-(3-(R)-aminopiperidin-1-yl)-7-but-2-ynyl-3-methyl-1-(4-methyl-quinazolin-2-ylmethyl)-3,7-dihydropurine-2,6-dione (BI 1356), a highly potent, selective, long-acting, and orally bioavailable DPP-4 inhibitor for the treatment of type 2 diabetes. *J. Med. Chem.* **50**, 6450–6453.
172 LG Life Sciences (2005) WO 2 006 104 356, Korea-prior. 1.4.2005.
173 LG Life Sciences (2010) WO 2 012 060 590, Korea-prior. 1.11.2010.
174 Dong-A Pharm. (2007) WO 2 008 130 151, Korea-priority date 19.4.2007.
175 Dong-A Pharm. (2009) WO 2 010 114 291, Korea-prior. 30.3.2009.
176 Dong-A Pharm. (2007) WO 2 010 114 292, Korea-prior. 19.4.2007.
177 Kim, H.J. et al. (2011) Discovery of DA-1229: a potent, long acting dipeptidyl peptidase-4 inhibitor for the treatment of type 2 diabetes. *Bioorg. Med. Chem. Lett.* **21**, 3809–3812.
178 Gilroy, C.A., Luginbuhl K.M., Chilkoti, A. (2016) Controlled release of biologics for the treatment of type 2 diabetes. *J. Contr. Rel.* **240**, 151–164.
179 Manandhar, B., Ahn, J.-M. (2015) Glucagon-like peptide-1 (GLP-1) analogues: recent advances, new possibilities and therapeutic implications. *J. Med. Chem.* **58**, 1020–1037.
180 Lorenz, M., Evers A., Wagner M. (2013) Recent progress and future options in the development of GLP-1 receptor agonists for the treatment of diabesity. *Bioorg. Med. Chem. Lett.* **23**, 4011–4018.
181 Meier, J.J. (2012) GLP-1 receptor agonists for the individualized therapy of type 2 diabetes mellitus. *Nat. Rev. Endocrinology* **8**, 728–742.
182 Madsbad, S. (2016) Review of head-to-head comparisons of glucagon-like peptide-1 receptor agonists. *Diab. Ob. Metab.* **18**, 317–332.
183 J. Eng (1993) US 5 424 286, US-prior. 24.5.1993.
184 Amylin Pharm. Inc. (1996) US 6 858 576, US-prior. 8.8.1996.
185 Amylin Pharm. Inc. (1997) US 6 956 026, US-prior. 7.1.1997.
186 Amylin Pharm. Inc. (2000) US 6 902 744, US-prior. 14.1.2000.
187 Amylin Pharm. Inc. (2000) US 6 872 700, US-prior. 14.1.2000.
188 Göke, R. et al. (1993) Exendin-4 Is a high potency agonist and truncated exendin-(9-39)-amide an antagonist at the glucagon-like peptide 1-(7-36)-amide receptor of insulin-secreting β-cells. *J. Biol. Chem.* **268**, 19650–19655.
189 Novo Nordisk (1993) WO 9 517 510, DK-prior. 23.12.1993.
190 Novo Nordisk (1996) WO 9 808 871, DK-prior. 30.8.1996.
191 Novo Nordisk (1996) US 6 458 924, DK-prior. 30.8.1996.
192 Novo Nordisk (1996) US 7 235 627, DK-prior. 30.8.1996.
193 Novo Nordisk (1998) WO 9 943 707, DK-prior. 27.2.1998.
194 Knudsen, L.B. et al. (2000) Potent derivatives of glucagon-like peptide-1 with pharmacokinetic properties suitable for once daily administration. *J. Med. Chem.* **43**, 1664–1669.
195 Zealand Pharma (1999) WO 2 001 004 156, EP-prior. 12.7.1999, US, 9.8.1999.
196 Zealand Pharma (2002) US 7 544 657, US-prior. 2.10.2002.
197 Glaxo-SmithKline (2011) WO 2 012 109 429, US-prior. 9.2.2011.
198 Bush, M.A. et al. (2009) Safety, tolerability, pharmacodynamics and pharmacokinetics of albiglutide, a long-acting glucagon-like peptide-1 mimetic, in healthy subjects. *Diabetes Obes. Metab.* **11**, 498–505.
199 Matthews, J.E. et al. (2008) Pharmacodynamics, pharmacokinetics, safety, and tolerability of albiglutide, a long-acting glucagon-like peptide-1 mimetic, in patients with type 2 diabetes. *J. Clin. Endocr. & Metab.* **93**, 4810–4817.
200 Glaesner, W. et al. (2010) Engineering and characterization of the long-acting glucagon-like peptide-1 analogue LY2189265, an Fc fusion protein. *Diab. Metab. Res. Rev.* **26**, 287–296.
201 Novo Nordisk (2005) WO 2 006 097 537, DK-prior. 18.3.2005.
202 Lau, J. et al. (2015) Discovery of the once-weekly glucagon-like peptide-1 (GLP-1) analogue semaglutide. *J. Med. Chem.* **58**, 7370–7380.
203 Koepsell, H. (2017) The Na$^+$-D-glucose cotransporters SGLT1 and SGLT2 are targets for the treatment of diabetes and cancer. *Pharm. & Therap.* **170**, 148–165.
204 Mudaliar, S., Polidori, D., Zambrowicz, B., Henry R.R. (2015) Sodium–glucose cotransporter inhibitors: Effects on renal and intestinal glucose transport from bench to bedside. *Diabetes Care* **38**, 2344–2353.
205 Washburn, W.N. (2014) Forxiga (Dapagliflozin), a potent selective SGLT2 inhibitor for the treatment of diabetes. *Ann. Rep. Med. Chem.* **49**, 363–382.
206 Bristol-Myers Squibb (1999) US 6 414 126, US-prior. 12.10.1999.
207 Bristol-Myers Squibb (1999) US 6 515 117, US-prior. 12.10.1999.
208 Bristol-Myers Squibb (2003) US 7 375 213, US-prior. 3.1.2003.
209 Bristol-Myers Squibb (2007) US 7 851 502, US-prior. 22.3.2007.
210 Bristol-Myers Squibb (2007) US 8 221 786, US-prior. 22.3.2007.
211 Bristol-Myers Squibb (2007) US 8 361 972, US-prior. 22.3.2007.
212 Bristol-Myers Squibb (2007) US 8 716 251, US-prior. 22.3.2007.
213 Meng, W. et al. (2008) Discovery of dapagliflozin: A potent, selective renal sodium-dependent glucose cotransporter 2 (SGLT2) inhibitor for the treatment of type 2 diabetes. *J. Med. Chem.* **51**, 1145–1149.
214 Tanabe Pharma Corp. (2003) WO 2 005 012 326, JP-prior. 1.8.2003.
215 Mitsubishi Tanabe Pharma Corp. (2006) WO 2 008 069 327, JP-prior. 4.12.2006.
216 Mitsubishi Tanabe Pharma Corp. (2006) US 7 943 582, JP-prior. 4.12.2006.
217 Janssen Pharmac., Mitsubishi Tanabe Pharma Corp. (2007) WO 2 009 035 969, US-prior. 10.9.2007.
218 Janssen Pharmac. (2011) WO 2 012 140 120, US-prior. 13.4.2011.
219 Nomura, S. et al. (2010) Discovery of canagliflozin, a novel C-glucoside with thiophene ring, as sodium-dependent glucose cotransporter 2 inhibitor for the treatment of type 2 diabetes mellitus. *J. Med. Chem.* **53**, 6355–6360. *tab. Res. Rev.* **26**, 287–296.
220 Boehringer Ingelheim (2005) WO 2 006 117 359, EU-prior. 3.5.2005, 19.8.2005.
221 Boehringer Ingelheim (2004) US 7 579 449, DE-prior. 16.3.2004, 18.8.2004, 16.12.2004.
222 Boehringer Ingelheim (2004) WO 2 006 120 208, DE-prior. 10.5.2005, 23.8. 2005, 15.9.2005.

223 Boehringer Ingelheim (2005) US 7 713 938, EU-prior. 3.5.2005, 19.8.2005.

224 Wang, X. et al. (2014) Efficient synthesis of empagliflozin, an inhibitor of SGLT-2, utilizing an AlCl3-promoted silane reduction of a β-glycopyranoside. *Org. Lett.* **16**, 4090–4093.

225 Pfizer (2008) US 8 080 580, US-prior. 24.8.2008.

226 Pfizer (2009) WO 2 010 023 594, US-prior. 28.8.2008, 21.7.2009.

227 Merck Sharp Dohme Ltd (2013) WO 2 014 159 151, US-prior. 14.4.2013, 7.10.2013, 5.12.2013.

228 Mascitti, V. et al. (2011) Discovery of a clinical candidate from the structurally unique dioxa-bicyclo[3.2.1]octane class of sodium-dependent glucose cotransporter 2 inhibitors. *J. Med. Chem.* **54**, 2952–2960.

229 Bernhardson, D. et al. (2014) Development of an early-phase bulk enabling route to sodium-dependent glucose cotransporter 2 inhibitor ertugliflozin. *Org. Proc. Res. Dev.* **18**, 57–65.

230 Tanabe Seyaku Co. (2003) WO 2 005 012 326, US-prior. 1.8.2003.

231 Tanabe Seyaku Co. (2003) US 20 050 233 988, US-prior. 1.8.2003.

232 Astellas Pharma, Kotobuki Pharmac. (2006) WO 2 007 114 475, JP-prior. 5.4.2006.

233 Astellas Pharma, Kotobuki Pharmac. (2006) WO 2 008 075 736, US-prior. 21.12.2006.

234 Imamura, M. et al. (2012) Discovery of ipragliflozin (ASP1941): A novel C-glucoside with benzothiophene structure as a potent and selective sodium glucose co-transporter 2 (SGLT2) inhibitor for the treatment of type 2 diabetes mellitus. *Bioorg. Med. Chem.* **20**, 3263–3279.

235 Guardado-Mendoza, R., Prioletta, A., Jiménez-Ceja, L.M., Sosale, A., Folli, F. (2013) The role of nateglinide and repaglinide, derivatives of meglitinide, in the treatment of type 2 diabetes mellitus. *Arch. Med. Sci.* **9**, 936–943.

236 Thomae GmbH (1985) US 5 216 167, DE-prior. 30.12.1983, 25.6.1985. 1.7.1985.

237 Thomae GmbH (1991) WO 9 300 337, EP-prior. 21.6.1991.

238 Grell, W. et al. (1998) Repaglinide and related hypoglycemic benzoic acid derivatives. *J. Med. Chem.* **41**, 5219–5246.

239 Ajinomoto (1985) EP 196 222, JP-prior. 27.3.1985.

240 Shinkai, H. et al. (1989) N-(cyclohexylcarbonyl)-D-phenylalanines and related compounds. A new class of oral hypoglycemic agents. *J. Med. Chem.* **32**, 1436–1441.

241 Amylin Pharmaceuticals (1991) WO 9 310 146, US-prior. 19.11.1991.

242 Science Union (1966) FR 1 517 587, UK-prior. 15.4.1966.

243 Science Union (1966) DE 1 593 991, UK-prior. 15.4.1966.

244 Science Union (1966) US 3 607 909, US-prior. 15.4.1966.

245 Kissei Pharmaceuticals (1991) EP 507 534, JP-prior. 30.3.1991.

246 Yamaguchi, T. et al. (1997) Preparation of optically active succinic acid derivatives. I. Optical resolution of 2-benzyl-3-(cis-hexahydroisoindolin-2-ylcarbonyl)-propionic acid. *Chem. Pharm. Bull.* **45**, 1518–1520.

247 Yamaguchi, T. et al. (1998) Preparation of optically active succinic acid derivatives. II. Efficient and practical synthesis of KAD-1229. *Chem. Pharm. Bull.* **46**, 337–340.

248 Ayerst, McKenna & Harrison Inc. (1981) EP 59 596, CA-prior. 15.10.1981.

249 Ayerst, McKenna & Harrison Inc. (1981) US 4 391 825, CA-prior. 15.10.1981.

250 Ayerst, McKenna & Harrison Inc. (1981) US 4 568 693, CA-prior. 2.3.1981.

251 Ayerst, McKenna & Harrison Inc. (1981) US 4 600 724, CA-prior. 2.3.1981.

252 Ayerst, McKenna & Harrison Inc. (1981) US 4 705 882, CA-prior. 2.3.1981.

253 Ayerst, McKenna & Harrison Inc. (1981) US 4 946 987, CA-prior. 2.3.1981.

254 Sestanj, K. et al. (1984) N-[[5-(trifluoromethyl)-6-methoxy-1-naphthalenyl]thioxomethyl]-N-methylglycine (tolrestat), a potent, orally active aldose reductase inhibitor. *J. Med. Chem.* **27**, 255–256.

Vitamins (A11)

Axel Kleemann, Hanau, Germany

1. Introduction

Vitamins are essential dietary nutrients, which are required in diet in only small amounts and therefore are also called micronutrients. They have to be taken up with the diet and cannot be synthesized by the human body. The normal balanced food provides the necessary requirement of vitamins, this being the reason that cases of malnutrition with vitamin deficiency in most countries of the world are rather rare, and their use as drugs is far smaller than as feed and food additives. There are two groups of vitamins, namely fat- and water-soluble ones. Many of the vitamins contained mainly in vegetables and fruits are small molecules, play an important and vital role in human metabolism, and are enzyme cofactors or cofactor precursors.

This applies to all water-soluble vitamins and the fat-soluble vitamin K.

There are numerous excellent reviews on vitamins, their discovery history, biochemical role, production, and use. Among them, references [1–5] were chosen and are recommended as entry for deeper studies, especially the vitamin articles of Ullmann's Encyclopedia [1].

In the following chapters, vitamins are reviewed according to Chapter A11 of the Anatomical Therapeutic Chemical classification system (ATC). Unlike in A11, only vitamin monosubstances are referenced, and no combinations or multivitamins. Furthermore, some vitamins are classified in other ATC chapters, so vitamin B_{12} under ATC number B03BA in Antianemic preparations, and vitamin K under ATC number B02BA in Antihemorrhagics.

Ullmann's Pharmaceuticals. Edited by Axel Kleemann and Bernhard Kutscher.
© 2022 Wiley-VCH GmbH
Set ISBN: 978-3-527-34252-5

2. A11C Vitamins A (Retinols) and D (Calciferols)

2.1. A11CA Vitamin A, Plain (Table 1)

Table 1. A11CA Vitamin A, plain

INN (Synonyms) [Brand names] ATC	Structure (Remarks)	[CAS No.] Formula MW, g/mol	Target (if known) Medical use Formulation DDD	Originator Approval (country/year) Production method [References]
Retinol (vitamin A; vitamin A₁; axerophthol; *all-trans*-retinol) [numerous generic preparations] A11CA01; D10AD02; R01AX02; S01XA02 **WHO essential medicine**		[68-26-8] C₂₀H₃₀O 286.46 acetate: [127-47-9] C₂₂H₃₂O₂ 328.50 propionate: [7069-42-3] C₂₃H₃₄O₂ 342.52 palmitate: [79-81-2] C₃₆H₆₀O₂ 524.87	fat-soluble essential vitamin; needed for eyesight and maintenance of the skin; measured in IU (1 IU is equivalent to 0.3 µg). treatment of vitamin A deficiency and fortification of food. cps. 30 000, 50 000 IU; ointment nose 1250 IU; ointment eye 250 IU. 50 000 IU O, P.	discovery 1913 by E. McCollum and M. Davis [6]. first industrial process was developed by Roche in the 1950s; key intermediate for all industrial syntheses is β-ionone; world production >7500 t/a: 75% for animal feed, 25% for food and pharmaceuticals [1, 2] synthesis: Roche [7]; BASF: [8]
β-Carotene (beta-carotene; betacarotene) [Carotaben, Provatene, Solatene] A11CA02; D02BB01		[7235-40-7] C₄₀H₅₆ 536.89	provitamin A; abundant in fungi, plants (e.g., carrots), and fruits; converted enzymatically to vitamin A by β-carotene 15,15'-monooxygenase to retinal and reduction by retinal reductase. mainly used as dietary supplement. cps. 25 mg.	discovery and isolation from carrots by F. Wackenroder; structure elucidation by P. Karrer. synthesis [9] industrial processes: Roche: [10, 11]; BASF: [8, 12]

2.2. A11CC Vitamin D and Analogues (Table 2)

Table 2. A11CC Vitamin D and analogues

INN (Synonyms) [Brand names] ATC	Structure (Remarks)	[CAS No.] Formula MW, g/mol	Target (if known) Medical use Formulation DDD	Originator Approval (country/year) Production method [References]
Ergocalciferol (calciferol, vitamin D$_2$; viosterol; oleovitamin D$_2$) [Deltalin, Drisdol, Calcidol] A11CC01 **WHO essential medicine**		[50-14-6] C$_{28}$H$_{44}$O 396.66	antirachitic vitamin, fat soluble. med. use in vitamin D deficiency, treatment of hypocalcemia, and in dialysis-dependent renal failure; also used as dietary supplement and food additive. cps. 50 000 iu.	discovery: H. STEENBOCK, T. NELSON (1924)/Wisconsin Alumni semisynthesis (from ergosterol by irradiation) [13–16]
Dihydrotachysterol (AT 10; antitetany substance 10; dichystrolum) [Antitanil, Calcamine, Dygratyl, Dihydral, Hytakerol, Parterol, Tachyrol] A11CC02		[67-96-9] C$_{28}$H$_{46}$O 398.68	vitamin D analogue, acts as bone density conservation agent (calcium regulator). cps. 0.125, 0.5 mg; concentrate 0.2 mg/mL. 1 mg O.	Winthrop synthesis (reduction of tachysterol) [17]
Alfacalcidol (1α-hydroxycholecalciferol; 1α-hydroxy-vitamin D$_3$) [Alfarol, Alpha, EinsAlpha, Etalpha, One-Alpha] A11CC03		[41294-56-8] C$_{27}$H$_{44}$O$_2$ 400.65	vitamin D analogue. med. use in management of hypocalcemia, secondary hyperparathyroidism, and osteodystrophy in patients with chronic renal failure. cps. 0.25, 1 μg; oral drops 2 μg/mL; inj. 2 μg/mL. 1 μg O, P.	Wisconsin Alumni Research Foundation USA/1977; GB/1978 synthesis (from cholesterol) [18–21]
Calcitriol (1α,25-dihydroxy-cholecalciferol; 1α,25-dihydroxyvitamin D$_3$; 1,25-DHCC; Ro-21-5535) [Calcijex, Decostriol, Osteotriol, Rocaltrol, Silkis] A11CC04; D05AX03		[32222-06-3] C$_{27}$H$_{44}$O$_3$ 416.65	active form of vitamin D. med. use for treatment of hypocalcemia–hypoparathyroidism, osteomalacia, rickets, renal osteodystrophy, chronic kidney disease, osteoporosis, and for prevention of corticosteroid-induced osteoporosis. amp. 0.5, 1, 2 μg/mL; cps. 0.25, 0.5 μg; tbl. 0.25, 0.5 μg. 1 μg O, P.	discovery by HOLICK, DELUCA et al., Univ. of Wisconsin−Madison [22, 23] USA/1978 synthesis (e.g., from pregnenolone or stigmasterol via 1α,25-dihydroxy-7-dehydrocholesterol) [22–25]

(continued)

Table 2. (*Continued*)

INN (Synonyms) [Brand names] ATC	Structure (Remarks)	[CAS No.] Formula MW, g/mol	Target (if known) Medical use Formulation DDD	Originator Approval (country/year) Production method [References]
Colecalciferol (cholecalciferol; vitamin D$_3$) [Dekristol, D3-Vicotrat, Oleovit D$_3$, Vigantol] A11CC05 **WHO essential medicine**		[*67-97-0*] C$_{27}$H$_{44}$O 384.65	physiological form of vitamin D. med. use for treatment of vitamin D deficiency, familial hypophosphatemia, hypoparathyroidism, and Fanconi syndrome. colecalciferol is synthesized in the skin from 7-dehydrocholesterol under UV irradiation and is converted in the liver into calcifediol and in the kidneys into calcitriol. amp. 1.25, 2.5 mg/mL; cps. 0.5 mg; drops 0.5 mg/mL; tbl. 0.01, 0.25, 5 mg.	E. Merck DE/1929 (Vigantol) synthesis (from 7-dehydrocholesterol) [26, 27]
Calcifediol (calcidiol; 25-hydroxy-cholecalciferol; 25-hydroxy-vitamin D$_3$; 25-HCC; U-32070E) [Calderol, Dedrogyl, Didrogyl, Hidroferol, Rayaldee] A11CC06		[*19356-17-3*] C$_{27}$H$_{44}$O$_2$ 400.65	metabolite of colecalciferol, formed by cholecalciferol 25-hydroxylase in the liver, and precursor of calcitriol, which is formed in the kidneys. Circulating form of vitamin D$_3$ and basis for the blood test to determine the vitamin D status. med. use for treatment of vitamin D deficiency, familial hypophosphatemia, and hypoparathyroidism. cps. 30 µg; drops 0.15, 0.45 mg/mL.	discovery 1968 in the lab of H.F. DeLuca [28, 29] USA/1979 (calderol) semisynthesis (from, e.g., 25-hydroxycholesterol or pregnenolone) [28–31]

3. A11D Vitamin B$_1$

3.1. A11DA Vitamin B$_1$, Plain (Table 3)

Table 3. A11DA Vitamin B$_1$, plain

INN (Synonyms) [Brand names] ATC	Structure (Remarks)	[CAS No.] Formula MW, g/mol	Target (if known) Medical use Formulation DDD	Originator Approval (country/year) Production method [References]
Thiamine (thiamin, vitamin B$_1$; aneurin) [Beatine, Benerva, Betabion, Betaline, Bewon, Metabolin, Vitaneuron] A11DA01 **WHO essential medicine**		[*59-43-8*] C$_{12}$H$_{17}$ClN$_4$OS 300.81 hydrochloride: [*67-03-8*] C$_{12}$H$_{17}$ClN$_4$OS ·HCl 337.26 mononitrate: [*532-43-4*] C$_{12H}$H$_{17}$N$_5$O$_4$S 327.37	water-soluble vitamin of the so-called B-complex, an essential micronutrient; precursor of thiamine pyrophosphate (TPP), which is important for some reactions in carbohydrate metabolism (with pyruvate dehydrogenase, α-ketoglutarate dehydrogenase, and transketolase). deficiency of thiamine causes beriberi disease and Wernicke encephalopathy. amp. 25, 100, 200 mg; tbl. 10, 100, 250, 300 mg; powder.	first isolation by JANSEN and DONATH in 1926 [32] first total synthesis by WILLIAMS [33] USA/1937 synthesis (by Roche and Merck & Co.) [32–36]
Benfotiamine (BTMP; *S*-benzoyl-thiamine-*O*-monophosphate) [Benalgis, Benfogamma, Biotamin, Milgamma, Vitanevrin] A11DA03		[*22457-89-2*] C$_{19}$H$_{23}$N$_4$O$_6$PS 466.45	vitamin B$_1$ derivative for treatment of diabetic neuropathy. tbl. 50, 150, 300 mg.	Sankyo synthesis [37]
Fursultiamine (thiamine tetrahydrofurfuryl disulfide; TTFD) [Adventan, Alinamin F, Benlipoid, Diteftin, Lipothiamine, Judolor) A11DA04		[*804-30-8*] C$_{17}$H$_{26}$N$_4$O$_3$S$_2$ 398.54	vitamin B$_1$ derivative for treatment of vitamin B$_1$ deficiency (e.g., beriberi) and neurotropic pain. amp. 5, 10, 25, 50, 100 mg; tbl. 5, 25, 50 mg.	Takeda JP/1965 synthesis [38]
Thiamine diphosphate (cocarboxylase; thiamine pyrophosphate; ThPP; TPP) [Berolase, Bivitasi, Cocalose, Cocarvit, Nutrase, Pyrolase] A11DA06		[*154-87-0*] C$_{12}$H$_{19}$ClN$_4$ O$_7$P$_2$S 460.76	cofactor of several enzymes (e.g., pyruvate dehydrogenase and pyruvate decarboxylase). med. use in vitamin B$_1$ deficiency. amp. 25, 50 mg.	Roche synthesis isolation from yeast: [39–42]

4. A11G Ascorbic Acid (Vitamin C)

4.1. A11GA Ascorbic Acid (Vitamin C), Plain (Table 4)

Table 4. A11GA Ascorbic acid (vitamin C), plain

INN (Synonyms) [Brand names] ATC	Structure (Remarks)	[CAS No.] Formula MW, g/mol	Target (if known) Medical use Formulation DDD	Originator Approval (country/year) Production method [References]
Ascorbic acid (L-ascorbic acid; vitamin C; 3-oxo-L-gulofuranolactone; L-threo-hex-2-enonic acid γ-lactone; antiscorbutic vitamin) [Cebion, Hermes Cevitt, Pascorbin, Xitix, and numerous other brand names] A11GA01 **WHO essential medicine**		[50-81-7] $C_6H_8O_6$ 176.12 calcium salt: [5743-27-1] $C_{12}H_{14}CaO_{12}$ sodium salt: [134-03-2] $C_6H_7NaO_6$ 198.11	water-soluble vitamin with potent antioxidant activity, found in citrus fruits and green vegetables; not formed in human body and has to be added by diets. its deficiency causes scurvy; important for formation of collagen. tbl. 100, 200, 500 mg; f.c. tbl. 500, 1000 mg; effervescent tbl. 1 g; amp. 100 mg/1 mL; 750 mg/5 mL; 300 mg/2 mL. very important as food additive, mainly in order to retard oxidation and enzymatic browning.	first isolation: [43, 44] synthesis: [45] first industrial synthesis: [46] fermentation to convert D-sorbitol into 2-keto-L-gulonic acid: [47]

5. A11H Other Plain Vitamin Preparations

For vitamin K, phylloquinone, ATC number B02BA, please refer to → Antithrombotics (B01) and Antihemorrhagics (B02). For vitamin B_{12}, cobalamins, ATC number B03BA, please refer to the vitamin article "Vitamins, 13. Vitamin B_{12} (Cobalamins)" of Ullmann's Encyclopedia online [78].

5.1. A11HA Other Plain Vitamin Preparations (Table 5)

Table 5. A11HA Other plain vitamin preparations

INN (Synonyms) [Brand names] ATC	Structure (Remarks)	[CAS No.] Formula MW, g/mol	Target (if known) Medical use Formulation DDD	Originator Approval (country/year) Production method [References]
Nicotinamide (niacinamide; nicotinic acid amide; vitamin B_3; vitamin PP) [Nicobion, Papulex] A11HA01 **WHO essential medicine**		[98-92-0] $C_6H_6N_2O$ 122.13	constituent of the coenzyme NADH/NAD$^+$, which has important function in physiological oxidation–reduction reactions, e.g., glycolysis, citric acid cycle, and electron transport chain, in use for treatment of pellagra, which is caused by nicotinamide deficiency; important dietary supplement and additive for animal feed. cps. 500 mg; tbl. 200 mg.	isolation/discovery: [48] synthesis (hydrolysis of 3-cyanopyridine): [49, 50] 3-cyanopyridine process from 3-methylpyridine by ammonoxidation: [51]
Pyridoxine (vitamin B_6; adermin; pyridoxol) [Hexobion, Pyridox] A11HA02 The vitamin B_6 group comprises the biochemically interconvertible compounds: pyridoxine (large scale production), pyridoxine-5′-phosphate, pyridoxal, pyridoxal-5′-phosphate, pyridoxamine, and pyridoxamine-5′-phosphate. **WHO essential medicine**		[65-23-6] $C_8H_{11}NO_3$ 169.18 hydrochloride: [58-56-0] $C_8H_{11}NO_3 \cdot HCl$ 205.64	water-soluble vitamin, part of the B_6 complex is converted to pyridoxal 5-phosphate in the body, which acts as coenzyme for biosynthesis of amino acids, neurotransmitters, etc. It is found in fruits, vegetables, and grain. medical use for treatment and prevention of pyridoxine deficiency, sideroblastic anemia, pyridoxine-dependent epilepsy, and isoniazid-induced peripheral neuropathy. Combin. with doxylamine is used as treatment for morning sickness in pregnant women; also in use as additive for food and feed. tbl. 20, 40, 50 mg (as hydrochloride).f.c. tbl. 300 mg; amp. 25 mg/2 mL.	first isolation was achieved simultaneously by five groups in 1938 [2], and synthesis in 1939 by two independent groups [2]. 1) synthesis of Merck & Co.: [52] 2) [53] industrial syntheses (Diels–Alder reaction with oxazoles): [54, 55]. estimated world production: 7000–8500 t/a

(continued)

Table 5. (*Continued*)

INN (Synonyms) [Brand names] ATC	Structure (Remarks)	[CAS No.] Formula MW, g/mol	Target (if known) Medical use Formulation DDD	Originator Approval (country/year) Production method [References]
α-Tocopherol (vitamin E; D-α-tocopherol; (2R,4'R,8'R)-α-tocopherol; most bioactive naturally occurring form of vitamin E) [mainly used in pharmaceutical preparations as acetate. e.g., Optovit and Mowivit Vitamin E] A11HA03 The β-, γ-, and δ-tocopherols are also naturally occurring and are minor components in admixture with α-tocopherol and can be converted into the α-product by methylation processes.	α-tocopherol β-tocopherol γ-tocopherol δ-tocopherol	[59-02-9] $C_{29}H_{50}O_2$ 430.72 acetate: [58-95-7] $C_{31}H_{52}O_3$ 472.75 D,L-α-tocopherol: [10191-41-0] $C_{29}H_{50}O_2$ 430.72 acetate: [52225-20-4] $C_{31}H_{52}O_3$ 472.75 β-tocopherol: [16698-35-4] $C_{28}H_{48}O_2$ 416.69 γ-tocopherol: [54-28-4] $C_{28}H_{48}O_2$ 416.69 δ-tocopherol: [119-13-1] $C_{27}H_{46}O_2$ 402.66	antioxidant, the acetate is indicated for dietary supplementation in patients with vitamin E deficiency. numerous combination products. 0.2 g O. P.	first isolation from wheat germ by H. M. EVANS in 1936 [56]; structure in 1937 by E. FERNHOLZ [57]; Synthesis of D,L-α-tocopherol in 1938 by P. KARRER [58]. isolation of natural α-tocopherol from soy deodorizer distillates and synthesis of D,L-α-tocopherol by different routes (main producers are BASF, DSM, and Chinese companies). (all-*rac*-) or D,L-α-tocopherol is industrially produced in an amount of about 30 000 *t/a* and mainly used as feed additive and antioxidant in food. review literature: [1–3]

Riboflavin
(vitamin B$_2$;
lactoflavine; [lactoflavin]
[Beflavin, Flavaxin]
A11HA04
**WHO essential
medicine**

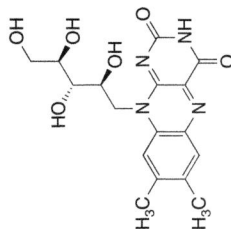

[83-88-5]
C$_{17}$H$_{20}$N$_4$O$_6$
376.37

water-soluble essential human nutrient, precursor, and constituent of the coenzymes FMN and FAD—both present in all living cells; yeast is the richest natural source. treatment of riboflavin deficiency; dietary supplement.
tbl. 1, 2, 5, 10, 100 mg; amp. 10 mg/2 mL; 20 mg/mL.

first isolation in 1933 by R. KUHN et al. [59]
structure and synthesis in 1935 by P. KARRER et al. [60] and 1934 by R. KUHN et al. [61, 62].
several techn. syntheses, e.g., by M. TISHLER et al. of Merck & Co. [63, 64], are of historical interest.
produced only by fermentation (BASF, DSM, Hubei Guangji Pharm.; in total ca. 400 t/a) [2].

Biotin
(vitamin B$_7$; vitamin H)
[Bio-H-Tin, Deacura, Natubiotin]
A11HA05

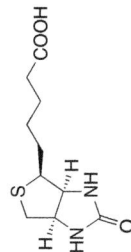

[58-85-5]
C$_{10}$H$_{16}$N$_2$O$_3$S
244.31

water-soluble B-vitamin, coenzyme for multiple carboxylase enzymes which are important for fatty acid synthesis, branched-chain amino acid catabolism, and gluconeogenesis.
treatment of biotin deficiency; component of vitamin comb. preparations; also in use as food and feed additive.
tbl. 2.5, 5, 10 mg.

first isolation from egg yolk in 1936 [65], structure determination in 1942 [66], and first total synthesis at Merck labs. [67]
actual world market 100 t/a.
review: [68]

Pyridoxal phosphate
(codecarboxylase;
coenzyme B$_6$; PLP; P5P)
[Pyromijin, Vitazechs]
A11HA06

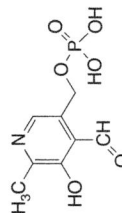

[54-47-7]
C$_8$H$_{10}$NO$_6$P
247.14

active coenzyme form of vitamin B$_6$, which is important for all-*trans*-amination reactions and L-amino acid decarboxylations.
cps. 50 mg.

first synthesis by GUNSALUS et al. in 1944
synthesis [69]

(continued)

Table 5. (*Continued*)

INN (Synonyms) [Brand names] ATC	Structure (Remarks)	[CAS No.] Formula MW, g/mol	Target (if known) Medical use Formulation DDD	Originator Approval (country/year) Production method [References]
Inositol (*myo*-inositol; *meso*-inositol) A11HA07		[87-89-8] $C_6H_{12}O_6$ 180.16 monophosphate: [573-35-3] $C_6H_{13}O_9P$	carbocyclic sugar alcohol, abundant in brain and other mammalian tissue, occurs widely in animals and plants (e.g. as phytic acid), and is considered as pseudovitamin. ingredient of OTC products. nutrient supplement in special dietary foods and infant formula.	first isolation in 1850 from heart muscle by SCHERER. produced from corn steep liquor [70, 71].
Tocofersolan (tocophersolan; TPGS, tocopheryl polyethylene glycol 1000 succinate) [Vedrop] A11HA08 orphan drug		[9002-96-4]	water-soluble form of D-α-tocopherol (vitamin E). treatment of vitamin E deficiency due to digestive malabsorption in pediatric patients with congenital chronic cholestasis or hereditary chronic cholestasis, from birth up to 18 years of age. oral sol. 50 mg/mL. 0.2 g O (as tocopherol)	Recordati Rare Diseases EU/2008 [72] Review: [73]
Dexpanthenol (panthenol; D-panthenol; pantothenol; pantothenyl alcohol) [Alcopan, Bepanthen, Cozyme, Ilopan, Intrapan, Pantenyl, Panthoderm] A11HA30 (Respiratory Disorders, 2. Nasal Preparations and Decongestants for Topical Use (R01))		[81-13-0] $C_9H_{19}NO_4$ 205.25	moisturizer and humectant, improves wound healing in pharmaceutical and cosmetic preparations. In organisms it is quickly oxidized to pantothenic acid, and so it acts as provitamin B_5. component in combination products.	Roche synthesis (from pantolactone and 3-hydroxypropylamine) [74]

Calcium pantothenate
(calpanate; pantholin)
A11HA31

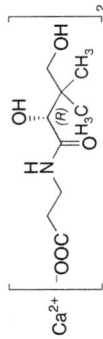

[137-08-6]
$C_{18}H_{32}CaN_2O_{10}$
476.54
pantothenic acid:
[79-83-4]
$C_9H_{17}NO_5$
219.24

calcium salt of water-soluble vitamin B_5, a component of coenzyme A (CoA) and a growth factor and plays a role in various metabolic functions.
besides medical use feed and food additive.

Merck & Co.; Roche; Upjohn synthesis [75, 76]

Pantethine
[Lipodel, Pantetina. Panthecin, Pantomin, Pantosin]
A11HA32

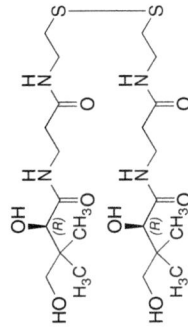

[16816-67-4]
$C_{22}H_{42}N_4O_8S_2$
554.72

lowering elevated LDL-cholesterol and triglycerides.
in use as dietary supplement in USA. Intermediate in CoA production.
cps. 300. 450. 600 mg.

Parke Davis
[77]

List of Abbreviations

amp.	ampoule(s)
ATC	Anatomical Therapeutic Chemical (classification system)
cps	capsule(s)
DDD	defined daily dose
f.c.	film coated
inj.	injection
IU	international unit
O	oral (DDD)
OTC	over-the-counter (drug)
P	parenteral (DDD)
tbl.	tablet(s)
TPP	thiamine pyrophosphate

References

1 Bonrath, W., et al. (2019) Vitamins, 1. Introduction, *Ullmann's Encyclopedia of Industrial Chemistry*, Wiley-VCH Verlag, Weinheim, Electronic Release, September.

2 Eggersdorfer, M., et al. (2012) *Angew. Chem.* **124**, 13134–13165.

3 Isler, O. and Brubacher, G. (1982) *Vitamine I Fettlösliche Vitamine*, G. Thieme Verlag, Stuttgart.

4 Isler, O., et al. (1988) *Vitamine II Wasserlösliche Vitamine*, G. Thieme Verlag, Stuttgart.

5 Devlin, T.M., (2006), *Textbook of Biochemistry: with Clinical Correlations*, Wiley-Liss, 2006; Chaney, S.G. Chap. 28, p. 1096–1120: Principles of Nutrition II: Micronutrients.

6 McCollum, E.V. and Davis, M. (1913) *J. Biol. Chem.* **15**, 167.

7 Roche:Isler, O., et al. (1947) *Helv. Chim. Acta* **30**, 1911; (1949) **32**, 489.

8 BASF:Pommer, H. (1977) *Angew. Chem.* **89**, 437.

9 (a) Karrer, P., et al. (1929) *Helv. Chim. Acta* **12**, 1142. (b) Karrer, P., et al. (1930) *Helv. Chim. Acta* **13**, 1084.

10 Isler, O. (1956) *Angew. Chem.* **68**, 547.

11 Roche, (1959) US 2 917 539, CH-prior.1955.

12 Pommer, H. (1960) *Angew. Chem.* **72**, 911.

13 Steenbock, H. and Nelson, T. (1924) *J. Biol. Chem.* **62**, 209.

14 Fischer, M. (1978) *Angew. Chem.* **90**, 17.

15 UCLAF, (1954) US 2 693 475, FR-prior.1949.

16 UCLAF, (1956) US 2 757 182, FR-prior.1949.

17 Winthrop, (1941) US 2 228 491, DE-prior.1938.

18 Wisconsin Alumni, (1973) US 3 741 996, USA-prior.1971.

19 Wisconsin Alumni, (1975) US 3 929 770, USA-prior.1973.

20 Wisconsin Alumni, (1977) US 4 022 768, USA-prior.1976.

21 Wisconsin Alumni, (1980) US 4 195 027, USA-prior.1978.

22 Holick, M.F., et al. (1971) *Biochemistry* **10**, 2799.

23 Wisconsin Alumni, (1972) US 3 697 559, USA-prior.1971.

24 Res. Inst. Med. and Chem. Cambridge/Hesse, R., (1988), US 4 772 433, GB-prior.1981.

25 Roche, (1982) US 4 310 467, USA-prior.1980.

26 UCLAF, (1955) US 2 707 710, FR-prior.1949.

27 Nisshin Flour Milling, (1972) US 3 661 939, JP-prior.1969.

28 Ponchon, G., et al. (1969) *J. Clin. Invest.* **48**, 1273.

29 Wisconsin Alumni, (1971) US 3 565 924, USA-prior.1968.

30 Blunt, J.W. and DeLuca, H.F. (1969) *Biochemistry* **8**, 671.

31 Upjohn, (1974) US 3 833 622, USA-prior.1969.

32 Jansen, B.C.P. and Donath, W.F. (1926) *Chem. Weekbl.* **23**, 201.

33 Williams, R.R. and Cline, J.K. (1936) *J. Am. Chem. Soc.* **58**, 1504.

34 Todd, A.R. and Bergel, F. (1937) *J. Chem. Soc.* **364**.

35 Todd, A.R. and Bergel, F. (1937) *J. Chem. Soc.* **26**.

36 Andersag, H. and Westphal, K. (1937) *Ber. Dtsch. Chem. Ges. A* **70**, 2035.

37 Sankyo, (1962) DE 1 130 811, JP-prior.1959.

38 Takeda, (1962) US 3 016 380, JP-prior.1957.

39 Lohmann, K. and Schuster, P. (1937) *Biochem. Z.* **294**, 183.

40 Weijlard, J. and Tauber, H. (1938) *J. Am. Chem. Soc.* **60**, 2263.

41 Merck & Co., (1940) US 2 188 323, USA-prior.1938.

42 E. Merck AG, (1961) US 2 991 284, DE-prior.1957.

43 Szent-Györgyi, A. (1928) *Biochem. J.* **22**, 1387.

44 Haworth, W.N. and Szent-Györgyi, A. (1933) *Nature* **131**, 24.

45 Reichstein, T., et al. (1933) *Helv. Chim. Acta* **16**, 1019.

46 Roche: Reichstein, T. and Grüssner, A. (1934) *Helv. Chim. Acta* **17**, 311, 510.

47 Roche, (1997) EP 518 136; EP-prior.1991.

48 von Euler, H., et al. (1936) *Z. Physiol. Chem.* **237**, 1801; (1936), *Z. Physiol. Chem.* **240**, 113.

49 Distillers Comp., (1959), US 2 904 552; GB-prior.1955.

50 Shimizu, S., et al. (2000) Pyridine and Pyridine Derivatives, *Ullmann's Encyclopedia of Industrial Chemistry*, Wiley-VCH Verlag, Weinheim, Electronic Release, June.

51 Beschke, H., et al. (1977) *Chem. Ztg.* **101**, 384.

52 Harris, S.A. and Folkers, K. (1939) *J. Am. Chem. Soc.* **61**, 1245, 3307.

53 Kuhn, R. and Westphal, O. (1939) *Naturwissenschaften* **27**, 469.

54 Bonrath, W. et al., (2020) Vitamins, 10. Vitamin B_6, *Ullmann's Encyclopedia of Industrial Chemistry*, Wiley-VCH Verlag, Weinheim, Electronic Release, January.

55 Eggersdorfer, M., et al. (2012) *Angew. Chem. Int. Ed.* **51**, 12960.

56 Evans, H.M., et al. (1936) *J. Biol. Chem.* **113**, 319.

57 Fernholz, E. (1937) *J. Am. Chem. Soc.* **59**, 1154; (1938), J. Am. Chem. Soc. **60**, 700.

58 Karrer, P., et al. (1938) *Helv. Chim. Acta* **21**, 520, 820.

59 Kuhn, R., et al. (1933) *Ber. Dtsch. Chem. Ges.* **66**, 576–580.

60 Kuhn, R., et al. (1935) *Helv. Chim. Acta* **18**, 426–429.

61 Kuhn, R., et al. (1934) *Ber. Dtsch. Chem. Ges.* **67**, 1125–1130.

62 Kuhn, R., et al. (1935) *Ber. Dtsch. Chem. Ges.* **68**, 625–634.

63 Tishler, M., et al. (1947) *J. Am. Chem. Soc.* **69**, 1487–1492.

64 Merck & Co, (1944) US 2 350 376, USA-prior.1941.

65 Kögl, F. and Tönnis, B. (1936) *Hoppe Seyler's Z. Physiol. Chem.* **242**, 43.

66 du Vigneaud, V., et al. (1942) *J. Biol. Chem.* **146**, 475; 487.

67 Harris, J.A., et al. (1943) *Science* **97**, 447.

68 De Clercq, P.J. (1997) Biotin : A Timeless Challenge for Total Synthesis, *Chem. Rev.* **97**, 1755–1792.

69 Gunsalus, I.C., et al. (1944) *J. Biol. Chem.* **155**, 685; J. Biol. Chem. **161**, 311; J. Biol. Chem. **161**, 743.

70 American Cyanamid, (1947) US 2 414 365, USA-prior.1942.

71 Hoglan, F.A. and Bartow, E. (1940) *J. Am. Chem. Soc.* **62**, 2397.

72 Eastman Kodak, (1954) US 2 680 749, USA-prior.1951.

73 Yang, C., et al. (2018) *Theranostics* **8**, 464–485.

74 Roche, (1946) US 2 413 077, CH-prior.1942.

75 Stiller, E.T., et al. (1940) *J. Am. Chem. Soc.* **62**, 1785–1790.

76 Upjohn, (1958) US 2 845 456, USA-prior.1954.

77 Wittle, E.L., et al. (1953) *J. Am. Chem. Soc.* **75**, 1694.

78 Hohmann, H.-P., et al. (2020) Vitamins, 13. Vitamin B_{12} (Cobalamins), *Ullmann's Encyclopedia of Industrial Chemistry*, Wiley-VCH Verlag, Weinheim, Electronic Release, March.

Anabolic Agents for Systemic Use (A14)

AXEL KLEEMANN, Hanau, Germany

ULLMANN'S ENCYCLOPEDIA OF INDUSTRIAL CHEMISTRY

1. Introduction

In the following chapters, the different drug groups are reviewed and arranged according to their Anatomical Therapeutic Chemical Classification (ATC) of the WHO, in which they are arranged according to their indication and chemical structure [1].

Since its isolation and first synthesis in 1935, the principal secreted androgen and male sex hormone testosterone is clinically used in varied cases of androgen deficiency and potency disorders [2a]. Besides its androgen activity, testosterone also shows anabolic effects, i.e., stimulation of protein synthesis and muscle growth. In the 1950s and 1960s, numerous synthetic variations of the testosterone structure were undertaken in order to develop drugs with more anabolic and less androgenic properties for treatment of different catabolic conditions [3]. These anabolics led to excessive use by bodybuilders, weightlifters, and other athletes with the aim to increase physical performance in sports. Within a rather short period, heavy misuse of these anabolic androgenic steroids (AAS) in sports (doping) and private life situations became evident, and most of the marketed anabolic drugs were withdrawn by the pharmaceutical companies. Medical use of the AAS was rather limited and commercially insignificant, especially after they became controlled substances in USA and many other countries. In addition, they exert serious adverse side effects, e.g., on heart and liver. But this does not mean that these drugs are not available without prescription; a simple search in the internet demonstrates the contrary. In many countries, possession of or dealing with AAS is illegal because of the abuse potential. All AAS are on the list of the World Anti-Doping Agency (WADA) in order to prohibit their use in sports [2b].

The anabolic steroids are classified in the ATC list under A14; they are listed in Tables 1 and 2 with the exception of quinbolone (A14AA06) and oxabolone cipionate (A14AB03), which did and do not play any commercial role. Testosterone itself is classified under G03BA03 as androgen, see → Genitourinary System and Sex Hormones (G)

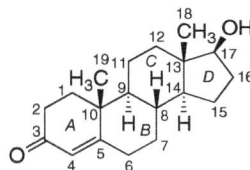

Testosterone [*58-22-0*]; for steroid structure numbering.

The APIs in Tables 1 and 2 have the following underlying structural modifications of the testosterone molecule.

- alkylation in 17α-position: metandienone, stanozolol, oxymetholone, oxandrolone, norethandrolone, chlorodehydromethyltestosterone, and ethylestrenol
- demethylation in position 19: nandrolone and ethylestrenol
- 5α-reduction: androstanolone, stanozolol, metenolone, oxymetholone, prasterone, and oxandrolone

Ullmann's Pharmaceuticals. Edited by Axel Kleemann and Bernhard Kutscher.
© 2022 Wiley-VCH GmbH
Set ISBN: 978-3-527-34252-5/ DOI: 10.1002/14356007.w02_w05

The two listed drugs, quinbolone (A14AA06) and oxaboloncipionate (A14AB03), did not play any role in the past and so were not included in Tables 1 and 2.

An excellent and exhaustive overview about all aspects of AAS is given in Wikipedia [4].

2. A14A Anabolic Steroids

2.1. A14AA Androstan Derivatives (Table 1)

Table 1. A14AA Androstan derivatives

INN (Synonyms) [Brand names] ATC	Structure (Remarks)	[CAS-No.] Formula MW, g/mol	Target (if known) Medical use Formulation DDD	Originator Approval (country/year) Production method [References]
Androstanolone (stanolone; androstan-17β-ol-3-one; DHT; 4-dihydrotestosterone) [Anabolex, Andractim, AndroGel-DHT, Androlone, Apeton, Gelovit, Neotrol, Ophtovital, Pesomax, Stanaprol] A14AA01; G03BB02 USA controlled substance (schedule III, CSA) WADA antidoping list (→ Genitourinary System and Sex Hormones (G)) wfm in many countries		[521-18-6] $C_{19}H_{30}O_2$ 290.45	androgen; anabolic steroid; biologically most active form of testosterone. medical use for treatment of low testosterone levels in men. hydroalcoholic gel 2.5%; tbl. sublingual 25 mg.	Schering Corp. USA/1953 synthesis [5–7]
Stanozolol (androstanazole; stanazol; NSC-43193; Win-14833) [Stromba, Strombaject, Winstrol] A14AA02 controlled substance (schedule III, CSA) WADA antidoping list wfm in many countries		[10418-03-8] $C_{21}H_{32}N_2O$ 328.50	androgen; anabolic steroid. med. use to promote weight gain in cachexia and debilitating diseases. tbl. 2, 5, 15 mg. 5 mg O, 3.5 mg P.	Sterling Drug/Winthrop USA/1962; DE/1962 synthesis (from mestanolone) [8–10]
Metandienone (methandienone; methandrostenolone; 1-dehydro-17α-methyltestosterone) [Danabol, Dianabol, Dianavit, Nabolin, Nerobol, Stenolon] A14AA03 controlled substance (schedule III, CSA) WADA antidoping list wfm in many countries		[72-63-9] $C_{20}H_{28}O_2$ 300.44	weak androgenic, strong anabolic steroid; med. use to promote weight gain in cachexia and debilitating diseases. tbl. 2 mg, 5 mg; ointment 10 mg/g. 5 mg O.	Ciba USA/1958 synthesis (from methyltestosterone) [11, 12]

(continued)

Table 1. (Continued)

INN (Synonyms) [Brand names] ATC	Structure (Remarks)	Target (if known) Medical use Formulation DDD	[CAS-No.] Formula MW, g/mol	Originator Approval (country/year) Production method [References]
Metenolone (methenolone; methylandrostenolone) [Primobolan/-Depot, Nibal, Primonabol] A14AA04 controlled substance (schedule III, CSA) WADA antidoping list wfm in many countries		androgen; anabolic steroid. treatment of anemia due to bone marrow failure, wasting syndromes, e.g., after surgery, osteoporosis, and sarcopenia. 17-acetate (Primobolan): amp. 20, 50, 100 mg/1 mL; tbl. 5 mg, 25 mg; 17-enanthate (Primobolan-depot): amp. 100 mg.	[153-00-4] $C_{20}H_{30}O_2$ 302.46 17-acetate: [434-05-9] $C_{22}H_{32}O_3$ 344.50 17-enanthate: [303-42-4] $C_{27}H_{42}O_3$ 414.63	Schering AG USA/1962 (enanthate) synthesis (from androstenolone) [13, 14]
Oxymetholone (hydroxymetholone) [Adroyd, Anapolon, Anadrol, Pardroyd, Plenastril, Protanabol, Nastenon, Synasteron] A14AA05 controlled substance (schedule III, CSA) WADA antidoping list wfm in many countries		androgen; anabolic steroid. treatment of anemias caused by deficient red cell production. tbl. 5, 10, 50 mg.	[434-07-1] $C_{21}H_{32}O_3$ 332.48	Syntex USA/1961; UK/1961 synthesis (from mestanolone) [15, 16]
Prasterone (dehydroepiandrosterone; DHEA: androstenolone) [Aylistormer, Diandrone, Gynodian, Intrarosa] A14AA07: G03EA03; G03XX01 USA: no controlled substance (OTC medicine!) WADA antidoping list (→ Genitourinary System and Sex Hormones (G))		endogenous steroid hormone precursor for androgen as well as for estrogen sex steroid hormones; most abundant circulating steroid in humans. treatment of DHEA deficiency (e.g., in older people); androgen in menopausal hormone therapy (with enanthate), atrophic vaginitis, and cervical dilation during childbirth (with sodium sulfate). vaginal suppos. 6.5 mg (prasterone); cps. 200 mg (Diandrone): amp. 200 mg enanthate + 4 mg estradiol valerate/1 mL (Gynodian).	[53-43-0] $C_{19}H_{28}O_2$ 288.43 enanthate: [23983-43-9] $C_{26}H_{40}O_3$ 400.60 sulfate: [651-48-9] $C_{19}H_{28}O_5S$ 368.49 sodium sulfate: [1099-87-2] $C_{19}H_{27}NaO_5S$ 390.47	discovery and isolation from male urine by BUTENANDT in 1934 [17] and synthesis from cholesterol in 1935 independently by BUTENANDT [18] and RUZICKA [19] DE/1978; USA/1980 (OTC) USA/2016 (Intrarosa) synthesis (from diosgenin via 16-dehydropregnenolon acetate) [17–23].

Name	Structure	CAS / Formula / MW	Description	Origin
Oxandrolone (SC-11585) [Anatrophill, Anavar, Antitriol, Lonavar, Oxandrin, Protavar, Vasorome] A14AA08 controlled substance (schedule III, CSA) WADA antidoping list	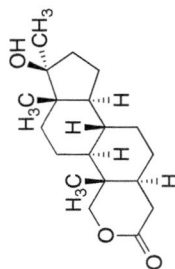	[53-39-4] $C_{19}H_{30}O_3$ 306.45	weak androgen; strong anabolic steroid. adjunctive therapy to promote weight gain after weight loss due to surgery, chronic infections, and severe trauma. tbl. 2.5 mg, 10 mg.	Searle USA/1964 synthesis [24, 25]
Norethandrolone (CB-822; ethylnandrolone; ethylnortestosterone) [Nilevar, Pronabol] A14AA09 controlled substance (schedule III, CSA) wfm in most countries	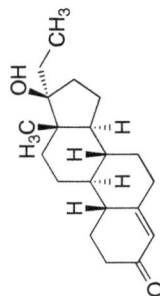	[52-78-8] $C_{20}H_{30}O_2$ 302.46	androgen; anabolic steroid. was used for treatment of muscle wasting. tbl. 10 mg.	Searle USA/1956 synthesis (selective hydrogenation of norethisterone) [26–28]
Chlorodehydromethyltestosterone (CDMT; 4-chloro-17β-hydroxy-17α-methylandrosta-1,4-dien-3-one) [Oral Turinabol] A14AA10	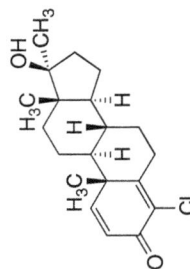	[2446-23-3] $C_{20}H_{27}ClO_2$ 334.88	androgenic–anabolic steroid. was "famous" doping substance for East German athletes.	Jenapharm GDR (DDR)/1965 synthesis production stop in 1994. [29]
Clostebol (4-chlorotestosterone) [Macrobin, Megagrisevit-Mono, Steranabol] A14AA11 controlled substance (schedule III, CSA) WADA antidoping list wfm in Germany in 2003	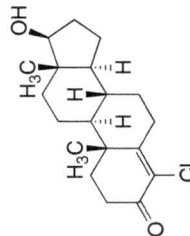	[1093-58-9] $C_{19}H_{27}ClO_2$ 322.87 acetate: [855-19-6] $C_{21}H_{29}ClO_3$ 364.91	androgen; anabolic steroid. was used for treatment of testosterone deficiency. misuse in body building and for doping in sports. tbl. 15 mg; vials 10 mg/1.5 mL; cream 0.5%.	Farmitalia synthesis (from testosterone) [30]

2.2. A14AB Estren Derivatives (Table 2)

Table 2. A14AB Estren derivatives

INN (Synonyms) [Brand names] ATC	Structure (Remarks)	[CAS-No.] Formula MW, g/mol	Target (if known) Medical use Formulation DDD	Originator Approval (country/year) Production method [References]
Nandrolone (19-nortestosterone) [Deca-Durabolin, Durabolin, Dynabolon, Duramin, Keratyl, Laurabolin, Nortestonate, Retabolil, Superanabolon] A14AB01 controlled substance (schedule III, CSA) WADA antidoping list		[*434-22-0*] $C_{18}H_{26}O_2$ 274.40 decanoate: [*360-70-3*] $C_{28}H_{44}O_3$ 428.66 dodecanoate: [*26490-31-3*] $C_{30}H_{48}O_3$ 456.71 phenpropionate: [*62-90-8*] $C_{27}H_{34}O_3$ 406.57	androgen and anabolic steroid; application in the form of esters. treatment of patients in catabolic state; management of anemia of renal insufficiency. decanoate: vials of 2 mL with 100 mg/mL; vials of 1 mL with 200 mg/mL; amp. 25, 50, 100 mg/mL.	Organon USA/1959 synthesis (from estradiol-3-methyl ether/Birch reduction) [31, 32]; esters: [33]
Ethylestrenol (äthylestrenol) [Orgaboline, Orabolin, Maxibolin] A14AB02 controlled substance (schedule III, CSA) wfm in all countries		[*965-90-2*] $C_{20}H_{32}O$ 288.48	androgen and anabolic steroid. was used for promotion of weight gain and treatment of anemia; only limited use as veterinary drug. tbl. 2 mg.	Organon USA/1964 synthesis (e.g., from nandrolone) [34, 35]

List of Abbreviations

AAS	anabolic–androgenic steroid
amp.	ampoule
API	active pharmaceutical ingredient
ATC	Anatomical Therapeutic Chemical Classification
cps.	capsule
CSA	United States Controlled Substances Act
OTC	over-the-counter (available without prescription)
tbl.	tablet
WADA	World Anti-Doping Agency
wfm	withdrawn from market
WHO	World Health Organization

References

1 WHO Collaborating Centre for Drug Statistics Methodology (2019) *Guidelines for ATC Classification and DDD Assignment*, WHO Collaborating Centre for Drug Statistics Methodology, Oslo, Norway.

2 (a) Nieschlag, E. and Behre, H.M. (1998) Pharmacology and Clinical Uses of Testosterone, in *Testosterone – Action, Deficiency, Substitution*, (eds. E. Nieschlag and H.M. Behre), 2nd edn, Springer Verlag, Berlin Heidelberg, pp. 293–328. (b) Schänzer, W. (1998) Abuse of Androgens and Detection of Illegal Use, in *Testosterone – Action, Deficiency, Substitution*, (eds. E. Nieschlag and H.M. Behre), 2nd edn, Springer Verlag, Berlin Heidelberg, pp. 545–565.

3 Chapter 41, Androgens – Testosterone and Other Androgens (2011) in Goodman & Gilman's The Pharmacological Basis of Therapeutics (ed. P.J. Snyder), 12th edn, McGraw-Hill, London, pp. 1195–1207.

4 Wikipedia: Anabolic steroid: https://en.wikipedia.org/wiki/Anabolic_steroid (accessed 28 December 2020).

5 Butenandt, A., et al. (1935) Hydrogenation of Testosterone, *Chem. Ber.* **68**, 2097.

6 Ruzicka, L., et al. (1937) From dehydroepiandrosterone, *Helv. Chim. Acta* **20**, 1557.

7 From 3,17-androstandione: Schering Corp., (1960) US 2 927 921, USA-prior.1952, 1954.

8 Clinton, R.O., et al. (1959) *J. Am. Chem. Soc.* **81**, 1513.

9 Clinton, R.O., et al. (1961) *J. Am. Chem. Soc.* **83**, 1478.

10 Sterling Drug, (1962) US 3 030 358, USA-prior.1959.

11 Vischer, F., et al. (1955) *Helv. Chim. Acta* **38**, 1502.

12 Ciba, (1959) US 2 900 398, CH-prior.1956.

13 Schering AG, (1958) DE 1 023 764, DE-prior.1957.

14 Schering AG, (1961) DE 1 152 100, DE-prior.1960.

15 Ringold, H.J., et al. (1959) *J. Am. Chem. Soc.* **81**, 427.

16 Syntex, (1959) DE 1 070 632; Mex-prior.1956.

17 Butenandt, A. and Tscherning, K. (1934) *Z. Physiol. Chem.* **229** (167), 192.

18 Butenandt, A., et al. (1935) *Z. Physiol. Chem.* **237**, 57.

19 Ruzicka, L. and Wettstein, A. (1935) *Helv. Chim. Acta* **18**, 986.

20 Parke, D. (1943) US 2 335 616, USA-prior.1941.

21 Farbwerke Hoechst, (1957) US 2 783 252, DE-prior.1953.

22 Schmidt-Thomé, J. (1955) *Chem. Ber.* **88**, 895.

23 Rosenkranz, G., et al. (1956) *J. Org. Chem.* **520**, 21.

24 Searle, (1964) US 3 128 283, Mex-prior.1961.

25 Pappo, R. and Jung, C.J. (1962) *Tetrahedron Lett.* **9**, 365.

26 Colton, F.B., et al. (1957) *J. Am. Chem. Soc.* **79**, 1123.

27 Searle, (1955) US 2 721 871, USA-prior.1954.

28 Searle, (1954) US 2 691 028, USA-prior.1952.

29 Schwarz, S., et al. (1999) The Steroid Story of Jenapharm, *Steroids* **64**, 439–445.

30 Farmitalia, (1960) US 2 953 582, IT-prior.1956.

31 Wilds, A.L. and Nelson, N.A. (1953) *J. Am. Chem. Soc.* **75**, 5366.

32 Syntex, (1956) US 2 774 777, USA-prior.1952.

33 Organon, (1961) US 2 998 423, NL-prior.1958.

34 Organon, (1959) US 2 878 267, NL-prior.1957.

35 Organon, (1963) US 3 112 328, NL-prior.1956.

Other Alimentary Tract and Metabolism Products (A16)

AXEL KLEEMANN, Hanau, Germany

ULLMANN'S ENCYCLOPEDIA OF INDUSTRIAL CHEMISTRY

1. Introduction

In the following chapters, the different drug groups are reviewed and arranged according to their Anatomical Therapeutic Chemical Classification (ATC) of the WHO in which they are arranged according to their indication and chemical structure [1]. It has to be mentioned that only drug substances (APIs; active pharmaceutical ingredients) with clearly defined structures are referenced here. This is the reason why the chapter A16AB, comprising enzymes, has been omitted.

This group A16 comprises all products acting on the alimentary tract and metabolism which cannot be classified in the preceding subgroups of ATC Group A; see compilation below. A16AA contains amino acids and derivatives, which are medically used in various metabolic deficiency states. The drugs of group A16AX build a mixture of partly older established drugs (e.g., thioctic acid) and mostly of orphan drugs for rare diseases with quite different indications and chemical structures.

Interlinks to preceding subgroups of ATC Group Alimentary Tract and Metabolism (A):

→ Stomatological Preparations (A01)
→ Drugs for Acid-Related Disorders (A02)
→ Drugs for Functional Gastrointestinal Disorders (A03)
→ Antiemetics and Antinauseants (A04)
→ Bile and Liver Therapy (A05)
→ Drugs for Constipation (A06)
→ Antidiarrheals, Intestinal Antiinflammatory and Antiinfective Agents (A07)
→ Anti-Obesity Drugs (A08)
→ Antidiabetic Drugs (A10)
→ Anabolic Agents for Systemic Use (A14)

For vitamins (ATC Group A11), Ullmann's Encyclopedia provides several articles, which are not arranged according to the ATC classification, and included interlinks to articles on single vitamins.

Ullmann's Pharmaceuticals. Edited by Axel Kleemann and Bernhard Kutscher.
© 2022 Wiley-VCH GmbH
Set ISBN: 978-3-527-34252-5/ DOI: 10.1002/14356007.w18_w02

2. A16AA Amino Acids and Derivatives (Table 1)

Table 1. A16AA Amino acids and derivatives

INN (Synonyms) [Brand names] ATC	Structure (Remarks)	[CAS-No.] Formula MW, g/mol	Target (if known) Medical use Formulation DDD	Originator Approval (country/year) Production method [References]
Levocarnitine (L-carnitine; (2R)-3-carboxy-2-hydroxypropyl)trimethyl-ammonium hydroxide) [Biocarn, L-Carn, Carnitor, Entomin, Levocarnil, Nefrocarnit] A16AA01		[541-15-1] $C_7H_{15}NO_3$ 161.20 hydrochloride: [6645-46-1] $C_7H_{15}NO_3 \cdot HCl$ 197.66 D,L-carnitine: [406-76-8] D,L-hydrochloride: [461-05-2]	natural (physiological) substance that the body needs for energy, essential cofactor of fatty acid metabolism. treatment of carnitine deficiency, e.g., in the case of dialysis. amp. 200 mg/mL for i.v. inj.; oral sol. 1 g/10 mL; tbl. 330 mg: cps. 250, 330 mg; syrup 1 g/3.3 mL. 2 g O, P.	Sigma-Tau USA/1985 synthesis or fermentation; enantioselective synthesis from glycerol: [1] enzymatic synthesis from γ-butyrobetaine: [2] synthesis from (S)-3-hydroxybutyrolactone: [3]
S-Adenosyl methionine (ademetionine; SAMe) [Donamet, Gumbaral, S-Amet, Samyr, Transmetil] A16AA02		[29908-03-0] $C_{15}H_{22}N_6O_5S$ 398.44 1,4-butanedisulfonate (Donamet): [200393-05-1] $2 C_{15}H_{22}N_6O_5S$ $\cdot 3 C_4H_{10}S_2O_6$ 1451.6 disulfate ditosylate (Gumbaral): [55722-12-8] $C_{15}H_{22}N_6O_5S$ $\cdot 2 C_7H_8O_3S$ $\cdot 2 H_2SO_4$ 938.98	physiologic methyl group donor in enzymatic transmethylation reactions. medical use as antirheumatic drug in osteoarthritis and antidepressant and has been used as hepatoprotectant; dietary supplement in USA. f.c. tbl. 200 mg ademetionine (= 384 mg disulfate ditosylate); tbl. 500 mg butanedisulfonate).	Bioresearch IT/1979; DE/1989; USA/1999 (dietary suppl.) fermentation (of yeast with addition of methionine) [4, 5]
Glutamine (L-glutamine; Gln; Q; glutamic acid 5-amide) [Cebrogen, Endari, Glumin, Levoglutamina, Stimulina] A16AA03		[56-85-9] $C_5H_{10}N_2O_3$ 146.15	proteinogenic nonessential α-amino acid with a variety of physiological functions. used as medical food in the case of certain catabolic states, is a source of energy for the nervous system. medical use to reduce the acute complications of sickle cell disease. nutraceutical, micronutrient. pack with oral powder 5 g.	first isolation from beet roots in 1877 by E. SCHULZE fermentation (by *Brevibacterium flavum*) or isolation from sugar beet juice. reviews: [6] numerous patents, e.g., [7]

Cysteamine

(mercaptamine;
2-aminoethanethiol, L-1573)
[Cystagon, Cystaran,
Procysbi, Cystadrops,
Dropcys]
A16AA04; S01XA21

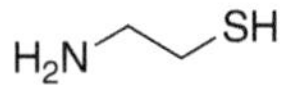

H_2N ⁀ SH

[60-23-1]
C_2H_7NS
77.15
hydrochloride:
[156-57-0]
$C_2H_7NS·HCl$
113.60
bitartrate:
[27761-19-9]
$C_2H_7NS·C_4H_6O_6$
227.23

biosynthesis by decarboxylation of cysteine; component of coenzyme A. treatment of nephropathic cystinosis, cystinuria, and corneal cysteine crystal accumulation in cystinosis patients. antidote to acetaminophen.
cps. 50, 150 mg free base (as bitartrate); sol. for ophthalmic use 0.44%.

Chiesi
USA/1994; DE/1997
synthesis (from aziridine and H_2S): [8, 9]
enterically coated cysteamine: [10]

Carglumic acid

(carbamylglutamic acid;
ureidoglutaric acid)
[Carbaglu, Ucedane]
A16AA05

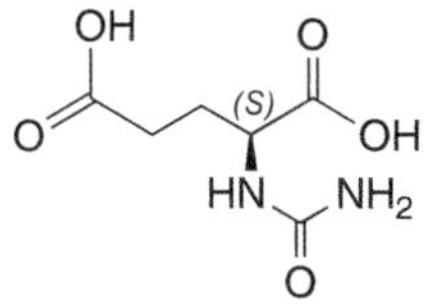

[structure: glutamic acid backbone, labeled (S), with OH, O, HN, NH₂ groups]

[1188-38-1]
$C_6H_{10}N_2O_5$
190.16

activator of carbamylphosphate synthetase (urea cycle).
treatment of hyperammonemia in patients with N-acetylglutamate synthase deficiency.
tbl. 200 mg.

Recordati
USA/2010; EU/2017
synthesis [11]

3. A16AX Various Alimentary Tract and Metabolism Products (Table 2)

Table 2. A16AX Various alimentary tract and metabolism products

INN (Synonyms) [Brand names] ATC	Structure (Remarks)	[CAS-No.] Formula MW, g/mol	Target (if known) Medical use Formulation DDD	Originator Approval (country/year) Production method [References]
Thioctic acid (α-lipoic acid; 6.8-thioctic acid) [Alpan, Alpha-Lipogamma, α-Vibolex, Biletan, Thioctacid, Thioctan, Tioctan, Thiogamma, Unilipon] A16AX01	(D-form)	racemate (D,L form): [1077-28-7] $C_8H_{14}O_2S_2$ 206.32 d form, ((R) form): [1200-22-2]	cofactor of pyruvate dehydrogenase complex and some other enzymes; many foods contain thioctic acid bound to lysine in proteins. medical use for treatment of diabetic neuropathy; in USA OTC as nutritional supplement and antioxidant. f.c. tbl. 200, 600 mg; cps. 300, 600 mg; vials 50 mL with sol. for infusion 600 mg; amp. 24 mL with sol. for infusion 600 mg (all formulations contain racemate). 0.6 g O, P.	L. J. Reed/Research Corp.; Lederle Labs. DE/1966 synthesis [12–14]
Anethole trithione (ADT; 3-(p-anisyl)trithione; trithioanethole) [Felviten, Mucinol, Stalor, Sulfarlem] A16AX02		[532-11-6] $C_{10}H_8OS_3$ 240.35	choleretic agent (stimulates bile secretion), is also used in the treatment of xerostomia. f.c. tbl. 12.5, 25 mg.	B. Böttcher (Germany) synthesis (from anethole and sulfur) [15–18]
Sodium phenylbutyrate (4-phenylbutyric acid sodium salt) [Ammonaps, Buphenyl, Pheburane] A16AX03 orphan drug		[1716-12-7] $C_{10}H_{11}NaO_2$ 186.19	treatment of hyperammonemia in patients with urea cycle disorders (genetic enzyme deficiencies), elimination of excess nitrogen. The drug is also an inhibitor of class I and II histone deacetylases. tbl. 5 g; granules 483 mg/g. 20 g O.	Eurocept International/Ucyclyd EU/2013; USA/1996 synthesis [19]
Nitisinone (NTBC) [Orfadin, Nityr] A16AX04 orphan drug		[104206-65-7] $C_{14}H_{10}F_3NO_5$ 329.23	reversible inhibitor of 4-hydroxyphenylpyruvate dioxygenase (HPPD). treatment of hereditary tyrosinemia type 1 (HT-1) as an adjunct to dietary restriction of tyrosine and phenylalanine. cps. 2, 5, 10, 20 mg. 20 mg O.	Swedish Orphan International EU/2005; USA/2002 synthesis [20–22]

Zinc acetate
[Galzin, Wilzin]
A16AX05

H₃C structure (zinc acetate)

[557-34-6]
$C_4H_6O_4Zn$
183.50
dihydrate:
[5970-45-6]
$C_4H_6O_4Zn \cdot 2 H_2O$
219.53

zinc is an essential element (food supplement); astringent. treatment of common cold; as daily supplement it inhibits the absorption of copper and is used for treatment of Wilson's disease.
cps. 25, 50 mg (as zinc).
0.15 g O.

synthesis
[23]

Miglustat
(OGT-918; SC-48334;
N-butylmoranoline)
[Zavesca]
A16AX06

structure

(2*R*,3*R*,4*R*,5*S*)-1-butyl-2-(hydroxymethyl)-3,4,5-piperidinetriol
(*N*-butyldeoxynojirimycin)

[72599-27-0]
$C_{10}H_{21}NO_4$
219.28

competitive and reversible inhibitor of glucosylceramide synthase. treatment of Gaucher disease type I (GD 1) and progressive neurological complications in people with Niemann–Pick disease.
tbl. 100 mg.
0.3 g O.

Oxford GlycoSciences;
Actelion
EU/2002; USA/2003
synthesis
[24, 25]

Sapropterin
(6*R*)-L-*erythro*-tetrahydro-biopterin; dapropterin; *R*-THBP; SUN-0588)
[Biopten, Kuvan]
A16AX07

structure

[62989-33-7]
$C_9H_{15}N_5O_3$
241.25
dihydrochloride:
[69056-38-8]
$C_9H_{15}N_5O_3$
$\cdot 2$ HCl
314.17

cofactor for hydroxylase enzymes (tryptophan, phenylalanine, and tyrosine hydroxylases) and nitric oxide synthase and also ether lipid oxidase.
medical use for patients with tetrahydrobiopterin deficiency and in phenylketonuria along with dietary measures.
tbl. 100 mg.
0.7 g O.

BioMarin
EU/2008; USA/2007
synthesis (hydrogenation of biopterin)
[26]
medical use for phenylketonuria:
[27, 28]

Teduglutide
(glucagon-like peptide-2 analogue; ALX-0600)
[Gattex, Revestive]
A16AX08
orphan drug

His-Gly-Asp-Gly-Ser-Phe-Ser-Asp-Glu-Met-Asn-Thr-Ile-Leu-Asp-Asn-Leu-Ala-Ala-Arg-Asp-Phe-Ile-Asn-Trp-Leu-Ile-Gln-Thr-Lys-Ile-Thr-Asp

[197922-42-2]
$C_{164}H_{252}N_{44}O_{55}S$
3752.13

treatment of short bowel syndrome. lyophilized powder for s.c. inj.
5 mg/vial.
5 mg P.

Allelix Biopharmaceutical
EU/2012; USA/2012
fermentation (rec. *E. coli*):
[29, 30]
solid phase synthesis: [31]
medical use (short bowel syndrome):
[32]

Glycerol phenylbutyrate
(HPN-100)
[Ravicti]
A16AX09
orphan drug

structure

[611168-24-2]
$C_{33}H_{38}O_6$
530.66

treatment of inborn urea cycle disorders (nitrogen-binding agent). oral liquid 1.1 g/mL.
16 g O.

Hyperion
Therapeutics/Horizon Pharma
USA/2013; EU/2015
synthesis
[33]

(continued)

Table 2. (Continued)

INN (Synonyms) [Brand names] ATC	Structure (Remarks)	[CAS-No.] Formula MW, g/mol	Target (if known) Medical use Formulation DDD	Originator Approval (country/year) Production method [References]
Eliglustat (Genz-99067) [Cerdelga] A16AX10 orphan drug		[49833-29-5] $C_{23}H_{36}N_2O_4$ 404.55 tartrate: [928659-70-5] $(C_{23}H_{36}N_2O_4)_2$ $\cdot C_4H_6O_6$ 959.19	inhibitor of glucosylceramide synthase, substrate of CYP2D6 and CYP3A4. treatment of Morbus Gaucher type 1. cps. 84 mg (tartrate). 0.168 g O.	Genzyme USA/2014; EU/2015; CH/2020 synthesis [34–36]
Sodium benzoate (E211) A16AX11 USA: GRAS substance		[532-32-1] $C_7H_5NaO_2$ 144.10	sodium benzoate binds amino acids, which leads to excretion of these amino acids and decrease of ammonia levels. used for treatment of urea cycle disorders: together with phenylbutyrate it is used to treat hyperammonemia. the substance is also in use as preservative (E number E211).	N.A. synthesis
Trientine (triethylenetetramine; TECZA; TETA) [Clovique, Cuprior, Cufence, Metalite, Syprine] A16AX12		[112-24-3] $C_6H_{18}N_4$ 146.24 dihydrochloride: [38260-01-4] $C_6H_{18}N_4 \cdot 2\,HCl$ 219.15	chelating agent, which is used to remove copper in patients with Wilson's disease. f.c. tbl. 150 mg; cps. 250, 300 mg (dihydrochloride).	N.A. USA/1985; EU/2017 synthesis (from ethylene diamine) [37]
Uridine triacetate (triacetyluridine; PN-401) [Vistogard, Xuriden] A16AX13 orphan drug		[4105-38-8] $C_{15}H_{18}N_2O_9$ 370.31	prodrug of uridine. treatment of hereditary orotic aciduria (Xuriden) and emergency treatment of patients who receive an overdose of the anticancer drugs fluorouracil or capecitabine (Vistogard). oral granules 2 g for $60–120\,mg\,kg^{-1}\,d^{-1}$.	Wellstat Therapeutics USA/2015; EU/2009 synthesis [38, 39]

Migalastat
(DDIG; AT-1001:
1-deoxygalactonojirimycin)
[Galafold]
A16AX14
orphan drug

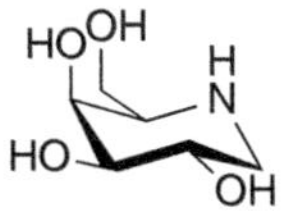

[108147-54-2]
$C_6H_{13}NO_4$
163.17
hydrochloride:
[75172-81-5]
$C_6H_{13}NO_4 \cdot HCl$
199.4

inhibitor of α-galactosidase
A (α-GalA).
treatment of Fabry disease.
cps. 150 mg (hydrochloride).

Amicus Therapeutics
EU/2016; USA/2018
synthesis (from D-galactose)
[40–43]

Telotristat ethyl
(LX-1032; LX-1606)
[Xermelo]
A16AX15
orphan drug

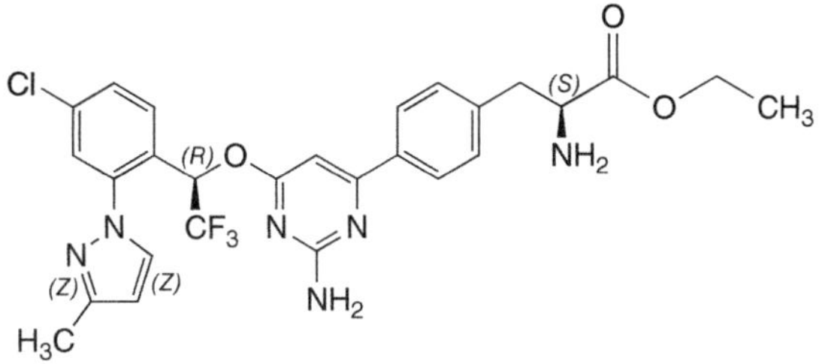

[1033805-22-9]
$C_{27}H_{26}ClF_3N_6O_3$
574.99
telotristat etiprate
(salt with hippuric
acid):
[1137608-69-5]
$C_{27}H_{26}ClF_3N_6O_3$
$\cdot C_9H_9NO_3$
754.16
telotristat (acid):
[1033805-28-5]
$C_{25}H_{22}ClF_3N_6O_3$
546.94

tryptophan hydroxylase inhibitor
(prodrug of telotristat), inhibits
production of serotonin.
treatment (in combination with
somatostatin analogue) of adults with
diarrhea associated with carcinoid
syndrome.
tbl. 250 mg (as etiprate).

Lexicon
EU/2017; USA/2017
synthesis
[44–46]

List of Abbreviations

amp.   ampoules
API   active pharmaceutical ingredient
ATC   Anatomical Therapeutic Chemical
           classification
cps.   capsules
f.c.   film-coated
GRAS   generally recognized as safe
inj.   injection
i.v.   intravenous
O   oral
OTC   over-the-counter drug
P   parenteral
prior.   priority (date)
s.c.   subcutaneous
sol.   solution
tbl.   tablets

References

1 Marzi, M., et al. (2000) *J. Org. Chem.* **65**, 6766.
2 Sigma-Tau, (1983) US 4 371 618, IT-prior.1980.
3 Samsung, (2002) US 6 342 034, KR-prior.1997.
4 Bioresearch, (1975) US 3 893 999, IT-prior.1972.
5 Bioresearch, (1976) US 3 954 726, IT-prior.1974.
6 Kaneko, T., et al. (1974) *Synthetic Production and Utilization of Amino Acids*, J. Wiley & Sons, New York, pp. 109–112.
7 Ajinomoto, (1975) US 3 886 039, JP-prior.1972.
8 Bestian, H., et al. (1950) *Ann.* **566**, 210–244.
9 I.G. Farben, (1941) DE 710 276, DE-prior.1939.
10 The Regents of the University of California, (2015) US 9 192 590, USA-prior.26.01.2007.
11 McIlwain, H. (1939) *Biochem. J.* **33**, 1942.
12 Bullock, M.W., et al. (1952) *J. Am. Chem. Soc.* **74**, 1868; 3455.
13 Research Corp., (1961) US 2 980 716, USA-prior.1954.
14 Degussa, (1987) US 4 705 867, DE-prior.1985.
15 Böttcher, B. and Lüttringhaus, A. (1947) *Ann.* **557**, 89.
16 Böttcher, B. (1952) DE 855 865, DE-prior.1942.
17 Böttcher, B. (1953) DE 869 799, DE-prior.1940.
18 Böttcher, B. (1953) DE 874 447, DE-prior.1944.
19 Brusilow, S., et al. (1980) *Science* **207** (4431), 659.
20 Zeneca, (1996) US 5 550 165, GB-prior.1991.
21 Swedish Orphan Biovitrum Int., (2016) US 9 301 932, SE-prior.23.06.2011.
22 Cycle Pharmaceuticals, (2019) US 10 328 029, GB-prior.03.01.2014.
23 Med. Use:Brewer, G.J., et al. (1987) *J. Lab. Clin. Med.* **109**, 526.
24 Bayer, (1987) US 4 639 436, DE-prior.1977.
25 Baxter, E.W. and Reitz, A.B. (1994) *J. Org. Chem.* **59**, 3175–3185.
26 Shiratori Pharm./Suntory, (1987), US 4 713 454, JP-prior.1985.
27 BioMarin, (2009) US 7 566 714, USA-prior.17.11.2003.
28 BioMarin, (2016) US 9 433 624, USA-prior.
29 Allelix, (1998) US 5 789 379, USA-prior.1995.
30 Allelix, (2001) US 6 184 201, USA-prior.1995.
31 NPS Pharm., (2013) EP 2 611 825, EP-prior. 30.08.2010.
32 NPS Pharm., (2010) US 7 847 061, EP-prior. 01.11.2004.
33 Hyperion, (2013) US 8 404 215, USA-prior.30.09.2011.
34 Univ. of Michigan/Genzyme, (2007) US 7 196 205, USA-prior.16.07.2001.
35 Univ. of Michigan/Genzyme, (2007) US 7 253 185, USA-prior.29.04.2002.
36 Univ. of Michigan/Genzyme, (2009) US 7 615 573, USA-prior.16.07.2001.
37 Walshe, J.M. (1979) *Prog. Clin. Biol. Res.* **34**, 271.
38 Pro-Neuron, (2001) US 6 258 795, USA-prior.28.10.1987.
39 Wellstat, (2010) US 7 776 838, USA-prior.28.10.1987.
40 Chacko, S. and Ramapanicker, R. (2015) *J. Org. Chem.* **80**, 4776.
41 Santoyo-Gonzalez, F., et al. (1998) *Synthesis* 1787.
42 Amicus, (2011) US 7 973 157, USA-prior.08.06.2005.
43 Amicus, (2018) US 10 076 514, USA-prior.22.03.2016.
44 Lexicon, (2009) US 7 553 840, USA-prior.12.12.2006.
45 Lexicon, (2011) US 7 968 729, USA-prior.24.08.2007.
46 Lexicon, (2014) US 8 653 094, USA-prior.28.09.2007.

Antithrombotics (B01) and Antihemorrhagics (B02)

Bernhard Kutscher, Maintal, Germany

1. Introduction

This article reviews the different drugs against blood disorders, with specific focus on antithrombotic (B01) and antihemorrhagic/hemostatic (B02) therapies. For an overview of cardiovascular drugs classified in the Anatomical Therapeutic Chemical Classification (ATC) group C, see → Cardiovascular System (C).

The blood system is a multifunctional transport medium of vital importance for the body of higher organisms. The composition of blood is very complex and includes various cell systems with peculiar functions, proteins, electrolytes, metabolites, and highly sophisticated mediator systems. Its undisturbed circulation in a ubiquitous system of blood vessels with fine ramifications throughout the whole organism, connections to the lymph vessels, and interfaces with the extravascular compartments is essential for the performance of all organs and enables numerous exchange processes. A sudden stop of blood circulation will cause the death of the organism within a short time.

Blood consists of various cell types, such as red blood cells or platelets, and plasma, the cell-free part of blood, containing numerous different proteins, each having certain and essential functions. Both cells and plasma can be used and further processed into medicines [1]. Drugs acting on the blood system are categorized by their ATC of the WHO, according to antithrombotics (B01), antihemorrhagics (B02), antianemics (B03), blood substitutes and perfusion solutions (B05), and other hematologics (B06). As the major focus of this article is on small synthetic molecules and therapeutic enzymes, large molecule biologics such as coagulation factors (ATC group B02BD) are not covered in this article.

Hemostasis is the human body's physiological response to blood vessel injury and subsequent prevention of hemorrhage [2]. This significant biological process involves a concerted coordination between blood-clotting proteins and platelets with the consequent formation of a clot, as repair of a damaged vascular tissue, or thrombus, which is a clot in a healthy blood vessel. This coagulation process

Ullmann's Pharmaceuticals. Edited by Axel Kleemann and Bernhard Kutscher.
© 2022 Wiley-VCH GmbH
Set ISBN: 978-3-527-34252-5/ DOI: 10.1002/14356007.w02_w03

involves local vasoconstriction which diminishes blood flow at the injury site as well as platelet plug formation. Subsequent secondary hemostasis implicates a series of enzymatic reactions between coagulation factors and cellular activity. These enzymatic reactions convert fibrinogen to fibrin which, together with the platelets, forms a thrombus [3]. Lastly, fibrinolysis is the biological mechanism which disperses the clot after the blood vessel has healed. During initiation after vascular injury, tissue factor (TF) is activated by proteases (blood-clotting proteins or factors), and the produced FVIIa/TF complex activates factor X to Xa that produces small amounts of thrombin, which subsequently activates platelets during the amplification phase and produces more thrombin. After that, fibrin production starts with clot formation and finally formation of a thrombus, thus producing the worldwide pathology called thrombosis. A thrombus formation, which obstructs arterial circulation, can end in acute myocardial infarction (AMI) or ischemic stroke. In venous circulation, deep vein thrombosis (DVT) can cause chronic leg pain, edema, and ulcers.

Thrombosis is a common causal pathology for cardiovascular disorders such as stroke, atrial fibrillation (AF), acute coronary syndrome (ACS), and venous thromboembolism (VTE), which is the most common cause of death and disability in the developed world [4]. Antithrombotic drugs are used for the prevention and treatment of thrombosis. Three classes target the thrombi components and pathology with agents including anticoagulants and antiplatelet drugs as well as fibrinolytic agents [5]. Anticoagulants slow down clotting, thereby reducing fibrin formation and preventing clots from forming/growing, and antiplatelet drugs prevent platelets from clumping. Fibrinolytic agents act when the thrombus is already formed.

The anticoagulants heparin and dicumarol were discovered by chance, long before their mechanism of action was well understood [6]. Unfractionated heparin (UFH) was first discovered in 1916 at Johns Hopkins University by a student investigating a clotting product from extracts of dog liver and heart.

In 1939, dicumarol, the precursor of warfarin, recognized as a promising therapeutic and rodenticide, was extracted by a biochemist at the University of Wisconsin. Heparin acts immediately and is given intravenously, whereas warfarin, approved in 1954 as the prototype vitamin K antagonist (VKA), is swallowed in tablet form, and its effect is delayed for days. Therefore, patients requiring anticoagulants who were admitted to a hospital were started on a heparin infusion, and discharged from the hospital on warfarin or other VKAs such as phencoumaron or acenocoumarol.

In the 1980s, low-molecular-weight heparins (LMWHs) were developed and produced by chemically splitting heparin into one-third of its original size. The products, such as enoxaparin or dalteparin, have fewer side effects than heparin, are more predictable in their anticoagulant response due to subcutaneous injection in a fixed dose, and replaced heparin in most indications as patients can be treated at home instead of in hospital.

With the biotechnology revolution, genetically engineered anticoagulant molecules that target specific clotting enzymes were designed. Anticlotting substances and their DNA were extracted from even exotic creatures such as ticks, leeches, snakes, or vampire bats and led to direct thrombin inhibitors such as lepirudin, bivalirudin (both based on hirudin, the anticlotting substance found in leeches), or argatroban, produced by chemical synthesis or recombinant techniques. In the 1990s, medicinal chemists began to fabricate small molecules designed to fit into the active site of clotting enzymes such as the indirect factor Xa inhibitor fondaparinux, a pentasaccharide analogue. However, all of these anticoagulants require parenteral administration. Hence, novel classes of small molecules and safer oral drugs were developed as direct FVIIa (dabigatran) or FXa inhibitors (rivaroxaban, apixaban, or betrixaban). With rivaroxaban as the first direct FXa inhibitor, this new and innovative therapeutic class is now called direct oral anticoagulants (DOACs) [7].

The first antiplatelet drug was aspirin (acetylsalicylic acid, ASA) (→ Analgesics (N02)), which has been used to relieve pain for decades. In 1980, researchers showed that aspirin in very low doses blocked an enzyme called cyclooxygenase (COX), responsible for platelet activation and required for platelet clumping [8]. Prostacyclin is one of the body's cardioprotective hormones and has a powerful antithrombotic role. Prostacyclin formation also requires activation of COX enzymes. Prostacyclin was developed and introduced into the market as epoprostenol, and further synthetic analogues such as iloprost or ciloprost were launched for the treatment of pulmonary arterial hypertension (PAH) ($\rightarrow$ Cardiovascular System (C)) [9].

The design of antiplatelet drugs directed to specific targets [10] of platelet activation even led to more potent therapies with, e.g., oral platelet aggregation inhibitors of the type of adenosine diphosphate P2Y12-receptor blocking thienopyridines (clopidogrel or ticlopidine) or PDE inhibitors such as dipyridamole or cilostazol. Glycoprotein platelet inhibitors and GPIIb/IIIa antagonists (abciximab, tirofiban, and eptifibatide) are intravenously available antiplatelet agents preventing platelet-to-platelet aggregation via the fibrinogen receptor. The thrombin receptor inhibitor voraxapar allows the targeting of yet another (third) pathway of platelet activation. Combinations of potent P2Y12 inhibitors such as clopidogrel, prasugrel, or ticagrelor with aspirin as dual antiplatelet therapy are considered as long-term standard of care for antithrombotic treatment, but even triple therapies in combination with DOACs are under clinical investigation [11].

Fibrinolytics or thrombolytic enzymes are mainly serine or metalloproteases having a direct or indirect mode of action for fibrinolysis. First-generation plasminogen activators, namely streptokinase and urokinase (or urokinase-type plasminogen activator), activate free circulatory plasminogen to plasmin and also degrade fibrinogen and other clotting factors in addition to fibrin. The driving force behind the development of the second-generation plasminogen activators was targeted thrombolysis, since the first-generation plasminogen activators showed nonspecific fibrin degradation and caused systemic fibrinolysis with the concomitant destruction of the hemostatic proteins, leading to hemorrhage. The development of second-generation plasminogen activators such as anistreplase, saruplase, and tissue-plasminogen activator (t-PA) partially controlled this problem. The third generation includes t-PA variants that have been engineered to improve the structural and functional properties such as longer half-life, resistance to inhibitors, and safety with the recombinant drugs alteplase, retaplase, and tenecteplase [12].

The incidence of venous thrombosis (VT) is 3% of the general population. In the United States, 100 000–300 000 people die from VT every year [13]. Admissions for ACS in the United States accounted for >1.3 million unique hospitalizations in 2006, with annual costs estimated as US$54 821 per patient [14]. Antithrombotic therapy has had an enormous impact in several clinical ways. Heparin has made bypass surgery and dialysis possible by blocking clotting in external tubings, and antithrombotics reduced the risk of blood clots in leg veins by more than 70%. And most importantly, it has markedly reduced death from heart attacks and the risk of stroke with atrial fibrillation (AF). Data collected by IQVIA showed that the global market sales of antithrombotic drugs were near to USD 26×10^9 in 2018 and still estimated to grow [13]. Aspirin remains the most frequently prescribed antiplatelet drug followed by P2Y12 blockers [10].

Bleeding disorder is the inability to form a normal clot after exposure to trauma or injury, which can lead to extensive bleeding that can be life threatening. Bleeding disorder may occur due a defect, mostly inherited, or a deficiency in one of the 13 natural blood-clotting factors. The most common bleeding disorders are hemophilia, caused by a deficiency or defect of factor VIII (hemophilia A) and factor IX (hemophilia B), or von Willebrand disease with the impairment of a protein called von Willebrand factor (VWF). An antihemorrhagic (or hemostatic) agent is defined as a substance that promotes hemostasis and stops bleeding [15]. A styptic is a specific type of antihemorrhagic that works by contracting tissue to seal injured blood vessels. Hemostatic agents used in medicine have various mechanisms of action and applications: Systemic drugs work by inhibiting fibrinolysis or by promoting coagulation; locally acting hemostatic agents work by causing vasoconstriction, promoting platelet aggregation, or stimulating clotting in a matrix. These agents are used during surgical procedures or even for emergency bleeding control in combat situation [16] and are categorized as hemostats, sealants, dressings, and adhesives [17].

The first class includes synthetic drugs such as tranexamic acid and aminocaproic acid as well as the naturally occurring protein aprotinin as fibrinolysis inhibitors. Carbazochrome is an oral antibleeding agent that stops the blood flow by increasing platelet aggregation and inducing platelets to form a plug.

Phylloquinone as vitamin K is used to treat bleeding disorders in patients with vitamin K deficiency. Fibrinogen concentrates, as well as protein concentrates, of coagulation factors such as FVIII complement bleeding management. The second class of hemostatic agents comprises locally acting agents such as cellulose, collagen, and gelatin as well as thrombin and thrombin-combination products. Passive hemostatic agents provide a structure for platelets to be aggregated and to form clots. For example, collagen-based agents can be applied to the bleeding site as powder, paste, or sponge, making direct contact between collagen and blood, which stimulates platelet aggregation. Active hemostatic agents such as thrombin have a biological activity in the coagulation cascade to induce clot formation at the site of bleeding. Thrombin can be used in combination with passive agents such as gelatin in order to increase effectiveness with more bleeding control. Fibrin sealants or glues are made of two components, thrombin and human fibrinogen, converting the latter into fibrin so that clotting is initiated and the mixture solidified [18]. Medicines derived from blood and plasma play an important role in medical therapy, particularly in modern surgery. The global hemostatic agents market was valued at USD 5.3×10^9 in 2018 and estimated to further grow [19].

Thrombocytopenia is a condition of an unusually low level of platelets in the blood and results from an imbalance between the production and destruction of platelets often seen in patients with advanced liver disease [20]. Thrombocytopenia is associated with several medical disorders including aplastic anemia, myelodysplasia, and idiopathic thrombocytopenic purpura (ITP). ITP can cause mucosal, skin, or intracranial bleeding. From 2006 to 2012, the number of ITP-related hospitalizations increased steadily by 30%, and the average length of stay was 6 days. The first two thrombopoietin receptor agonists (TPO-RAs), eltrombopag [21] and romiplostim, were licensed in 2008 for the treatment of ITP. Both bind to the TPO receptor, resulting in increased platelet production. However, romiplostim is a peptibody binding directly to the TPO-binding site, whereas eltrombopag is an orally bioavailable small molecule which binds to the transmembrane site. Two other oral nonpeptide TPO-RAs, avatrombopag and lusutrombopag, were recently approved to increase platelet counts in the outpatient setting and thus to avoid platelet transfusion [22, 23]. Another oral drug and first-in-class spleen tyrosine kinase (Syk) inhibitor, fostamatinib, was approved in 2018 for treating immune ITP [24, 25].

2. B Blood and Blood-Forming Organs, B01 Antithrombotics

2.1. B01AA Vitamin K Antagonists (Table 1)

Table 1. B01AA vitamin K antagonists

INN (Synonyms) [Brand names] ATC	Structure (Remarks)	[CAS-No.] Formula MW, g/mol	Target (if known) Medical use Formulation DDD	Originator Approval (country/year) Production method [References]
Dicoumarol (bishydroxycoumarin) [Miradon; Sintrom] B01AA01		[66-76-2] $C_{19}H_{12}O_6$ 336.30	agent that works by interfering with the metabolism of vitamin K, in biochemical experiments also used as inhibitor of reductases. oral anticoagulant. tabl. 25, 50, and 100 mg. 0.1 g O.	Univ. Wisconsin/Abbott US/1944 synthetic [26] first discovered in wet sweet-clover hay
Phenindione [Dindevan, Fenindion, Hedulin] B01AA02		[83-12-5] $C_{15}H_{10}O_2$ 222.23	vitamin K antagonist that reduces vitamin K reductase. anticoagulant similar to warfarin, but with hypersensitivity reactions. tabl. 10, 25, and 50 mg. 0.1 g O.	Merrell/Sanofi-Aventis US/1952 synthetic [27]
Warfarin [Coumadin; Miraban, Jantoven] B01AA03 **WHO essential medicine**		[81-81-2] $C_{19}H_{16}O_4$ 308.33 sodium salt: [129-06-6] $C_{19}H_{15}NaO_4$ 330.32	blocks vitamin K reductase, and subsequently clotting factors such as II, VII, IX, and X have decreased clotting ability. anticoagulant used for prevention of stroke. amp. 2 mg/mL; tabl. 1, 2, 2.5, 3, 4, 5, 6, 7.5, and 10 mg (as sodium salt). 7.5 mg O, P.	Bristol Myers Squibb US/1954 synthetic [28, 29]
Phenprocoumon [Falithrom, Marcumar; Liquamar] B01AA04		[435-97-2] $C_{18}H_{16}O_3$ 280.32	vitamin K antagonist as derivative of coumarin that inhibits coagulation by blocking synthesis of clotting factors II, VII, IX, and X. anticoagulant used for prophylaxis and treatment of thromboembolic disorders. f.c. tabl. 3 mg; tabl. 3 mg. 3 mg O.	Organon US/1957 synthetic [30, 31]
Acenocoumarol (acenokumarin, nicoumalone) [Ascumar, Sintrom] B01AA07		[152-72-7] $C_{19}H_{15}NO_6$ 353.33	vitamin K antagonist and coumarin derivative that inhibits coagulation by blocking synthesis of clotting factors II, VII, IX, and X. anticoagulant for treatment and prevention of thromboembolic disorders. tabl. 1 and 4 mg. 5 mg O.	Geigy/Paladin Labs 1952 synthetic [32, 33]

(continued)

Table 1. (*Continued*)

INN (Synonyms) [Brand names] ATC	Structure (Remarks)	[CAS-No.] Formula MW, g/mol	Target (if known) Medical use Formulation DDD	Originator Approval (country/year) Production method [References]
Ethyl bis-coumacetate [Tromexan] B01AA08		[548-00-5] $C_{22}H_{16}O_8$ 408.36	vitamin K antagonist that reduces vitamin K reductase. anticoagulant and antithrombotic. tabl. 300 mg. 0.6 g O.	synthetic [34]
Clorindione (G 25766) [Indaliton] B01AA09		[1146-99-2] $C_{15}H_9ClO_2$ 256.69	vitamin K antagonist that reduces vitamin K reductase. anticoagulant.	synthetic [35]
Diphenadione [Dipaxin] B01AA10 wfm		[82-66-6] $C_{23}H_{16}O_3$ 340.38	vitamin K antagonist that reduces vitamin K reductase. anticoagulant, rodenticide. tabl. 20 and 50 mg.	Upjohn US/1955 synthetic [36]
Tioclomarol [Apegmone] B01AA11		[22619-35-8] $C_{22}H_{16}Cl_2O_4S$ 447.34	vitamin K antagonist that reduces vitamin K reductase. anticoagulant and rodenticide. tabl. 4 mg.	Lipha synthetic [37]
Fluindione [Previscan] B01AA12		[957-56-2] $C_{15}H_9FO_2$ 240.23	vitamin K antagonist that reduces vitamin K reductase. anticoagulant. tabl. 20 mg.	Merck Serono synthetic [38]

2.2. B01AB Heparin Group (Table 2)

Table 2. B01AB Heparin group

INN (Synonyms) [Brand names] ATC	Structure (Remarks)	[CAS-No.] Formula MW, g/mol	Target (if known) Medical use Formulation DDD	Originator Approval (country/year) Production method [References]
Heparin [Calciparine, Caprocin, Eparical, Esberiven, Heparin Sodium, Multiparin, Pemiroc] B01AB01, C05BA03, and S01XA14 **WHO essential medicine**	(unfractionated heparin (UFH) is a heterogeneous preparation of anionic, sulfated glycosaminoglycan polymers with masses ranging from 3000 to 30 000 Da)	[9005-49-6] $[C_{24}H_{38}N_2O_{35}S_5]_x$ sodium salt: [9041-08-1] calcium salt: [37270-89-6] magnesium salt: [54479-70-8]	naturally occurring anticoagulant released from mast cells; binds reversibly to antithrombin III (ATIII) and greatly accelerates the rate at which ATIII inactivates coagulation enzymes thrombin (factor IIa) and factor Xa. UFH is different from low-molecular-weight heparin (LMWH) in the following ways: the average molecular weight of LMWH is about 4.5 kDa, whereas it is 15 kDa for UFH. Unfractionated heparin is more specific than LMWH for thrombin. anticoagulant, antithrombotic. amp. 12 500, 20 000, 50 000, and 60 000 iu; cream, eye drops, and eye ointment 30 000, 60 000, and 150 000 iu (as sodium salt or calcium salt); inj. 1000 iu/1 mL (as sodium salt); ointment 500 iu/1 mL; syringe 5000 and 7500 iu; vial 5000 iu/5 mL, 10 000 iu/10 mL, 20 000 iu/20 mL, 50 000 iu/50 mL, and 100 000 iu/100 mL (as calcium salt). 10 000 iu P.	Vitrum AB 1936 isolated from bovine or pig lung or intestinal mucosa [39–41]
Antithrombin III (AT; Org-10849) [ATryn, Antithrombin II NF, Thrombate III] B01AB02	Single-chain glycoprotein with 432 amino acids	[9000-94-6]	glycoprotein that accounts for major antithrombin activity of normal plasma and inhibits several other enzymes. thrombolytic. indicated for patients with hereditary antithrombin deficiency. powder for solution. 21 000 iu P.	Grifols/Baxter 1991 human plasma derived GTC Biotherapeutics US/2009 recombinant rhAT produced in genetically altered goats [42, 43]

(continued)

Table 2. (*Continued*)

INN (Synonyms) [Brand names] ATC	Structure (Remarks)	[CAS-No.] Formula MW, g/mol	Target (if known) Medical use Formulation DDD	Originator Approval (country/year) Production method [References]
Dalteparin sodium (edelparin, Kabi 2165, FR-860) [Fragmin] B01AB04	(low-molecular-mass fragment of heparin)	[9041-08-1]	antithrombotic, anticoagulant. sol. in amp.. graduated syringe, and multidose vial. 2500 iu P anti-Xa.	Kabi/Pharmacia-Pfizer US/1994 semisynthetic [44, 45]
Enoxaparin [Clexane, Lovenox] B01AB05 **WHO essential medicine**	(low-molecular-mass fragment of heparin)	[679809-58-6]	anticoagulant, platelet aggregation inhibitor, prophylaxis of deep vein thrombosis, and treatment of deep thrombosis. a) concentration 100 mg/mL in prefilled syringes with 30 and 40 mg and graduated syringes with 60, 80, and 100 mg and multiple-dose vials with 300 mg/3 mL b) concentration 150 mg/mL in graduated prefilled syringes with 120 and 150 mg. 2000 iu P anti-Xa.	Sanofi-Aventis semisynthetic from heparin [46]
Nadroparin calcium [Fraxiparine, Fraxodi] B01AB06	low-molecular-mass heparin (structure see above)	[9005-49-6]	anticoagulant and antithrombotic. i.v. and sc. 2850 iu P anti-Xa.	Sanofi-Synthelabo/Aspen depolymerization and fractionation of heparin [47]
Parnaparin sodium (OP 2123) [Fluxum, Minidalton, Tromboparin, Zoltan] B01AB07	low-molecular-mass heparin (structure see above)	[91449-79-5]	anticoagulant and antithrombotic. 3200 iu P anti-Xa.	Alfa/Opocrin IT depolymerization and fractionation of heparin [48]
Reviparin sodium [Clivarin, Clivarina, Clivarine] B01AB08	low-molecular-mass heparin (structure see above)	[9041-08-1]	anticoagulant and antithrombotic. prefilled syringe. 1430 iu P anti-Xa.	Abbott 1995 depolymerization and fractionation of heparin [49]

Name	Structure	CAS No.	Properties / Use	Source / References
Danaparoid sodium (Org-10172) [Orgaran] B01AB09	heparin sulfate, approx. 84%, R^1, $R^3 = SO_3H/H$, $R^2 = SO_3H/Ac$, $X^1 = COOH$ and $X^2 = H$ or $X^1 = H$ and $X^2 = COOH$ dermatan sulfate, approx. 12% chondroitin-4/6 sulfates, approx. 4% (low-molecular-mass heparinoid, mixture of sodium salts of the shown compounds)	[83513-48-8]	anticoagulant, platelet aggregation inhibitor, and antithrombotic. inj. sol.: amp. 0.6 mL with 1250 iu activity per mL 1500 iu P anti-Xa.	Akzo/Organon US/2001 isolated from pig intestine mucosa [50, 51]
Tinzaparin sodium [Innohep] B01AB10	low-molecular-mass heparin (structure see above)	[9041-08-1]	anticoagulant and antithrombotic. i.v. 2000 iu/mL and syringe 3500 iu P anti-Xa.	Leo Pharma US/2000 depolymerization and fractionation of heparin [52]
Sulodexide (SDX, KRX-101) [Aterina, Sulodexide Gelcaps, Sulonex, Vessel] B01AB11	glucosaminoglycan sulfate as highly purified mixture of low-molecular-mass heparin and dermatan sulfate (structures see above)	[57821-29-1]	anticoagulant and antithrombotic. oral and i.v. 500 iu O, P.	Alfa Wassermann/Keryx extracted from porcine intestinal mucosa [53, 54]
Bemiparin sodium [Badyket, Ivor, Hibor, Zibor] B01AB12	ultra-low-molecular-mass heparin (structure see above)	[91449-79-5]	anticoagulant and antithrombotic. 2500 iu P.	Laboratorios Rovi [55]
Certoparin sodium [Embolex, Sandoparin, Troparin] B01AB13	low-molecular-mass heparin (structure see above)	[9005-49-6]	anticoagulant and antithrombotic. 3000 iu P anti-Xa.	Novartis/Sandoz [56]

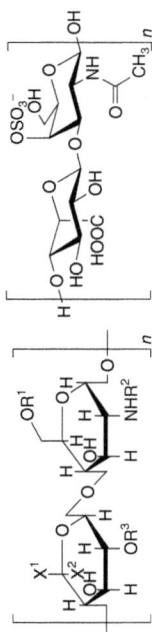

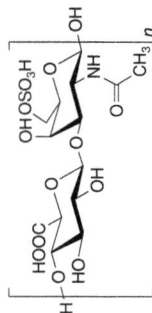

2.3. B01AC Platelet Aggregation Inhibitors excl. Heparins (Table 3)

Table 3. B01AC Platelet aggregation inhibitors excl. heparins

INN (Synonyms) [Brand names] ATC	Structure (Remarks)	[CAS-No.] Formula MW, g/mol	Target (if known) Medical use Formulation DDD	Originator Approval (country/year) Production method [References]
Ditazole [Ageroplas] B01AC01		[18471-20-0] $C_{19}H_{20}N_2O_3$ 324.37	platelet aggregation inhibitor and NSAID with analgesic activity. anticoagulant. caps. 400 and 500 mg.	Angelini ES synthetic [57]
Cloricromen (AD-6) [Assogen, Proendotel] B01AC02		[68206-94-0] $C_{20}H_{26}ClNO_5$ 395.88 hydrochloride salt: [74697-28-2] $C_{20}H_{26}ClNO_5 \cdot HCl$ 432.34	platelet aggregation inhibitor. coronary vasodilator and antithrombotic. amp. 30 mg/5 mL; cps. 100 mg; vial 30 mg.	Fidia IT/1991 synthetic [58, 59]
Picotamide [Plactidil] B01AC03		[32828-81-2] $C_{21}H_{20}N_4O_3$ 376.42 tartrate salt: [86247-87-2] $C_{21}H_{20}N_4O_3$ $\cdot x\ C_4H_6O_6$ hydrate: [80530-63-8] $C_{21}H_{20}N_4O_3 \cdot H_2O$ 394.43	platelet aggregation inhibitor with dual mechanism of action as thromboxane antagonist and thromboxane synthase inhibitor. anticoagulant and fibrinolytic. tabl. 300 mg (as hydrate).	Novartis synthetic [60, 61]
Clopidogrel (SR-25990C) [Iscover, Plavix] B01AC04 **WHO essential medicine**		[120202-66-6] $C_{16}H_{16}ClNO_2S \cdot$ H_2SO_4 419.91 (+)-base: [113665-84-2] $C_{16}H_{16}ClNO_2S$ 321.83	prodrug of a platelet aggregation inhibitor. used to reduce the risk of myocardial infarction. f.c. tabl. 75 mg (as hydrogen sulfate). 75 mg O.	Sanofi US/1997 synthetic [62–65]
Ticlopidine [Ticlid, Tiklyd] B01AC05		[55142-85-3] $C_{14}H_{14}ClNS$ 263.79 hydrochloride salt: [53885-35-1] $C_{14}H_{14}ClNS \cdot HCl$ 300.25	prodrug of a platelet aggregation inhibitor. cps. 250 mg; drg. 250 mg; f.c. tabl. 250 mg (as hydrochloride); tabl. 250 mg. 0.5 mg O.	Sanofi DE/1990 synthetic [66–68]

Table 3. (*Continued*)

INN (Synonyms) [Brand names] ATC	Structure (Remarks)	[CAS-No.] Formula MW, g/mol	Target (if known) Medical use Formulation DDD	Originator Approval (country/year) Production method [References]
Acetylsalicylic acid (acidum acetylsalicylicum, aspirin, ASA, ASS) [Aggrenox, Aspirin] B01AC06; A01AD05; M01BA03; N02BA01; N02BA51 **WHO essential medicine**		[*50-78-2*] $C_9H_8O_4$ 180.16 sodium salt: [*493-53-8*] $C_9H_7NaO_4$ 202.14 lysine salt: [*62952-06-1*] $C_9H_8O_4 \cdot$ $C_6H_{14}N_2O_2$ 326.35	cyclooxygenase (COX) inhibitor. analgesic, antipyretic, antirheumatic, and platelet aggregation inhibitor; used for reducing the risk of myocardial infarction and prevention of thromboembolism after hip surgery. cps. 325 and 500 mg; enteric tabl. 100 mg; powder; suppos. 125, 150, 300, 500, and 750 mg; tabl. 50, 75, 100, 300, and 500 mg. 1 tablet O.	Bayer 1899 synthetic [69, 70]
Dipyramidole [Aggrenox, Asasantine, Corosan, Persantine] B01AC07		[*58-32-2*] $C_{24}H_{40}N_8O_4$ 504.64	phosphodiesterase inhibitor. coronary vasodilator and platelet aggregation inhibitor. amp. 10 mg/2 mL; cps. 75 mg; drg. 25 and 75 mg; f.c. tabl. 75 mg; powder 12.5%; s.r. cps. 150 mg; tabl. 12.5, 25, and 100 mg. 0.4 g O; 0.2 g P.	Boehringer Ingelheim US/1999 synthetic [71–73]
Carbasalate calcium (calcium carbaspirin) [Cardiosolupsan, Flogesic, Iromin, Solupsan] B01AC08; N02BA15		[*5749-67-7*] $C_{18}H_{16}CaO_8 \cdot$ CH_4N_2O 458.44	cyclooxygenase (COX) inhibitor as chelate of calcium acetylsalicylate and urea. analgesic and platelet aggregation inhibitor. tabl. 500 mg. 1 tabl. O.	Bayer synthetic [74, 75]
Epoprostenol (PGI$_2$, prostacyclin) [Flolan, Veletri] B01AC09		[*35121-78-9*] $C_{20}H_{32}O_5$ 352.47 monosodium salt: [*61849-14-7*] 374.45	natural prostaglandin, acts as physiological antagonist of platelet aggregation; anticoagulant, vasodilator, and platelet aggregation inhibitor. used to treat pulmonary arterial hypertension (PAH). vial 0.5 and 1.5 mg; vial (lyo.) 0.5 mg (as sodium salt). 38 μg P.	Glaxo Wellcome US/2000 synthetic [76–79]

(*continued*)

Table 3. (*Continued*)

INN (Synonyms) [Brand names] ATC	Structure (Remarks)	[CAS-No.] Formula MW, g/mol	Target (if known) Medical use Formulation DDD	Originator Approval (country/year) Production method [References]
Indobufen [Ibustrin] B01AC10		[63610-08-2] $C_{18}H_{17}NO_3$ 295.34	cyclooxygenase (COX) inhibitor. anti-inflammatory, antithrombotic. amp. 200 mg (as sodium salt); tabl. 100 and 200 mg.	Pharmacia & Upjohn 1984 synthetic [80, 81]
Iloprost (ciloprost, E-1030, SH-401, ZK-36374) [Ilomedin, Endoprost, Ventavis] B01AC11		[78919-13-8] $C_{22}H_{32}O_4$ 360.49 trometamol salt: [73873-87-7] $C_{26}H_{43}NO_7$ 481.63	synthetic prostaglandin analogue of prostacyclin PGI2; vasodilator and platelet aggregation inhibitor. used to treat PAH. amp. 50 µg/0.5 mL, 100 µg/1 mL (as trometamol salt). 0.15 mg inhalation; 50 µg P.	Bayer/Schering 1992 synthetic [82–84]
Sulfinpyrazone [Anturan, Apo-sulfin-pyrazone] B01AC12; M04AB02		[57-96-5] $C_{23}H_{20}N_2O_3S$ 404.48	uricosuric medication and reduces platelet aggregation by inhibiting degranulation of platelets which reduces ADP and thromboxane. tabl.	Novartis 1959 synthetic [85]
Abciximab (c7E3) [ReoPro] B01AC13	Fab fragment of the chimeric human–murine monoclonal antibody 7E3	[143653-53-6] $C_{2101}H_{3229}N_{551}$ $O_{673}S_{15}$ 47455.4	immunoglobulin G (human–mouse monoclonal clone) antihuman glycoprotein IIb/IIIa receptor antagonist also binding to vitronectin receptor. platelet aggregation inhibitor and antianginal. for use in angioplasty. vial 10 mg/5 mL. 25 mg P.	Lilly/Centocor US/1994 biofermentation at Janssen Biologics [86–88]
Anagrelide (BL-4162A, BMY-26538-01) [Agrylin, Xagrid] B01AC14		[68475-42-3] $C_{10}H_7Cl_2N_3O$ 256.09 hydrochloride salt: [8579-51-4] $C_{10}H_7Cl_2N_3O \cdot$ HCl 292.55	phosphodiesterase III inhibitor that reduces platelet counts. antithrombotic. cps. 0.5 and 1 mg (as hydrochloride hydrate).	Shire US/1998 synthetic [89, 90]
Aloxiprin [Palaprin, Superpirin] B01AC15; N02BA02 wfm		[9014-67-9]	cyclooxygenase (COX) inhibitor as aluminum acetylsalicylate complex. analgesic and antiplatelet agent. tabl. 400, 450, and 600 mg.	synthetic [91]

Table 3. (*Continued*)

INN (Synonyms) [Brand names] ATC	Structure (Remarks)	[CAS-No.] Formula MW, g/mol	Target (if known) Medical use Formulation DDD	Originator Approval (country/year) Production method [References]
Eptifibatide (C 68-22, SB-1, Sch-60936, intrifiban) [Integrilin] B01AC16		[*188627-80-7*] $C_{35}H_{49}N_{11}O_9S_2$ 831.98	cyclic heptapeptide as GPIIb/III receptor antagonist derived from venom of pygmy rattlesnake. fibrinogen receptor antagonist and platelet aggregation inhibitor. vial for inj. 20 mg/10 mL and 75 mg/10 mL. 0.2 g P.	GSK/Schering Corp. synthetic [92, 93]
Tirofiban (L 700462, MK 383) [Aggrastat] B01AC17		[*144494-65-5*] $C_{22}H_{36}N_2O_5S$ 440.61 hydrochloride salt: [*142373-60-2*] $C_{22}H_{36}N_2O_5S \cdot$ HCl 477.07	glycoprotein (GP) IIb/IIIa receptor antagonist. platelet aggregation inhibitor. sol. for inj. 0.05 and 0.25 mg/mL; vial 50 mL and 0.25 mg/mL. 10 mg P.	Merck, Sharp & Dohme 1998 synthetic [94, 95]
Triflusal [Aflen, Disprin, Grendis, Triflux] B01AC18		[*322-79-2*] $C_{10}H_7F_3O_4$ 248.16	salicylate derivative that blocks cyclooxygenase (COX) and phosphodiesterase (PDE). platelet aggregation inhibitor. caps. and oral solution. 0.6 g O.	Uriach labs. ES/1981 synthetic [96, 97]
Beraprost (MDL-201229) [Beracle, Berasil, Berasus, Procylin] B01AC19		[*88430-50-6*] $C_{24}H_{30}O_5$ 398.49	prostacyclin analogue. vasodilator and antiplatelet agent also used for PAH. tabl. 20 µg.	Toray/Astellas/Kaken JP/2007 synthetic [98]
Treprostinil (uniprost, BW-15AU, LRX-15, 15 AU81) [Remodulin, Trevyent, Tyvaso] B01AC21		[*81846-19-7*] $C_{23}H_{34}O_5$ 390.52 sodium salt: [*289480-64-4*] $C_{23}H_{33}NaO_5$ 412.50	synthetic prostaglandin analogue of prostacyclin PGI$_2$. treatment of pulmonary hypertension and peripheral vascular disease. vial 20 mL: 1.0, 2.5, 5.0, and 10 mg/mL. 4.3 mg P.	United Therapeutics 2002 synthetic [99]

(*continued*)

Table 3. (*Continued*)

INN (Synonyms) [Brand names] ATC	Structure (Remarks)	[CAS-No.] Formula MW, g/mol	Target (if known) Medical use Formulation DDD	Originator Approval (country/year) Production method [References]
Prasugrel (CS-747, LY-640315) [Efient, Effient] B01AC22		[*150322-43-3*] $C_{20}H_{20}FNO_3S$ 373.45 hydrochloride salt: [*389574-19-0*] $C_{20}H_{20}FNO_3S \cdot$ HCl 409.91	purinergic P2Y12 ADP receptor antagonist. antithrombotic, platelet inhibitor, and vasodilator. tabl. 5 and 10 mg (as hydrochloride). 10 mg O.	E. Lilly/Daiichi Sankyo EU/2009 synthetic [100]
Cilostazol [Pletal] B01AC23		[*73963-72-1*] $C_{20}H_{27}N_5O_2$ 369.46	selective inhibitor of PDE 3 increasing cAMP. platelet aggregation inhibitor. tabl. 50 and 100 mg. 0.2 g O.	Otsuka US/1999 synthetic [101]
Ticagrelor (AZD-6140) [Brilinta, Brilique, Possia] B01AC24		[*274693-27-5*] $C_{23}H_{28}F_2$ N_6O_4S 522.6	purinergic P2Y12 receptor antagonist. antithrombotic. oral tabs., 90 mg. 0.18 g O.	Astra Zeneca EU/2010; US/2011 synthetic [102, 103]
Cangrelor (ARC69931MX) [Kengreal, Kengrexal] B01AC25		[*63706-06-7*] $C_{17}H_{25}Cl_2F_3$ $N_5O_{12}P_3S_2$ 776.4 tetrasodium salt: [*63706-36-3*] $C_{17}H_{21}Cl_2F_3N_5$ $Na_4O_{12}P_3S_2$ 864.3	purinergic P2Y12 receptor antagonist. antiplatelet and antithrombotic agent. powder for i.v. infusion; 50 mg/vial containing cangrelor tetrasodium. 50 mg P.	Medicines Company 2015 synthetic [104, 105]
Vorapaxar (SCH 530348) [Zontivity] B01AC26		[*618385-01-6*] $C_{29}H_{33}FN_2O_4$ 492.6 sulfate salt: [*705260-08-8*] $C_{29}H_{33}FN_2O_4 \cdot$ H_2SO_4 590.7	thrombin receptor antagonist. antiplatelet agent. tabl.; eq. to 2.08 mg base as sulfate.	MSD US/2014; EU/2015 synthetic [106, 107]

Table 3. (*Continued*)

INN (Synonyms) [Brand names] ATC	Structure (Remarks)	[CAS-No.] Formula MW, g/mol	Target (if known) Medical use Formulation DDD	Originator Approval (country/year) Production method [References]
Selexipag (NS-304; ACT-293987; prodrug of ACT-333679) [Uptravi] B01AC27	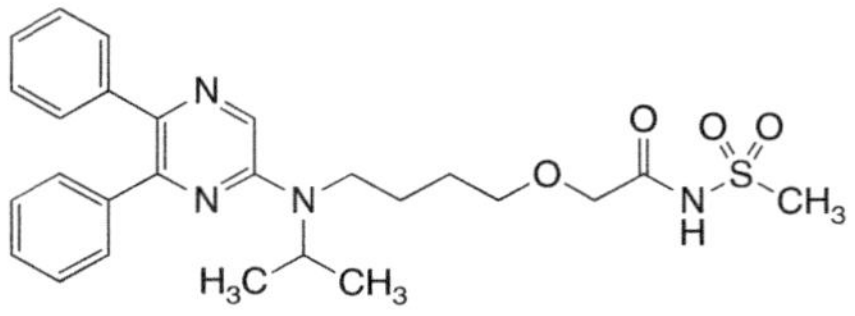	[*475086-01-2*] $C_{26}H_{32}N_4O_4S$ 496.6	prostacyclin receptor agonist for treatment of pulmonary arterial hypertension (PAH). antiplatelet and antihypertensive agent. tabl.; 0.2, 0.4, 0.6, 0.8, 1.2, 1.4, 1.6, and 1 mg.	Actelion Pharma EU/2016 synthetic [108, 109]

2.4. B01AD Enzymes (Table 4)

Table 4. B01AD Enzymes

INN (Synonyms) [Brand names] ATC	Structure/Amino acid sequence (Remarks)	[CAS-No.] Formula MW, g/mol	Target (if known) Medical use Formulation DDD	Originator Approval (country/year) Production method [References]
Streptokinase [Streptase, Varidase] B01AD01, B06AA55 **WHO essential medicine**	MKNYLSFGMFALLFALTFGTVNSVQAIAGPE-WLLDRPSVNNSQLVVSVAGTVEGTNQDIS LKFFEIDLTSRPAHGGKTE-QGLSPKSKPFATDSGAMSHKLEKADLLKAIQEQLIANVHSN DDYFEVIDFASDATITDRNGKVYFADKDGSVTLPTQPVQEFLLSGHVRVR-PYKEKPIQNQ AKSVDVEYTVQFTPLNPDDDFRPGLKDTKLLKTLAIGDTITSQEL-LAQAQSILNKNHPGY TIYERDSSIVTHDNDIFRTILPMDQEFTYRVKNREQAYRINKKSGLNEEINNT-DLISEKY YVLKKGEKPYDPFDRSHLKLFTIKYVDVDTNELLKSEQLLTASERNLDFRD-LYDPRDKAK LLYNNLDAFGIMDYTLTGKVEDNHDDTNRIITVYMGKRPEGENASYHLAYD-KDRYTEEER EVYSYLRYTGTPIPDNPNDK	[9002-01-1]	thrombolytic enzyme that activates while binding the human plasminogen to produce plasmin. Plasmin is produced in the blood to break down fibrin, the major constituent of blood thrombi. thrombolytic, fibrinolytic, and plasminogen activator. buccal tabl. 10×10^3 iu; vial 100×10^3 iu, 250×10^3 iu, 600×10^3 iu, 750×10^3 iu, and 1.5×10^6 iu. 1.5×10^6 iu P.	CSL Behring isolated from *Streptococci* spp. bacteria [110–112]
Alteplase (t-PA) [Activase, Actilyse, Cathflo] B01AD02, S01XA13 **WHO essential medicine**	SYQVICRDEKTQMIYQQHQSWLRPVLRSNR-VEYCWCNSGRAQCHSVPVKSCSEPRCFNGG TCQQALYFSDFVC-QCPEGFAGKCCEIDTRATCYEDQGISYRGTWSTAESGAECTNWNSSA LAQKPYSGRRPDAIRLGLGNHNYCRNPDRDSKPWCYVFKAGKYSSEFCST-PACSEGNSDC YFGNGSAYRGTHSLTESGASCLPWNSMILIGKVYTAQNPSAQALGLGKHNY-CRNPDGDAK PWCHVLKNRRLTWEYCDVPSCSTCGLRQYSQPQFRIKGGLFADIASHP-WQAAIFAKHRRS PGERFLCGGILISSCWILSAAHCFQERFPPHHLTVILGRTYRVVPGEEEQK-FEVEKYIVH KEFDDDTYDNDIALLQLKSDSSRCAQESSVVRTVCLPPADLQLPDWTECELS-GYGKHEAL SPFYSERLKEAHVRLYPSSRCTSQHLLNRTVTDNMLCAGDTRSGGPQANLH-DACQGDSGG PLVCLNDGRMTLVGIISWGLGCGQKDVPGVYTKVTNYLDWIRDNMRP	[105857-23-6] $C_{2569}H_{3928}$ $N_{746}O_{781}S_{40}$ 59042.3	serine protease that acts as tissue-plasminogen activator. thrombolytic agent used to treat myocardial infarctions as it binds to fibrin in a thrombus and initiates fibrinolysis. powder for injection solution vial 50 mg/50 mL or 100 mg/100 mL. 0.1 g P.	Genentech/Roche US/1987 Boehringer Ingelheim EU/1988 recombinant production in CHO cells [113]

Name	Sequence	CAS	Description	Production
Anistreplase (anisoylated plasminogen streptokinase activator complex) [Eminase] B01AD03	SYQVICRDEKTQMIYQQHQSWLRPVLRSNR-VEYCWCNSGRAQCHSVPVKSCSEPRCFNGG TCQQALYFSDFVCQCPEGFAGKCCEIDTRATCYEDQGISYRGTWSTAESGAECTNWNSSA LAQKPYSGRRPDAIRLGLGNHNYCRNPDRDSKPWCYVFKAGKYSSEFCSTPACSEGNSDC YFGNGSAYRGTHSLTESGASCLPWNSMILGKVYTAQNPSAQALGLGKHNYCRNPDGDAK PWCHVLKNRRLTWEYCDVPSCSTCGLRQYSQPQFRIKGGLFADIASHPWQAAIFAKHRRS PGERFLCGGILJSSCWILSAAHCFQERFPPHHLTVILGRTYRVVPGEEEQKFEVEKYIVH KEFDDDTYDNDIALLQLKSDSSRCAQESSVVRTVCLPPADLQLPDWTECELSGYGKHEAL SPFYSERLKEAHVRLYPSSRCTSQHLLNRTVTDNMLCAGDTRSGGPQANLHDACQGDSGG PLVCLNDGRMTLVGIISWGLGCGQKDVPGVYTKVTNYLDWIRDNMRP	[81669-57-0]	complex of purified human plasminogen and bacterial streptokinase. thrombolytic agent activated after administration by hydrolysis. powder for solution 30 iu. 30 iu P.	Beecham/Wulfing Pharma US/1997 recombinant production in CHO cells [114]
Urokinase [Abbokinase, Corase, Kinlytic, Rheothromb, Uroninase] B01AD04	KPSSPPEELKFQCGQKTLRPRFKIIGGEFTTI-ENQPWFAAIYRRHRGGSVTYVCGGSLMS PCWVISATHCFIDYPKKEDYIVYLGRSRLNSNTQGEMKFEVENLLLHKDYSADTLAHHND IALLKIRSKEGRCAQPSRTIQTICLPSMYNDPQFGTSCEITGFGKENSTDYLYPEQLKMT VVKLISHRECQQPHYYGSEVTTKMLCAADPQWKTDSCQGDSGGPLVCSLQGRMTLTGIVS WGRGCALKDKPGVYTRVSHFLPWIRSHTKEENGLAL	[9039-53-6]	plasminogen activator. fibrinolytic. vial 5000 iu/mL, 250×10^3 iu/5 mL, 2500 iu, 50×10^3 iu, 60×10^3 iu, 100×10^3 iu. 120×10^3 iu, 240×10^3 iu. 250×10^3 iu, 500×10^3 iu, and 600×10^3 iu. 3×10^6 iu P.	Microbix Biosystems US/1988 recombinant production with neonatal kidney cells [115, 116]
Fibrinolysin (serum tryptase, plasmin, bovine plasmin) [Elase, Fibrinolysin human, Fibrolan] B01AD05	fibrinolysin heavy chain DLLDDYVNTQGASLLSLSRKNLAGRSVEDCAAKCEEETDFVCRAFQYHSKEQQCVVMAEN SKNTPVFRMRDVILYEKRIYLLECKTGNGQTYRGTTAETKSGVTCQKWSATSPHVPKFSP EKFPLAGLEENYCRNPDNDENGPWCYTTDPDKRYDYCDIPECEDKCMHCSGENYEGKIAK TMSGRDCQAWDSQSPHAHGYIPSKFPNKNLKMNYCRNPDGEPRPWCFTTDPQKRWEFCDI PRCTTPPPSSGPKYQCLKGTGKNYGGTVAVTESGHTCQRWSEQTPHKHNRTPENFPCKNL EENYCRNPNGEKAPWCYTTNSEVRWEYCTIPSCESSPLSTERMDVPVPPEQTPVPQDCYH GNGQSYRGTSSTTTTGRKCQSWSSMTPHRHLKTPENYPNAGLTMNYCRNPDADKSPWCYT TDPRVRWEFCNLKKCSETPEQVPAAPQAPGVENPEADCMIGTGKSYRGKKATTVAGVPC QEWAAAQEPHQHSIFTPETNPQSGLERNYCRNPDGDVNGPWCYTMNPRKPFDYCDVPQCES SFDCGKPKVEPKKCSGR >Fibrinolysin light chain IVGGCVSKPHSWPWQVSLRSSRHFCGGTLISPKWVLTAAHCLDNILALSFYKVILGAHN EKVREQSVQEIPVSRLFREPSQADIALLKLSRPAIITKEVIPACLPPPNYMVAARTECY1 TGWGETQGTTGEGLLKEAHLPVIENKVCNRNEYLDGRVKPTELCAGHLIGGTDSCQGDSG GPLVCFEKDKYILQGVTSWGLGCARPNKPGVYVRVSPYVPWIEETMRRN	[9004-09-5]	fibrinolytic that attacks and inactivates fibrin molecules occurring in undesirable exudates on the surface of the human body or on human mucosa. assists in healing of minor burns, superficial wounds and hematomas, ulcers, and surgical wounds. ointment 10 mg/l g (1%); vial 25 iu.	Pfizer extracted from bovine plasma or bacterial cultures [117–119]
Brinase [Brinolase] B01AD06	n.a.	[9000-99-1]	fibrinolytic enzyme. thrombolytic.	derived from *Aspergillus oryzae* [120]

(continued)

Table 4. (Continued)

INN (Synonyms) [Brand names] ATC	Structure/Amino acid sequence (Remarks)	[CAS-No.] Formula MW, g/mol	Target (if known) Medical use Formulation DDD	Originator Approval (country/year) Production method [References]
Reteplase [Rapilysin, Retavase] B01AD07	SYQGNSDCYFGNGSAYRGTHSLTESGASCLPWNSMILIGKVYTAQNPSAQALGLGKHNYC RNPDGDAKPWCHVLKNRRLTWEYCDVPSCSTCGLRQYSOPQFRIKGGLFADIASHPWQAA IFAKHRRSPGERFLCGGILISSCWILSAAHCFQERFPPHHLTVILGRTYRVVPGEEEQKF EVEKYIVHKEFDDTYDNDIALLQLKSDSSRCAQESSVVRTVCLPPADLQLPDWTECELS GYGKHEALSPFYSERLKEAHVRLYPSSRCTSQHLLNRTVTDNMLCAGDTRSGGPQANLHD ACQGDSGGPLVCLNDGRMTLVGIISWGLGCGQKDVPGVYTKVTNYLDWIRDNMRP	[133652-38-7] C_{1736}H_{2671}N_{499} O_{522}S_2 39589.6	recombinant nonglycosylated form of human plasminogen activator similar to alteplase. thrombolytic agent. kit 1.81 mg/mL, powder for solution. 20 iu P.	Centocor/Roche US/1996 recombinant production in *E. coli* or CHO [121]
Saruplase [Prolyse] B01AD08	411 amino acids	[99149-95-8]	fibrinolytic enzyme closely related to urokinase. thrombolytic.	Abbott [122]
Ancrod [Arwin, Viprinex] B01AD09 wfm	n.a.	[9046-56-4]	thrombin-like serine protease venom derived from pit viper. defibrogenating agent, anticoagulant. liquid for injection.	Knoll Pharma/Abbott 1986 venom from pit viper [123]
Drotrecogin alfa [Xigris] B01AD10 wfm	Heavy chain LIDGKMTRRGDSPWQVVLLDSKKKLACGAVLIHPSWVLTAAHCMDESKKLLVRLGEYDLR RWEKWELDLDIKEVFVHPNYSKSTTDNDIALLHLAQPATLSQTIVPICLPDSGLAERELN QAGQETLVTGWGYHSSREKEAKRNRTFVLNFIKIPVVPHNECSEVMSNMVSENMLCAGIL GDRQDACEGDSGGPMVASFHGTWFLVGLVSWGEGCGLLHNYGVYTKVSRYLDWI- HGHIRD KEAPQKSWAP >Light chain SKHVDGDQCLVLPLEHPCASLCCGHGTCIXGIGSFSCDCRSGWEGRFCQREVSFLNCSLD NGGCTHYCLEEVGWRRCSCAPGYKLGDDLLQCHPAVKFPCGRPWKRMEKKRSHL	[98530-76-8] C_{1786}H_{2779}N_{509} O_{519}S_{29} 55000	activated human protein C and serine protease that inhibits factor Va and VIIIa as well as plasminogen activator inhibitor. antithrombotic and anticoagulant. vial 5 and 20 mg. 40 mg P.	E. Lilly US/2001 recombinant [124]
Tenecteplase [Metalyse, TNKase] B01AD11	SYQVICRDEKTQMIYQQHQSWLRPVLRSNRVEYCWCNSGRAQCHSVPVKSCSEPRCFNGG TCQQALYFSDFVCQCPEGFAGKCCEIDTRATCYEDQGISYRGNWSTAESGAECTNWQSSA LAQKPYSGRRPDAIRLGLGNHNYCRNPDRDSKPWCYVFKAGKYSSEFCSTPACSEGNSDC YFGNGSAYRGTHSLTESGASCLPWNSMILIGKVYTAQNPSAQALGLGKHNYCRNPDGDAK PWCHVLKNRRLTWEYCDVPSCSTCGLRQYSOPQFRIKGGLFADIASHPWQAAIFAAAAAS PGERFLCGGILISSCWILSAAHCFQERFPPHHLTVILGRTYRVVPGEEEQKFEVEKYIVH KEFDDTYDNDIALLQLKSDSSRCAQESSVVRTVCLPPADLQLPDWTECELSGYGKHEAL SPFYSERLKEAHVRLYPSSRCTSQHLLNRTVTDNMLCAGDTRSGGPQANLHDACQGDSGG PLVCLNDGRMTLVGIISWGLGCGQKDVPGVYTKVTNYLDWIRDNMRP	[191588-94-0] C_{2561}H_{3919}N_{747} O_{781}S_{40} 58951.2	glycoprotein with 527 amino acids that acts like tissue-plasminogen activator (tPA); derived from native tPA by modifications at three sites of the protein structure. thrombolytic. i.v. powder for solution 50 mg. 40 mg P.	Roche-Genentech/ Boehringer Ingelheim 2001 recombinant production in CHO cells [125, 126]
Protein C [Ceprotin] B01AD12	n.a.		endogenous plasma protein that is the natural precursor of an anticoagulant serine protease. anticoagulant. injection powder for solution 100iu/mL, 500iu/5 mL and 1000iu/10 mL.	Baxter US/2001 isolated from blood plasma [127]

2.5. B01AE Direct Thrombin Inhibitors (Table 5)

Table 5. B01AE Direct thrombin inhibitors

INN (Synonyms) [Brand names] ATC	Structure/Amino acid sequence (Remarks)	Target (if known) Medical use Formulation DDD	[CAS-No.] Formula MW, g/mol	Originator Approval (country/year) Production method [References]
Desirudin (63-desulfohirudin) [Iprivask, Revasc] B01AE01	Val - Val - Tyr - Thr - Asp - Cys - Thr - Glu - Ser - Gly^{10} Gln - Asn - Leu - Cys - Leu - Cys - Glu - Gly - Ser - Asn^{20} Val - Cys - Gly - Gln - Gly - Asn - Lys - Cys - Ile - Leu^{30} Gly - Ser - Asp - Gly - Glu - Lys - Asn - Gln - Cys - Val^{40} Thr - Gly - Glu - Gly - Thr - Pro - Lys - Pro - Gln - Ser^{50} His - Asn - Asp - Gly - Asp - Phe - Glu - Glu - Ile - Pro^{60} Glu - Glu - Tyr - Leu - Gln^{65}	direct human thrombin inhibitor. analogue of naturally occurring anticoagulant hirudin extracted from leeches; antithrombotic. inj. powder 15 and 30 mg. 30 mg P.	[120993-53-5] $C_{287}H_{440}N_{80}O_{110}S_6$ 6963.4	Novartis/Canyon Pharma EU/2004; US/2010 recombinant production in yeast strain 1456 [128]
Lepirudin [Refludan] B01AE02 wfm	LVYTDCTESGQNLCLCEGSNVCGQGNKCILGSDGEKNQCVTGEGTPKPQS-HNDGDFEEIP EEYLQ	direct thrombin inhibitor. analogue of naturally occurring anticoagulant hirudin. inj. sol. concentrate 20 mg. 0.25 g P.	[138068-37-8] $C_{288}H_{448}N_{80}O_{110}S_6$ 6963.43	Bayer/Celgene EU/1997 recombinant production in yeast cells [129]
Argatroban (argipidine, MQPA, DK-7419, MCl-9038, MD-805, OM-805) [Argatra, Novastan, Slonnon] B01AE03		direct and selective thrombin inhibitor. antithrombotic, anticoagulant. 100 mg/mL: 2.5 mL. vial. 0.2 g P.	[74863-84-6] $C_{23}H_{36}N_6O_5S$ 508.64 monohydrate: [141396-28-3] $C_{23}H_{36}N_6O_5S \cdot H_2O$ 526.66	Mitsubishi 2005 synthetic [130, 131]
Melagatran [Exanta] B01AE04		thrombin and serine protease inhibitor as active metabolite of ximelagatran. anticoagulant.	[159776-70-2] $C_{22}H_{31}N_5O_4$ 429.51	Astra Zeneca synthetic [132]

(continued)

Table 5. (*Continued*)

INN (Synonyms) [Brand names] ATC	Structure/Amino acid sequence (Remarks)	Target (if known) Medical use Formulation DDD	[CAS-No.] Formula MW, g/mol	Originator Approval (country/year) Production method [References]
Ximelagatran (H376/95) [Exanta, Exarta] B01AE05 wfm		thrombin inhibitor and prodrug of melagatran. oral anticoagulant. oral, 24 and 36 mg. 48 mg O.	[192939-46-1] $C_{24}H_{35}N_5O_5$ 473.57 monohydrate: [260790-58-7] $C_{24}H_{35}N_5O_5 \cdot H_2O$ 491.59 monobromide: [260790-59-8] $C_{24}H_{35}N_5O_5 \cdot HBr$ 554.49	Astra Zeneca 2004 synthetic [133–135]
Bivalirudin (hirulog-1/8, BG-8967) [Angiox, Angiomax] B01AE06	D-Phe-Pro-Arg-Pro-Gly-Gly-Gly-Asn-Gly-Asp-Phe-Glu-Glu-Ile-Pro-Glu-Glu-Tyr-Leu	direct thrombin inhibitor as congener of the naturally occurring drug hirudin. antithrombotic and anticoagulant. powder for inj., vial 250 mg (lyophilized product). 0.25 g P.	[128270-60-0] $C_{98}H_{138}N_{24}O_{33}$ 2180.32	Nycomed 2008 synthetic [136–138]
Dabigatran etexilate (BIBR 1048, BIBR 953) [Pradaxa] B01AE07 **WHO essential medicine**		direct thrombin inhibitor that binds to the active site and prevents thrombin-mediated activation of coagulation factors. antithrombotic, anticoagulant used to prevent blood clots following hip and knee surgery. caps. 75 and 100 mg; tabl. 75 and 100 mg. 0.3 g O.	[211915-06-9] $C_{34}H_{41}N_7O_5$ 627.75 dabigatran: [211914-51-1] $C_{25}H_{25}N_7O_3$ 471.52	Boehringer Ingelheim EU/2008; US/2010 synthetic [139–141]

2.6. B01AF Direct Factor Xa Inhibitors (Table 6)

Table 6. B01AF Direct factor Xa inhibitors

INN (Synonyms) [Brand names] ATC	Structure (Remarks)	[CAS-No.] Formula MW. g/mol	Target (if known) Medical use Formulation DDD	Originator Approval (country/year) Production method [References]
Rivaroxaban (Bay 59-7939) [Xarelto] B01AF01		[366789-02-8] $C_{19}H_{18}ClN_3O_5S$ 435.89	selective and direct inhibitor of factor Xa. antithrombotic, oral anticoagulant used for treatment of deep vein thrombosis (DVT). tabl. 20 mg. 20 mg O.	Bayer/Johnson & Johnson US/2011 synthetic [142, 143]
Apixaban (BMS-562247) [Eliquis] B01AF02		[503612-47-3] $C_{25}H_{25}N_5O_4$ 459.5	direct inhibitor of factor Xa. anticoagulant used to treat and prevent blood clots as well as DVT and to prevent stroke in people with atrial fibrillation (AT). oral f.c. tabl.; 2.5 and 5 mg. 10 mg O.	Bristol Myers Squibb US/2012 synthetic [144–146]
Edoxaban (DU-176b) [Lixiana, Savaysa] B01AF03		[480449-70-5] $C_{24}H_{30}ClN_7O_4S$ 548.1 hydrochloride salt: [480448-29-1] $C_{24}H_{30}ClN_7O_4S \cdot HCl$ 584.6 tosylate salt: [480449-71-6] $C_{24}H_{30}ClN_7O_4S \cdot C_7H_8O_3S$ 738.3	direct inhibitor of factor Xa. antithrombotic and anticoagulant used for the prevention of stroke and embolism. tabl.: 15 and 30 mg as tosylate. 60 mg O.	Daiicji Sankyo JP/2011; US; EU 2015 synthetic [147–149]
Betrixaban (PRT-054021) [Bevyxxa] B01AF04		[330942-05-7] $C_{23}H_{22}ClN_5O_3$ 451.1 maleate salt: [936539-80-9] 568.0	direct factor Xa inhibitor (DOAC). anticoagulant for prevention of DVT. caps.: 40 and 80 mg.	Portola Pharmaceuticals US/2017 synthetic [150–153]

2.7. B01AX Other Antithrombotic Agents (Table 7)

Table 7. B01AX Other antithrombotic agents

INN (Synonyms) [Brand names] ATC	Structure (Remarks)	[CAS-No.] Formula MW, g/mol	Target (if known) Medical use Formulation DDD	Originator Approval (country/year) Production method [References]
Defibrotide [Defitelio, Noravid, Procicide] B01AX01	Polydeoxyribonucleotide, sodium salt (mixture of single-stranded oligonucleotides)	[83712-60-1]	antithrombotic, cholinergic channel modulator, stimulates fibrinolysis. used to treat veno-occlusive disease of the liver of patients having had a bone marrow transplant. vial 200 mg, 80 mg/mL concentrate for infusion. 1.75 g P.	Crinos-Gentium/Jazz Pharmaceuticals IT/1986; EU/2014; US/2016 isolation from bovine lung or intestinal pig mucosa [154, 155]
Chondroitin sulfate B (dermatan sulfate) [Remaxazon, Theraflex] B01AX04; M01AX25	(sulfated glycosaminoglycan composed of alternating *N*-acetylgalactosamine and glucuronic acid)	[24967-93-9] sodium salt: [39455-18-0]	used for treating osteoarthritis. caps. and tabl. 500 mg; patch.	Bayer US/2015 extraction from cartilaginous cow and pig tissues [156]
Fondaparinux sodium (Org-31540, SR-90107A) [Arixtra] B01AX05		[114870-03-0] $C_{31}H_{43}N_3Na_{10}O_{49}S_8$ 1728.09 free acid: [104993-28-4] $C_{31}H_{53}N_3O_{49}S_8$ 1508.27	selective and synthetic pentasaccharide factor Xa inhibitor. anticoagulant for prevention of deep vein thrombosis and pulmonary embolism. prefilled syringe 2.5 mg/0.5 mL for s.c. application 2.5 mg P.	Sanofi-Organon/GSK US/2001; EU/2008 synthetic [157, 158]
Pentosan polysulfate sodium (Bay-946, Hoe-946) [Elmiron, Hemoclar] B01AX07; C05BA04; G05BX04		[116001-96-8] free acid: [37300-21-3]	low-molecular-weight heparin-like compound that binds to fibroblast growth factors (FGFs). anticoagulant with fibrinolytic effects. caps. 100 mg.	Ortho Mc Neil/Janssen Pharmaceuticals US/1996 semisynthetic [159]

3. B Blood and Blood-Forming Organs, B02 Antihemorrhagics

3.1. B02A Antifibrinolytics

3.1.1. B02AA Amino Acids (Table 8)

Table 8. B02AA Amino acids

INN (Synonyms) [Brand names] ATC	Structure (Remarks)	[CAS-No.] Formula MW (g/mol)	Target (if known) Medical use Formulation DDD	Originator Approval (country/year) Production method [References]
Aminocaproic acid (epsilcapramin) [Amicar, Caprolisin, Hemocid] B02AA01		[60-32-2] $C_6H_{13}NO_2$ 131.18	binds to lysine-binding sites within plasminogen/plasmin molecule and interferes with the ability to lyse fibrin clots. antifibrinolytic and plasmin inhibitor. gran. 98.6%; inj. flask 250 mg/mL; syrup 25%; tabl. 500 and 1000 mg. 16 g O, P.	Clover/Xanodyne US/1964 synthetic [160–162]
Tranexamic acid (TXA) [Cyklokapron, Exacyl, Lysteda, Transamin] B02AA02 **WHO essential medicine**		[1197-18-8] $C_8H_{15}NO_2$ 157.21	synthetic analogue of lysine that serves as antifibrinolytic via binding 4–5 lysine receptor sites on plasminogen. antifibrinolytic, hemostatic used to treat or prevent blood loss after surgery, tooth removal, or major trauma. amp. 250 mg/5 mL, 500 mg/5 mL, 5%, 1%; cps. 250 and 500 mg; f.c. tabl. 500 mg; gran. 50%; syrup 5%; tabl. 250 and 500 mg. 2 g O, P.	Daiichi/Xanodyne Pharms. JP/1965; US/2009 synthetic [163–165]
Aminomethylbenzoic acid [Pamba] B02AA03		[56-91-7] $C_8H_9NO_2$ 151.16	antifibrinolytic and hemostatic agent. 0.25 g O.	Nycomed DE synthetic [166]

3.1.2. B02AB Proteinase Inhibitors (Table 9)

Table 9. B02AB Proteinase inhibitors

INN (Synonyms) [Brand names] ATC	Structure (Remarks)	[CAS-No.] Formula MW, g/mol	Target (if known) Medical use Formulation DDD	Originator Approval (country/year) Production method [References]
Aprotinin [Artiss, Tisseel, Trasylol] B02AB01	globular 58 amino acid polypeptide	[9087-70-1] $C_{284}H_{432}N_{84}O_{79}S_7$ 6511.51	also known as bovine pancreatic trypsin inhibitor (BPTI); competitive inhibitor of several serine proteases such as specifically trypsin, chymotrypsin, plasmin, and kallikrein; inhibition of factor XIIa and as result both intrinsic pathways of coagulation and fibrinolysis are inhibited. anticoagulant and hemostatic for prophylactic use to reduce perioperative blood loss. intravenous solution with 10×10^3–200×10^3 iu and topical solution. 500×10^3 iu P.	Bayer/Nordic Group 1959 isolated from bovine pancreas and lung or gene expression in mammalian cells [167–170]
Alfa-1 antitrypsin (A1AT, AAT) [Aralast, Glassia, Prolastin, Zemaira] B02AB02	394 amino acid glycoprotein	[9041-92-3] $C_{200}H_{3130}N_{514}O_{601}S_{10}$ 44324.5	protease inhibitor glycoprotein; inhibition of elastase, plasmin, and thrombin. therapy of AAT deficiency (AATD). vials 1, 4, and 5 g. for i.v. dose 60 mg/kg once weekly. 0.6 g P.	Bayer US/1987 isolation from blood plasma [171, 172]
Camostat (FOY-305) [Foipan, Libilister, Pancrel] B02AB03		[59721-28-7] $C_{20}H_{22}N_4O_5$ 398.42 monomesylate salt: [59721-29-8] $C_{20}H_{22}N_4O_5 \cdot CH_4O_3S$ 494.53	serine protease inhibitor. trypsin inhibitor for treatment of chronic pancreatitis or liver fibrosis. gran. 200 mg; tabl. 100 mg.	Ono JP/1985 synthetic [173, 174]

3.2. B02B Vitamin K and Other Hemostatics

3.2.1. B02BA Vitamin K (Table 10)

Table 10. B02BA Vitamin K

INN (Synonyms) [Brand names] ATC	Structure (Remarks)	[CAS-No.] Formula MW, g/mol	Target (if known) Medical use Formulation DDD	Originator Approval (country/year) Production method [References]
Phytomenadione (phylloquinone, phytonadione, vitamin K$_1$) [Kativ N, Kaywan, Konakion, Mephyton] B02BA01 **WHO essential medicine**		[84-80-0] C$_{31}$H$_{46}$O$_2$ 450.71	vitamin K1 found in food, cofactor for the formation of coagulation factors. antihemorrhagic vitamin as diet supplement against bleeding disorders. amp. 1 mg/0.5 mL, 1 mg/mL, 10 mg/1 mL, 20 mg/3 mL, 50 mg/5 mL; chewing drg. 10 mg; cps. 10 mg, 20 mg; powder 1%; sol. 2 mg/0.2 mL, 20 mg/mL; syrup 20 mg/mL; tabl. 5 and 10 mg. 20 mg O, P.	Hospira US/1983 synthetic [175–178]
Menadione (menaphthone, menaquinone, vitamin K$_3$) [Bilkapy] B02BA02		[58-27-5] C$_{11}$H$_8$O$_2$ 172.18	precursor of various types of vitamin K antihemorrhagic vitamin (prothrombogenic). tabl. 2 mg. 10 mg O: 0.2 mg P.	Lehning FR synthetic [179, 180]

3.2.2. B02BB Fibrinogen (Table 11)

Table 11. B02BB Fibrinogen

INN (Synonyms) [Brand names] ATC	Structure (Remarks)	[CAS-No.] Formula MW, g/mol	Target (if known) Medical use Formulation DDD	Originator Approval (country/year) Production method [References]
Fibrinogen human (factor I) [Artiss, Evicel, Fibryna, Riastab, TachoSil] B02BB01	soluble plasma glycoprotein with 3410 amino acids	[9001-32-5]	physiological substrate for the three enzymes plasmin, factor VIIIa, and thrombin. indicated for patients with acute bleeding episodes due to congenital fibrinogen deficiency. i.v. and topical solution, 1 g.	CSL Behring US/2009 fractionated plasma product [181]

3.2.3. B02BC Local Hemostatics (Table 12)

Table 12. B02BC Local hemostatics

INN (Synonyms) [Brand names] ATC	Structure (Remarks)	[CAS-No.] Formula MW, g/mol	Target (if known) Medical use Formulation DDD	Originator Approval (country/year) Production method [References]
Absorbable gelatin sponge [Gelfoam, Gelita, Gelaspon, Stypro] B02BC01			medical device intended for application to bleeding surfaces as a hemostatic. antihemorrhagic, sometimes soaked with buprenorphine. sponge size 100.	Baxter 1945 produced from purified porcine skin [182]
Oxidized cellulose [Oxycel, Surgicel, Traumastem] B02BC02			water-insoluble derivative of cellulose used to control postsurgical bleeding. antihemorrhagic that provides matrix for clotting initiation.	Ethicon 1947 synthetic, produced from cellulose [183]
Tetragalacturonic acid hydroxy-methylester B02BC03	(oligosaccharide)	[53008-15-4] $C_{28}H_{42}O_{29}$ 842.6	antihemorrhagic.	produced by hydrolysis of pectic acid with yeast polygalac-torunase [184]

Table 12. (*Continued*)

INN (Synonyms) [Brand names] ATC	Structure (Remarks)	[CAS-No.] Formula MW, g/mol	Target (if known) Medical use Formulation DDD	Originator Approval (country/year) Production method [References]
Adrenalone [Hemorrodine, Stryphnasal] B02BC05, A01AD06		[99-45-6] $C_9H_{11}NO_3$ 181.19 hydrochloride salt: [62-13-5] $C_9H_{11}NO_3 \cdot HCl$ 217.65	keto-analogue of adrenaline; adrenergic agonist. sympathomimetic, vasoconstrictor, and hemostyptic. 60 mg/stick.	1907 synthetic [185, 186]
Thrombin (fibrinogenase) [Thrombinar, Thrombostat] B02BC06	enzyme consisting of 622 amino acids	[9002-04-4]	serine protease also called coagulation factor II; converts soluble fibrinogen into insoluble strands of fibrin as well as catalyzing other coagulation reactions. vasoconstrictor and hemostatic. powder for solution, 5000 or 1000 or 5000 iu/vial for topical use.	Pfizer 1951 isolation from bovine material or recombinant production [187]
Collagen [Carticel, MACI] B02BC07	triple-helix protein aggregated to fibrils, main structural protein in extracellular matrix of connective tissues		guides fibroblasts. artificial skin substitute used for wound healing in surgery.	Vericel/Genzyme US/1997 derived from bovine, porcine, or equine sources [188]
Calcium alginate [Fibracol, Seasorb, Sorbsan, Ultraplast] B02BC08	(hydrophilic and anionic polysaccharides present in the cell wall of brown algae)	[9005-35-5] $(C_{12}H_{14}CaO_{12})_n$	hemostatic surgical dressing.	Wallace 1969 produced from seaweed [189]
Epinephrine (adrenaline) [Adrenaclick, Anapen, Epipen, Fastjekt, Suprarenin, Twinject] B02BC09, A01AD01, C01CA24, R03AA01	(natural hormone)	[51-43-4] $C_9H_{13}NO_3$ 183.21 hydrochloride salt: [55-31-2] $C_9H_{13}NO_3 \cdot HCl$ 219.67 tartrate salt: [51-42-3] $C_9H_{13}NO_3$ $\cdot C_4H_6O_6$ 333.29	neurotransmitter binding to adrenergic receptors. sympathomimetic, vasoconstrictor used to treat anaphylaxis and cardiac arrest. amp. 0.05 mg/10 mL, 1 mg/1 mL, 2.05 mg/2.05 mL (as hydrochloride); eye drops 1.25%, 2 mg/mL, 5 mg/mL; eye ointment 1 mg/g (as tartrate); syringe 1 mg/1 mL and pen injector.	Hoechst 1900 synthetic [186, 190]

(*continued*)

Table 12. (*Continued*)

INN (Synonyms) [Brand names] ATC	Structure (Remarks)	[CAS-No.] Formula MW, g/mol	Target (if known) Medical use Formulation DDD	Originator Approval (country/year) Production method [References]
Thromboplastin (TPL, thrombokinase) B02BC12	protein consisting of 295 amino acids		tissue coagulation factor III found in plasma aiding blood coagulation through catalyzing the conversion of prothrombin into thrombin.	derived from placental sources [191]
Polyglycolic acid (PGA) [Assucryl, Dexon] B02BC13	(biodegradable thermoplastic polymer)	[26124-63-5] $(C_2H_2O_2)_n$	absorbable suture used in surgery.	American Cyanamide 1962 synthetic [192]
Gelatin [Floseal, Orabase, Surgiflo] B02BC14	mixture of peptides and proteins	[9000-70-3]	chemotactic properties on fibroblasts. hemostatic as it provides a physical framework within which clotting may occur, sealant. injection vial 4.5 mg/mL and paste for topical or oral use.	Baxter/ConvaTec 1999 derived from collagen by hydrolysis [193]

3.2.4. B02BX Other Systemic Hemostatics (Table 13)

Table 13. B02BX Other systemic hemostatics

INN (Synonyms) [Brand names] ATC	Structure (Remarks)	[CAS-No.] Formula MW, g/mol	Target (if known) Medical use Formulation DDD	Originator Approval (country/year) Production method [References]
Etamsylate (ethamsylate) [Altodor, Cyclonamine, Dicynene, Dicynone, Eselin, Haemostop] B02BX01		[88-46-0] $C_6H_6O_5S$ 190.18 diethylammonium salt: [2624-44-4] $C_6H_6O_5S \cdot$ $C_4H_{11}N$ 263.31	promotor of platelet adhesion. hemostatic (capillary protective). amp. 250 mg/2 mL; tabl. 250, 500 mg.	Sanofi IT synthetic [194]
Carbazochrome [Adona, Adrenoxyl, Fleboside, Sumlin, Tazin] B02BX02		[69-81-8] $C_{10}H_{12}N_4O_3$ 236.23	promotes clotting via platelet aggregation and adhesion. antihemorrhagic, hemostatic. amp 5 mg/1 mL; gran. 10%; inj. sol. 1.5 mg/3.6 mL, 50 mg/10 mL; tabl. 2.5, 10, 30 mg.	Beecham 1953 synthetic [195, 196]
Batroxobin [Botropase, Defibrase, Reptilase] B02BX03	serine protease with 231 amino acids isolated from the venom of a pit viper, *Bothrops atrox*	[9039-61-6]	serine protease defibrinogating hemostatic agent. anticoagulant, fibrinolytic. amp. 10 iu/1 mL; 20 iu.	Pentapharma EU/1954 recombinant production [197, 198]
Romiplostim (AMG 531) [Nplate] B02BX04	dimer Fc-peptide fusion protein with each 269 amino acids	[2167639-76-9] $C_{2634}H_{4086}N_{722}$ $O_{790}S_{18}$	fusion protein analogue of thrombopoietin, a hormone that regulates platelet production. treatment of chronic idiopathic (immune) thrombocytopenic purpura (ITP). vial, 250 μg for s.c. injection 30 μg P.	Amgen US/2008 recombinant [199]

(continued)

Table 13. (*Continued*)

INN (Synonyms) [Brand names] ATC	Structure (Remarks)	[CAS-No.] Formula MW, g/mol	Target (if known) Medical use Formulation DDD	Originator Approval (country/year) Production method [References]
Eltrombopag (SB-497115-GR) [Promacta, Revolade] B02BX05		[496775-61-2] $C_{25}H_{22}N_4O_4$ 442.48 olamine salt: [96775-62-3] $C_{29}H_{36}N_6O_6$ 564.64	selective thrombopoietin receptor agonist that increases platelet count. antithrombocyto-penic used to treat idiopathic thrombocy-topenic purpura (ITP). tabl. 25, 50, and 75 mg and oral suspension. 50 mg O.	GlaxoSmithKline US/2008 synthetic [200, 201]
Emicizumab (ACE 910) [Hemlibra] B02BX06	humanized monoclonal modified immunoglobulin G4 (IgG4) antibody with a bispecific antibody structure	[1610943-06-0] $C_{6434}H_{9940}N_{1724}$ $O_{2047}S_{45}$ 145 639.0	mimics the function of coagulation factor VIII. humanized bispecific antibody for the treatment of hemophilia A. solution 30 mg/mL for sc injection.	Roche/Chugai US/2019 recombinant production in CHO cells [202]
Lusutrombopag (S-888711) [Mulpleta] B02BX07		[1110766-97-6] $C_{29}H_{32}Cl_2N_2$ O_5S 591.6	thrombopoietin receptor agonist. improvement of thrombocytope-nia. f.c. tabl. 3 mg.	Shionogi JP/2015; US/2019 synthetic [203, 204]
Avatrombopag (AKR-501; E-5501; YM-477) [Doptelet] B02BX08		[570406-98-3] $C_{29}H_{34}Cl_2N_6$ O_3S_2 649.7 maleate salt: [677007-74-8] $C_{29}H_{34}Cl_2N_6$ $O_3S_2 \cdot C_4H_4O_4$ 765.7	thrombopoietin receptor agonist. treatment of thrombocytope-nia. tabl. 20 mg base as maleate.	AkaRx US/2018 synthetic [205]
Fostamatinib (R788; NSC-745942; prodrug of tamatinib) [Tavalisse] B02BX09		[025687-58-4] $C_{23}H_{24}FN_6O_9P$ ·2 Na 624.4 free acid: [901119-35-5] $C_{23}H_{26}FN_6O_9P$ 580.5	SYK tyrosine kinase inhibitor. treatment of autoimmune thrombocytope-nia, antirheumatic. tabl. 100 and 150 mg base.	Rigel Pharma. US/2019 synthetic [206, 207]

List of Abbreviations

AATD	Alpha-1 antitrypsin deficiency
ACS	acute coronary syndrome
AF	atrial fibrillation
AMI	acute myocardial infarction
amp.	ampule(s)
ASA	acetylsalicylic acid/aspirin
ATIII	antithrombin III
CHO	Chinese hamster ovary (cells)
cps.	capsules
COX	cyclooxygenase
DDD	defined daily dose
DOAC	direct acting oral anticoagulant
DVT	deep vein thrombosis
drg.	dragee
eff.	effervescent
FGF	fibroblast growth factor
FVIIa	factor VII activated
FVIII	factor VIII
FXa	factor X activated
f.c.	film coated
GPIIa	glycoprotein IIa
gran.	granules
i.m.	intramuscular
inj.	injection
i.v.	intravenous
INN	international nonproprietary name
ITP	idiopathic thrombocytopenic purpura
LMWH	low-molecular-weight heparin
mL	milliliter (cubic centimeter)
MW	molecular mass
N	nasal
n.a.	not available
NSAID	nonsteroidal anti-inflammatory drug
O	oral
P	parenteral
PAH	pulmonary arterial hypertension
PDE	phosphodiesterase
P2Y12	purinergic P2Y12 chemoreceptor for adenosine diphosphate
sol.	solution
s.r.	slow release
suppos.	suppositories
susp.	suspension
SYK	spleen tyrosine kinase
synth.	synthesis
tabl.	tablets
TF	tissue factor
t-PA	tissue-plasminogen activator

TPO	thrombopoietin
TPO-RA	thrombopoietin receptor agonist
iu	international units according to the WHO Expert Committee on Biological Standardization
UFH	unfractionated heparin
VKA	vitamin K antagonists
VT	venous thrombosis
VTE	venous thromboembolism
VWF	von Willebrand factor
wfm	withdrawn from market
WHO	World Health Organization

References

1 Farrugia, A. and Cassar, J. (2012) *Blood Transfus.* **10** (3), 273–278.
2 Palta, S., et al. (2014) *Ind. J. Anaesth.* **58** (5), 515–523.
3 Smith, S. (2009) *J. Vet. Emerg. Crit. Care* **19** (3), 3–10.
4 Felgin, V., et al. (2003) *Lancet Neurol.* **2**, 43–53.
5 Longo, D., et al. (2011) *Harrison's Principles of Internal Medicine*, 18th edn, McGraw-Hill, New York.
6 Rao, P., et al. (2017) *Blood Rev.* **31**, 205–211.
7 Zacconi, T.C., (2018) Intechopen, DOI: 10.5772/Intechopen.76518
8 Hall, R., et al. (2011) *Anesth. Analg.* **112** (2), 292–318.
9 Moncada, S., et al. (1976) *Nature* **263** (5579), 663–665.
10 Gremmel, T., et al. (2018) *Res. Pract. Thromb. Haemost.* **2**, 439–449.
11 George, S., et al. (2019) *Clin. Drug Investig.* **39**, 495–502.
12 Dagar, V.K. and YP, A.K. (2017) *Bioengineered* **8** (4), 331–358.
13 Fan, P., et al. (2018) *J. Thorac. Dis.* **10** (3), 2011–2015.
14 Johnson, S.S., et al. (2011) *J. Occup. Environ. Med.* **53**, 2–7.
15 Ghareep, H. and Karaman, R. (2015) *Commonly Used Drugs*, Chap. 6, Nova Science Publishers.
16 Benett, B.L. (2011) *Wilderness Environ. Med.* **28**, 539–549.
17 Galanakis, I., et al. (2011) *Rev. Urol.* **13** (3), 131.
18 Scarano, A., et al. (2013) *Int. J. Immunopathol. Pharmacol.* **26** (4), 847–854.
19 Grand View Research Market Report February (2019).
20 Ghanima, W., et al. (2018) *Haematologica* **104** (6), 1112.
21 Erickson-Miller, C.I., et al. (2009) *Stem Cells* **27**, 424–430.
22 Nilles, K.M., et al. (2019) *Hematol. Commun.* **3** (11), 1423.
23 Al-Samkari, H., et al. (2019) *Ther. Adv. Hematol.* **10**, 1–13.
24 Bussel, J., et al. (2018) *Am. J. Hematol.* **93** (7), 921–930.
25 Connell, N.T. and Berliner, N. (2019) *Blood* **133** (19), 2027–2030.
26 Kresge, N., et al. (2005) *J. Biol. Chem.* **280** (8), e5.
27 Naisbitt, D.J., et al. (2005) *J. Pharmacol. Exp. Ther.* **313** (3), 1058–6529.
28 Wisconsin Alumni Research, US 2 427 578 (16.9.1947; prior. 2.4.1945).
29 Wisconsin Alumni Research, US 3 077 481 (12.2.1963; appl. 21.2.1961).
30 Hoffmann-La Roche, (1955) US 2 701 804 (CH-prior. 1952).
31 Wisconsin Alumni Research, US 3 239 529 (8.3.1966; appl. 1.3.1962).
32 Ivanov, I.C., et al. (1990) *Arch. Pharm. Ber. Dtsch. Pharm. Ges. (APBDAJ)* **323**, 521.

33 Geigy, (1953) US 2 648 682 (CH-prior. 1950).
34 Spójené farmaceutické Zovody, (1949) US 2 482 510.
35 Lund, E. (1957) *Acta Med Scand.* **157** (1), 39–41.
36 Upjohn, (1954) US 2 672 483 (prior. 1951).
37 Lipha, US 3 574 234 (6.4.1971; F-prior. 13.12.1966, 13.11.1967).
38 Mentre, F., et al. (1998) *Clin. Pharmacol. Ther.* **63** (1), 64–78.
39 Hind, H.G. (1963) *Manuf. Chem. (MACSAS)* **34**, 510.
40 Southern California Gland Co., US 2 884 358 (28.4.1959; appl. 22.4.1957).
41 Uclaf, US 2 989 438 (20.6.1961; appl. 29.12.1958).
42 Menache, D., et al. (1992) *Transfusion* **32** (6), 580–588.
43 Adiguzel, C., et al. (2009) *Clin. Appl. Thromb. Hemost.* **15** (6), 645–651.
44 Lindahl, U., et al. (1979) *Proc. Natl. Acad. Sci. USA (PNASA6)* **76**, 3198.
45 Kabi, A.B., US 4 303 651 (1.12.1981; appl. 4.1.1980; S-prior. 8.1.1979).
46 Rhône-Poulenc Rorer, US 5 389 618 (14.2.1995; F-prior. 26.6.1990).
47 Shafiq, N., et al. (2006) *Pharmacology* **78** (3), 136–143.
48 Maugeri, N., et al. (2007) *Thromb. Haemost.* **97** (6), 965–973.
49 Yusuf, S., et al. (2005) *JAMA* **293** (4), 427–435.
50 Akzo, EP 66 908 (15.12.1982; appl. 7.5.1982; NL-prior. 21.5.1981).
51 Akzo, US 4 438 108 (20.3.1984; appl. 12.5.1982; NL-prior. 21.5.1981).
52 Friedel, H.A., et al. (1994) *Drugs* **48** (4), 638–640.
53 Cosmi, B., et al. (2003) *Thromb. Res.* **109** (5), 333–339.
54 Opocrin, EP1833492 (19.9.2007; appl. 27.10.2005).
55 Chapman, T.A., et al. (2003) *Drugs* **63** (21), 2357–2377.
56 Rüdiger, R. (1999) *Dtsch. Artztebl.* **96** (44), A-2840.
57 Caprino, L., et al. (1977) *Haemostasis* **6** (5), 310–317.
58 Fidia, US 4 296 039 (20.10.1981; I-prior. 17.11.1977).
59 Weizmann, C., Bergmann, E. and Sulzbacher, M. (1950) *J. Org. Chem. (JOCEAH)* **15**, 918.
60 Selleri, R., et al. (1971) *Chim. Ther. (CHTPBA)* **6**, 203.
61 Manetti Roberts, FR 2 100 850 (appl. 30.6.1971; I-prior. 1.7.1970).
62 Sanofi, US 4 847 265 (11.7.1989; F-prior. 17.2.1987, 27.11.1987).
63 Elf Sanofi, EP 281 459 (appl. 16.2.1988; F-prior. 17.2.1987).
64 Sanofi, WO 9 839 322 (appl. 5.3.1998; F-prior. 5.3.1997).
65 Garcia, M.J. and Azerad, R. (1997) *Tetrahedron: Asymmetry (TASYE3)* **8** (1), 85.
66 Centre Etud. Ind. Pharm., DE 2 404 308 (prior. 30.1.1974).
67 Centre Etud. Ind. Pharm., US 4 051 141 (27.9.1977; F-prior. 1.2.1973).
68 Maffrand, J.P. and Eloy, F. (1974) *Eur. J. Med. Chem. (EJMCA5)* **9**, 483.
69 *Ullmanns Encykl. Tech. Chem.*, 3. Aufl., Vol. 13, 90.
70 Norwich Pharmacal, US 3 235 583 (15.2.1966; appl. 22.7.1964).
71 Thomae, DE 1 116 676 (appl. 1955).
72 Thomae, GB 807 826 (appl. 1956; D-prior. 1955).
73 Thomae, US 3 031 450 (24.4.1962; D-prior. 1959).
74 Parrott, E.L. (1962) *J. Pharm. Sci. (JPMSAE)* **51**, 897.
75 Lee Labs., US 2 003 374 (1935; appl. 1932).
76 Corey, E.J., et al. (1976) *J. Am. Chem. Soc. (JACSAT)* **99**, 2006.
77 Nicolaou, K.C., et al. (1977) *Lancet (LANCAO)* 1058.
78 Burroughs Wellcome, US 4 539 333 (3.9.1985; GB-prior. 11.5.1976, 17.8.1976, 3.9.1976).
79 Burroughs Wellcome, US 4 883 812 (28.11.1989; GB-prior. 17.8.1976, 3.9.1976).
80 Nannini, G., et al. (1973) *Arzneim.-Forsch. (ARZNAD)* **23**, 1090.
81 Carlo Erba, US 4 118 504 (3.10.1978; I-prior. 10.11.1970).
82 Skuballa, W. and Vorbrueggen, H. (1981) *Angew. Chem. (ANCEAD)* **93** (12), 1080.
83 Schering AG, DE 2 845 770 (D-prior. 19.10.1978).
84 Schering AG, DE 3 839 155 (D-prior. 17.11.1988).
85 Schrader, B.J., et al. (1990) *Clin. Pharm.* **9** (2), 118–124.
86 Knight, D.M., et al. (1995) *Mol. Immunol. (MOIMD5)* **32**, 1271.
87 Gold, H.K., et al. (1989) *Circulation Suppl. (CISUAQ)* **80** (4) Abst.), 1063.
88 Centocor, WO 9 512 412 (11.5.1995; USA-prior. 5.11.1993).
89 Bristol Myers Co., US 3 932 407 (13.1.1976; USA-prior. 4.2.1972).
90 Bristol Myers Co., US 4 208 521 (17.6.1980; USA-prior. 31.7.1978).
91 Cummings, A.J., et al. (1963) *J. Pharm. Pharmacol. (JPPMAB)* **15**, 56.
92 COR Therap., WO 9 015 620 (appl. 15.6.1990; USA-prior. 20.2.1990).
93 Scarborough, R.M., et al. (1993) *J. Biol. Chem. (JBCHA3)* **268**, 1066–1073.
94 Egbertson, M.S., et al. (1994) *J. Med. Chem. (JMCMAR)* **37**, 2537.
95 Merck & Co., US 5 292 756 (8.3.1994; USA-prior. 27.9.1990, 30.8.1991).
96 Murdioch, D., et al. (2006) *Drugs* **66** (5), 671–692.
97 Expermed, WO2015051811 (16.4.2015; appl. 10.10.2013).
98 Melian, E.B., et al. (2002) *Drugs* **62** (1), 107–133.
99 United Therapeutics Corp., US 6 441 245 (27.8.2002; USA-prior. 24.10.1997).
100 Ube Ind., EP 542 411 (12.8.1998; J-prior. 9.9.1991).
101 Teva Pharma., US 20 020 099 213 (25.7.2002; USA-prior. 20.3.2001).
102 Springthorpe, B., et al. (2007) *Bioorg. Med. Chem. Lett.* **17**, 6013–6018.
103 AstraZeneca, US 6 251 910 (26.6.2001; appl. 21.9.1998; SE-prior. 22.7.1997).
104 Fisons PLC, WO 9 418 216 (18.8.1994; appl. 8.2.1994; GB-prior. 10.2.1993).
105 Medicines Co., US 8 759 316 (24.6.2014; appl. 28.6.2013; USA-prior. 10.11.2010).
106 Chackalamannil, S., et al. (2008) *J. Med. Chem.* **51**, 3061–3064.
107 Schering Corp., US 6 326 380 (4.12.2001; appl. 7.4.2000; USA-prior. 25.11.1997).
108 Asaki, T., et al. (2015) *J. Med. Chem.* **58** (18), 7128–7137.
109 Nippon Shinyaku, US 7 205 302 (27.5.2004; appl. 23.10.2003; JP-prior. 26.4.2001).
110 Merck & Co., (1954) US 2 666 729 (appl. 1951).
111 American Cyanamid, US 2 701 227 (1955; appl. 1951).
112 Behringwerke, US 3 063 913 (13.11.1962; D-prior. 29.12.1959).
113 Mc Cartney, P. (2019) *JAMA* **321**, 56–68.
114 Hannaford, P., et al. (1995) *Br. J. Gen. Pract.* **45** (393), 175–179.
115 Sobel, G.W., et al. (1952) *Am. J. Physiol. (AJPHAP)* **171**, 768.
116 Ortho (1960) US 2 961 382 (appl. 1957).
117 Parke Davis, (1953) US 2 624 691 (appl. 1946).
118 Ortho Pharmaceutical, US 3 136 703 (9.6.1964; prior. 1.10.1957, 22.4.1958).
119 Cutter Labs., US 3 234 106 (8.2.1966; appl. 3.12.1962).
120 Frisch, E.P. (1972) *J. Clin. Pathol.* **25** (7), 654–655.

121 Boehringer Mannheim, CA2107476 (17.10.1992; USA-prior. 16.4.1991).

122 Moser, M., et al. (1999) *Exp. Opin. Investig. Drugs* **8** (3), 329–335.

123 Hennerici, M.G., et al. (2006) *Lancet* **368** (9550), 1871–1878.

124 Lilly, E., CA2139468 (6.7.1995; USA-prior. 5.1.1994).

125 Parsons, M., et al. (2012) *New Engl. J. Med.* **399** (12), 1099–1107.

126 Genentech, CA1341432 (17.6.2003; USA-prior. 20.5.1988).

127 Esmon, C.T., et al. (1987) *Adv. Exp. Med. Biol.* **214**, 47–54.

128 Matheson, A.J. and Goa, K.L. (2000) *Drugs* **60** (3), 679–700.

129 Lubenow, N., et al. (2005) *J. Thromb. Haemostat.* **3** (11), 2428–2436.

130 Mitsubishi Chem. Ind., EP 8 746 (11.8.1982; appl. 22.8.1979; USA-prior. 31.8.1978).

131 Mitsubishi Chem. Corp., US 5 925 760 (20.7.1999; appl. 13.3.1992; J-prior. 7.8.1996).

132 Gustafsson, D., et al. (1998) *Thromb. Haemost.* **79**, 110–118.

133 Gustafsson, D., et al. (2004) *Nat. Rev. Drug Discovery (NRDDAG)* **3**, 649.

134 Seebach, D., et al. (1990) *Liebigs Ann. Chem. (LACHDL)* 687.

135 Astra, WO 9 723 499 (3.7.1997; appl. 17.12.1996; GB-prior. 21.12.1995).

136 Okayama, T., et al. (1996) *Chem. Pharm. Bull. (CPBTAL)* **44** (7), 1344–1350.

137 Biogen, US 5 196 404 (23.3.1993; appl. 6.7.1990; USA-prior. 18.8.1989).

138 Novetide, WO 20 070 033 383 (22.3.2007; appl. 14.9.2006; USA-prior. 14.9.2005).

139 Hanel, N., et al. (2002) *J. Med. Chem. (JMCMAR)* **45**, 1757–1766.

140 Boehringer Ingelheim, WO 9 837 075 (27.8.1998; appl. 16.2.1998; D-prior. 18.2.1997).

141 Boehringer Ingelheim, WO 2 007 071 743 (28.6.2007; D-prior. 20.12.2005).

142 Roehrig, S. et al.: J. Med. Chem. (JMCMAR) **48**, 5900–5908 (2005).

143 Bayer AG, WO 200 147 919 (5.7.2001; appl. 11.12.2000; D-prior. 24.12.1999).

144 Pinto, D.J.P., et al. (2007) *J. Med. Chem.* **50**, 5339.

145 Pinto, D.J.P., et al. (2006) *Bioorg. Med. Chem.* **16**, 4141–4147.

146 BMS, US 6 413 980 (2.7.2002; appl. 22.12.1999; USA-prior. 23.12.1998).

147 Yoshikawa, K., et al. (2009) *Bioorg. Med. Chem.* **17**, 8221.

148 Haginoya, N., et al. (2004) *J. Med. Chem.* **47**, 5167.

149 Daiichi Seiyaku, WO 0 174 774 (11.10.2001; appl. 5.4.2001; JP-prior. 5.4.2000).

150 Zhang, P. (2009) *Bioorg. Med. Chem. Lett.* **19** (8), 2179–2185.

151 COR Therapeutics, US 6 376 515 (23.4.2002; appl. 28.2.2001; USA-prior. 29.2.2000).

152 Millenium Pharms., US 8 524 907 (3.9.2013; appl. 1.11.2007; USA-prior. 2.11.2006).

153 Portola Pharms., US 8 946 269 (3.2.2015; appl. 31.8.2011; USA-prior. 18.3.2011).

154 Dalle, J.H., et al. (2016) *Biol. Blood Marrow Transplant.* **22** (3), 400–409.

155 Crinos Ind., US4693995 (15.9.1987; appl. 14.2.1985; IT-prior. 16.2.1984).

156 Verges, J., et al. (2004) *Proc. West. Pharmacol. Soc.* **47**, 50–53.

157 Linhardt, R.J. (2003) *J. Med. Chem. (JMCMAR)* **46**, 2551–2564.

158 Choay, US 4 818 816 (4.4.1989; F-prior. 28.4.1981).

159 Jerebtsova, M., et al. (2007) *Am. J. Physiol. Heart Circ.* **292** (2), 743–750.

160 Gabriel, S., et al. (1899) *Ber. Dtsch. Chem. Ges. (BDCGAS)* **32**, 1266.

161 Galat, A., et al. (1946) *J. Am. Chem. Soc. (JACSAT)* **68**, 2729.

162 American Enka Corp., US 2 453 234 (1948; NL-prior. 1946).

163 Einhorn, A. and Ladisch, C. (1900) *Justus Liebigs Ann. Chem. (JLACBF)* **310** (194).

164 Levine, M. and Sedlecky, R. (1959) *J. Org. Chem. (JOCEAH)* **24**, 115.

165 Daiichi Seiyaku, DAS 1 443 755 (appl. 23.12.1964; J-prior. 24.12.1963).

166 Heidrich, R., et al. (1978) *J. Neurol.* **219** (1), 83–85.

167 Kunitz, M., et al. (1936) *J. Gen. Physiol.* **19** (6), 991–1007.

168 Bayer AG, DE 2 748 295 (prior. 27.10.1977).

169 Bayer AG, DE 3 339 693 (15.5.1985; appl. 3.11.1983).

170 Mahdy, A.M. (2004) *Br. J. Anaesth.* **93** (6), 842–858.

171 Axelson, U., et al. (1965) *Am. J. Human Genet.* **17** (6), 466–472.

172 Parr, D.G., et al. (2017) *Drud Des. Devel. Ther.* **11**, 2149–2162.

173 Ono Pharmac., DOS 2 548 886 (appl. 31.10.1975; J-prior. 1.11.1974).

174 Ono Pharmac., US 4 021 472 (3.5.1977; J-prior. 1.11.1974).

175 Fieser, L.F. (1939) *J. Am. Chem. Soc. (JACSAT)* **61**, 2559, 3467.

176 Hirschmann, R., et al. (1954) *J. Am. Chem. Soc. (JACSAT)* **76**, 4592.

177 Isler, O. and Doebel, K. (1939) *Helv. Chim. Acta (HCACAV)* **22**, 945; (1954) **37**, 225.

178 Roche, (1943) US 2 325 681 (CH-prior. 1939).

179 Fieser, L.F., et al. (1939) *J. Am. Chem. Soc. (JACSAT)* **61**, 2559, 3216.

180 Velsicol, (1946) US 2 402 226 (appl. 1943).

181 Acharya, S.S., et al. (2008) *Haemophilia* **114**, 1151–1158.

182 Shenoi, P.M. (1973) *Proc. R. Soc. Med.* **66** (2), 193–196.

183 Oto, A., et al. (1999) *Am. J. Roentgenol.* **172** (6), 1481–1484.

184 Demain, D.L., et al. (1954) *Arch. Biochem. Biophys.* **51** (1), 114–121.

185 *Stolz: Ber. Dtsch. Chem. Ges. (BDCGAS)* **37**, 4152 (1904).

186 Hoechst, (1903) DRP 152 814.

187 Crawley, J.T., et al. (2007) *J. Thromb. Heamost.* **5**, 95–101.

188 Singh, O., et al. (2011) *J. Cutanous Aesth. Surg.* **4** (1), 12–16.

189 Szekalska, M., et al. (2016) *Int. J. Polym. Sci.* 769703.

190 Tullar, B.F. (1948) *J. Am. Chem. Soc. (JACSAT)* **70**, 2067.

191 van der Besselaar, A.M., et al. (2010) *Biologicals* **38** (2), 430–436.

192 Gilding, D.K., et al. (1979) *Polymer* **20** (12), 1559–1464.

193 Coenen, M., et al. (2006) *Equine Vet. J. Suppl.* **36**, 606–610.

194 Lab. OM S.A., GB 895 709 (appl. 31.12.1959; CH-prior. 28.1.1959).

195 Soc. Belge de l'Azote et des Prod. Chim., (1950) US 2 506 294 (B-prior. 1943).

196 Labaz, GB 806 908 (appl. 1957; USA-prior. 1956).

197 Pentapharm, US 3 849 252 (19.11.1974; CH-prior. 18.1.1971).

198 Pentapharm, DOS 2 201 993 (appl. 17.1.1972; CH-prior. 18.1.1971).

199 Kuter, D.J., et al. (2008) *Lancet* **371** (9610), 395–404.

200 SmithKline Beecham, WO 0 189 457 (29.11.2001; appl. 24.5.2001; USA-prior. 25.5.2000).

201 SmithKline Beecham, WO 03 098 992 (4.12.2003; appl. 21.5.2003; USA-prior. 22.5.2002).

202 Shima, M., et al. (2016) *New Engl. J. Med.* **374** (21), 2044–2053.

203 Shionogi, US 7 601 746 (13.10.2013; appl. 22.2.2007; JP-prior. 12.8.2003).

204 Flick, A.C., et al. (2017) *J. Med. Chem.* **60**, 6480–6515.

205 Astellas Pharma, US 7 638 536 (29.12.2009; appl. 15.1.2003; JP-prior. 18.1.2002).

206 Hart, R., et al. (2015) *Org. Process Res. Dev.* **19**, 537–542.

207 Rigel Pharms., WO 2 005 012 294 (10.2.2005; appl. 30.7.2004; USA-prior. 30.7.2003).

Cardiovascular System (C)

AXEL KLEEMANN, Hanau, Germany

Ullmann's Pharmaceuticals. Edited by Axel Kleemann and Bernhard Kutscher.
© 2022 Wiley-VCH GmbH
Set ISBN: 978-3-527-34252-5/ DOI: 10.1002/14356007.a05_289.pub3

1. Introduction

In the following chapters, the different cardiovascular (CV) drug classes are reviewed and arranged according to the Anatomical Therapeutical Chemical Classification (ATC) of the WHO, in which they are arranged according to the organs or organ systems on which they act, and their chemical, pharmacological, and therapeutic properties. For each drug, the ATC classification number is given, except in the very rare cases where such an ATC numbering is still pending. As the new drugs are numbered consecutively in the ATC classification system, the new chemical entities (NCEs) that are approved each year can always be added easily. By this procedure it will be possible to keep the Ullmann's chapters

on pharmaceuticals up-to-date without writing new articles every few years.

Biopharmaceuticals (e.g., antibodies, polypeptides, proteins ...) are not included in the tables, only so-called "small molecules". The references for each drug include the basic patents and publications, in which the relevant synthetic or biosynthetic processes for production are described. In addition, the reader is referred to the standard reference source [1] and/or its online version, which is updated 1–2 times per year and so contains the most recent drug substances with their syntheses.

The syntheses of older cardiovascular drugs can also often be found in the respective older Ullmann's chapters (where it deemed appropriate, some text explanations were taken over from this earlier chapter).

For detailed informations about pharmacology, pharmacokinetics, metabolism, mechanism of action, toxicology, interactions with other drugs etc., the reader is referred to [2, 3].

The general review publications and the review references directly under the headlines of certain chapters or tables have been carefully selected especially for the use of chemists, who want to be directed to interesting overviews. Drugs that have been withdrawn from the market or drugs that have become economically insignificant are not included, even if they are still contained in the ATC list. In the tables of the particular chapters, the individual drug substances are characterized by their INN, synonyms, brand or trade names, ATC number, structures, CAS-no., formula, molecular mass, mode of action, medical use, year of approval, formulation, DDD (Defined Daily Dose), originator and references as shown here:

INN	Structure	[CAS-No.]	Target (if	Originator(s)
(Synonyms)	(Remarks)	Formula	known)	Approval
[Brand		MW (g/mol)	Medical use	(country/year)
names]			Formulation	Production
ATC			DDD	method
				[References]

Cardiovascular drugs are also classified in the ATC groups **B01** (Antithrombotic agents) and **B02** (Antihemorrhagics).

According to WHO figures for the year 2012, cardiovascular disease (CVD) was number one cause of mortality worldwide with 17.5 million deaths (= 46% of all noncommunicable disease (NCD) deaths), an estimated 7.4 million due to CHD and 6.7 million due to stroke. Worldwide, a decline in CVD mortality rates was seen since 2000, highest decline in high-income countries. 58% of cardiovascular deaths are attributable to high blood pressure. Other risk factors are smoking, and unhealthy diets [4].

Although CVDs are a globally increasing burden and still the leading cause of death, the global market of CV drugs declined considerably within the last five years. The main reason is that almost all effective antihypertensives including ACE inhibitors and ATII antagonists ("sartans") and also the lipid-lowering statins have lost patent protection and exclusivity, and face heavy generic competition (e.g., Pfizer's Lipitor®/atorvastatin with peak sales of $ 13 × 10^9$ in 2011, AstraZeneca's Crestor®/Rosuvastatin with peak sales over $ 7 × 10^9$ before 2016, and Novartis' Diovan®/valsartan with peak sales over $ 6 × 10^9$). In parallel there has been a steep decline in CV drug development since the 1990s, resulting in only few NCEs reaching the market. Pharmaceutical companies have experienced a number of late-stage failures of once-promising CVD drug candidates, often after very expensive phase III clinical trials each with more than 20 000 patients. This, and also higher hurdles for registration, diminished the interest to develop new CVD drugs. Instead, some companies lay focus on dual or triple combinations, e.g., of sartans with calcium channel blockers and/or diuretics and statins, and look for synergistic effects. Quite a number of such combinations are now in clinical phase III [5, 6].

The global market for antihypertensive drugs including sartans, ACE inhibitors, calcium antagonists, betablocker, diuretics, fixed-dose combinations, and others was estimated for the year 2015 at US $ 28.2 × 10^9$. It is projected to decline to $ 24.9 × 10^9$ in 2020 [7].

2. C Cardiovascular System

2.1. C01 Cardiac Therapy

2.1.1. C01A Cardiac Glycosides

2.1.1.1. C01AA Digitalis Glycosides (Table 1)

Table 1. C01AA Digitalis glycosides

INN (Synonyms) [Brand names] ATC	Structure (Remarks)			[CAS-No.] Formula MW (g/mol)	Target (if known) Medical use Formulation DDD	Originator(s) Approval (country/year) Production method [References]

	R^1	R^2	R^3			
β-Acetyldigoxin [Digotab, Novodigal, Stillacor] C01AA02	H	CH_3CO	OH	[5355-48-6] $C_{43}H_{66}O_{15}$ 822.99	Na^+,K^+-ATPase inhibitor (prodrug)/ cardiotonic. tbl. 0.1, 0.2 mg. 0.5 mg, O.	Beiersdorf/ Sandoz semisynth. from digoxin [8, 9]
Digitoxin [Digimerck, Digitaline] C01AA04	H	H	H	[71-63-6] $C_{41}H_{64}O_{13}$ 764.95	Na^+,K^+-ATPase inhibitor/CHF and atrial fibrillation. tbl. 0.05, 0.1 mg; amp. 0.1, 0.25 mg/1 mL. 0.1 mg, O, P.	Sandoz extraction from *Digitalis purpurea* [10]
Digoxin [Lanicor, Lanoxin] C01AA05 **WHO essential medicine**	H	H	OH	[20830-75-5] $C_{41}H_{64}O_{14}$ 780.96	Na^+,K^+-ATPase inhibitor/CHF and atrial fibrillation. tbl. 0.125, 0.25 mg; solution 0.05 mg/1 mL. 0.25 mg, O, P.	Wellcome Found. extraction from *Digitalis purpurea* (*foxglove*). [11]
Metildigoxin **(Medigoxin)** [Lanitop, Cardiolan] C01AA08	H	CH_3	OH	[30685-43-9] $C_{42}H_{66}O_{14}$ 794.98	Na^+,K^+-ATPase inhibitor/CHF and atrial fibrillation. tbl. 0.05, 0.1, 0.15 mg; amp. 0.65 mg/1 mL. 0.2 mg, O, P.	Boehringer Mannheim semisynth. from digoxin [12]

2.1.2. C01B Antiarrhythmics, Class I and III

Although many antiarrhythmic drugs exert multiple effects that contribute to their clinical actions, in the ATC code system they are classified according to their major action into

Class I:	Na^+ channel block
Class Ia:	$T_{recovery}$ 1–10 s
Class Ib:	$T_{recovery}$ <1 s
Class Ic:	$T_{recovery}$ >10 s
$T_{recovery}$:	rate of recovery from drug-induced block under physiological conditions

Class II:	β blockade (see C07A)
Class III:	K^+ channel block (action potential prolongation)
Class IV:	Ca^{2+} channel block (see C08D)

Here under C01B only antiarrhythmics of class I and III are referenced.

Pharmacology and clinical application of these drugs are reviewed in detail by K. J. SAMPSON and R. S. KASS in "Goodman & Gilman's" [13].

2.1.2.1. C01BA Antiarrhythmics, Class Ia (Table 2)

Table 2. C01BA Antiarrhythmics, class Ia

INN (Synonyms) [Brand names] ATC	Structure (Remarks)	[CAS-No.] Formula MW (g/mol)	Target (if known) Medical use Formulation DDD	Originator(s) Approval (country/year) Production method [References]
Quinidine (Chinidin) [Kinidin Durules, Cardioquin, Cordichin, Quinora] C01BA01		[56-54-2] $C_{20}H_{24}N_2O_2$ 324.42 Gluconate: [7054-25-3] $C_{20}H_{24}N_2O_2 \cdot C_6H_{12}O_7$ 520.58 Hydrogen sulfate: [747-45-5] $C_{20}H_{24}N_2O_2 \cdot H_2SO_4$ 422.5 Sulfate: [50-54-4] $(C_{20}H_{24}N_2O_2)_2 \cdot H_2SO_4$ 746.93	blocks cardiac and sodium channels on neuronal cell membranes and ATP-sensitive potassium channels/antiarrhythmic agent to treat atrial flutter/fibrillation and ventricular arrhythmias; treatment of malaria. tbl. 100, 200, 300 mg; vials 800 mg gluconate/10 mL. 1200 mg, O.	Boehringer Mannheim DE/1918 (?) isomerization of quinine [14, 15]
Disopyramide (H-3292; SC-7031; SC-13957) [Isorythm, Lispine, Norpace, Ritmodan, Rythmodan] C01BA03	(racemate)	[3737-09-5] $C_{21}H_{29}N_3O$ 339.48 Phosphate: [22059-60-5] $C_{21}H_{29}N_3O \cdot H_3PO_4$ 437.48	block of cardiac sodium channels/ treatment of ventricular tachycardia. amp. 50 mg/10 mL; cps. 100, 150 mg. 400 mg, O, P.	Searle synth. [16, 17]
Ajmaline (Rauwolfine) [Gilurytmal, Ritmos, Tachmalin] C01BA05		[4360-12-7] $C_{20}H_{26}N_2O_2$ 326.44 Hydrochloride: [4410-48-4] $C_{20}H_{26}N_2O_2 \cdot HCl$	block of cardiac sodium channels/ treat of ventricular tachycardia. amp. 50 mg/10 mL. 50 mg, P.	S. and R.H. Siddiqui/India isolation from roots of *Rauwolfia serpentine (L.) Benth.* [18]

2.1.2.2. C01BB Antiarrythmics, Class Ib (Table 3)

Table 3. C01BB Antiarrythmics, class Ib

INN (Synonyms) [Brand names] ATC	Structure (Remarks)	[CAS-No.] Formula MW (g/mol)	Target (if known) Medical use Formulation DDD	Originator Approval (country/year) Production method [References]
Lidocaine (Lignocaine) [Xylocaine, Xylocard, Lidoderm] C01BB01 and C05AD01, D04AB01, L01BB02, R02AD02, S01HA07, S02DA01.		[137-58-6] $C_{14}H_{22}N_2O$ 234.34 Hydrochloride: [73-78-9] $C_{14}H_{22}N_2O \cdot HCl$ 270.80	cardiac sodium channel blocker/antiarrythmic and anesthetic. solution for inj. 10, 20 mg/1 mL and many different formulations for oral, nasal, topical applications.	Astra AB SE/1948 synth. [19]
WHO essential medicine Mexiletine (Ko-1173) [Mexitil] C01BB02	(racemate)	[31828-71-4] $C_{11}H_{17}NO$ 179.26 Hydrochloride: [5370-01-4] $C_{11}H_{17}NO \cdot HCl$ 215.72	cardiac sodium channel blocker/antiarrythmic agent; treatment of muscle stiffness from myotonic dystrophy. cps. 150, 200, 250 mg; amp. 250 mg/10 mL. 800 mg, O, P.	Boehringer Ing. DE/1979 synth. [20]

2.1.2.3. C01BC Antiarrhythmics, Class Ic (Table 4)

Table 4. C01BC Antiarrythmics, class Ic

INN (Synonyms) [Brand names] ATC	Structure (Remarks)	[CAS-No.] Formula MW (g/mol)	Target (if known) Medical use Formulation DDD	Originator Approval (country/year) Production method [References]
Propafenone (SA-79) [Rythmol, Rytmonorm] C01BC03	(racemate)	[*54063-53-5*] $C_{21}H_{27}NO_3$ 341.45 Hydrochloride: [*34183-22-7*] $C_{21}H_{27}NO_3 \cdot HCl$ 377.91	cardiac sodium channel and β-adrenergic receptor blocker/antiarrythmic agent and β-blocker. f.c. tbl. 150, 300 mg; cps. 225, 325, 425 mg. 500 mg, O.	Helopharm synth. [21]
Flecainide (R-818) [Tambocor] C01BC04	(racemate)	[*54143-55-4*] $C_{17}H_{20}F_6N_2O_3$ 414.35 Acetate: [*54143-56-5*] $C_{17}H_{20}F_6N_2O_3 \cdot C_2H_4O_2$ 474.40	cardiac sodium channel blocker/antiarrythmic for treatment of ventricular and supraventricular tachycardia and atrial fibrillation. amp. 50 mg/5 mL; tbl. 50, 100 mg. 500 mg, O, P.	Riker DE/1982; USA/1985 synth. [22, 23]
Lorcainide (R-15889; Ro13-1042/001) [Remivox, Lopantrol, Lorivox] C01BC07		[*59729-31-6*] $C_{22}H_{27}ClN_2O$ 370.92 Hydrochloride: [*58934-46-6*] $C_{22}H_{27}ClN_2O \cdot HCl$ 407.38	cardiac sodium channel blocker with prolonged duration of action/treatment of ventricular tachycardia. tbl. 100 mg.	Janssen synth. [24]

2.1.2.4. C01BD Antiarrhythmics, Class III (Table 5)

Table 5. C01BD Antiarrythmics, class III

INN (Synonyms) [Brand names] ATC	Structure (Remarks)	[CAS-No.] Formula MW (g/mol)	Target (if known) Medical use Formulation DDD	Originator Approval (country/year) Production method [References]
Amiodarone (L-3428) [Cordarone, Pacerone, Cordarex, Ancaron] C01BD01		[1951-25-3] $C_{25}H_{29}I_2NO_3$ 645.32 Hydrochloride: [19774-82-4] $C_{25}H_{29}I_2NO_3 \cdot HCl$ 681.78	potassium and sodium channel blocker/ antiarrythmic in case of ventricular tachycardia and fibrillation and atrial fibrillation. amp. 150 mg/3 mL; tbl. 100, 200, 400 mg. 200 mg, O, P.	Labaz USA/1974 synth. [25]
Dofetilide (UK-68798) [Tikosyn] C01BD04		[115256-11-6] $C_{19}H_{27}N_3O_3S_2$ 441.57	cardiac potassium channel blocker/atrial flutter and fibrillation. cps. 125, 250, 500 µg.	Pfizer USA/2000 synth. [26, 27]
Ibutilide (U-70226E) [Corvert] C01BD05		[122647-31-8] $C_{20}H_{36}N_2O_3S$ 384.59 Fumarate: [122647-32-9] $C_{20}H_{36}N_2O_3S \cdot C_4H_4O_4$ 885.24	potassium channel blocker/ treatment of atrial flutter and atrial fibrillation. vials 1 mg/10 mL. 1 mg, P.	Upjohn USA/1996 synth. [28, 29]
Dronedarone (SR-33589) [Multaq] C01BD07		[141626-36-0] $C_{31}H_{44}N_2O_5S$ 556.77 Hydrochloride: [141625-93-6] $C_{31}H_{44}N_2O_5S \cdot HCl$ 593.22	multi channel blocker (Na, K, Ca)/ treatment of atrial fibrillation and atrial flutter. tbl. 400 mg. 800 mg, O.	Sanofi USA/2009 synth. [30]

2.1.2.5. C01BG Antiarrhythmics, Class I and III (Table 6)

Table 6. C01BG Other antiarrhythmics, class I and III

INN (Synonyms) [Brand names] ATC	Structure (Remarks)	[CAS-No.] Formula MW (g/mol)	Target (if known) Medical use Formulation DDD	Originator Approval (country/year) Production method [References]
Vernakalant (MK-6621; RSD-1235) [Brinavess] C01BG11		[794466-70-9] C$_{20}$H$_{31}$NO$_4$ 349.47 Hydrochloride: [748810-28-8] C$_{20}$H$_{31}$NO$_4$·HCl 385.93	dual sodium/potassium channel blocker/acute conversion of atrial fibrillation. vials for inj. 200 mg/10 mL, 500 mg/25 mL. 200 mg, P.	Cardiome Pharma EU/2010 synth. [31]

2.1.3. C01C Cardiac Stimulants Excluding Cardiac Glycosides

2.1.3.1. C01CA Adrenergic and Dopaminergic Agents (Table 7)

Table 7. C01CA Adrenergic and dopaminergic agents

INN (Synonyms) [Brand names] ATC	Structure (Remarks)	[CAS-No.] Formula MW (g/mol)	Target (if known) Medical use Formulation DDD	Originator Approval (country/year) Production method [References]
Etilefrine [Cardanat, Effortil, Thomasin] C01CA01	((R,S) form)	[709-55-7] C$_{10}$H$_{15}$NO$_2$ 181.24 Hydrochloride: [943-17-9] C$_{10}$H$_{15}$NO$_2$·HCl 217.70	α- and β-adrenergic receptor stimulant; positive inotrope and chronotrope action/antihypotensive. drops 5 mg/1 mL; 7.5 mg/1 mL; tbl. 5, 10, 25 mg. 50 mg O, P.	Boehringer Ingelheim DE/1949 synth. [32, 33, 34]
Norepinephrine (Noradrenaline) [Arterenol, Levophed] C01CA03	((R)-(−) form)	[51-41-2] C$_8$H$_{11}$NO$_3$ 169.18 Hydrochloride: [329-56-6] C$_8$H$_{11}$NO$_3$·HCl 205.64 D-Tartrate: [69815-49-2] C$_8$H$_{11}$NO$_3$·C$_4$H$_6$O$_6$ 319.27	α$_1$-, β$_1$- and β$_3$-adrenergic receptor stimulant/septic anaphylactic shock. amp. 1.22 mg/1 mL (as hydrochloride); vials 25 mL. 6 mg P.	Sterling Drug synth. [35, 36]
Dopamine (ASL-279) [Cardiosteril, Dopastat, Dynatra, Inovan, Revivan] C01CA04 **WHO essential medicine**		[51-61-6] C$_8$H$_{11}$NO$_2$ 153.18 Hydrochloride: [62-31-7] C$_8$H$_{11}$NO$_2$·HCl 189.64	D$_1$, D$_2$, β$_1$ and at higher doses also α$_1$ and α$_2$ adrenergic receptor stimulant/correction of haemodynamic imbalances in shock situations. amp. 50 and 200 mg/5 mL. 500 mg P.	Wellcome Labs. USA/1973 synth. [37, 38, 39]

Table 7. (*Continued*)

INN (Synonyms) [Brand names] ATC	Structure (Remarks)	[CAS-No.] Formula MW (g/mol)	Target (if known) Medical use Formulation DDD	Originator Approval (country/year) Production method [References]
Phenylephrine [Neosynephrin, Nostril, Vibrocil, Visadron] C01CA06 and R01AA04, R01AB01	((R)-(−) form)	[59-42-7] $C_9H_{13}NO_2$ 167.21 Hydrochloride: [61-76-7] $C_9H_{13}NO_2 \cdot HCl$ 203.67	α_1-adrenergic receptor agonist/decongestant (substitute for pseudoephedrine), mydriatic. solution 10 mg/mL 4 mg P.	H. Legerlotz synth. [40, 41]
Dobutamine (CPD. 81929) [Dobuject, Dobutrex] C01CA07	((R,S) form)	[34368-04-2] $C_{18}H_{23}NO_3$ 301.39 Hydrochloride: [49745-95-1] $C_{18}H_{23}NO_3 \cdot HCl$ 337.85	β_1-adrenergic receptor stimulant; positive inotropic/septic or cardiogenic shock; increases cardiac output in cases of CHF. vials 280, 560 mg/50 mL. 500 mg P.	Eli Lilly USA/1978 synth. [42]
Dopexamine (FPL-60278) [Dopacard, Dopagard] C01CA14		[86197-47-9] $C_{22}H_{32}N_2O_2$ 356.51 Dihydrochloride: [86484-91-5] $C_{22}H_{32}N_2O_2 \cdot 2HCl$ 429.43	stimulant of β_2-adrenergic and peripheral dopamine D1 and D2 receptors/positive inotropic agent. amp. 50 mg/5 mL for infus. 0.5g P.	Fisons; Teva GB/1996 synth. [43]
Midodrine (ST-1085) [Gutron, ProAmatine, Amatine] C01CA17	((R,S) form)	[42794-76-3] $C_{12}H_{18}N_2O_4$ 254.29 Hydrochloride: [3092-17-9] $C_{12}H_{18}N_2O_4 \cdot HCl$ 290.75	α_1-adrenergic receptor agonist/antihypertensive in case of orthostatic hypotension and dysautonomia. tbl. 2.5, 5 mg; sol.1% for drops. 30 mg O.	Lentia (ÖSW, Linz) USA/1996 synth. [44]
Fenoldopam (SKF 82526-J) [Corlopam] C01CA19	((R,S) form)	[67227-56-9] $C_{16}H_{16}ClNO_3$ 305.76 Mesylate: [67227-57-0] $C_{16}H_{16}ClNO_3 \cdot CH_4O_3S$ 401.87 Hydrochloride: [181217-39-0] $C_{16}H_{16}ClNO_3 \cdot HCl$ 342.22	selective D_1 receptor partial agonist/ antihypertensive agent (improves renal perfusion) vials for inj. 10 mg/mL	SmithKline USA/1997 synth. [45, 46]
Cafedrine [Akrinor (comb.w/ Theodrenaline)] C01CA21		[58166-83-9] $C_{18}H_{23}N_5O_3$ 357.41 Hydrochloride: [3039-97-2] $C_{18}H_{23}N_5O_3 \cdot HCl$ 393.88	in combination with theodrenaline positive inotropic effect/application in anesthesiology to increase blood pressure during surgery, especially in spinal anesthesia, e.g., in cesarean section. amp. 200 mg + 10 mg theodrenalin/2 mL	Degussa DE/1963 synth. [47]

Table 7. (*Continued*)

INN (Synonyms) [Brand names] ATC	Structure (Remarks)	[CAS-No.] Formula MW (g/mol)	Target (if known) Medical use Formulation DDD	Originator Approval (country/year) Production method [References]
Theodrenaline [Akrinor (comb. w/ Cafedrine)] C01CA23	((*R,S*) form)	[*13460-98-5*] C$_{17}$H$_{21}$N$_5$O$_5$ 375.39 Hydrochloride: [*2572-61-4*] C$_{17}$H$_{21}$N$_5$O$_5$·HCl 411.85	cardiotonic/s cafedrine	Degussa DE/1963 synth. [48]
Epinephrine (Adrenaline) [Epifrin, EpiPen, Glaucon, Suprarenin] C01CA24 and A01AD01, B02BC09, R01AA14, R03AA01, S01EA01.	((*R*)-(−) form)	[*51-43-4*] C$_9$H$_{13}$NO$_3$ 183.21 Hydrochloride: [*55-31-2*] C$_9$H$_{13}$NO$_3$·HCl 219.67 D-Tartrate: [*51-42-3*] C$_9$H$_{13}$NO$_3$·C$_4$H$_6$O$_6$ 333.29	α- and β-adrenoceptor agonist with positive inotrope and chronotrope action/vasoconstrictor for combination with local anesthetics; treatment of anaphylactic shocks and cardiac arrest. amp. 0.5, 1 mg/1 mL. EpiPen Auto-injector 0.15 and 0.3mg. 0.5 mg, P.	Farbwerke Hoechst DE/1906 synth. [49, 50]
WHO essential medicine **Amezinium metilsulfate** (LU-1631) [Regulton, Supratonin] C01CA25	x CH$_3$OSO$_3^-$	[*30578-37-1*] C$_{11}$H$_{12}$N$_3$O·CH$_3$O$_4$S 313.33	indirect sympathomimetic, inhibits metabolism of noradrenaline/ treatment of orthostatic hypotension. tbl. 10 mg 30 mg, O.	BASF DE/1984 synth. [51, 52]
Ephedrine (L-(−)-Ephedrine) [Wick MediNait] C01CA26	((1*R*,2*S*) form)	[*299-42-3*] C$_{10}$H$_{15}$NO 165.24 Hydrochloride: [*50-98-6*] C$_{10}$H$_{15}$NO·HCl 201.70 Sulfate: [*134-72-5*] C$_{10}$H$_{15}$NO·1/2H$_2$SO$_4$ 428.55	indirect sympath- omimetic/bronchodilator and vasoconstrictor, mostly in combinations used as anti-cough preparations (in many countries withdrawn from market); prevention of low blood pressure during spinal anesthesia. amp. 10, 25, 40, 50 mg/mL. 50 mg, P.	Knoll AG (?) semisynthetic (via hydrogenation of phenylacetylcarbinol, made by fermentation of sugar with yeast in presence of benzaldehyde, in presence of methylamine; natural alkaloid, first isolation from *Ephedra vulgaris* in 1885 by N. Nagai.) [53]

2.1.3.2. C01CE Phosphodiesterase Inhibitors (Table 8)

Table 8. C01CE Phosphodiesterase inhibitors

INN (Synonyms) [Brand names] ATC	Structure (Remarks)	[CAS-No.] Formula MW (g/mol)	Target (if known) Medical use Formulation DDD	Originator Approval (country/year) Production method [References]
Amrinone (Inamrinone; Win-40680) [Wincoram, Inocor, Vesistol] C01CE01		[60719-84-8] $C_9H_9N_3O$ 187.20	PDE-3 inhibitor; positive inotrope and chronotrope/treatment of CHF. amp.50 mg/10 mL, 100 mg/20 mL. 500 mg, P.	Winthrop/Sterling Drug DE/1984 synth. [54, 55]
Milrinone (Win-47203) [Corotrop, Primacor] C01CE02		[78415-72-2] $C_{12}H_9N_3O$ 211.22	PDE-3 inhibitor; positive inotrope and chronotrope/treatment of CHF. amp. 1 mg/1 mL; 10 mg/10 mL. 50 mg, P.	Winthrop/Sterling Drug NL/1989 synth. [56, 57, 58]
Enoximone (RMI-17043; MDL-17043) [Perfan] C01CE03		[77671-31-9] $C_{12}H_{12}N_2O_2S$ 248.31	selective PDE-3 inhibitor; positive inotropic and vasodilating activity/ treatment of CHF. amp. 100 mg/20 mL. 1000 mg, P.	Richardson-Merrell/ Merrell-Dow FR/1988 synth. [59, 60]

2.1.3.3. C01CX Other Cardiac Stimulants (Table 9)

Table 9. C01CX Other cardiac stimulants

INN (Synonyms) [Brand names] ATC	Structure (Remarks)	[CAS-No.] Formula MW (g/mol)	Target (if known) Medical use Formulation DDD	Originator Approval (country/year) Production method [References]
Levosimendan (R-Simendan; (–)-OR 1259) [Simdax] C01CX08	((R)-(–) form)	[141505-33-1] $C_{14}H_{12}N_6O$ 280.29	calcium sensitizer, increases cardiac contractility without rising intracellular Ca; positive inotropic/ treatment of acutely decompensated CHF. amp. 12.5 mg/5 mL; 25 mg/10 mL. 11 mg, P.	Orion SE/2000; DE/2013 synth. [61, 62]

2.1.4. C01D Vasodilators Used in Cardiac Diseases

2.1.4.1. C01DA Organic Nitrates (Table 10)

Table 10. C01DA Organic nitrates

INN (Synonyms) [Brand names] ATC	Structure (Remarks)	[CAS-No.] Formula MW (g/mol)	Target (if known) Medical use Formulation DDD	Originator Approval (country/year) Production method [References]
Nitroglycerine (Glyceryl trinitrate) [Minitrans, Nitroderm, Nitrolingual] C01DA02 **WHO essential medicine**		[*55-63-0*] $C_3H_5N_3O_9$ 227.09	active metabolite is the potent natural vasodilator nitric oxide/treatment of angina pectoris. different tbl., ointments, sol. for i.v. infusion, transdermal patches and sprays. 5 mg, O., TD; 2.5 mg oral aerosol.	W. Murrell/UK UK/1878 synth. [63]
Pentaerythrityl tetranitrate (PETN) [Pentalong, Peritrate, Mycardol, Pentral, Dilcoran, Prevangor, Subicard, Vasodiatol] C01DA05		[*78-11-5*] $C_5H_8N_4O_{12}$ 316.14	(see nitroglycerine C01DA02) tbl. 50, 80 mg. 120 mg, O.	Warner-Chilcott USA/1896 synth. [64]
Isosorbide dinitrate (ISDN) [Isoket, Isocard, Isordil, Dilatrate] C01DA08		[*87-33-2*] $C_6H_8N_2O_8$ 236.14	(see nitroglycerine C01DA02) tbl. 10, 20, 40 mg; s.r. tbl. 20, 40, 60, 80 mg; sublingual tbl. 5 mg; sol. for inj. 0.1%. 60 mg, O.	Schwarz Pharma synth. [65]
WHO essential medicine **Isosorbide mononitrate** (ISMN) [Elantan, Imdur, Ismo, Mono Mack, Monoket] C01DA14		[*16051-77-7*] $C_6H_9NO_6$ 191.14	(see nitroglycerine C01DA02); main metabolite of isosorbide dinitrate. tbl. 20, 40, 60 mg; s.r. tbl. 40, 100 mg; s.r. cps. 40, 50, 60 mg. 40 mg, O.	American Home Products synth. [66]

2.1.4.2. C01DX Other Vasodilators Used in Cardiac Diseases (Table 11)

Table 11. C01DX Other vasodilators used in cardiac diseases

INN (Synonyms) [Brand names] ATC	Structure (Remarks)	[CAS-No.] Formula MW (g/mol)	Target (if known) Medical use Formulation DDD	Originator Approval (country/year) Production method [References]
Trapidil (AR-12008) [Rocornal, Avantrin] C01DX11		[*15421-84-8*] $C_{10}H_{15}N_5$ 205.27	vasodilator and PDGF antagonist; non selective phosphodiesterase inhibitor/treatment in CHF and after myocard infarction. cps. 200 mg.	VEB Deutsches Hydrierwerk Rodleben GDR/1971; JP/1979;DE/1992. synth. [67]
Molsidomine (SIN-10) [Corvaton, Corvasal] C01DX12	(sydnone imine type compound)	[*25717-80-0*] $C_9H_{14}N_4O_4$ 242.24	nitric oxide donor/treatment of angina pectoris s.r.tbl. 8 mg. 8 mg, O.	Takeda; Cassella DE/1977 synth. [68, 69, 70]

2.1.5. C01E Other Cardiac Preparations

2.1.5.1. C01EA Prostaglandins (Table 12)

Table 12. C01EA Prostaglandins

INN (Synonyms) [Brand names] ATC	Structure (Remarks)	[CAS-No.] Formula MW (g/mol)	Target (if known) Medical use Formulation DDD	Originator Approval (country/year) Production method [References]
Alprostadil (Prostaglandin E_1; PGE1; U-10136) [Caverject, Minprog, MUSE, Pridax, Prostavasin, Viridal] C01EA01; G04BE01 **WHO essential medicine**	(endogenous vasodilator substance)	[*745-65-3*] $C_{20}H_{34}O_5$ 354.48	physiological prostaglandin with vasodilatory properties; inhibits platelet aggregation and relaxes smooth muscle of ductus arteriosus/ maintain patency of ductus arteriosus; treatment of erectile dysfunction. amp. 10, 20 µg powder for inj. or infus. sol.; concentrate with 500 µg for infus. sol.; urethral suppositories. 500 µg, P.	Upjohn; Pfizer USA/1981 synth. [71, 72]

2.1.5.2. C01EB Other Cardiac Preparations (Table 13)

Table 13. C01EB Other cardiac preparations

INN (Synonyms) [Brand names] ATC	Structure (Remarks)	[CAS-No.] Formula MW (g/mol)	Target (if known) Medical use Formulation DDD	Originator Approval (country/year) Production method [References]
Adenosine (Adenine riboside) [Adenoscan, Adrekar, Adenocard, Adenocor, Krenosin] C01EB10	(endogenous purine nucleoside)	[58-61-7] $C_{10}H_{13}N_5O_4$ 267.25	acts on adenosine receptors (4 subtypes), which can either stimulate or inhibit adenylate cyclase activity; vasodilator/antiarrythmic agent in supraventricular tachycardia and vasodilator in myocardia scintigraphy. vials with 30 mg/10 mL; amp. 6 mg/2 mL, 5 mg/1 mL. 15 mg, P.	G. V. R. Born about 1970 extraction from RNA hydrolyzates [73, 74]
Ivabradine (S-16257-2) [Coralan, Corlentor, Procoralan] C01EB17		[155974-00-8] $C_{27}H_{36}N_2O_5$ 468.59 Hydrochloride: [148849-67-6] $C_{27}H_{36}N_2O_5 \cdot HCl$ 505.06	selective inhibitor of the pacemaker I_f current/ symptomatic treatment of chronic stable angina pectoris in patients with normal sinus rhythm who cannot take β-blockers; decreases death rate in CHF patients. f.c.tbl. 5, 7.5 mg. 10 mg, O.	Servier DE/2006 synth. [75]
Ranolazine (CVT-303; KEG-1295; RS-43285) [Ranexa] C01EB18		[95635-56-6] $C_{24}H_{33}N_3O_4$ 427.55 Dihydrochloride: [95635-56-6] $C_{24}H_{33}N_3O_4 \cdot 2HCl$ 500.47	inhibitor of voltage-gated sodium channels in heart muscles/treatment of chronic angina pectoris. s.r.tbl. 500 mg. 1500 mg, O.	Syntex/CV Therapeutics USA/2006 synth. [76]
Regadenosone (CVT-3146) [Lexiscan, Rapiscan] C01EB21		[313348-27-5] $C_{15}H_{18}N_8O_5$ 390.36 Monohydrate: [875148-45-1] $C_{15}H_{18}N_8O_5 \cdot H_2O$ 408.38	A_{2A} adenosine receptor agonist; coronary vasodilator/ used for cardiac stress testing instead of adenosine (radionuclide myocardial perfusion imaging). amp. 0.4 mg/5 mL. 0.4 mg, P.	CV Therapeutics USA/2008 synth. from guanosine. [77]
Meldonium [Mildronate] C01EB22		[86426-17-7] $C_6H_{14}N_2O_2$ 147.19	treatment of coronary artery disease (ischemia), mainly in Eastern European countries; it was added 2016 to the WADA list of doping agents because of its metabolic modulator properties of hormones. cps. 250 mg.	Grindeks (LV) LV/1970 (?) synth. [78]

2.2. C02 Antihypertensives

2.2.1. C02A Antiadrenergic Agents, Centrally Acting

2.2.1.1. C02AB Methyldopa (Table 14)

Table 14. C02AB Methyldopa

INN (Synonyms) [Brand names] ATC	Structure (Remarks)	[CAS-No.] Formula MW (g/mol)	Target (if known) Medical use Formulation DDD	Originator Approval (country/year) Production method [References]
Methyldopa (MK-351) [Aldomet, Aldoril, Dopamet, Dopegyt, Medopren, Presinol] C02AB01 **WHO essential medicine**	HO, HO (S) COOH, CH₃, NH₂	[555-30-6] $C_{10}H_{13}NO_4$ 211.22	competitive enzyme inhibitor of L-Dopa decarboxylase/management of pregnancy-induced hypertension. f.c.tbl. 125, 250, 500 mg. 1000 mg, O, P.	Merck & Co. USA/1964 synth. [79, 80, 81]

2.2.1.2. C02AC Imidazoline Receptor Agonists (Table 15)

Table 15. C02AC Imidazoline receptor agonists

INN (Synonyms) [Brand names] ATC	Structure (Remarks)	[CAS-No.] Formula MW (g/mol)	Target (if known) Medical use Formulation DDD	Originator Approval (country/year) Production method [References]
Clonidine (ST-155) [Catapres, Catapresan, Clorpres, Haemiton, Isoglaucon, Paracefan] C02AC01; N02CX02; S01EA04		[4205-90-7] $C_9H_9Cl_2N_3$ 230.10 Hydrochloride: [4205-91-8] $C_9H_9Cl_2N_3 \cdot HCl$ 266.56	α_2-adrenergic agonist/antihypertensive; treatment of ADHD and Tourette syndrome. tbl. 0.075, 0.15, 0.3 mg, amp. 0.15 mg/1 mL; 0.75 mg/5 mL;eye drops 0.625, 1.25/1 mL; transdermal patch. 0.45 mg, O, P.	Boehringer Ing. DE/1966 synth. [82]
Guanfacine [Estulic, Tenex, Intuniv] C02AC02		[29110-47-2] $C_9H_9Cl_2N_3O$ 246.10 Hydrochloride: [29110-48-3] $C_9H_9Cl_2N_3O \cdot HCl$ 282.56	α_2-adrenergic receptor agonist/antihypertensive; treatment of ADHD. tbl. 0.5, 1, 2 mg. 3 mg, O.	Dr. A. Wander FR/1981; USA/1987; USA/2010 (ADHD, Intuniv) synth. [83].
Moxonidine (BDF-5895) [Cynt, Normatens, Physiotens] C02AC05		[75438-57-2] $C_9H_{12}ClN_5O$ 241.68	α_2-adrenergic and imidazoline I_1-receptor agonist/ antihypertensive. f.c.tbl. 0.2, 0.3, 0.4 mg. 0.3 mg, O.	Beiersdorf DE/1991 synth. [84]
Rilmenidine (Oxaminozolin; S-3341) [Albarel, Hyperium, Iterium, Tenaxum] C02AC06		[54187-04-1] $C_{10}H_{16}N_2O$ 180.25 Dihydrogen phosphate [85409-38-7] $C_{10}H_{16}N_2O \cdot H_3PO_4$ 278.25	mainly imidazoline I_1-receptor agonist and weaker α_2-adrenergic receptor agonist/antihypertensive. tbl. 1 mg.	Science Union synth. [85]

2.2.2. C02C Antiadrenergic Agents, Peripherally Acting

2.2.2.1. C02CA Alpha-Adrenoreceptor Antagonists (Table 16)

Table 16. C02CA Alpha-adrenoreceptor antagonists

INN (Synonyms) [Brand names] ATC	Structure (Remarks)	[CAS-No.] Formula MW (g/mol)	Target (if known) Medical use Formulation DDD	Originator Approval (country/year) Production method [References]
Prazosin (CP-12299-1) [Minipress] C02CA01		[19216-56-9] $C_{19}H_{21}N_5O_4$ 383.41 Hydrochloride: [19237-84-4] $C_{19}H_{21}N_5O_4 \cdot HCl$ 419.87	α_1-adrenergic receptor blocker/ antihypertensive agent; also used to treat benign prostate hyperplasia (BPH). tbl. 0.5, 1, 2, 5 mg; cps. 1, 2, 5 mg. 5 mg, O.	Pfizer USA/1971 synth. [86]
Indoramin (Wy-21901) [Baratol, Doralese, Vidora, Wydora] C02CA02		[26844-12-2] $C_{22}H_{25}N_3O$ 347.46 Hydrochloride: [38821-52-2] $C_{22}H_{25}N_3O \cdot HCl$ 383.92	α_1-adrenergic receptor blocker/ antihypertensive agent; also used to treat BPH. tbl. 20, 25, 50 mg.	Wyeth synth. [87, 88]
Doxazosin (UK-33274-27) [Alfadil, Cardura, Diblocin, Zoxan] C02CA04		[74191-85-8] $C_{23}H_{25}N_5O_5$ 451.58 Methanesulfonate: [77883-43-3] $C_{23}H_{25}N_5O_5 \cdot CH_4O_3S$ 547.59	α_1-adrenergic receptor blocker/ antihypertensive agent; also used to treat BPH. tbl. 0.5, 1, 2, 4 mg. 4 mg, O.	Pfizer USA/1990 synth. [89]
Urapidil (B-66256) [Ebrantil, Eupressyl, Hypotrit, Mediatensyl] C02CA06		[34661-75-1] $C_{20}H_{29}N_5O_3$ 387.48 Hydrochloride: [64887-14-5] $C_{20}H_{29}N_5O_3 \cdot HCl$ 423.95	α_1-adrenergic receptor antagonist and 5-HT$_{1A}$ receptor agonist/antihypertensive agent. cps. 30, 60, 90 mg; amp. (i.v.) 25 mg/5 mL; 50 mg/10 mL. 120 mg, O, 50 mg, P.	Byk Gulden DE/1977 synth. [90, 91]
Bunazosin (E-643) [Andante, Detantol] C02CA07		[80755-51-7] $C_{19}H_{27}N_5O_3$ 373.46 Hydrochloride: [52712-76-2] $C_{19}H_{27}N_5O_3 \cdot HCl$ 409.92	α_1-adrenergic receptor blocker/ antihypertensive agent; also used to treat BPH and glaucoma. tbl. 0.5, 1, 3 mg; s.r.tbl. 3, 6 mg; eye drops 0.01%. 6 mg, O.	Eisai JP/1985 synth. [92]
Terazosin (Abbott-45975) [Flotrin, Heitrin, Hytrin] C02CA08 and G04CA03		[63590-64-7] $C_{19}H_{25}N_5O_4$ 387.44 Hydrochloride: [63074-08-08] $C_{19}H_{25}N_5O_4 \cdot HCl$ 423.90	α_1-adrenergic receptor blocker/ antihypertensive agent; also used to treat BPH. tbl. 1,2, 5, 10 mg. 5 mg, O.	Abbott USA/1984 synth. [93]

2.2.3. C02D Arteriolar Smooth Muscle, Agents Acting on

2.2.3.1. C02DA Thiazide Derivatives (Table 17)

Table 17. C02DA Thiazide derivatives

INN (Synonyms) [Brand names] ATC	Structure (Remarks)	[CAS-No.] Formula MW (g/mol)	Target (if known) Medical use Formulation DDD	Originator Approval (country/year) Production method [References]
Diazoxide (SRG-95213) [Eudemine, Hyperstat, Hypertonalum, Proglicem, Proglycem] C02DA01; V03AH01		[*364-98-7*] $C_8H_7ClN_2O_2S$ 230.68	potassium channel opener, decreasing calcium influx; inhibits secretion of insulin from pancreas; relaxation of smooth muscles/antihypertensive; treatment of hypoglycemia. cps. 25, 100 mg; amp. 300 mg/20 mL. 300 mg, P.	Schering Corp. synth. [94, 95]

2.2.3.2. C02DB Hydrazinophthalazine Derivatives (Table 18)

Table 18. C02DB Hydrazinophthalazine derivatives

INN (Synonyms) [Brand names] ATC	Structure (Remarks)	[CAS-No.] Formula MW (g/mol)	Target (if known) Medical use Formulation DDD	Originator Approval (country/year) Production method [References]
Dihydralazine [Depressan, Nepresol] C02DB01		[*484-23-1*] $C_8H_{10}N_6$ 190.21 Hydrogen sulfate: [*7327-87-9*] $C_8H_{10}N_6 \cdot H_2SO_4$ 288.29	smooth muscle relaxant; peripheral vasodilator/antihypertensive. tbl. 25, 50 mg; vials 25 mg powder for inj. sol. 75 mg, O; 25 mg, P.	Ciba synth. [96, 97]
Hydralazine (C-5968) [Apresoline, Hydrapres, Pertenso] C02DB02 **WHO essential medicine**		[*86-54-4*] $C_8H_8N_4$ 160.18 Hydrochloride: [*304-20-1*] $C_8H_8N_4 \cdot HCl$ 196.64	smooth muscle relaxant; peripheral vasodilator/antihypertensive. tbl. 10, 25, 50 mg. 100 mg, O.	Ciba synth. [98, 99]

2.2.3.3. C02DC Pyrimidine Derivatives (Table 19)

Table 19. C02DC Pyrimidine derivatives

INN (Synonyms) [Brand names] ATC	Structure (Remarks)	[CAS-No.] Formula MW (g/mol)	Target (if known) Medical use Formulation DDD	Originator Approval (country/year) Production method [References]
Minoxidil (U-10858) [Alopexy, Loniten, Lonolox, Regaine, Rogaine] C02DC01; D11AX01		[38304-91-5] C₉H₁₅N₅O 209.25	potassium channel activator; NO-donor/antihypertensive; treatment of androgenic alopecia. tbl. 2.5, 10 mg; topical sol. or gel 2%, 5%. 20 mg, O.	Upjohn USA/1979 synth. [100, 101]

2.2.4. C02K Other Antihypertensives

2.2.4.1. C02KX Antihypertensives for Pulmonary Arterial Hypertension (Table 20)

Table 20. C02KX Antihypertensives for pulmonary arterial hypertension

INN (Synonyms) [Brand names] ATC	Structure (Remarks)	[CAS-No.] Formula MW (g/mol)	Target (if known) Medical use Formulation DDD	Originator Approval (country/year) Production method [References]
Bosentan (Ro-47-0203) [Tracleer] C02KX01		[147536-97-8] C₂₇H₂₉N₅O₆S 551.62 Monohydrate: [157212-55-0] C₂₇H₂₉N₅O₆S·H₂O 569.64	endothelin receptor antagonist/ pulmonary arterial hypertension. tbl. 62.5, 125 mg. 250 mg,O.	Roche; Actelion USA/2001 synth. [102]
Ambrisentan (BSF-208075; LU-208075) [Letairis, Volibris] C02KX02	((S) form)	[177036-94-1] C₂₂H₂₂N₂O₄ 378.42	endothelin receptor (type A) antagonist/ pulmonary arterial hypertension. f.c.tbl. 5, 10 mg. 7.5 mg, O.	BASF; GSK USA/2007 synth. [103–105]
Macitentan (ACT-064992) [Opsumit] C02KX04		[441798-33-0] C₁₉H₂₀Br₂N₆O₄S 588.28	endothelin receptor antagonist/ pulmonary arterial hypertension. f.c.tbl. 10 mg. 10 mg, O.	Actelion USA/2013 synth. [106, 107]

Table 20. (*Continued*)

INN (Synonyms) [Brand names] ATC	Structure (Remarks)	[CAS-No.] Formula MW (g/mol)	Target (if known) Medical use Formulation DDD	Originator Approval (country/year) Production method [References]
Sildenafil (UK-92480) [Revatio, Viagra] C02KX06; G04BE03		[*139755-83-2*] $C_{22}H_{30}N_6O_4S$ 474.59 Citrate: [*171599-83-0*] $C_{22}H_{30}N_6O_4S \cdot C_6H_8O_7$ 666.71	inhibitor of c-GMP-specific PDE5/pulmonary arterial hypertension (Revatio) and erectile dysfunction (Viagra). f.c.tbl. 20 mg; vials 10 mg/12.5 mL (Revatio); tbl. 50, 100 mg (Viagra). 60 mg, O.	Pfizer USA/1999 (Viagra) USA/2005 (Revatio) synth. [108, 109]
Tadalafil (IC-351; GF-196960) [Adcirca, Cialis] C02KX07; G04BE08	((6*R*,12a*R*) form)	[*171596-29-5*] $C_{22}H_{19}N_3O_4$ 389.41	inhibitor of cGMP-specific PDE5/pulmonary arterial hypertension (Adcirca) and erectile dysfunction (Cialis). f.c.tbl. 5, 10, 20 mg. 40 mg, O.	Icos; Eli Lilly DE/2002; USA/2003 (Cialis). USA/2009 (Adcirca) synth. [110, 111]
Riociguat (Bay 63-2521) [Adempas] C02KX05		[*625115-55-1*] $C_{20}H_{19}FN_8O_2$ 422.42	s-GC activity agonist/pulmonary arterial hypertension. tbl. 0.5, 1, 1.5, 2, 2.5 mg. 4.5 mg, O.	Bayer EU/2013 synth. [112, 113]
Iloprost (Ciloprost, E-1030, SH-401, ZK-36374) [Ilomedin, Ventavis, Endoprost] C02KX08		[*78919-13-8*] $C_{22}H_{32}O_4$ 360.49 Trometamol salt: [*73873-87-7*] $C_{22}H_{32}O_4 \cdot C_4H_{11}NO_3$ 481.63	prostacyclin PGI$_2$ analog dilating blood vessels/pulmonary arterial hypertension. amp. 10 μg, 20 μg/1 mL for nebulizer; 30 μg, inhalation.	Schering AG DE/1992 synth. [114, 115]
Selexipag (NS-304; ACT-293987; prodrug of ACT-333679) [Uptravi] C02KX09; B01AC27		[*475086-01-2*] $C_{26}H_{32}N_4O_4S$ 496.6	prostacyclin receptor agonist/pulmonary arterial hypertension. tbl. 0.2, 0.4, 0.6, 0.8, 1, 1.2, 1.4 and 1.6 mg. 1.8 mg, O.	Actelion; Nippon Shinyaku USA/2015; EU/2016 synth. [116–118]

2.3. C03 Diuretics [119]

2.3.1. C03A Low-Ceiling Diuretics, Thiazides

2.3.1.1. C03AA Thiazides, Plain (Table 21)

Table 21. C03AA Thiazides, plain

INN (Synonyms) [Brand names] ATC	Structure (Remarks)	[CAS-No.] Formula MW (g/mol)	Target (if known) Medical use Formulation DDD	Originator Approval (country/year) Production method [References]
Bendroflumethiazide (Bendrofluazide) [Aprinox, Berkozide, Corzide, Dociretic] C03AA01		[73-48-3] $C_{15}H_{14}F_3N_3O_4S_2$ 421.42	see hydrochlorothiazide C03AA03/antihypertensive. tbl. 2.5, 5 mg; cps. 1.25, 2.5 mg. 2.5 mg, O.	Loevens Kemiske Fabrik synth. [120, 121]
Hydrochlorothiazide (HCT; HCTZ) [Esidrex, Esidrix, Hydrodiuril] C03AA03 **WHO essential medicine**		[58-93-5] $C_7H_8ClN_3O_4S_2$ 297.74	reduction of sodium readsorption in the distal convoluted tubule/ treatment of hypertension, CHF, and edema. tbl. 12.5, 25 mg. 25 mg, O.	Ciba USA/1959 synth. [122, 123]
Chlorothiazide [Diuril] C03AA04		[58-94-6] $C_7H_6ClN_3O_4S_2$ 295.73	see hydrochlorothiazide C03AA03/ treatment of hypertension and edema. tbl. 250, 500 mg. 500 mg, O.	Merck & Co. synth. [124, 125]
Trichlormethiazide [Achletin, Anistadin, Diu-Hydrin, Esmalorid] C03AA06		[133-67-5] $C_8H_8Cl_3N_3O_4S_2$ 380.66	see hydrochlorothiazide C03AA03/ treatment of hypertension and edema. tbl. 2, 4 mg. 4 mg, O.	Ciba synth. [126]

2.3.2. C03B Low-Ceiling Diuretics, Excluding Thiazides

2.3.2.1. C03BA Sulfonamides, Plain (Table 22)

Table 22. C03BA Sulfonamides, plain

INN (Synonyms) [Brand names] ATC	Structure (Remarks)	[CAS-No.] Formula MW (g/mol)	Target (if known) Medical use Formulation DDD	Originator Approval (country/year) Production method [References]
Clopamide (DT-327) [Adurix, Brinaldix, Viskaldix] C03BA03		[636-54-4] $C_{14}H_{20}ClN_3O_3S$ 345.85	inhibitor of sodium chloride symporter at the proximal convoluted tubule (PCT) of the nephron; thiazide diuretic like/antihypertensive; antiedema. drg. 2.5, 5 mg. 10 mg, O.	Sandoz synth. [127, 128]
Chlorthalidone (Chlortalidone; G-33182) [Hygroton] C03BA04		[77-36-1] $C_{14}H_{11}ClN_2O_4S$ 338.77	see clopamide C03BA03/ antihypertensive, antiedema. tbl. 25, 50, 100 mg. 25 mg, O.	Geigy synth. [129, 130]
Xipamide (Bei-1293) [Aquaphor, Aquaphoril, Diurexan, Lumitens] C03BA10		[14293-44-8] $C_{15}H_{15}ClN_2O_4S$ 354.81	see clopamide C03BA03/ antihypertensive, antiedema. tbl. 10, 20, 40 mg. 20 mg, O.	Beiersdorf synth. [131]
Indapamide (S-1520; SE-1520) [Natrilix] C03BA11		[26807-65-8] $C_{16}H_{16}ClN_3O_3S$ 365.84	see clopamide C03BA03/ antihypertensive, antiedema. s.r.tbl. 1.25, 2.5 mg. 2.5 mg, O.	Science Union synth. [132]

2.3.3. C03C High-Ceiling Diuretics

2.3.3.1. C03CA Sulfonamides, Plain (Table 23)

Table 23. C03CA Sulfonamides, plain

INN (Synonyms) [Brand names] ATC	Structure (Remarks)	[CAS-No.] Formula MW (g/mol)	Target (if known) Medical use Formulation DDD	Originator Approval (country/year) Production method [References]
Furosemide (LB-502; Frusemide) [Diusemide, Lasix] C03CA01 **WHO essential medicine**		[54-31-9] $C_{12}H_{11}ClN_2O_5S$ 330.75	loop diuretic (inhibitor of NKCC2, the Na-K-Cl cotransporter in the Henle loop)/treatment of hypertension and edema. tbl. 20, 40, 125, 250, 500 mg; s.r.cps. 30 mg; solution for infusion 250 mg/25 mL 40 mg, O, P.	Hoechst DE/1966 synth [133, 134]
Bumetanide (PF-1593; Ro-10-6338) [Burinex, Bumex] C03CA02		[28395-03-1] $C_{17}H_{20}N_2O_5S$ 364.42	loop diuretic/treatment of edema and heart failure (CHF). tbl. 1 mg; amp. 1 mg/2 mL. 1 mg, O, P.	Loevens Kemiske Fabrik CH/1974 synth. [135, 136]
Piretanide (Hoe-118) [Arelix, Eurelix] C03CA03		[55837-27-9] $C_{17}H_{18}N_2O_5S$ 362.41	loop diuretic/treatment of edema and heart failure (CHF), and hypertension. tbl. 3, 6 mg. 6 mg, O.	Hoechst synth. [137, 138]
Torasemide (Torsemide; AC-4464; BM-02015) [Demadex, Torem, Unat] C03CA04		[56211-40-6] $C_{16}H_{20}N_4O_3S$ 348.43 Sodium salt: [72810-59-4] $C_{16}H_{20}N_4O_3S \cdot x$ Na	loop diuretic/treatment of edema and heart failure (CHF), and hypertension. tbl. 2.5, 5, 10, 20, 200 mg; amp. 10, 20 mg. 15 mg, O, P.	A.Christiaens; Boehringer Mannheim DE/1993 synth. [139, 140]

2.3.3.2. C03CC Aryloxyacetic Acid Derivatives (Table 24)

Table 24. C03CC Aryloxyacetic acid derivatives

INN (Synonyms) [Brand names] ATC	Structure (Remarks)	[CAS-No.] Formula MW (g/mol)	Target (if known) Medical use Formulation DDD	Originator Approval (country/year) Production method [References]
Etacrynic acid (Ethacrynic acid) [Edecrin, Hydromedin] C03CC01		[58-54-8] $C_{13}H_{12}Cl_2O_4$ 303.14 Sodium salt: [6500-81-8] $C_{13}H_{11}Cl_2NaO_4$ 325.12	loop diuretic, inhibits NKCC2 in the thick ascending loop of Henle and the macula densa/treatment of edema associated with CHF, cirrhosis and renal disease, and other kinds of edema. tbl. 25, 50 mg (free acid); vials with 53.6 mg (sodium salt) for inj. sol. 50 mg O, P.	Merck & Co. synth. [141, 142]

2.3.4. C03D Potassium-Sparing Agents

2.3.4.1. C03DA Aldosterone Antagonists (Table 25)

Table 25. C03DA Aldosterone antagonists

INN (Synonyms) [Brand names] ATC	Structure (Remarks)	[CAS-No.] Formula MW (g/mol)	Target (if known) Medical use Formulation DDD	Originator Approval (country/year) Production method [References]
Spironolactone (SC-9420) [Aldactone, Osyrol] C03DA01		[52-01-7] $C_{24}H_{32}O_4S$ 416.58	diuretic, androgen receptor antagonist/treatment of CHF, edema and symptoms of hyperandrogenism. tbl. 25, 50, 100 mg; cps. 100 mg. 75 mg, O.	Searle USA/1959 semisynth. [143–145]
Canrenone (**Potassium canrenoate**; **Aldadiene**) [Aldactone, Contaren, Luvion, Phanurane, Soldactone, Spiroletan] C03DA02; C03DA03		[976-71-6] $C_{22}H_{28}O_3$ 340.46 Potassium canrenoate: [2181-04-6] $C_{22}H_{29}KO_4$ 396.57	active metabolite of spironolactone; steroidal antimineralocorticoid/diuretic, antiedema. amp. 200 mg. 400 mg, P.	Searle semisynth. [146, 147]
Eplerenone (SC-66110; CGP-30083) [Inspra, Salara] C03DA04		[107724-20-9] $C_{24}H_{30}O_6$ 414.50	aldosteronantagonist, antimineralocorticoid; potassium-sparing diuretic/ treatment of CHF; antihypertensive. f.c.tbl. 25, 50 mg. 50 mg, O.	Ciba-Geigy; Searle USA/2003 semisynth. [148–150]

2.3.4.2. C03DB Other Potassium-Sparing Agents (Table 26)

Table 26. CD03DB Other potassium-sparing agents

INN (Synonyms) [Brand names] ATC	Structure (Remarks)	[CAS-No.] Formula MW (g/mol)	Target (if known) Medical use Formulation DDD	Originator Approval (country/year) Production method [References]
Amiloride (MK-870) [Midamor, Moduretic] C03DB01		Base: [2609-46-3] $C_6H_8ClN_7O$ 229.63 Hydrochloride dihydrate: [17440-83-4] $C_6H_8ClN_7O\cdot HCl\cdot2 H_2O$ 302.12	blocker of epithelium sodium channel and so inhibiting sodium reabsorption in the late distal convoluted tubules/antihypertensive, often in combination with a thiazide diuretic. tbl. 2.5, 5, 10 mg. 10 mg, O.	Merck & Co.. synth. [151]
Triamterene (NSC-77625; SKF-8542) [Dyazide, Dyrenium, Dytac, Maxzide/comb.] C03DB02		[396-01-0] $C_{12}H_{11}N_7$ 253.27	see amiloride C03DB01/potassium-sparing diuretic in comb. with HCT; antihypertensive. tbl. 30, 50, 75 mg; cps. 50, 100 mg. 100 mg, O.	Smith Kline & French synth. [152, 153]

2.3.5. C03X Other Diuretics

2.3.5.1. C03XA Vasopressin Antagonists (Table 27)

Table 27. C03XA Vasopressin antagonists

INN (Synonyms) [Brand names] ATC	Structure (Remarks)	[CAS-No.] Formula MW (g/mol)	Target (if known) Medical use Formulation DDD	Originator Approval (country/year) Production method [References]
Tolvaptan (OPC-41061) [Samsca] C03XA01		[150683-30-0] $C_{26}H_{25}ClN_2O_3$ 448.95	selective V_2 receptor antago- nist/hyponatremia. tbl. 15, 30 mg. 30 mg, O.	Otsuka Pharmaceutical EU/2009 synth. [154, 155]
Conivaptan (CI-1025; YM-087; YM-35087) [Vaprisol] C03XA02		[210101-16-9] $C_{32}H_{26}N_4O_2$ 498.59 Hydrochloride: [168626-94-6] $C_{32}H_{26}N_4O_2\cdot HCl$ 535.04	dual V_{1a} and V_2 receptor antago- nist/hyponatremia. solution for i.v. infusion 20 mg/100 mL.	Yamanouchi USA/2004 Synth. [156, 157]

2.4. C04 Peripheral Vasodilators

2.4.1. C04A Peripheral Vasodilators

2.4.1.1. C04AA 2-Amino-1-phenylethanol Derivatives (Table 28)

Table 28. C04AA 2-Amino-1-phenylethanol derivatives

INN (Synonyms) [Brand names] ATC	Structure (Remarks)	[CAS-No.] Formula MW (g/mol)	Target (if known) Medical use Formulation DDD	Originator Approval (country/year) Production method [References]
Isoxsuprine [Duvadilan, Vasodilan, Vasoplex] C04AA01	(racemic)	[395-28-8] $C_{18}H_{23}NO_3$ 301.39 Hydrochloride: [579-56-6] $C_{18}H_{23}NO_3 \cdot HCl$ 337.85	β_2-receptor agonist, relaxation of uterine and vascular smooth muscle/treatment of premature labor (tocolytic) and use as vasodilator. amp. 5, 10 mg/1 mL; tbl. 10, 20 mg; cps. 40 mg. 60 mg, O, P.	Philips synth. [158]

2.4.1.2. C04AB Imidazoline Derivatives (Table 29)

Table 29. C04AB Imidazoline derivatives

INN (Synonyms) [Brand names] ATC	Structure (Remarks)	[CAS-No.] Formula MW (g/mol)	Target (if known) Medical use Formulation DDD	Originator Approval (country/year) Production method [References]
Phentolamine [Regitin, Regitine, Rogitine; OraVerse] C04AB01		[50-60-2] $C_{17}H_{19}N_3O$ 281.36 Hydrochloride: [73-05-2] $C_{17}H_{19}N_3O \cdot HCl$ 317.82 Mesylate: [65-28-1] $C_{17}H_{19}N_3O \cdot CH_4O_3S$ 377.47	α_1- and α_2-receptor blocker/antihypertensive and local anesthetic reversal agent. amp. 10 mg/1 mL; vials 5 mg; patr./ cartridge 400 µg/1.7 mL (OraVerse). 10 mg, O, P.	Ciba synth. [159, 160]

2.4.1.3. C04AD Purine Derivatives (Table 30)

Table 30. C04AD Purine derivatives

INN (Synonyms) [Brand names] ATC	Structure (Remarks)	[CAS-No.] Formula MW (g/mol)	Target (if known) Medical use Formulation DDD	Originator Approval (country/year) Production method [References]
Pentoxifylline (Oxpentifylline) [Trental, Torental] C04AD03		[6493-05-6] $C_{13}H_{18}N_4O_3$ 278.31	nonselective PDE inhibitor/ intermittent claudication resulting from peripheral artery disease. f.c.tbl. 400 mg; s.r.cps. 400, 600 mg; s.r.tbl. 400, 600 mg; amp. 100 mg/5 mL, 300 mg/15 mL. 1000 mg, O; 300 mg,P.	Chem. Werke Albert Synth. [161–163]

2.4.1.4. C04AE Ergot Alkaloids (Table 31)

Table 31. C04AE Ergot alkaloids

INN (Synonyms) [Brand names] ATC	Structure (Remarks)	[CAS-No.] Formula MW (g/mol)	Target (if known) Medical use Formulation DDD	Originator Approval (country/year) Production method [References]
Ergoloid mesylate (Codergocrine mesylate; Dihydroergotoxine mesylate; CCK-179) [DCCK, Hydergine, Circanol, Lysergin, Orphol, Progeril] C04AE01; N06DX07	Dihydroergocornine R=CH(CH$_3$)$_2$ Dihydroergocristine R=CH$_2$C$_6$H$_5$ Dihydro-α-ergocryptine R=CH$_2$CH(CH$_3$)$_2$ Dihydro-β-ergocryptine R=CH(CH$_3$)CH$_2$CH$_3$	[8067-24-1] formula and MW unspecified.	mechanism not clear/vasodilator, sympatholytic agent; nootropic, has been used to treat dementia. tbl. 1, 2 mg; amp. 0.3 mg/1 mL; 1.5 mg/5 mL. 3 mg, O, P.	Sandoz Semisynth. (hydrogenation of Ergotoxine). [164, 165]
Nicergoline (MNE; FI-6714) [Sermion, Ergobel, Nicergolin] C04AE02		[27848-84-6] $C_{24}H_{26}BrN_3O_3$ 484.39 Tartrate: [32222-75-6] $C_{24}H_{26}BrN_3O_3 \cdot x C_4H_6O_6$ unspecified	selective αA1-adrenergic receptor antagonist; decreasing vascular resistance and increasing arterial blood flow/ treatment of cerebral thrombosis and atherosclerosis. f.c.tbl. 10, 20, 30 mg; cps. 5, 10, 15, 30 mg; vials 4 mg/4 mL.	Farmitalia semisynth. from methyl lysergate. [166, 167]

2.4.1.5. C04AX Other Peripheral Vasodilators (Table 32)

Table 32. C04AX Other peripheral vasodilators

INN (Synonyms) [Brand names] ATC	Structure (Remarks)	[CAS-No.] Formula MW (g/mol)	Target (if known) Medical use Formulation DDD	Originator Approval (country/year) Production method [References]
Phenoxybenzamine [Dibenzyran, Dibenyline, Dibenzyline] C04AX02	(racemate)	[59-96-1] $C_{18}H_{22}ClNO$ 303.83 Hydrochloride: [63-92-3] 340.29	non selective irreversible α-adrenergic receptor antagonist/antihypertensive especially in cases of pheochromocytoma. cps. 1, 5, 10 mg. 30 mg, O.	Smith Kline & French (SKF) synth. [168]
Vincamine [Oxybral SR, Cetal, Oxygeron, Pervincamine; these drugs are no longer used] C04AX07		[1617-90-9] $C_{21}H_{26}N_2O_3$ 354.45	peripheral vasodilator, increases blood flow to the brain. tbl. 10 mg; s.r.cps. 30,60 mg.	(E.Schlittler et al.); Richter Gedeon extraction from *Vinca minor L. (periwinkle)* or semisynth. from tabersonine. [169]
Piribedil (ET-495; EU-4200) [Clarium, Trivastal] C04AX13		[3605-01-4] $C_{16}H_{18}N_4O_2$ 298.35 Mesylate: [52293-23-9] $C_{16}H_{18}N_4O_2 \cdot CH_4O_3S$ 394.45	adrenergic receptor antagonist, dopamine receptor D_4-antagonist, dopamine receptor D_2- and D_3-agonist/treatment of Parkinson's disease and intermittent claudication. s.r.tbl. 50 mg.	Science Union synth. [170, 171]
Naftidrofuryl (Nafronyl) [Dusodril, Naftilong, Di-Actane, Gevatran, Praxilene] C04AX21		[31329-57-4] $C_{24}H_{33}NO_3$ 383.53 Hydrogen oxalate: [3200-06-4] $C_{24}H_{33}NO_3 \cdot C_2H_2O_4$ 473.57	selective antagonist of $5\text{-}HT_2$ receptor/ treatment of peripheral and cerebral vascular disorders and claudication. cps. 100 mg; f.c.tbl. 200 mg. 600 mg, O.	Lipha synth. [172]
Papaverine [Papacon, Para-Time SR, Pavacort; Artegodan, Panergon] C04AX38; A03AD01; GE04BE02.		[58-74-2] $C_{20}H_{21}NO_4$ 339.39 Hydrochloride: [61-25-6] $C_{20}H_{21}NO_4 \cdot HCl$ 375.85	inhibitor of cAMP-phosphodiesterase (PDE)/ clinical use as cerebral and coronary vasodilator in subarachnoid hemorrhage and coronary bypass surgery. s.r.cps. 150, 300 mg; amp. 30 mg/mL.	G. Merck (discovery in 1848/opium alkaloid) synth. [173, 174]
Moxaverine [Eupaverin, Kollateral] C04AX42; A03AD30; G04BE02.		[10539-19-2] $C_{20}H_{21}NO_2$ 307.39 Hydrochloride: [1163-37-7] $C_{20}H_{21}NO_2 \cdot HCl$ 343.85	inhibitor of cAMP-phosphodiesterase (PDE)/ peripheral vasodilator and spasmolytic. drg. 100, 150 mg; amp. 150 mg/5 mL.	Orgamol synth. [175]

2.5. C05 Vasoprotectives

2.5.1. C05A Agents for Treatment of Hemorrhoids and Anal Fissures for Topical Use

2.5.1.1. C05AA Corticosteroids
All in C05AA listed corticosteroids are described in D07AA, D07AB, and D07AC (Dermatologicals): hydrocortisone, prednisolone, betamethasone, fluorometholone, fluocortolone, dexamethasone, fluocinolonacetonide, fluocinonide, triamcinolone.

2.5.1.2. C05A Local Anesthetics
All in C05AD listed local anesthetics are described in N01BB Nervous System, Anesthetics, Local: lidocaine, tetracaine, benzocaine, cinchocaine, procaine, oxetacaine, pramocaine, polidocanol, quinisocaine.

2.6. C07 Beta Blocking Agents

2.6.1. C07A Beta Blocking Agents

2.6.1.1. C07 Beta Blocking Agents, Nonselective (Table 33)

Table 33. C07AA Beta blocking agents, nonselective

INN (Synonyms) [Brand names] ATC	Structure (Remarks)	[CAS-No.] Formula MW (g/mol)	Target (if known) Medical use Formulation DDD	Originator Approval (country/year) Production method [References]
Oxprenolol (Ba-39089) [Coretal, Laracor, Trasicor, Trasitensin] C07AA02	(racemate)	[6452-71-7] $C_{15}H_{23}NO_3$ 265.35 Hydrochloride [6452-73-9] $C_{15}H_{23}NO_3 \cdot HCl$ 301.81	antagonist of β_1- and β_2-adrenergic receptors/antihypertensive and treatment of angina pectoris (antianginal). f.c.tbl. 20, 40, 80, 160 mg. 160 mg, O.	Ciba synth. [176]
Pindolol (LB-46) [Visken] C07AA03	(racemate)	[13523-86-9] $C_{14}H_{20}N_2O_2$ 248.33	antagonist of β_1- and β_2-adrenergic receptors/antihypertensive and antianginal. tbl. 1, 2.5, 5, 10, 15 mg; s.r.cps. 5, 15 mg. 15 mg, O.	Sandoz synth. [177]
Propranolol (AY-64043; ICI-45520; NSC-91523) [Dociton, Inderal] C07AA05 **WHO essential medicine**	(racemate)	[525-666-6] $C_{16}H_{21}NO_2$ 259.35 Hydrochloride [318-98-9] $C_{16}H_{21}NO_2 \cdot HCl$ 295.81	antagonist of β_1- and β_2-adrenergic receptors/antihypertensive, antianginal, antiarrythmic. tbl.10, 20, 25,40, 80 mg; s.r.cps. 60, 80, 120, 160 mg. 160 mg, O.	ICI USA/1967 synth. [178–180]
Timolol (MK-950) [Blocadren, Ophtamolol, Timacor, Timoptic, Timoptol] C07AA06; S01ED01 **WHO essential medicine**	((S)-(–) form)	[26839-75-8] $C_{13}H_{24}N_4O_3S$ 316.43 Hydrogen maleate [26921-17-5] $C_{13}H_{24}N_4O_3S \cdot C_4H_4O_4$ 432.50	antagonist of β_1- and β_2-adrenergic receptors/antihypertensive, antianginal, and antiglaucoma. tbl. 5, 10, 20 mg; eye drops. 20 mg, O.	C.E.Frosst & Co.; MSD USA/1978 synth. [181, 182]

Table 33. (*Continued*)

INN (Synonyms) [Brand names] ATC	Structure (Remarks)	[CAS-No.] Formula MW (g/mol)	Target (if known) Medical use Formulation DDD	Originator Approval (country/year) Production method [References]
Sotalol (MJ-1999) [Sotacor, Sotalex, Betapace, Darob, Rentibloc, Rytmobeta] C07AA07	(racemate)	[3930-20-9] $C_{12}H_{20}N_2O_3S$ 272.37 Hydrochloride: [959-24-0] $C_{12}H_{20}N_2O_3S \cdot HCl$ 308.83	antagonist of β_1- and β_2-adrenergic receptors/antihypertensive and antiarrhythmic (class III). tbl. 40, 80, 160, 240 mg; amp. 40 mg/4 mL. 160 mg, O.	Bristol-Myers GB/1974; DE/1975; USA/1992. synth. [183]
Nadolol (SQ-11725) [Corgard, Corzide, Solgol] C07AA12	((R,S)-cis form)	[42200-33-9] $C_{17}H_{27}NO_4$ 309.41	antagonist of β_1- and β_2-adrenergic receptors/antihypertensive, antianginal. tbl. 20, 30, 40, 60,80, 120 mg. 160 mg, O.	Squibb synth. [184, 185]
Carteolol (OPC-1085; Abbott 43326) [Arteoptic, Endak, Mikelan, Teoptic] C07AA15	(racemate)	[51781-06-7] $C_{16}H_{24}N_2O_3$ 292.38 Hydrochloride: [51781-21-6] $C_{16}H_{24}N_2O_3 \cdot HCl$ 328.84	antagonist of β_1- and β_2-adrenergic receptors with intrinsic sympathomimetic activity/antianginal, antiglaucoma. tbl. 2.5, 5, 20 mg; eye drops 1%, 2%. 10 mg, O.	Otsuka synth. [186, 187]
Metipranolol (VUFB-6453) [Betamet, Betamann, Betanol, Beta-Ophtiole Disorat, Glauline, Turoptin] C07AA18; S01ED04.	(racemate)	[22664-55-7] $C_{17}H_{27}NO_4$ 309.41 Hydrochloride: [36592-77-5] $C_{17}H_{27}NO_4 \cdot HCl$ 345.87	antagonist of β_1- and β_2-adrenergic receptors/antihypertensive, mainly used as antiglaucoma agent. tbl. 20 mg; eye drops 0.1%, 0.3%, 0.6%.	Spofa synth. [188]
Penbutolol (HOE-893d; HOE-39-893d) [Betapressin, Levatol, Paginol] C07AA23	((S) form)	[38363-40-5] $C_{18}H_{29}NO_2$ 291.44 Sulfate: [38363-32-5] $(C_{18}H_{29}NO_2)_2 \cdot H_2SO_4$ 680.95	antagonist of β_1- and β_2-adrenergic receptors with intrinsic sympathomimetic activity/antihypertensive and antianginal. f.c.tbl. 20, 40 mg. 40 mg, O.	Hoechst USA/1987 synth. [189, 190]

2.6.1.2. C07AB Beta Blocking Agents, Selective (Table 34)

Table 34. C07AB Beta blocking agents, selective

INN (Synonyms) [Brand names] ATC	Structure (Remarks)	[CAS-No.] Formula MW (g/mol)	Target (if known) Medical use Formulation DDD	Originator Approval (country/year) Production method [References]
Metoprolol (CGP-2175; H93/26) [Beloc, Lopresor, Seloken, Toprol] C07AB02	(racemate)	[37350-58-6] $C_{15}H_{25}NO_3$ 267.37 Tartrate: [56392-17-7] $(C_{15}H_{25}NO_3)_2 \cdot$ $C_4H_6O_6$ 684.82	selective β_1-adrenoceptor blocker/antihypertensive, antianginal, treatment of tachycardia and CHF. f.c.tbl. 20, 40, 50, 100 mg; s.r.tbl. 120, 200 mg; amp. 5 mg/5 mL. 150 mg, O.	AB Hässle USA/ 1978 synth. [191, 192]
Atenolol (ICI-66082) [Tenoretic, Tenormin, Prenormine, Uniloc] C07AB03	(racemate)	[29122-68-7] $C_{14}H_{22}N_2O_3$ 266.34	selective β_1-adrenoceptor blocker/antihypertensive, antianginal, and treatment of tachycardia. f.c.tbl. 25, 50, 100 mg. 75 mg, O.	ICI USA/ 1976 synth. [193, 194]
Acebutolol (M&B 17803A; IL-17803A) [Acecor, Acetanol, Prent, Sectral] C07AB04	(racemate)	[37517-30-9] $C_{18}H_{28}N_2O_4$ 336.43 Hydrochloride: [34381-68-5] $C_{18}H_{28}N_2O_4 \cdot HCl$ 372.89	selective β_1-adrenoceptor blocker with intrinsic sympathomimetic activity/ antihypertensive and antiarrhythmic agent. tbl. 200, 400, 500 mg; cps. 100, 200 mg. 400 mg, O.	May & Baker synth. [195, 196]
Betaxolol (SLD-212; SL-75,212) [Betoptic, Betoptima, Kerlone] C07AB05; S01ED02	(racemate)	[63659-18-7] $C_{18}H_{29}NO_3$ 307.43 Hydrochloride: [63659-19-8] $C_{18}H_{29}NO_3 \cdot HCl$ 343.90	selective β_1-adrenoceptor blocker/antihypertensive and antiglaucoma. tbl. 5, 10, 20, 25 mg; eye drops 0.25, 0.5%. 20 mg, O.	Synthelabo FR/1983; USA/ 1985 synth. [197, 198]
Bisoprolol [Concor, Detensiel, Emconcor, Monocor, Soprol, Zebeta] C07AB07	(racemate)	[66722-44-9] $C_{18}H_{31}NO_4$ 325.45 Fumarate: [104344-23-2] $(C_{18}H_{31}NO_4)_2 \cdot$ $C_4H_4O_4$ 766.97	selective β_1-adrenoceptor blocker/antihypertensive, antianginal, and treatment of tachycardia. f.c.tbl. 1.25, 2.5, 3.75, 5, 7.5, 10mg. 10 mg, O.	E. Merck AG DE/1986 synth. [199, 200]
Celiprolol (ST-1396) [Celectol, Corliprol, Selectol] C07AB08	(racemate)	[56980-93-9] $C_{20}H_{33}N_3O_4$ 379.50 Hydrochloride: [57470-78-7] $C_{20}H_{33}N_3O_4 \cdot HCl$ 415.96	selective β_1-adrenoceptor blocker and partial β_2-agonist/antihypertensive, antianginal and management of CHF. f.c.tbl. 100, 200 mg. 200 mg, O.	Chemie Linz DE/1983 synth. [201].

Table 34. (*Continued*)

INN (Synonyms) [Brand names] ATC	Structure (Remarks)	[CAS-No.] Formula MW (g/mol)	Target (if known) Medical use Formulation DDD	Originator Approval (country/year) Production method [References]
Esmolol (ASL-8052) [Brevibloc] C07AB09	(racemate)	[81147-92-4] $C_{16}H_{25}NO_4$ 295.38 Hydrochloride: [81161-17-3] $C_{16}H_{25}NO_4 \cdot HCl$ 331.84	selective β_1-adrenoceptor blocker with rapid onset and very short duration of action/treatment of acute supraventricular tachycardia. soln. for infus. 10 mg/mL; vials with 100 mL.	AB Hässle; American Hospital Supply (AHS) USA/ 1987 synth. [202–205]
Nebivolol (R-65824) [Bystolic, Lobivon, Nebilet, Silostar] C07AB12	((±) form of (*RSSS* and *SRRR* enantiomers)	[99200-09-6] $C_{22}H_{25}F_2NO_4$ 405.44 Hydrochloride: [152520-56-4] $C_{22}H_{25}F_2NO_4 \cdot HCl$ 441.90	selective β_1-adrenoceptor blocker; nitric oxide mediated vasodilator/antihypertensive. tbl. 2.5, 5, 10 mg. 5 mg, O.	Janssen DE/1997 synth. [206]

2.6.1.3. C07AG Alpha and Beta Blocking Agents (Table 35)

Table 35. C07AG Alpha and beta blocking agents

INN (Synonyms) [Brand names] ATC	Structure (Remarks)	[CAS-No.] Formula MW (g/mol)	Target (if known) Medical use Formulation DDD	Originator Approval (country/year) Production method [References]
Labetalol (AH-5158A; Sch-15719W) [Amipress, Ipolab, Labelol, Laprocol, Normodyne, Presdate, Trandate] C07AG01	(racemic mixture, 4 stereoisomers)	[36894-69-6] $C_{19}H_{24}N_2O_3$ 328.41 Hydrochloride [32780-64-6] $C_{19}H_{24}N_2O_3 \cdot HCl$ 364.87	mixed α_1-/β-adrenoceptors antagonist /antihypertensive. tbl. 50, 100, 200, 300, 400 mg; amp. 50, 100 mg (5 mg/mL). 600 mg, O.	Allen & Hanburys synth. [207, 208]
Carvedilol (BM-14190) [Coreg, Dilatrend, Eucardic, Kredex, Querto] C07AG02	(racemate)	[72956-09-3] $C_{24}H_{26}N_2O_4$ 406.48	mixed α_1-/β-adrenoceptors antagonist/antihypertensive and treatment of CHF. tbl. 3.125, 6.25, 12.5, 25 mg. 37.5 mg, O.	Boehringer Mannheim USA/ 1995 synth. [209]

2.7. C08 Calcium Channel Blockers

2.7.1. C08C Selective Calcium Channel Blockers with Mainly Vascular Effects

2.7.1.1. C08CA Dihydropyridine Derivatives (Table 36)

Table 36. C08CA Dihydropyridine derivatives

INN (Synonyms) [Brand names] ATC	Structure (Remarks)	[CAS-No.] Formula MW (g/mol)	Target (if known) Medical use Formulation DDD	Originator Approval (country/year) Production method [References]
Amlodipine (UK-48340) [Amloclair, Norvasc, Caduet] C08CA01 **WHO essential medicine**	(racemate)	[88150-42-9] $C_{20}H_{25}ClN_2O_5$ 408.88 Besilate: [111470-99-6] $C_{20}H_{25}ClN_2O_5 \cdot$ $C_6H_6O_3S$ 567.06	angioselective Ca channel blocker/antihypertensive and antianginal. tbl. 5, 10 mg. 5 mg, O.	Pfizer USA/ 1990 synth. [210, 211]
Felodipine (H-154/82) [Agon, Flodil, Hydac, Modip, Plendil, Prevex, Splendil, Munobal] C08CA02	(racemate)	[72509-76-3] $C_{18}H_{19}Cl_2NO_4$ 384.26	Ca channel blocker/antihypertensive and antianginal (stable angina p.). s.r.tbl. 2.5, 5, 10 mg. 5 mg, O.	AB Hässle USA/ 1991 synth. [212]
Isradipine (Isrodipine; PN-200-110) [DynaCirc, Icaz, Lomir, Prescal, Vascal] C08CA03	(racemate)	[75695-93-1] $C_{19}H_{21}N_3O_5$ 371.39	Ca channel blocker/antihypertensive and antianginal (stable angina p.). cps. 2.5, 5 mg; s.r.tbl. 5, 10 mg. 5 mg, O.	Sandoz GB/1989; USA/1989. synth. [213]
Nicardipine (RS-69216; YC-93) [Antagonil, Cardene, Lincil, Nimicor, Perdipine, Vasodin, Vasonase] C08CA04	(racemate)	[55985-32-5] $C_{26}H_{29}N_3O_6$ 479.53 Hydrochloride: [54527-84-3] $C_{26}H_{29}N_3O_6 \cdot HCl$ 515.99	Ca channel blocker/antihypertensive and antianginal (stable angina p.). s.r.tbl. 20, 30, 40 mg; tbl. 10, 20, 30 mg; amp. 2, 10, 25 mg. 90 mg, O, P.	Yamanouchi USA/1988 synth. [214, 215]

Table 36. (*Continued*)

INN (Synonyms) [Brand names] ATC	Structure (Remarks)	[CAS-No.] Formula MW (g/mol)	Target (if known) Medical use Formulation DDD	Originator Approval (country/year) Production method [References]
Nifedipine (Bay a 1040) [Adalat, Adipine, Aprical, Cardilate, Coracten, Cordilan, Corinfar, Corotrend, Nifecard, Procardia, Tensipine, Zenusin] C08CA05 **WHO essential medicine**		[*21829-25-4*] $C_{12}H_{18}N_2O_6$ 346.34	Ca channel blocker/antihypertensive and antianginal (stable angina p.) and treatment of Raynaud-syndrome. s.r.tbl. 20, 30, 60, 90 mg; cps. 5, 10 mg; tbl. 30, 60, 90 mg; vials 5 mg/50 mL (inj.) 30 mg, O, P.	Bayer DE/1975; USA/1981. synth. [216, 217]
Nimodipine (Bay e 9736) [Nimotop, Admon, Periplum] C08CA06	(racemate)	[*66085-59-4*] $C_{21}H_{26}N_2O_7$ 418.45	Ca channel blocker/antihypertensive and use in preventing complications in subarachnoid hemorrhage (vasospasm). f.c.tbl. 30 mg; cps. 30 mg; vials 10 mg/50 mL. 300 mg, O; 50 mg, P.	Bayer USA/1985 synth. [218–220]
Nisoldipine (Bay k 5552) [Baymycard, Sular, Syscor, Zadipina] C08CA07	(racemate)	[*63675-72-9*] $C_{20}H_{24}N_2O_6$ 388.42	Ca channel blocker/antihypertensive and antianginal (chronic. stable angina p.). f.c.tbl. 5, 10 mg; s.r.tbl. 10, 20, 30 mg; 8.5, 17, 34 mg. 20 mg, O.	Bayer DE/1990 synth. [221]
Nitrendipine (Bay e 5009) [Baypress, Bayotensin] C08CA08	(racemate)	[*39562-70-4*] $C_{18}H_{20}N_2O_6$ 360.37	Ca channel blocker/antihypertensive. tbl. 10, 20 mg; oral soln. 5 mg/mL. 20 mg, O.	Bayer DE/1985 synth. [222, 223]
Lacidipine (GR-43659X; GX-1048; SN-305) [Caldine, Lacipil, Midotens, Motens] C08CA09	(*E* isomer)	[*103890-78-4*] $C_{26}H_{33}NO_6$ 455.55	Ca channel blocker/antihypertensive. tbl. 2, 4 mg. 4 mg, O.	Glaxo DE, UK/1991 synth. [224, 225]

Table 36. (*Continued*)

INN (Synonyms) [Brand names] ATC	Structure (Remarks)	[CAS-No.] Formula MW (g/mol)	Target (if known) Medical use Formulation DDD	Originator Approval (country/year) Production method [References]
Nilvadipine (FR-34235; FK-235; SKF-102362) [Escor, Nilvadil] C08CA10	(racemate)	[*75530-68-6*] $C_{19}H_{19}N_3O_6$ 385.38	Ca channel blocker/ antihypertensive. s.r.cps. 8, 16 mg; tbl. 2, 4 mg. 8 mg, O.	Fujisawa DE/1989 synth. [226–228]
Manidipine (Franidipine) [Calslot, Iperten, Manyper, Vascoman] C08CA11	(racemate)	[*89226-50-6*] $C_{35}H_{38}N_4O_6$ 610.71 Dihydrochloride: [*89226-75-5*] $C_{35}H_{38}N_4O_6 \cdot 2HCl$ 683.63	Ca channel blocker/ antihypertensive. tbl. 5, 10, 20 mg. 10 mg, O.	Takeda JP/1990 synth. [229, 230]
Lercanidipine (Rec-15-2375; R-75) [Carmen, Corifeo, Lercan, Zanidip] C08CA13	(racemate)	[*100427-26-7*] $C_{36}H_{41}N_3O_6$ 611.74 Hydrochloride: [*132866-11-6*] $C_{36}H_{41}N_3O_6 \cdot HCl$ 648.20	Ca channel blocker/ antihypertensive tbl. 10, 20 mg. 10 mg, O.	Recordati DE/1997 synth. [231, 232]
Cilnidipine (FRC-8653) [Atelec, Cilacar, Cilaheart] C08CA14	(racemate)	[*132203-70-4*] $C_{27}H_{28}N_2O_7$ 492.52	Ca channel blocker/ antihypertensive. tbl. 10 mg. 10 mg, O.	Fujirebio JP/1991 synth. [233]
Benidipine (KW-3049) [Coniel] C08CA15		[*105979-17-7*] $C_{28}H_{31}N_3O_6$ 505.57 Hydrochloride: [*91599-74-5*] $C_{28}H_{31}N_3O_6 \cdot HCl$ 542.03	Ca channel blocker/ antihypertensive. tbl 2,4,8 mg.	Kyowa Hakko JP/1991 synth. [234, 235]

Table 36. (*Continued*)

INN (Synonyms) [Brand names] ATC	Structure (Remarks)	[CAS-No.] Formula MW (g/mol)	Target (if known) Medical use Formulation DDD	Originator Approval (country/year) Production method [References]
Clevidipine [Clevelox, Cleviprex] C08CA16	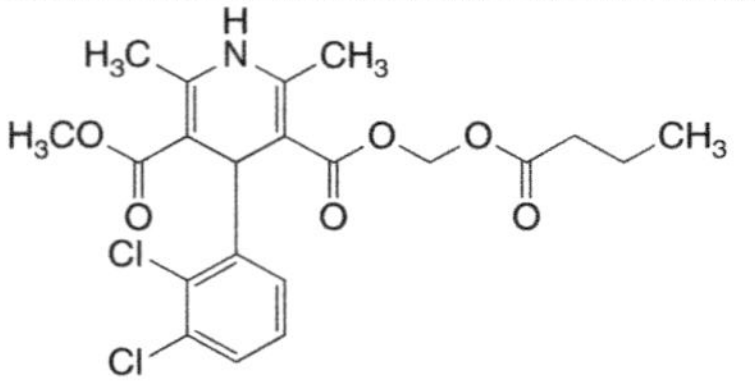 (racemate)	[*167221-71-8*] $C_{21}H_{23}Cl_2NO_6$ 456.32	Ca channel blocker/ ultrashort acting in vascular smooth muscle and reducing mean arterial pressure, e.g., in perioperative patients. vials 25 mg/50 mL. 120 mg, P.	Astra (AstraZeneca) USA/2008 synth. [236]

2.7.2. C08D Selective Calcium Channel Blockers with Direct Cardiac Effects

2.7.2.1. C08DB Benzothiazepine Derivatives (Table 37)

Table 37. C08DB Benzothiazepine derivatives

INN (Synonyms) [Brand names] ATC	Structure (Remarks)	[CAS-No.] Formula MW (g/mol)	Target (if known) Medical use Formulation DDD	Originator Approval (country/year) Production method [References]
Diltiazem (CRD-401) [Cardizem, Dilacor, Dilzem, Tiazac] C08DB01	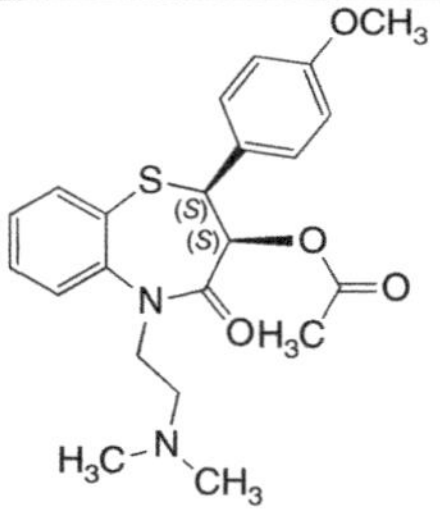 ((2*S-cis*) form)	[*42399-41-7*] $C_{22}H_{26}N_2O_4S$ 414.53 Hydrochloride: [*33286-22-5*] $C_{22}H_{26}N_2O_4S \cdot HCl$ 450.99	Ca channel blocker/ antihypertensive, antianginal, antiarrhythmic (class IV). s.r.tbl. 90, 120, 180 mg; cps. 180, 240 mg. 240 mg, O.	Tanabe Seiyaku USA/1982 synth. [237–239]

2.8. C09 Agents Acting on the Renin-Angiotensin System [240, 241]

2.8.1. C09A ACE Inhibitors, Plain

2.8.1.1. C09AA ACE Inhibitors, Plain (Table 38)

Table 38. C09AA ACE inhibitors, plain

INN (Synonyms) [Brand names] ATC	Structure (Remarks)	[CAS-No.] Formula MW (g/mol)	Target (if known) Medical use Formulation DDD	Originator Approval (country/year) Production method [References]
Captopril (SQ-14225) [Acepril, Acepress, Capoten, Capozide, Lopirin, Tensobon] C09AA01		[62571-86-2] $C_9H_{15}NO_3S$ 217.29	first orally active ACE inhibitor/antihypertensive. tbl. 6.25, 12.5, 25, 50 mg; cps. 18.75 mg. 50 mg, O.	Squibb USA/1981 synth. [242–244]
Enalapril (MK-421) [Amprace, Enacard, Glioten, Innovace, Naprilene, Pres, Renitec, Renivace, Vaseretic, Vasotec, Xanef] C09AA02 **WHO essential medicine**		Base: [75847-73-3] $C_{20}H_{28}N_2O_5$ 376.45 Maleate: [76095-16-4] $C_{20}H_{28}N_2O_5 \cdot C_4H_4O_4$ 492.53 **Enalaprilat** = Diacid: [76420-72-9] $C_{18}H_{24}N_2O_5$ 348.40 Dihydrate: [84680-54-6] $C_{18}H_{24}N_2O_5 \cdot 2H_2O$ 384.43	ACE inhibitor/ antihypertensive tbl. 2.5, 5, 10, 20 mg. 10 mg, O. enalaprilat: amp. 1.25/1.25 mL. 10 mg, P.	Merck & Co. USA/1985 synth. [245–248]
Lisinopril (MK-521) [Acerbon, Alapril, Coric, Lisitril, Prinil, Prinivil, Prinzide, Vivatec, Zestril] C09AA03		[76547-98-3] $C_{21}H_{31}N_3O_5$ 405.50 Dihydrate: [83915-83-7] $C_{21}H_{31}N_3O_5 \cdot 2H_2O$ 441.53	ACE inhibitor/ antihypertensive. tbl. 2.5, 5, 10, 20, 40 mg. 10 mg, O.	Merck & Co. USA/1987 synth. [249, 250]
Perindopril (McN-A-2833; S-9490) [Aceon, Coversum, Coversyl, Procaptan] C09AA04		[82834-16-0] $C_{19}H_{32}N_2O_5$ 368.47 Erbumine salt (tert-Butylamine): [107133-36-8] $C_{19}H_{32}N_2O_5 \cdot C_4H_{11}N$ 441.61	ACE inhibitor/ antihypertensive. tbl. 2, 4, 8 mg. 4 mg, O.	ADIR USA/1999 synth. [251, 252]
Ramipril (Hoe-498) [Altace, Cardace, Delix, Pramace, Tritace Unipril, Vesdil] C09AA05		[87333-19-5] $C_{23}H_{32}N_2O_5$ 416.52	ACE inhibitor/ antihypertensive. tbl. 2.5, 5, 10 mg. 2.5 mg, O.	Hoechst USA/1989 synth. [253, 254]

Table 38. (*Continued*)

INN (Synonyms) [Brand names] ATC	Structure (Remarks)	[CAS-No.] Formula MW (g/mol)	Target (if known) Medical use Formulation DDD	Originator Approval (country/year) Production method [References]
Quinapril (CI-906; PD-109452-2) [Accupril, Accupro, Accuretic] C09AA06		[*85441-61-8*] $C_{25}H_{30}N_2O_5$ 438.52 Hydrochloride: [*82586-55-8*] $C_{25}H_{30}N_2O_5 \cdot HCl$ 474.99	ACE inhibitor/ antihypertensive. f.c.tbl. 5, 10, 20, 40 mg. 15 mg, O.	Warner-Lambert USA/1989 synth. [255, 256]
Benazepril (CGS-14824A) [Cibacen, Lotensin, Lotrel] C09AA07		[*86541-75-5*] $C_{24}H_{28}N_2O_5$ 424.50 Hydrochloride: [*86541-74-4*] $C_{24}H_{28}N_2O_5 \cdot HCl$ 460.96	ACE inhibitor/ antihypertensive. f.c.tbl. 5, 10, 20 mg; tbl. 2.5, 5, 10 mg. 7.5 mg, O.	Ciba-Geigy USA/1990 synth. [257, 258]
Cilazapril (Ro-31-3113) [Dynorm, Inhibace] C09AA08		[*88768-40-5*] $C_{22}H_{31}N_3O_5$ 417.51 Monohydrate: [*92077-78-6*] $C_{22}H_{31}N_3O_5 \cdot H_2O$ 435.52	ACE inhibitor/ antihypertensive. f.c.tbl. 0.5, 1, 2.5, 5 mg; tbl. 0.25, 0.5, 1 mg. 2.5 mg, O.	Roche USA/1990 synth. [259, 260]
Fosinopril (SQ-28555) [Dynacil, Fosinorm, Fositen, Monopril, Staril] C09AA09		[*98048-97-6*] $C_{30}H_{46}NO_7P$ 563.67 Sodium salt: [*88889-97-6*] $C_{30}H_{45}NNaO_7P$ 585.65	ACE inhibitor/ antihypertensive. tbl. 5, 10, 20, 40 mg. 15 mg, O.	Squibb USA/1991 synth. [261–264]
Trandolapril (RU-44570) [Gopten, Mavik, Odrik, Tarka] C09AA10		[*87679-37-6*] $C_{24}H_{34}N_2O_5$ 430.55 Hydrochloride: [*87725-72-2*] $C_{24}H_{34}N_2O_5 \cdot HCl$ 467.01	ACE inhibitor/ antihypertensive. tbl. 0.5, 1, 2, 4 mg; cps. 0.5, 1, 2 mg. 2 mg, O.	Hoechst USA/1993 synth. [265]
Spirapril (Sch-33844) [Quadropril, Setrilan] C09AA11		[*83647-97-6*] $C_{22}H_{30}N_2O_5S_2$ 466.62 Hydrochloride: [*94841-17-5*] $C_{22}H_{30}N_2O_5S_2 \cdot HCl$ 503.08	ACE inhibitor/ antihypertensive. tbl. 6 mg. 6 mg, O.	Schering Corp.; Sandoz DE/1995 synth. [266, 267]

Table 38. (*Continued*)

INN (Synonyms) [Brand names] ATC	Structure (Remarks)	[CAS-No.] Formula MW (g/mol)	Target (if known) Medical use Formulation DDD	Originator Approval (country/year) Production method [References]
Delapril (CV-3317; REV-6000A) [Adecut, Delaket] C09AA12		[*83435-66-9*] $C_{26}H_{32}N_2O_5$ 452.56 Hydrochloride: [*83435-66-9*] $C_{26}H_{32}N_2O_5 \cdot HCl$ 489.01	ACE inhibitor/ antihypertensive. tbl. 15, 30 mg. 30 mg, O.	Takeda JP/1989 synth. [268]
Moexipril (CI-925; RS-10085-197; SPM-925) [Fempress, Perdix, Univasc] C09AA13		[*103775-10-6*] $C_{27}H_{34}N_2O_7$ 498.58 Hydrochloride: [*82586-52-5*] $C_{27}H_{34}N_2O_7 \cdot HCl$ 535.04	ACE inhibitor/ antihypertensive. f.c.tbl. 7.5, 12.5, 15 mg. 15 mg, O.	Warner-Lambert USA/1995 synth. [269, 270]
Temocapril (RS-5142; CS-622) [Acecol] C09AA14		[*111902-57-9*] $C_{23}H_{28}N_2O_5S_2$ 476.62 Hydrochloride: [*110221-44-8*] $C_{23}H_{28}N_2O_5S_2 \cdot HCl$ 513.08	ACE inhibitor/ antihypertensive. tbl. 1, 2, 4 mg. 10 mg, O.	Sankyo JP/1994 synth. [271, 272]
Zofenopril (SQ-26991; MEN-8029) [Bifril, Zofenil] C09AA15		[*81872-10-81*] $C_{22}H_{23}NO_4S_2$ 429.55 Calcium salt: [*81938-43-4*] $C_{44}H_{44}CaN_2O_8S_4$ 897.16	ACE inhibitor/ antihypertensive. tbl. 7.5, 15, 30 mg. 30 mg, O.	Squibb A, CH/2000; DE/2008 Synth. [273]
Imidapril (TA-6366) [Tanatril] C09AA16		[*89371-37-9*] $C_{20}H_{27}N_3O_6$ 405.45 Hydrochloride: [*89396-94-1*] $C_{20}H_{27}N_3O_6 \cdot HCl$ 441.91	ACE inhibitor/ antihypertensive. tbl. 2.5, 5, 10 mg. 10 mg, O.	Tanabe Seiyaku JP/1993 synth. [274, 275]

2.8.2. C09C Angiotensin II Antagonists, Plain

2.8.2.1. C09CA Angiotensin II Antagonists, Plain (Table 39)

Table 39. C09CA Angiotensin II antagonists, plain

INN (Synonyms) [Brand names] ATC	Structure (Remarks)	[CAS-No.] Formula MW (g/mol)	Target (if known) Medical use Formulation DDD	Originator Approval (country/year) Production method [References]
Losartan (DuP-753; MK-954) [Cozaar, Hyzaar, Lorzaar] C09CA01		[114798-26-4] $C_{22}H_{23}ClN_6O$ 422.92 potassium salt: [124750-99-8] $C_{22}H_{22}ClKN_6O$ 461.01	angiotensin II type 1 receptor (AT$_1$)antagonist/antihypertensive. f.c.tbl. 12.5, 25, 50, 100 mg; powder for suspension 2.5 mg/mL. 50 mg, O.	DuPont; DuPont Merck USA/1994 synth. [276–278]
Eprosartan (SKF-108566) [Teveten] C09CA02		[137862-53-4] $C_{23}H_{24}N_2O_4S$ 424.52 Mesilate: [144143-96-4] $C_{23}H_{24}N_2O_4S \cdot CH_4O_3S$ 520.63	angiotensin II type 1 receptor (AT$_1$)antagonist/antihypertensive. f.c.tbl. 300, 400, 600 mg. 600 mg, O.	SmithKline Beecham DE/1997 synth. [279–281]
Valsartan (CGP-48933) [Diovan, Provas, Tareg] C09CA03		[137862-53-4] $C_{24}H_{29}N_5O_3$ 435.53	angiotensin II type 1 receptor (AT$_1$)antagonist/antihypertensive. tbl. 20, 40, 80, 160 mg; ccps. 80, 160 mg. 80 mg, O.	Ciba-Geigy DE/1996; USA/1996. synth. [282, 283]
Irbesartan (BMS-186295; SR-47436) [Aprovel, Avalide, Avapro, Karvea] C09CA04		[138402-11-6] $C_{25}H_{28}N_6O$ 428.54	angiotensin II type 1 receptor (AT$_1$)antagonist/antihypertensive. tbl. 75, 150, 300 mg. 150 mg, O.	Sanofi DE/1997 synth. [284, 285]

Table 39. (*Continued*)

INN (Synonyms) [Brand names] ATC	Structure (Remarks)	[CAS-No.] Formula MW (g/mol)	Target (if known) Medical use Formulation DDD	Originator Approval (country/year) Production method [References]
Candesartan (CV-11974; TCV-116) [Amias, Atacand, Blopress] C09CA06		[*139481-59-7*] $C_{24}H_{20}N_6O_3$ 440.46 Cilexetil: [*145040-37-5*] $C_{33}H_{34}N_6O_6$ 610.67	angiotensin II type 1 receptor (AT_1)antagonist/ antihypertensive. tbl. 2, 4, 8, 12, 16 mg. 8 mg, O.	Takeda SE/1997 synth. [286, 287]
Telmisartan (BIBR-277) [Kinzal, Kinzalmono, Micardis, Pritor] C09CA07		[*144701-48-4*] $C_{33}H_{30}N_4O_2$ 514.63	angiotensin II type 1 receptor (AT_1)antagonist/ antihypertensive. tbl. 20, 40, 80 mg. 40 mg, O.	Thomae (Boehringer Ing.) USA/1999 synth. [288, 289]
Olmesartan medoxomil (CS-866) [Benicar, Olmetec, Votum] C09CA08		[*144689-63-4*] $C_{29}H_{30}N_6O_6$ 558.60	angiotensin II type 1 receptor (AT_1)antagonist/ antihypertensive. tbl. 5, 20, 40 mg. 20 mg, O.	Sankyo USA/2002; DE/2002. synth. [290, 291]
Azilsartan medoxomil (TAK-536) [Edarbi] C09CA09		[*863031-21-4*] $C_{30}H_{24}N_4O_8$ 568.54 Kamedoxomil (potassium salt) [*863031-24-7*] $C_{30}H_{23}K N_4O_8$ 607.64	angiotensin II type 1 receptor (AT_1)antagonist/ antihypertensive. tbl. 40, 80 mg. 40 mg, O.	Takeda DE/2012 synth. [292, 293]

2.8.3. C09D Angiotensin II Antagonists, Combinations

2.8.3.1. C09DX Angiotensin II Antagonists, Other Combinations (Table 40)

Table 40. C09DX Angiotensin II antagonists, other combinations

INN (Synonyms) [Brand names] ATC	Structure (Remarks)	[CAS-No.] Formula MW (g/mol)	Target (if known) Medical use Formulation DDD	Originator Approval (country/year) Production method [References]
Sacubitril + Valsartan (LCZ-696; AHU-377) [Entresto] C09DX04 (comb. w/Valsartan, see C09CA03)		Sacubitril: [149709-62-6] C$_{24}$H$_{29}$NO$_5$ 411.49 Comb. w/Valsartan: [936623-90-4]	inhibitor of neprilysin (neutral endopeptidase)/antihypertensive for patients with CHF and a reduced LVEF. f.c.tbl. with cocrystals (1:1 molar) 24/26, 49/51 and 97/103 mg sacubitril/valsartan.	Novartis USA/2015; EU/2016. synth. [294–296]

2.8.4. C09X Other Agents Acting on the Renin-Angiotensin System

2.8.4.1. C09XA Renin Inhibitors (Table 41) [297, 298]

Table 41. C09XA Renin inhibitors

INN (Synonyms) [Brand names] ATC	Structure (Remarks)	[CAS-No.] Formula MW (g/mol)	Target (if known) Medical use Formulation DDD	Originator Approval (country/year) Production method [References]
Aliskiren (CGP-60536) [Rasilez, Tekturna] C09XA02		[173334-57-1] C$_{30}$H$_{53}$N$_3$O$_6$ 551.77 Hemifumarate: [173334-58-2] C$_{64}$H$_{110}$N$_6$O$_{16}$ 1219.61	direct inhibition of renin, the enzyme that cleaves angiotensinogen to angiotensin I/antihypertensive f.c.tbl. 150, 300 mg. 150 mg, O.	Ciba-Geigy/ Novartis; Speedel USA/2007 synth. [299–302]

2.9. C10 Lipid-Modifying Agents

2.9.1. C10A Lipid-Modifying Agents, Plain

2.9.1.1. C10AA HMG CoA Reductase Inhibitors (Table 42) [303, 304]

Table 42. C10AA HMG CoA reductase inhibitors

INN (Synonyms) [Brand names] ATC	Structure (Remarks)	[CAS-No.] Formula MW (g/mol)	Target (if known) Medical use Formulation DDD	Originator Approval (country/year) Production method [References]
Simvastatin (MK-733; Synvinolin) [Zocor, Vytorin] C10AA01 **WHO essential medicine**		[79902-63-9] $C_{25}H_{38}O_5$ 418.57	HMG CoA reductase inhibitor/ antilipemic. Tbl. 5, 10, 20, 40 mg; f.c.tbl. 5, 10, 20, 40, 80 mg. 30 mg, O.	Merck & Co. USA/1990 semisynth. from lovastatin [305, 306]
Lovastatin (MK-803; Mevinolin; Monacolin K) [Advicor, Altoprev, Mevacor, Mevinacor] C10AA02		[75330-75-5] $C_{24}H_{36}O_5$ 404.55	HMG CoA reductase inhibitor/antilipemic. tbl. 10, 20, 40 mg. 45 mg, O.	Merck & Co; Sankyo USA/1987; DE/1989. fermentation of *Aspergillus terreus* (Merck) or *Monascus ruber* (Sankyo) [307–309]
Pravastatin (CS-514; SQ-31000; Eptastatin) [Lipostat, Mevalotin, Pravachol, Pravacol, Selectin] C10AA03		[81093-37-0] $C_{23}H_{36}O_7$ 424.53 Sodium salt: [81131-70-6] $C_{23}H_{35}NaO_7$ 446.52	HMG CoA reductase inhibitor/antilipemic. tbl. 5, 10, 20 mg. 30 mg, O.	Sankyo JP/1989 microbial hydroxylation of mevastatin (= compactin) [310–312]
Fluvastatin (XU-62-320; XU-620; SRI-62320) [Cranoc, Lescol, Locol] C10AA04		Acid: [93957-54-1] $C_{24}H_{26}FNO_4$ 411.46 Sodium salt: [93957-55-2] $C_{24}H_{25}FNNaO_4$ 433.46	HMG CoA reductase inhibitor/antilipemic. s.r.tbl. 80 mg; cps. 20, 40 mg. 60 mg, O.	Sandoz GB/1994 synth. [313–315]

Table 42. (*Continued*)

INN (Synonyms) [Brand names] ATC	Structure (Remarks)	[CAS-No.] Formula MW (g/mol)	Target (if known) Medical use Formulation DDD	Originator Approval (country/year) Production method [References]
Atorvastatin (CI-981; YM-548) [Lipitor, Sortis, Tahor, Torvast, Caduet] C10AA05		Acid: [*134523-00-5*] $C_{33}H_{35}FN_2O_5$ 558.64 Ca-salt: [*134523-03-8*] $C_{66}H_{68}CaF_2N_4O_{10}$ 1155.36	HMG CoA reductase inhibitor/antilipemic. f.c.tbl. 10, 20, 40, 80 mg. 20 mg, O.	Warner-Lambert GB/1997 synth. [316, 317]
Rosuvastatin (S-4522; ZD-4522) [Crestor] C10AA07		Acid: [*287714-14-4*] $C_{22}H_{28}FN_3O_6S$ 481.55 Calcium salt: [*147098-20-2*] $C_{44}H_{54}CaF_2N_6O_{12}S_2$ 1001.15	HMG CoA reductase inhibitor/antilipemic. f.c.tbl. 5, 10, 20 mg. 10 mg, O.	Shionogi; AstraZeneca NL/2003 synth. [318, 319]
Pitavastatin (NK-104; Itavastatin) [Livalo, Livazo, Alipza] C10AA08		[*147511-69-1*] $C_{25}H_{24}FNO_4$ 421.47 1,5-Lactone (**Nisvastatin**): [*141750-63-2*] $C_{25}H_{22}FNO_3$ 403.45 Calcium salt: $C_{50}H_{46}CaF_2N_2O_8$ 881.00	HMG CoA reductase inhibitor/ antilipemic. tbl. 1, 2 mg. 2 mg, O.	Nissan Chem.; Kowa JP/2003 synth. [320, 321]

2.9.1.2. C10AB Fibrates (Table 43)

Table 43. C10AB Fibrates

INN (Synonyms) [Brand names] ATC	Structure (Remarks)	[CAS-No.] Formula MW (g/mol)	Target (if known) Medical use Formulation DDD	Originator Approval (country/year) Production method [References]
Bezafibrate (BM-15075) [Befizal, Bezalip, Bezatol, Cedur, Difaterol] C10AB02		[41859-67-0] $C_{19}H_{20}CINO_4$ 361.83	PPAR$_\alpha$ agonist; lipoprotein-lipase enhancer/antilipemic. f.c.tbl. 200 mg; drg. 200 mg; s.r.tbl. 400 mg. 600 mg, O.	Boehringer Mannheim DE/1977 synth. [322]
Gemfibrozil (CI-719) [Gevilon, Jezil, Lopid] C10AB04		[25812-30-0] $C_{15}H_{22}O_3$ 250.34	PPAR$_\alpha$ agonist; lipoprotein-lipase enhancer/antilipemic. f.c.tbl. 600, 900 mg. 1200 mg, O.	Parke Davis CH/1985 synth. [323]
Fenofibrate (LF-178; Procetofene) [Lipanthyl, Lipidil, Lipoclar, Lipofene, Lipsil, Lofibra, Nolipax, Secalip, Supralip, Tricor] C10AB05		[49562-28-9] $C_{20}H_{21}ClO_4$ 360.84	PPAR$_\alpha$ agonist; lipoprotein-lipase enhancer/antilipemic. f.c.tbl. 145, 160 mg; cps. 200 mg. 200 mg (micronized), O; 250 mg (normal), O.	Lab. Fournier; Orchimed FR/1975; CH/1977. synth. [324]

2.9.1.3. C10AC Bile Acid Sequestrants (Table 44)

Table 44. C10AC Bile acid sequestrants

INN (Synonyms) [Brand names] ATC	Structure (Remarks)	[CAS-No.] Formula MW (g/mol)	Target (if known) Medical use Formulation DDD	Originator Approval (country/year) Production method [References]
Colestyramine (Cholestyramine; MK-135) [Cholibar, Cuemid, Quantalan, Questran, Lipocol] C10AC01	Styrene-divinylbenzene copolymer containing quaternary ammonium groups (strongly basic anion exchange resin)	[11041-12-6]	basic ion exchanger with high affinity to bile acids/antilipemic powder 4 g; chewing tbl. 2 g 14 g, O	Bristol-Myers; Merck & Co. Synth. [325]
Colestipol hydrochloride (U-26597A) [Colestid] C10AC02	Copolymer of diethylenetriamine with 1-chloro-2,3-epoxypropane (basic anion exchange resin)	[37296-80-3]	basic ion exchanger with high affinity to bile acids/antilipemic. tbl. 1 g (micronized hydrochloride). 20 g, O.	Upjohn CH/1972 synth. [326]
Colesevelam hydrochloride (GT-31-104) [CholestaGel, Lodalis, Welchol] C10AC04	Allylamine polymer with 1-chloro-2,3-epoxypropane, [6-(allylamino)hexyl]-trimethylammonium chloride and N-allyldecylamine	[182815-44-7]	bile acid sequestrant; lowers total and LDL cholesterol and binds bile acids in the intestine/antilipemic agent. f.c.tbl. 625 mg. 3.75 g, O.	GelTex Pharmaceuticals USA/2003 synth. [327]

2.9.1.4. C10AX Other Lipid-Modifying Agents (Table 45)

Table 45. C10AX Other lipid-modifying agents

INN (Synonyms) [Brand names] ATC	Structure (Remarks)	[CAS-No.] Formula MW (g/mol)	Target (if known) Medical use Formulation DDD	Originator Approval (country/year) Production method [References]
Ezetimibe (Sch-58235) [Ezetrol, Inegy, Vytorin, Zetia] C10AX09		[163222-33-1] $C_{24}H_{21}F_2NO_3$ 409.43	inhibits absorption of cholesterol from the small intestine; exact mechanism unknown/antilipemic; can be applied as monotherapy or in combination with simvastatin. tbl. 10 mg; comb. w/simvastatin 10, 20, 40, or 80 mg. 10 mg, O.	Schering Plough; Merck & Co. USA/2002; DE/2002. synth. [328, 329]
Lomitapide (BMS-201038) [Juxtapid, Lojuxta] C10AX12		[182431-12-5] $C_{39}H_{37}F_6N_3O_2$ 693.72	inhibition of the microsomal triglyceride transfer protein (MTTP), which is necessary for VLDL assembly and secretion in the liver/antilipemic. cps. 5, 10, 20 mg. 40 mg, O.	BMS; Aegerion Pharmac. USA/2012; EU/2013. synth. [330, 331]

Abbreviations

ACE	angiotensin converting enzyme
ADHD	attention-deficit/hyperactivity disorder
amp.	ampule(s)
c-AMP	cyclic adenosine monophosphate
AT	angiotensin
ATC	anatomical therapeutic chemical (classification system)
ATP	adenosine triphosphate
BPH	benign prostate hyperblasia
cGMP	cyclic guanosine monophosphate
CAS	chemical abstracts service
CHD	coronary heart disease
CHF	congestive heart failure
cps.	capsule(s)
CVD	cardiovascular disease
DDD	defined daily dose (see WHO's ATC classification system)
drg.	dragees
f.c. tbl.	film coated tablet(s)
HCT	hydrochlorothiazide
HMG CoA	3-hydroxy-3-methyl-glutaryl coenzyme A
HT	5-hydroxytryptamine = serotonin
infus.	infusion
inj.	injection
iv.	intravenous
INN	international nonproprietary name
LDL	low-density lipoprotein
LVEF	left ventricular ejection fraction
MTTP	microsomal triglyceride transfer protein
NCD	noncommunicable disease
NCE	new chemical entity
NKCC2	Na-K-Cl cotransporter
O	oral
P	parenteral
patr.	patrone ($\triangleq$ cartridge)
PCT	proximal convoluted tubule
PDE	phosphodiesterase
PDGF	platelet derived growth factor
PGI$_2$	prostaglandin I$_2$
PPAR	peroxisome proliferator-activated receptor
RAS	renin-angiotensin system
s-GC	soluble guanylyl cyclase
Sol.	solution
s.r.tbl.	sustained release tablet(s)
Synth.	synthetic
tbl.	tablet(s)
V	vasopressin
VLDL	very low-density lipoprotein
WADA	world anti-doping agency
WHO	world health organization

References

General References and Reviews

Bagal, S.K., et al. (2013) Ion channels as therapeutic targets: a drug discovery perspective, *J. Med. Chem. (Perspective)*, **56**, 593–624.

Edmondson, S.D., et al. (2015) Cardiovascular and metabolic diseases: 50 years of progress, *Medicinal Chemistry Reviews (Ann. Reports in Med. Chem.) ACS*, **50**, 83–117.

Fiuzat, M. et al. (2016) Resourcing Drug Development Commensurate with its Public Health Importance, *JACC: Basic to Translational Science* 1, 5, 309–312 (open access article).

Finlay, H. (2014) Current Approaches to the Treatment of Atrial Fibrillation, *Annual Reports in Medicinal Chemistry*, **49**, 101–113.

Hwang, Th.J., et al. (2016) Temporal Trends and Factors Associated with Cardiovascular Drug Development, 1990-2012, *JACC: Basic to Translational Science* 1, 5, 301–308 (open access article).

Lundberg, J.O. et al. (2015) Strategies to increase nitric oxide signaling in cardiovascular disease, *Nature Rev. Drug Discov.* 14 623–641.

Nabel, E.G., (2003) Cardiovascular Disease (Lessons learned from monogenic cardiovascular disorders), *N. Engl. J. Med.* 349 60–72 (review article in Genomic Medicine [A.E. Guttmacher et al. (Eds.)].

Schade, D. and Plowright, A.T. (2015) Medicinal Chemistry Approaches to Heart Regeneration, *J. Med. Chem., (Perspective)*, **58** 9451–9479 (Cardiomyogenic agents).

Tamargo, J. and Lopez-Sendon, J. (2011) Novel therapeutic targets for the treatment of heart failure, *Nature Rev. Drug Discov.* **10** 536–555.

Topol, E.J. (2009) Past the wall in cardiovascular R&D, *Nature Rev. Drug Discov.* 8 259 (Editorial).

Specific References

1 Kleemann-Engel-Kutscher-Reichert (2009) Pharmaceutical Substances", 5th edn, Thieme Verlag Stuttgart.

2 Goodman & Gilman's (2011) The Pharmacological Basis of Therapeutics, 12th edn, McGraw Hill Medical, N.Y. et al.; Especially Section III, Modulation of Cardiovascular Function, chapters 25-31, pages 669–908.

3 Mutschler (2012) "Arzneimittelwirkungen, 10th edn, Wissenschaftliche Verlagsgesellschaft, Stuttgart, chapter 14, Herz-Kreislauf-System, p. 451–560.

4 WHO (2015) Health in 2015: from MDGs, Millennium Development Goals to SDGs, Sustainable Development Goals, ISBN 978 92 4 156511 0.

5 Mullard, A. (2016) Cardiovascular pipeline decline quantified, *Nature Rev. Drug Discov.* 15 669.

6 Plump, A. (2010) Accelerating the pulse of cardiovascular R&D, *Nature Rev, Drug Discov.* 9 823–824.

7 Adam A.M. et al. (2017) Trends in the market for antihypertensive drugs, Nature Rev. Drug Discov. Online Publication doi: 10.1038/nrd.2016.262, 10 March 2017.

8 Stoll, A. and Kreis, W. (1935) *Helv. Chim. Acta* 18 120.

9 Beiersdorf (1982) DE 2 316 269, DE-prior. 1973.

10 Sandoz (1937) DE 646 930, CH-prior. 1932.

11 Wellcome Foundation (1930) GB 337 091, GB-prior. 1929.

12 Boehringer (1970) US 3 538 078, DE-prior. 1968.

13 Goodman & Gilman's (2011) The Pharmacological Basis of Therapeutics, 12th edn, McGraw Hill Medical, N.Y. et al.; Section III, Modulation of Cardiovascular Function, chapter 29, pp. 815–848.

14 Doering, W.E. et al. (1949) J. Am. Chem. Soc. 69 1700.

15 Boehringer Mannheim (1953) DE 877 611, DE-prior. 1950.

16 Adelstein, G.W. (1973) J. Med. Chem. 16 309.

17 Searle (1965) US 3 225 054, USA-prior. 1961.

18 Siddiqui, S. and Siddiqui, R.H.S. (1931) J. Indian Chem. Soc. 8 667; 9 (1932) 539; 12 (1935) 37.

19 Astra AB (1948) US 2 441 498, SE-prior. 1943.

20 Boehringer Ing. (1976) US 3 954 872, DE-prior. 1967.

21 Helopharm (1971) DE 2 001 431, DE-prior. 1970.

22 Banitt, E.H. et al. (1977) J. Med. Chem. 20 821.

23 Riker (1975) US 3 900 481, USA-prior. 1974.

24 Janssen (1978) US 4 126 689, USA-prior. 1976.

25 Labaz (1966) US 3 248 401, DE-prior. 1961.

26 Cross, P.E. et al. (1990) J. Med. Chem. 33 1151.

27 Pfizer (1990) US 4 959 366, GB-prior. 1986.

28 Hester, J.B. et al. (1991) J. Med. Chem. 34 308.

29 Upjohn (1992) US 5 155 268, USA-prior. 1984.

30 Sanofi (1993) US 5 223 510, FR-prior. 1990.

31 Cardiome Pharma (06.06.2006) US 7 057 053, USA-prior. 06.10.2000.

32 Boehringer Ingelheim (1931) DE 520 079, DE-prior.1926.

33 Boehringer Ingelheim (1931) DE 522 790, DE-prior.1929.

34 Goto, T. et al. (1954) J. Pharm. Soc. Jpn. 74 318.

35 Tullar, B.F. (1948) J. Am. Chem. Soc. 70 2067.

36 Sterling Drug (1956) US 2 774 789, USA-prior. 1947.

37 Barger, G. and Ewins, A.J. (1910) J. Chem. Soc. 97 2253.

38 Wasre, E. and Sommer, H. (1923) Helv. Chim. Acta 6 54.

39 Schöpf, C. et al. (1934) Liebigs Ann. Chem. 513 196.

40 F. Stearns & Co. (1933) US 1 932 347, DE-prior.1927.

41 Bergmann, E.D. and Sulzbacher, M. (1951) J. Org. Chem. 16 84.

42 Eli Lilly (1976) US 3 987 200, US-prior.1972.

43 Fisons; Teva (1982) EP 72 061, GB-prior. 1981.

44 Lentia (1967) US 3 340 298, AT-prior. 1963.

45 Weinstock, J. et al. (1980) J. Med. Chem. 23 973.

46 SmithKline (1979) US 4 160 765, USA-prior. 1976.

47 Degussa (1962) US 3 029 239, DE-prior. 1954.

48 Degussa (1963) US 3 112 313, DE-prior. 1959.

49 Stolz, A.F. (1904) Chem. Ber. 37 4149.

50 Farbwerke (1904) DE 152 814, DE-prior. 1903.

51 Reicheneder, F. et al. (1981) Arzneim.-Forsch. 31 1529.

52 BASF (1971) US 3 631 038, DE-prior. 1969.

53 Bilhuber, E. (1934) US 1 956 950, DE-prior.1930.

54 Winthrop/Sterling Drug (1977) US 4 004 012, USA-prior. 1975.

55 Winthrop/Sterling Drug (1978) US 4 072 746, USA-prior. 1975.

56 Singh, B. (1985) Heterocycles 23 1479.

57 Winthrop/Sterling Drug (1982) US 4 312 875, USA-prior. 1980.

58 Winthrop/Sterling Drug (1982) US 4 313 951, USA-prior. 1980.

59 Schnettler, R.A. et al. (1982) J. Med. Chem. 25 1477.

60 Richardson-Merrell/Merrell-Dow (1982) US 4 405 635. USA-prior. 1979.

61 Orion (22.08.1990) EP 383 449 A2, GB-prior. 11.02.1989 (racemate).

62 Orion (23.07.1992) WO 1992012135A1, GB-prior. 03.01.1991 (Levosimendan).

63 Murrell, W. (1879) "Nitro-glycerine as a remedy for angina pectoris", Lancet 1 80.

64 Thieme, B. (1895) US 541 899, DE-prior. 1894.

65 Goldberg, L. (1948) Acta Physiol. Scand. 15 173.

66 American Home Products (1975) US 3 886 186, US-prior. 1971.

67 Tenor, E. and Ludwig, R. (1971) Pharmazie 26 534.

68 Masuda, K. et al. (1971) Chem. Pharm. Bull. 19 72.

69 Takeda (1973) US 3 769 283, JP-prior. 1966.

70 Cassella (1977) DE 2 532 897, DE-prior. 1975.

71 Corey, E.J. et al. (1969) J. Am. Chem. Soc. 91 535; (1970) 92 2586.

72 Schaaf, T.K. and Corey, E.J. (1972) J. Org. Chem. 37 2921.

73 Born, G.V.R. et al. (1964) Nature 202 761.

74 Review: Eltzschig, H.K. (2009) Anesthesiology 111 904–915.

75 Servier (1994) US 5 296 482, FR-prior. 1991.

76 Syntex (1986) US 4 567 264, USA-prior. 1983.

77 CV Therapeutics (11.06.2002) US 6 403 567, USA-prior. 22.06.1999.

78 Grindeks (LV) (1984) US 4 481 218 A, SU-prior. 1978.

79 Tristram, E.W. et al. (1964) J.Org. Chem. 29 2053.

80 Reinhold, D.F. et al. (1968) J. Org. Chem. 33 1209.

81 Merck & Co. (1964) US 3 158 648, USA-prior. 1961.

82 Boehringer Ingelheim (1965) US 3 202 660, DE-prior. 1961.

83 Dr. A. Wander (1972) US 3 632 645, CH-prior. 1967

84 Beiersdorf (1982) US 4 323 570, DE-prior. 1978.

85 Science Union (1976) US 3 988 464, FR-prior. 1972.

86 Pfizer (1970) US 3 511 836, USA-prior. 1966.

87 Archibald, J.L. et al. (1971) J. Med. Chem. 14 1054.

88 Wyeth (1970) US 3 527 761, GB-prior. 1967.

89 Pfizer (1980) US 4 188 390, GB-prior. 1977.

90 Klemm, K. et al. (1977) Arzneim.-Forsch. 27 1895.

91 Byk Gulden (1976) US 3 957 786, DE-prior. 1969.

92 Eisai (1975) US 3 920 636, JP-prior. 1972.

93 Abbott (1977) US 4 026 894, USA-prior. 1975.

94 Rubin, A.A. et al. (1961) Science 133 2067.

95 Schering Corp. (1967) US 3 345 365, USA-prior. 1960.

96 Druey, J. et al. (1951) Helv. Chim. Acta 34 195.

97 Ciba (1949) US 2 484 785, CH-prior. 1947.

98 Druey, J. et al. (1951) Helv. Chim Acta 34 204.

99 Ciba (1949) US 2 484 029, CH-prior. 1945.

100 McCall, J.M. et al. (1975) J. Org. Chem. 40 3304.

101 Upjohn (1968) US 3 382 247, USA-prior. 1965.

102 Roche (1994) US 5 292 740, CH-prior. 1991.

103 Riechers, H. et al. (1996) J. Med. Chem. 39 2123.

104 BASF (1997) US 5 703 017, DE-prior. 1993.

105 BASF (1999) US 5 932 730, DE-prior. 1994.

106 Bolli, H.M. et al. (2012) J. Med. Chem. 55 7849.

107 Actelion (22.08.2006) US 7 285 549, EP-prior. 18.12.2000.

108 Terrett, N.K. et al. (1996) Bioorg. Med. Chem. Lett. 6 1819.

109 Pfizer (1993) US 5 250 534, GB-prior. 1990.

110 Daugan, A.C. et al. (2003) J. Med. Chem. 46 4525; 4533.

111 Icos (12.01.1999) US 5 859 006, GB-prior. 21.01.1994.

112 Mittendorf, J. et al. (2009) ChemMedChem 4 853.

113 Bayer (19.06.2013) EP 2 604 608, 19.06.2013; EP-prior. 27.11.2009.

114 Skuballa, W. and Vorbrueggen, H. (1981) Angew. Chem. 93 1080.

115 Schering AG (1987) US 4 692 464, DE-prior. 1978.

116 Asaki, T. et al. (2015) *J. Med. Chem.* **58** 7128.
117 Nippon Shinyaku (27.05.2004) US 7 205 302, JP-prior. 26.04.2001.
118 Actelion (03.11.2015) US 9 173 881, WO-prior.13.08.2008.
119 Kuo, I.Y. and Ehrlich, B.E. (2012) Ion channels in renal disease, *Chem. Rev.* **112** 6353–6372.
120 Holdrege, C.T. et al. (1959) *J. Am. Chem. Soc.* **81** 4807.
121 Loevens Kemiske Fabriken (1968) US 3 392 168, GB-prior. 1958.
122 de Stevens, G. et al. (1958) *Experientia* **14** 463.
123 Ciba (1964) US 3 163 645, USA-prior. 1958.
124 Novello, F.C. et al. (1957) *J. Am. Chem. Soc.* **79** 2028.
125 Merck & Co. (1957) US 2 809 194, USA-prior. 1956.
126 Ciba (1959) BE 576 304, USA-prior. 1958.
127 Jucker, E. and Lindenmann, A. (1962) *Helv. Chim. Acta* **45** 2316.
128 Sandoz (1969) US 3 459 756, CH-prior. 1960.
129 Graf, W. et al. (1959) *Helv. Chim. Acta* **42** 1085.
130 Geigy (1962) US 3 055 904, CH-prior. 1957.
131 Beiersdorf (1971) US 3 567 777, DE-prior. 1965.
132 Science Union (1971) US 3 565 911, GB-prior. 1968.
133 Sturm, K. et al. (1966) *Chem. Ber.* **99** 328.
134 Hoechst (1962) US 3 058 882, DE-prior. 1959.
135 Feit, P.W. et al. (1971) *J. Med. Chem.* **14** 432.
136 Loevens Kemiske Fabrik (1974) US 3 806 534, GB-prior. 1968.
137 Merkel, W. et al. (1976) *Eur. J. Med. Chem.* **11** 399.
138 Hoechst (1977) US 4 010 273, DE-prior. 1974.
139 Delarge, J. (1988) *Arzneim.-Forsch.* **38** 1a.
140 Boehringer Mannheim (1977) US 4 018 929, GB-prior. 1974.
141 Schultz, E.M. et al. (1962) *J. Med. Pharm. Chem.* **5** 660.
142 Merck & Co, (1966) US 3 255 241, USA-prior. 1961
143 Cella, J.A. et al. (1959) *J. Org. Chem.* **24** 1109.
144 Tweit, R.C. et al. (1962) *J. Org. Chem.* **27** 3325.
145 Searle (1961) US 3 013 012, USA-prior. 1958.
146 Cella, J.A. et al. (1959) *J. Org. Chem.* **24** 1109.
147 Searle (1959) US 2 900 383, USA-prior. 1957.
148 Grob, J. et al. (1997) *Helv. Chim. Acta* **80** 566.
149 Ciba-Geigy (1985) US 4 559 332, CH-prior. 1983.
150 Searle (2001) US 6 331 622, USA-prior. 1995.
151 Merck & Co. (1967) US 3 313 813, USA-prior. 1962.
152 Spickett, R.G.W. and Timmis, G.M. (1954) *J. Chem. Soc.* **1954**, 2887.
153 Smith, Kline & French (1963) US 3 081 230, USA-prior. 1960.
154 Kondo, K. et al. (1999) *Bioorg. Med. Chem.* **7** 1743.
155 Otsuka Pharmaceutical (1993) US 5 258 510, JP-prior. 1989.
156 Matsuhisa, A. et al. (2000) *Chem. Pharm. Bull.* **48** 21.
157 Yamanouchi (1998) US 5 723 606, JP-prior. 1993.
158 Philips (1962) US 3 056 836, NL-prior. 1955.
159 Urech, E. et al. (1950) *Helv. Chim Acta* **33** 1386.
160 Ciba (1950) US 2 503 059, CH-prior. 1947.
161 Mohler, W. et al. (1966) *Arch. Pharm.* **299** 448.
162 Chem Werke Albert (1969) US 3 422 107, DE-prior. 1964.
163 Chem. Werke Albert (1973) US 3 737 433, DE-prior. 1964.
164 Stoll, A. and Hofmann, A. (1943) *Helv. Chim. Acta* **26** 2070.
165 Sandoz (1941) DE 883 153, CH-prior.1940.
166 Arcari, G. et al. (1972) *Experientia* **28** 819.
167 Farmitalia (1966) US 3 228 943, IT-prior. 1962.
168 Farmitalia (1952) US 2 599 000, USA-prior. 1950.
169 Schlittler, E. et al. (1953) *Helv. Chim. Acta* **36** 2017.
170 Regnier, G.J. et al. (1968) *J. Med. Chem.* **11** 1151.
171 Science Union (1967) US 3 299 067, GB-prior. 1963.
172 Lipha (1967) US 3 334 096, FR-prior. 1963.
173 Merck, G. (1848) *Ann. Chem.* **66** 125.
174 Pictet et al. (1909) *Compt. Rend.* **149** 210. (Synthesis)
175 Orgamol (1963) GB 1 030 022, CH-prior. 1962.
176 Ciba (1969) US 3 483 221, CH-prior. 1965.
177 Sandoz (1967) US 3 471 515, CH-prior. 1965.
178 Black, J.W. et al. (1964) *The Lancet* **283** 1080.
179 Crowther, A.F. and Smith, L.H. (1968) *J. Med. Chem.* **11** 1009.
180 ICI (1967) US 3 337 628, GB-prior. 1962.
181 Wasson, B.K. et al. (1972) *J. Med. Chem.* **15** 651.
182 C.E. Frosst & Co. (1972) US 3 655 663, USA-prior. 1968.
183 Uloth, R.H. et al. (1966) *J. Med. Chem.* **9** 88.
184 Condon, M.E. et al. (1978) *J. Med. Chem.* **17**, 913.
185 Squibb (1976) US 3 935 267, USA-prior. 1970.
186 Nakagawa, K. et al. (1974) *J. Med. Chem.* **17** 529.
187 Otsuka (1975) US 3 910 924, JP-prior. 1972.
188 Sopfa (1969) DE 1 668 964, PL-prior.1967.
189 Härtfelder, G. et al. (1972) *Arzneim.-Forsch.* **22** 930.
190 Hoechst (1970) US 3 551 493, DE-prior. 1967.
191 A.B. Hässle (1975) US 3 873 600, SE-prior. 1970.
192 A.B. Hässle (1976) US 3 998 790, USA-prior. 1971.
193 ICI (1972) US 3 663 607, GB-prior. 1969.
194 ICI (1974) US 3 836 671, GB-prior. 1969.
195 May & Baker (1973) US 3 726 919, GB-prior. 1967.
196 May & Baker (1974) US 3 857 952, GB-prior. 1967.
197 Synthelabo (1981) US 4 252 984, FR-prior. 1975.
198 Synthelabo (1982) US 4 311 708, FR-prior. 1975
199 Harting, J. et al. (1986) *Arzneim.-Forsch.* **36** 200.
200 E. Merck AG (1981) US 4 258 062, DE-prior. 1976.
201 Chemie Linz (1977) US 4 034 009, AT-prior. 1973.
202 Erhardt, P.W. et al. (1982) *J. Med. Chem.* **25** 1408.
203 AB Hässle (1982) EP 41 491, SE-prior. 1980.
204 AHS (1983) US 4 387 103, USA-prior. 1980.
205 AHS (1986) US 4 593 119, USA-prior. 1980.
206 Janssen (1987) US 4 654 362, USA-prior.1983.
207 Clifton, J.E. et al. (1982) *J. Med. Chem.* **25** 670.
208 Allen & Hanburys (1977) US 4 012 444, GB-prior. 1970.
209 Boehringer Mannheim (1985) US 4 503 067, DE-prior. 1978.
210 Arrowsmith, J.E. et al. (1986) *J. Med. Chem.* **29** 1696.
211 Pfizer (1986) US 4 572 909, GB-prior. 1982.
212 AB Hässle (1981) US 4 264 611, SE-prior. 1978.
213 Sandoz (1984) US 4 466 972, CH-prior. 1977.
214 Iwanami, M. et al. (1979) *Chem. Pharm. Bull.* **27** 1426.
215 Yamanouchi (1976) US 3 985 758, JP-prior. 1973.
216 Bayer (1969) US 3 485 847, DE-prior. 1967.
217 Bayer (1972) US 3 644 627, DE-prior. 1967.
218 Meyer, H. et al. (1981) *Arzneim. Forsch.* **31** 407;   (1983) **33** 106.
219 Bayer (1974) US 3 799 934, DE-prior. 1971.
220 Bayer (1976) US 3 932 645, DE-prior. 1971.
221 Bayer (1979) US 4 154 839, DE-prior. 1975.
222 Meyer, H. et al. (1981) *Arzneim.-Forsch.* **31**, 407.
223 Bayer (1974) US 3 799 934, DE-prior. 1971.
224 Glaxo (1989) US 4 801 599, IT-prior. 1984.
225 Glaxo (1991) US 5 011 848, IT-prior. 1984.
226 Miyamae, A. et al. (1986) *Chem. Pharm. Bull.* **34**, 3071.
227 Fujisawa (1982) US 4 338 322, GB-prior. 1978.
228 Fujisawa (1985) US 4 525 478, GB-prior. 1978.
229 Meguro, K. et al. (1985) *Chem. Pharm. Bull.* **33** 3787.
230 Takeda (1990) US 4 892 875, JP-prior. 1982.
231 Leonardi, A. et al. (1988) *Eur. J. Med. Chem.* **33** 399.
232 Recordati (1987) US 4 705 797, GB-prior. 1984.
233 Fujirebio (1987) US 4 672 068, JP-prior. 1984.
234 Muto, K. et al. (1988) *Arzneim.-Forsch.* **38** 1662.
235 Kyowa Hakko (1984) US 4 448 964, JP-prior. 1981.
236 Astra (AstraZeneca) (1999) US 5 856 346, SE-prior. 1999.
237 Kugita, H. et al. (1970) *Chem. Pharm. Bull.* **18** 2028, 2284; (1971) **19** 595.

238 Tanabe Sciyaku (1971) US 3 562 257, JP-prior. 1967.
239 Tanabe Sciyaku (1983) US 4 420 628, JP-prior. 1981.
240 Zaman, M.A. et al. (2002) Drugs targeting the Renin-Angio-tensin-Aldosterone System, *Nature Rev. Drug Discov.* **1** 621–636.
241 Perico, N. et al. (2008) Present and future drug treatments for chronic kidney diseases: Evolving targets in renoprotection, *Nature Rev. Drug Discov.* **7** 936–953.
242 Ondetti, M.A. et al. (1977) *Science* **196** 441.
243 Squibb (1977) US 4 046 889, USA-prior. 1976.
244 Squibb (1978) US 4 105 776, USA-prior. 1976.
245 Patchett, A.A. et al. (1980) *Nature* **288** 280.
246 Wyvratt, M.J. et al. (1984) *J. Org. Chem.* **49** 2816.
247 Merck & Co. (1983) US 4 374 829, USA-prior. 1978.
248 Merck & Co. (1984) US 4 472 380, USA-prior. 1978.
249 Wu, M.T. et al. (1985) *J. Pharm. Sci.* **74** 352.
250 Blacklock, T.J. et al. (1988) *J. Org. Chem.* **53** 836.
251 Vincent, M. et al. (1982) *Tetrahedron Lett.* **23** 1677.
252 ADIR (1985) US 4 508 729, FR-prior. 1980.
253 Teetz, V. et al. (1984) *Arzneim.-Forsch.* **34** 1399.
254 Hoechst (1991) US 5 061 722, DE-prior. 1981.
255 Klutchko, S. et al. (1986) *J. Med. Chem.* **29** 1953.
256 Warner-Lambert (1982) US 4 344 949, USA-prior. 1980.
257 Watthey, J.W.H. et al. (1985) *J. Med. Chem.* **28** 1511.
258 Ciba-Geigy (1983) US 4 410 520, USA-prior. 1981.
259 Attwood, M.R. (1986) *J. Chem. Soc. Perkin Trans.* **1** 1011.
260 Roche (1985) US 4 512 924, GB-prior. 1982.
261 Krapcho, J. et al. (1988) *J. Med. Chem.* **31** 1148.
262 Thottathil, J.K. et al. (1986) *Tetrahedron Lett.* **27** 151.
263 Thottathil, J.K. et al. (1986) *J. Org. Chem.* **51** 3140.
264 Squibb (1982) US 4 337 201, USA-prior. 1980.
265 Hoechst (1990) US 4 933 361, DE-prior. 1981.
266 Smith, E.M. et al. (1989) *J. Med. Chem.* **32** 1600.
267 Schering Corp. (1984) US 4 470 972, USA-prior. 1980.
268 Miyake, A. et al. (1986) *Chem. Pharm. Bull.* **34** 2852.
269 Klutchko, S. et al. (1986) *J. Med. Chem.* **29** 1953.
270 Warner-Lambert (1982) US 4 344 949, USA-prior. 1980.
271 Yanagisawa, H. et al. (1987) *J. Med. Chem.* **30** 1984.
272 Sankyo (1987) US 4 699 905, JP-prior. 1985.
273 Squibb (1982) US 4 316 906, USA-prior. 1978.
274 Hayashi, K. et al. (1989) *J. Med. Chem.* **32** 289.
275 Tanabe Seiyaku (1985) US 4 508 727, JP-prior. 1982.
276 Carini, D.J. et al. (1991) *J. Med. Chem.* **34** 2525.
277 Larsen, R.D. et al. (1994) *J. Org. Chem.* **59** 6391.
278 DuPont (1992) US 5 138 069, USA-prior. 1986.
279 Weinstock, J. et al. (1991) *J. Med. Chem.* **34** 1514.
280 Keenan, R.M. et al. (1993) *J. Med. Chem.* **36** 1880.
281 SmithKline Beecham (1993) US 5 185 351, USA-prior. 1989.
282 Bühlmayer, P. et al. (1994) *Bioorg. Med. Chem. Lett.* **4** 29.
283 Ciba-Geigy (1995) US 5 399 578, CH-prior. 1990.
284 Bernhart, C.A. et al. (1993) *J. Med. Chem.* **36** 3371.
285 Sanofi (1993) US 5 270 317, FR-prior. 1990.
286 Kubo, K.K. et al. (1993) *J. Med. Chem.* **36** 2343.
287 Takeda (1993) US 5 196 444, JP-prior. 1990.
288 Ries, U. et al. (1993) *J. Med. Chem.* **36** 4040.

289 Thomae (Boehringer Ing.) (1997) US 5 591 762, DE-prior. 1991.
290 Yanagisawa, H. et al. (1996) *J. Med. Chem.* **39** 323.
291 Sankyo (1997) US 5 616 599, JP-prior. 1991.
292 Kohara, Y. et al. (1996) *J. Med. Chem.* **39** 5228.
293 Takeda (1993) US 5 243 054, JP-prior. 1991.
294 Ksander, C.M. et al. (1989) *J. Med. Chem.* **32** 2519; (1995) **38** 1689.
295 Ciba-Geigy (1993) US 5 217 996 (Ciba-Geigy), USA-prior. 1992.
296 Novartis (23.12.2008) US 7 468 390, USA-prior. 17.01.2002.
297 Jensen, C. et al. (2008) Aliskiren: the first renin inhibitor for clinical treatment, *Nature Rev. Drug Discov.* **7** 399–410.
298 Webb, R.L. et al. (2010) Direct renin inhibitors as a new therapy for hypertension, *J. Med. Chem.* **53** 7490–7520.
299 Göschke, R. et al. (2007) *J. Med. Chem.* **50** 4818.
300 Maibaum, J. et al. (2007) *J. Med. Chem.* **50** 4832.
301 Ciba-Geigy (24.09.1996) US 5 559 111, CH-prior. 18.04.1994.
302 Speedel (07.11.2006) US 7 132 569, CH-prior. 29.07.1999.
303 Hudson, V. (2014) The dyslipidaemia market, *Nature Rev. Drug Discov.* **13** 807–808.
304 Tobert, J.A. (2003) Lovastatin and beyond: the history of the HMG-CoA reductase inhibitors, *Nature Rev. Drug Discov.* **2** 517–526.
305 Hoffmann, W.F. et al. (1986) *J. Med. Chem.* **29** 849.
306 Merck & Co. (1984) US 4 444 784, USA-prior. 1980.
307 Endo, A. (1979) *J. Antibiot.* **32** 852.
308 Merck & Co. (1979) US 4 231 938 USA-prior. 1979.
309 Sankyo (1982) US 4 323 648, JP-prior. 1979.
310 Serizawa, N. et al. (1983) *J. Antibiot.* **36** 604.
311 Sankyo (1982) US 4 346 227, JP-prior. 1980.
312 Mevastatin: Sankyo (1976) US 3 983 140, JP-prior. 1974.
313 Fuenfschilling, P.C. et al. (2007) *Org. Process Res. Dev.* **11** 13.
314 Sandoz (1988) US 4 739 073, USA-prior. 1982.
315 Sandoz (1994) US 5 354 772, USA-prior. 1982.
316 Roth, B.D. et al. (1991) *J. Med. Chem.* **34** 357.
317 Warner-Lambert (1993) US 5 273 995, USA-prior. 1989.
318 Watanabe, M. et al. (1997) *Bioorg. Med. Chem.* **5** 437.
319 Shionogi (1993) US 5 260 440, JP-prior. 1991.
320 Suzuki, M. et al. (1999) *Bioorg. Med. Chem. Lett.* **9** 2977.
321 Nissan (1991) US 5 011 930, JP-prior. 1987.
322 Boehringer Mannheim (1973) US 3 781 328, DE-prior. 1971.
323 Parke Davis (1972) US 3 674 836, USA-prior. 1968.
324 Orchimed (1977) US 4 058 552, CH-prior. 1969.
325 Merck & Co. (1968) US 3 383 281, USA-prior. 1958.
326 Upjohn (1971) US 3 692 895, USA-prior. 1968.
327 GelTex Pharmaceuticals (1997) US 5 693 675, USA-prior. 1994.
328 Clader, J.W. (2004) *J. Med. Chem.* **47** 1.
329 Schering Corp. (16.06.1998) US 5 767 115, USA-prior. 21.09.1993.
330 Bristol-Meyers Squibb (27.01.1998) US 5 712 279, USA-prior. 21.02.1995.
331 Univ. of Pennsylvania (31.12.2013) US 8 618 135, USA-prior. 05.03.2004.

Dermatologicals (D), 1. Introduction

ULLMANN'S ENCYCLOPEDIA OF INDUSTRIAL CHEMISTRY

Bernhard Kutscher, Maintal, Germany

1. Introduction

The following three articles review the different drugs against dermatological diseases. They are arranged according to the Anatomical Therapeutical Chemical (ATC) Classification System — a system of alphanumeric codes developed by the WHO for the classification of drugs and other medical products. Here, the active substances are divided into different groups according to the organ or system on which they act, and their therapeutic, pharmacological, and chemical properties. Each drug is given an ATC classification number, which for dermatologicals starts with a D. As new drugs are numbered consecutively in the ATC classification system, novel approved new chemical entities (NCEs) can be easily added. This will allow to keep the Ullmann Pharmaceutical chapters up to date without writing entirely new articles.

In the tables of the particular chapters, the individual drug substances are described by their international nonproprietary name (INN), synonyms, brand or trade names, ATC number, structure, CAS registry number, formula, molecular mass, mode of action, medical use, formulation, daily defined dose (DDD), originator, country and/or date of approval, production method, and references.

The references for each drug typically include the basic patents and publications in which the relevant synthetic or biosynthetic processes for production are described.

Dermatologicals are categorized into Antifungals (D01), Emollients and Protectives (D02), Antipruritics (D04), Antipsoriatics (D05), Antibiotics and Chemotherapeutics (D06), Corticosteroids (D07), Antiseptics and Disinfectants (D08), Anti-Acne Preparations (D10), and Other Dermatological Preparations (D11).

Further systemic anti-inflammatory or immunosuppressive drugs, janus kinase (JAK) inhibitors, or biologics against skin disorders including skin cancers are classified in different ATC categories, which can be found in other Ullmann Pharmaceuticals chapters (→ Antineoplastic and Immunomodulating Agents (L)) are not the subject of this article. Focus is on the so-called synthetic "small molecules," and hence homeopathy, plant-derived, or inorganic preparations are not included in the tables.

2. The Skin as Organ and Target for Drug Delivery

The human skin is composed of three layers of tissue: the outer protective thin layer called epidermis, the dermis, and hypodermis. The epidermis is primarily composed of keratinocytes, producing keratin as protective insoluble protein and melanocytes, producing melanin as pigment. Keratinocytes form several layers that constantly grow outward as the external cells die and flake off. This covering of dead skin is known as stratum corneum or horny

Ullmann's Pharmaceuticals. Edited by Axel Kleemann and Bernhard Kutscher.
© 2022 Wiley-VCH GmbH
Set ISBN: 978-3-527-34252-5/ DOI: 10.1002/14356007.a08_301.pub4

layer, a highly lipid-rich region that minimizes the ingress and egress of water, oxygen, and chemicals. The epidermis also harbors defensive Langerhans cells, which alert the body's immune system to viruses and other infectious agents. The epidermis is bonded to a deeper layer below known as the dermis, which gives to the organ its strength and elasticity, thanks to fibers of collagen and elastin. A network of nerve fibers and receptors pick up signals and feelings such as touch, temperature, and pain relaying them to the brain. The skin's base layer is the hypodermis that works as insulation and cushion from knocks and falls [1].

The human skin is an important administration route for both topical and systemic delivery of therapeutic molecules targeting various disorders [2]. Administration of the drug to the skin for systemic effect bypasses the effect of the first pass through the liver and is an easy and patient-friendly route of administration with some control of the delivery of the drug [3]. Transdermal products successfully target the systemic circulation using various dosage forms such as patches, gels, and ointments [4].

Topical dermatology drugs used for the treatment of skin diseases, e.g., antifungal, antiseptic, anti-inflammatory, or anti-allergic drugs as well as ultraviolet (UV) radiation blocking agents, can be formulated into liquid, solid, and semisolid dosage forms. Semisolids are the most common topical formulations and constitute approx. 80% of the global topical dermatology market with a wide range of dosage forms including creams, emulsions, ointments, pastes, and gels [5]. Semisolid formulations are primarily intended for surface action on the membranes, such as cornea, nasal mucosa, rectal tissue, vaginal tissue, and the internal lining of the ear.

3. Skin Diseases and Healthcare Implications

The skin is the largest organ of the human body ($1.5–2.0\,m^2$ in adults) that protects against external factors such as UV radiation and also one of the first lines of defense against microbial invasion [6].

Skin diseases are one of the most common health problems, affecting approximately one-third of the global population. Analysis of the burden of skin disease in the United States demonstrates that nearly 85 million Americans were seen by a physician for at least one skin disease in 2013, which led to estimated direct health care cost of 75×10^9 \$. The costs and prevalence of skin disease are comparable with or exceed other diseases with significant public health concerns, such as cardiovascular disease and diabetes [7].

Each year, more than 60×10^9 units of topical skin medications are sold worldwide, and the global dermatology market exceeds 20.4×10^9 \$ [8]. The most common disease categories and indications are acne, psoriasis, tinea infections, pruritis, rosacea, vitiligo, warts, and alopecia as well as seborrheic and atopic dermatitis (eczema) [9].

Healthy skin harbors a diverse range of bacteria, known as the skin microbiome, and depending on various factors this bacterial population can be protective or harmful. However, breaches within the skin whether accidental, e.g., trauma, burns, or insect bite, or intentional, e.g., due to surgical intervention, allow incursion of bacterial pathogens and can lead to skin and soft tissue infection (SSTI). SSTI is an extremely common infectious disease syndrome with an estimated 14.2 million ambulatory care attendances in the United States in 2005 [10].

4. Dermatological Drugs

Considerable progress has been made in the treatment of dermatological diseases [11]. Due to the wide range of skin diseases, dermatological drugs cover different chemical classes with distinct mechanism of action and combinations thereof.

After the introduction, the second article (→ Dermatologicals (D), 2. Antifungals (D01), Emollients and Protectives (D02), and Antipruritics (D04) for Dermatological Use) presents drugs used to treat common symptoms of skin disorders and dermatophytosis as well as pruritis with antifungals, antibiotics, antihistamines, local anesthetics, and photoprotectors.

The third article (→ Dermatologicals (D), 3. Antipsoriatics (D05), Antibiotics and

Chemotherapeutics (D06), and Corticosteroids (D07) for Dermatological Use) focuses on topical anti-inflammatory corticosteroids, retinoids, and psoralenes used to treat psoriasis and antibiotics as well as chemotherapeutics for topical use in skin disease.

The fourth and final article ($\rightarrow$ Dermatologicals (D), 4. Antiseptics and Disinfectants (D08), Anti-Acne Preparations (D10), and Other Dermatological Preparations (D11)) describes the antimicrobial activity spectrum of antiseptics and disinfectants as well as anti-acne drugs and other dermatologicals such as calcineurin inhibitors and drugs to treat male baldness, among others.

References

1 Bos, J.D. and Menardi, M.M.H.M. (2000) *Exp. Dermatol.* **9** (3), 165–194.
2 Chen, Y.C., et al. (2012) *Int. J. Nanomed.* **7**, 4409–4418.
3 Alkilani, A.Z., et al. (2015) *Pharmaceutics* **7**, 438–470.
4 Benson, H.A.E., et al. (2019) *Curr. Drug Deliv.* **16**, 444–460.
5 Manufacturing Chemist, (2017) https://www.manufacturing chemist.com/news/article_page/Trends_in_the_topical_delivery _of_dermatology_medications/129282 (accessed 7 April 2020)
6 Wilmer, E.N., et al. (2014) *Cutis* **94** (6), 285–292.
7 Lim, H.W., et al. (2017) *J. Am Acad. Dermatol.* **76**, 958–972.
8 Hellbrandt, S. and Marx, D. (2013) *Drug Dev. Deliv.* September.
9 Wysocki, A.B. (1999) *Nurs. Clin. North Am.* **34**, 777–797.
10 Findley, K., et al. (2014) *PloS Pathog.* **10**, e1004436.
11 WHO Report Drugs Used in Skin Diseases, (1997) https://apps .who.int/medicinedocs/en/d/Jh2918e/ (accessed 7 April 2020).

Dermatologicals (D), 2. Antifungals (D01), Emollients and Protectives (D02), and Antipruritics (D04) for Dermatological Use

ULLMANN'S
ENCYCLOPEDIA
OF INDUSTRIAL
CHEMISTRY

Bernhard Kutscher, Maintal, Germany

1. Introduction

The general topic of "Dermatologicals" is reviewed in this and in two further articles. Please see → Dermatologicals (D), 1. Introduction for a general introduction to the topic of dermatologic drug development [1].

Healthy skin harbors a diverse range of bacteria, known as the skin microbiome, and depending on various factors, this bacterial population can be protective or harmful. However, breaches within the skin whether accidental, e.g., trauma, burns, or insect bite, or intentional, e.g., due to surgical intervention, allow incursion of bacterial pathogens and can lead to skin and soft tissue infection (SSTI).

Dermatophytosis is generally defined as an infection of the hair, nails, or glabrous skin. These infections are caused by the keratinophilic fungi *Trichophyton* spp., *Microsporum* spp., and *Epidermophyton*, which have been recovered from both symptomatic and asymptomatic individuals (→ Antiinfectives for Systemic Use, 2. Antimycotics (Antifungals) and Antimycobacterials). Although dermatophytosis is generally not a life-threatening condition, these types of infections are among the most common infections worldwide, and their incidence continued to increase consistently due to the increasing number of immunocompromised patients [2].

Antifungal drugs are characterized by a variety of chemical structures and a broad range of mechanisms of action. The intervention typically consists in one of the four main classes of antifungals: azole derivatives applied orally and topically, such as ketoconazole, fluconazole, econazole, or clotrimazole; inhibitors of squalene epoxidase, e.g., terbinafine and naftifine; morpholine derivatives, amorolfine; or polyene antifungal drugs, i.e., nystatin and amphotericin B [3]. Azoles chemically belong to either imidazole or triazole classes and are exerting their antifungal effect by blocking the

Ullmann's Pharmaceuticals. Edited by Axel Kleemann and Bernhard Kutscher.
© 2022 Wiley-VCH GmbH
Set ISBN: 978-3-527-34252-5/ DOI: 10.1002/14356007.w08_w01

synthesis of ergosterol, an important component of the fungal cell membrane. Combinations of topical azoles and corticosteroids, e.g., miconazole plus hydrocortisone or clotrimazole plus betamethasone, are used in acute inflammatory mycoses.

Naftifine and terbinafine are synthetic allylamine derivatives, effective against dermatophytes and molds, by inhibiting squalene oxidase and thus decreasing ergosterol. Butenafine is a benzylamine structurally related to the allylamines and indicated for tinea pedis. Amorolfine is a morpholine derivative which possesses both fungistatic and fungicidal activity, effective against dermatophytes, yeast, and molds. Ciclopirox as a pyridone is structurally distinct to other antifungals and inhibits transmembrane transport of ions as substrates essential for cell metabolism. Nystatin, amphotericin B, and griseofulvin are systemic natural antibiotics active against candidiasis and tinea capitis. Terbinafine and fluconazole, approved after 1990, have a broad antifungal spectrum combined with fewer side effects and hence widely applied to many types of diseases.

Topical antifungals are also used for treatment of seborrheic dermatitis (SD), a common anti-inflammatory disease affecting scalp, face, and other seborrheic sites [4]. A variety of treatment options are available for SD including topical antifungals and corticosteroids as first-line agents in various formulations such as ointments, creams, gels, and shampoo [5]. Other therapeutic modalities include keratolytic salicylic acid, selenium sulfide, zinc pyrithione or tar.

Emollients are cosmetic preparations used for protecting, moisturizing, and lubricating the skin. These functions are normally performed by sebum produced by healthy skin. Emollients, e.g., petrolatum (white soft paraffin), castor, or silicon oils, prevent evaporation of water from skin by forming a coating [6]. Dimethicone as polydimethylsiloxane (PDMS) is a silicon oil used in cosmetic and consumer products, e.g., for treatment of head lice on the scalp [7]. Photoprotection is the biochemical process that helps organisms cope with molecular damage caused by sunlight. In humans, melanin is an efficient photoprotective substance that dissipates more than 99.9% of absorbed UV radiation as heat [8]. p-Aminobenzoic acid

(PABA) is a component of the vitamin B complex and a chemical UV light filter used in sunscreens, and therapeutically used in fibrotic skin disorders.

Betacarotene as vitamin A precursor and nutraceutical and the peptide drug afamelanotide are used in a rare disease called erythropoietic protoporphyria (EPP) for reduction of photosensitivity in patients with EPP and other photosensitivity disease [9]. EPP is an inherited metabolic disorder of the heme pathway causing severe phototoxicity in skin due to the buildup of a phototoxic chemical. Afamelanotide as melanocyte-stimulating hormone (MSH) analogue acts by increasing the levels of melanin in the skin and shields the skin [10].

Along with antibiotics, antihistamines are the most widely used systemic drugs in dermatology [11]. This is due to the major role played by histamines in common diseases such as pruritis, urticaria, dermatitis, and atopic eczema [12]. The mainstay of pharmacotherapy for pruritis and chronic urticaria is the administration of low-sedation third-generation anti-H1-histamines, e.g., loratadine, cetirizine, levocetirizine, and fexofenadine, which all have a low incidence of adverse events. Low-sedation antihistamines decrease the intensity of hives and pruritus in patients with mild chronic urticaria and are considered first-line therapy in relieving itching [13]. In addition, further first- and second-generation H1 antagonists, such as hydroxyzine, cyproheptadine, and diphenhydramine, are widely used in dermatological treatments alone or in combinations. The antipruritic effects may be primarily mediated by sedative action [14, 15].

Other anti-allergic drugs such as cromoglicic acid (→ Respiratory Disorders, 2. Nasal Preparations and Decongestants for Topical Use (R01)) are antagonizing histamine as mast cell stabilizers and can be used in topical formulations against dermatitis, conjunctivitis, and various ulcerations. Local anesthetics as neuromodulators, e.g., topical lidocaine, prilocaine, or benzocaine preparations, may provide pain and itching relief and were successfully used in pruritic conditions [16]. Further antipruritics such as camphor can be used for symptomatic treatment as baths or lotions against eczema or urticaria.

2. D01 Antifungals for Dermatological Use

2.1. D01A Antifungals for Topical Use

2.1.1. D01AA Antibiotics (Table 1)

Table 1. D01AA Antibiotics

INN (Synonyms) [Brand names] ATC	Structure (Remarks)	[CAS-No.] Formula MW, g/mol	Target (if known) Medical use Formulation DDD	Originator Approval (country/year) Production method [References]
Nystatin [Mycolog, Nystan, Nystatin] D01AA01; A07AA02; G01AA01	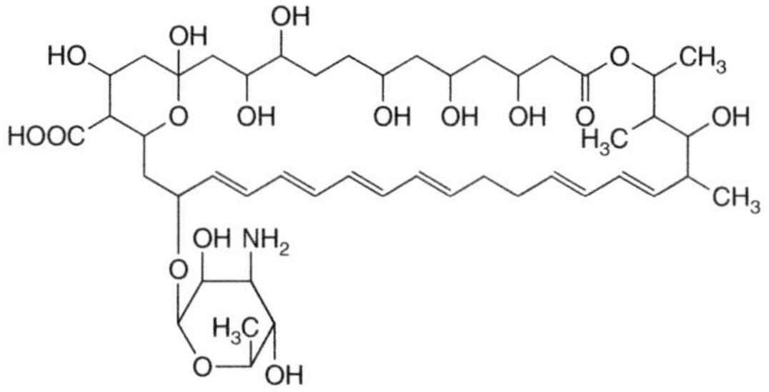	[1400-61-9] $C_{47}H_{75}NO_{17}$ 926.09	fungicidal antibiotic and antimycotic. cream 100 000 iu/g; drg. 500 000 iu; f.c. tabl. 500 000 iu/g; gel 100 000 iu/g; ointment 100 000 iu; pessaries 100 000 iu; susp. 100 000 iu/mL; tabl. 500 000 iu. 250 000 iu T.	BMS US/1998 from fermentation solutions of *Streptomyces noursei* [17, 18]
Natamycin (pimaricin) [Natafucin, Pimafucin] D01AA02; A01AB10; A07AA03; G01AA02; S01AA10	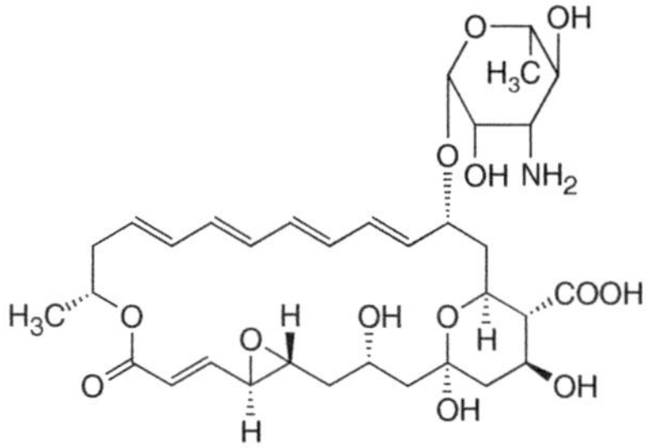	[7681-93-8] $C_{33}H_{47}NO_{13}$ 665.73	fungicidal antibiotic and antimycotic. cream 2 g/100 g; drg. 100 mg; eye drops 5%; eye ointment 1%; ointment 10 mg/g. 2%; tabl. 10 mg.	Gist/American Cyanamide from fermentation solutions of *Streptomyces natalensis* [19, 20]

(continued)

Table 1. (*Continued*)

INN (Synonyms) [Brand names] ATC	Structure (Remarks)	Target (if known) Medical use Formulation DDD	[CAS-No.] Formula MW, g/mol	Originator Approval (country/year) Production method [References]
Hachimycin (trichomycin) [Tricomycin] D01AA03; G01AA06; J0AA02	A B	antifungal agent.	[1394-02-1] $C_{116}H_{168}N_4O_{36}$ 2194.58	derived from *Streptomyces* [21]
Pecilocin [Variotin] D01AA04		fungicidal antibiotic. ointment 3.000 iu/g; topical sol. 1.500 iu/mL.	[19504-77-9] $C_{17}H_{25}NO_3$ 291.39	from culture of *Paecilomyces varioti* Bainier var. *antibioticus* [22]

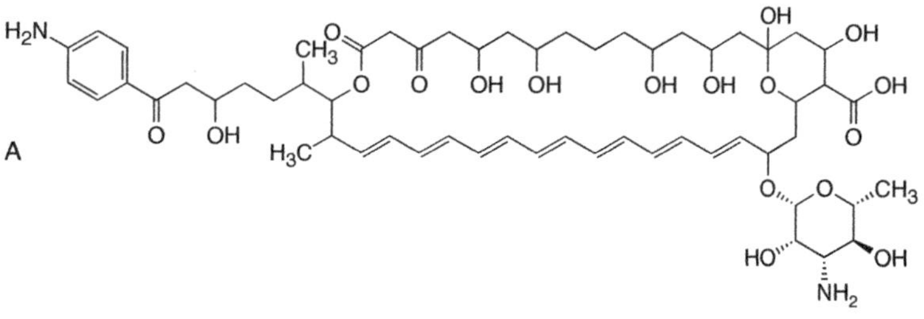

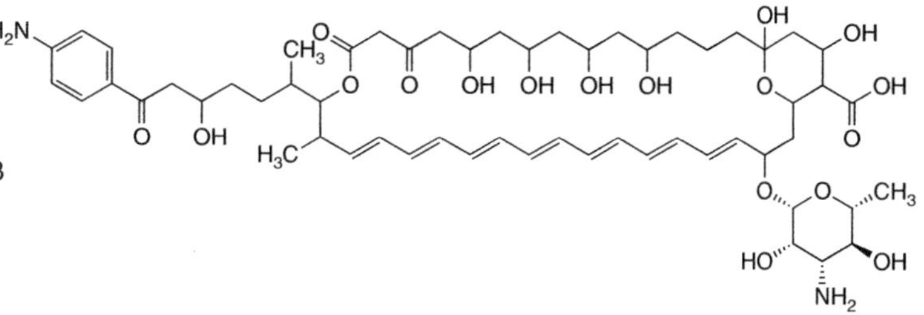

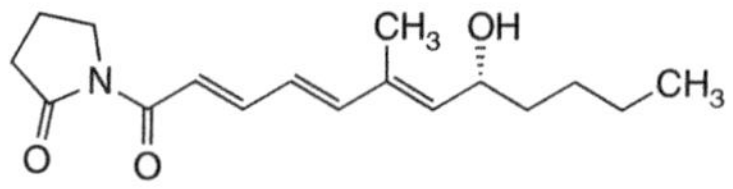

Mepartricin
(methylpartricin)
[Tricandil]
D01AA06; A01AB16;
G01AA09

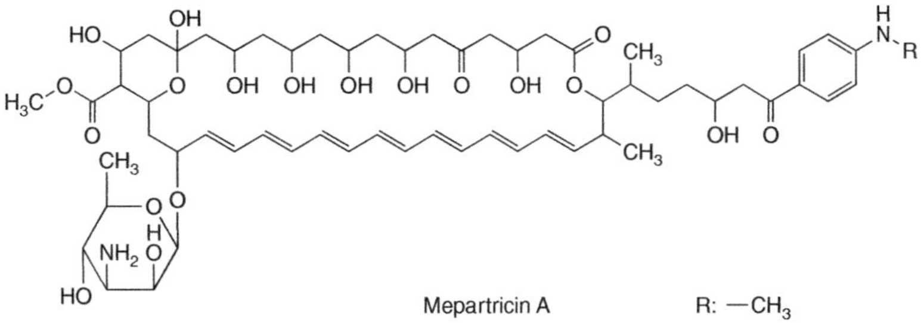

Mepartricin A   R: —CH₃
Mepartricin B   R: —H

[11121-32-7]
$C_{60}H_{88}N_2O_{19}$
1141.36

polyene antibiotic.
treatment of candidal and
trichomonal
gynecological infections,
treatment of benign
prostatic hypertrophy.
tabl. 50 000 iu/g, 40 mg;
vaginal cream 5000 iu/g;
vaginal tabl. 25 000 iu.

Spa
1975
fermentation of
Streptomyces
aureofaciens
[23, 24]

Pyrrolnitrin
[Micutrin, Micutrine]
D01AA07

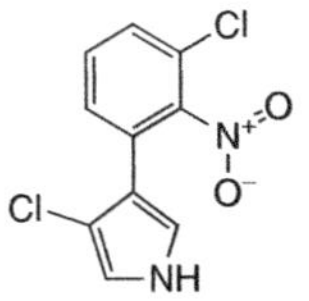

[1018-71-9]
$C_{10}H_6Cl_2N_2O_2$
257.08

antibiotic and antifungal.
cream 1%.

synthetic
[25, 26]

Griseofulvin
[Griseofulline, Grisetin,
Likuden, Poncine]
D01AA08; D01BA01

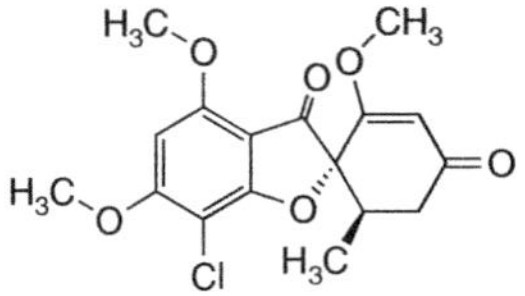

[126-07-8]
$C_{17}H_{17}ClO_6$
352.77

antifungal antibiotic and
antimycotic.
cps. 125, 250 mg; cream
5 g/100 g; tabl. 125, 165,
330, 500 mg.

from fermentation
solutions of
Penicillium patulum
[27, 28]

(continued)

Table 1. (*Continued*)

INN (Synonyms) [Brand names] ATC	Structure (Remarks)	[CAS-No.] Formula MW, g/mol	Target (if known) Medical use Formulation DDD	Originator Approval (country/year) Production method [References]
Amphotericin B [Abelcet, Ambisome, Fungilin, Fungizone, Mysteclin] D01AA10; A01AB04; A07AA07; G01AA03; J02AA01	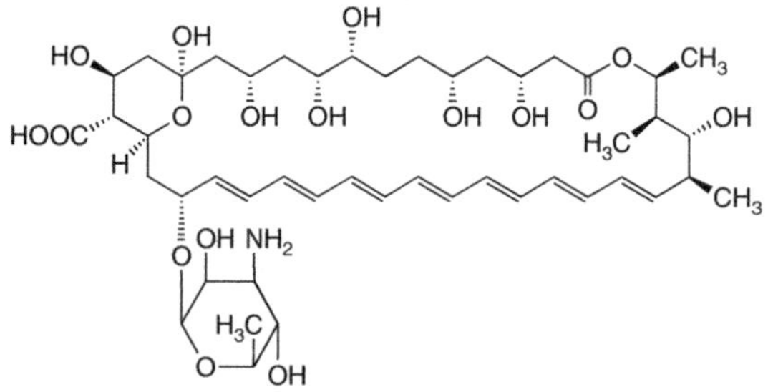	[*1397-89-3*] $C_{47}H_{73}NO_{17}$ 924.09	antimycotic and antibiotic. caramels 10 mg; cream 30 mg/g; ointment 30 mg/1 g; powder 50 mg; susp. 100 mg, 500 mg; syrup 10%; tabl. 10 mg, 100 mg; liposome-encapsulated amphotericin B in a complex with dimyristoyl phosphatidylcholine and dimyristoyl phosphatidylglycerol, vial 20 mL; vial 50 mg.	Gilead 1999 fermentation from *Streptomyces nodosus* [29]

2.1.2. D01AC Imidazole and Triazole Derivatives (Table 2)

Table 2. D01AC Imidazole and triazole derivatives

INN (Synonyms) [Brand names] ATC	Structure (Remarks)	[CAS-No.] Formula MW, g/mol	Target (if known) Medical use Formulation DDD	Originator Approval (country/year) Production method [References]
Clotrimazole [Canesten, Empecid, Lotrimin] D01AC01; A01AB18; G01AF02		[23593-75-1] $C_{22}H_{17}ClN_2$ 344.85	blocking the synthesis of ergosterol, an important component of the fungal cell membrane. antifungal and antimycotic. ointment 1%; pessaries 100, 200 mg, powder 10 mg/g (1%); 500 mg; sol. 1%, 10 mg; spray 10 mg/mL (1%); topical cream 1%, 10 mg; troche 10 mg; vaginal cream 10 mg (2%), 20 mg/mL (10%), 100 mg; vaginal tabl. 100, 200, 500 mg. 25 mg T.	Bayer/Schering 1973/1976 synthetic [30, 31]
Miconazole [Daktar, Gyno-Daktar, Daktarin, Fungoid] D01AC02; A01AB09; A07AC01; G01AF04; J02AB01; S02AA13 (→ Antiinfectives for Systemic Use, 2. Antimycotics (Antifungals) and Antimycobacterials)		[22916-47-8] $C_{18}H_{14}Cl_4N_2O$ 416.14 mononitrate: [22832-87-7] $C_{18}H_{14}Cl_4N_2O \cdot HNO_3$ 479.15	ergosterol biosynthesis inhibitor, topical antifungal and antimycotic. amp. 200 mg/20 mL; cream 1%, 2 g/100 g, 20 mg/g; lotion 1%; ointment 1%; oral gel 2%; powder 2 g/100 g, 20 mg/g (as mononitrate); sol. 20 mg/mL; suppos. 100 mg; tabl. 250 mg (as free base); vaginal cream 20 mg/g; vial 400 mg/40 mL. 40 mg T.	Janssen 1974 synthetic [32, 33]
Econazole [Ecorex, Epipevisone, Gyno-Pevaryl, Palavale] D01AC03; G01AF05		[27220-47-9] $C_{18}H_{15}Cl_3N_2O$ 381.69 mononitrate: [24169-02-6] $C_{18}H_{15}Cl_3N_2O \cdot HNO_3$ 444.70	ergosterol biosynthesis inhibitor. antifungal and antimycotic. cream 1%; lotion 1 g/100 g; pastes 10 mg; powder 1 g/100 g; sol. 1%; spray 1 g/100 g (as nitrate); suppos. 50 mg.	Janssen 1974 synthetic [34–36]
Chlormidazole clomidazole [Myco-polycid] D01AC04		[3689-76-7] $C_{15}H_{13}ClN_2$ 256.37	ergosterol biosynthesis inhibitor. antifungal for treatment of topical mycosis.	synthetic [37]
Isoconazole [Adestan, Travogen] D01AC05; G01AF07		[27523-40-6] $C_{18}H_{14}Cl_4N_2O$ 416.14 mononitrate: [24168-96-5] $C_{18}H_{14}Cl_4N_2O \cdot HNO_3$ 479.15	ergosterol biosynthesis inhibitor. antifungal and antimycotic. cream 1%; pessaries 100, 300, 600 mg (as nitrate); spray 10 mg/mL; vaginal tabl. 100, 300 mg.	Janssen EU/1979 synthetic [38–40]

(continued)

Table 2. (*Continued*)

INN (Synonyms) [Brand names] ATC	Structure (Remarks)	[CAS-No.] Formula MW, g/mol	Target (if known) Medical use Formulation DDD	Originator Approval (country/year) Production method [References]
Tiabendazole [Mintezol, Minzolum] D01AC06; P02CA02		[148-79-8] $C_{10}H_7N_3S$ 201.25	anthelminthic. chewing tabl. 500 mg; susp. 500 mg/5 mL; tabl. 500 mg.	synthetic [41]
Tioconazole [Mykontral, Trosyd, Trosyl] D01AC07; G01AF08		[65899-73-2] $C_{16}H_{13}Cl_3N_2OS$ 387.72	ergosterol biosynthesis inhibitor. antimycotic and topical antifungal. cream 10 mg/g; lotion 10 mg/g; ointment vaginal 6.5%; powder 1 g/100 g; spray 1 g/100 g. 20 mg T.	Pfizer/Combe US/1998 synthetic [42, 43]
Ketoconazole [Ketoderm, Nizoral, Terzolin, Triatop] D01AC08; G01AF11; J02AB02 (→ Antiinfectives for Systemic Use, 2. Antimycotics (Antifungals) and Antimycobacterials)		[65277-42-1] $C_{26}H_{28}Cl_2N_4O_4$ 531.44	ergosterol biosynthesis inhibitor. antimycotic. cream 2%; shampoo 2%; sol. 20 mg/mL; susp. 100 mg; tabl. 200 mg. 30 mg T.	Janssen/J&J US/1996 synthetic [44–46]
Sulconazole [Exelderm] D01AC09		[612318-90-9] $C_{18}H_{15}Cl_3N_2S$ 397.75	ergosterol biosynthesis inhibitor. antifungal. cream and solution.	synthetic [47]
Bifonazole (bifonazolum) [Amycor, Azolmen, Bifazol, Mycospor] D01AC10		[60628-96-8] $C_{22}H_{18}N_2$ 310.40 hydrochloride salt: [60629-09-6] $C_{22}H_{18}N_2 \cdot HCl$ 346.86 sulfate salt: [60629-08-5]	inhibitor of ergosterol biosynthesis in yeasts and dermatophytes. topical antimycotic. cream 1%; gel 10 mg; lotion 1%; powder 10 mg (1%); sol. 1%. 10 mg T.	Bayer synthetic [48, 49]
Oxiconazole [Fonx, Myfungar, Okinazole, Oxistat] D01AC11; G01AF17		[64211-45-6] $C_{18}H_{13}Cl_4N_3O$ 429.13 mononitrate: [64211-46-7] $C_{18}H_{13}Cl_4N_3O \cdot HNO_3$ 492.15	ergosterol biosynthesis inhibitor. antifungal and antimycotic. cream 1%; pessaries 600 mg (as mononitrate); powder 10 mg/g; sol. 1%; vaginal tabl. 100 mg, 600 mg. 10 mg T.	synthetic [50]

Table 2. (*Continued*)

INN (Synonyms) [Brand names] ATC	Structure (Remarks)	[CAS-No.] Formula MW, g/mol	Target (if known) Medical use Formulation DDD	Originator Approval (country/year) Production method [References]
Fenticonazole [Falvin, Fenizolan, Lomexin, Terlomexin] D01AC12; G01AF12		[72479-26-6] $C_{24}H_{20}Cl_2N_2OS$ 455.41 mononitrate: [73151-29-8] $C_{24}H_{20}Cl_2N_2OS$ ·HNO₃ 518.42	ergosterol biosynthesis inhibitor. antifungal and antimycotic. cream 2%, gel 2%; vaginal ovules 200 mg.	Recordati 1986 synthetic [51, 52]
Omoconazole [Fungisan, Fongamil, Fongarex] D01AC13; G01AF16		[74512-12-2] $C_{20}H_{17}Cl_3N_2O_2$ 423.73 mononitrate: [83621-06-1] $C_{20}H_{17}Cl_3N_2O_2$ ·HNO₃ 486.74	ergosterol biosynthesis inhibitor. topical antifungal and antimycotic. cream 10 mg/1 g.	synthetic [53, 54]
Sertaconazole [Monazol, Sertaderm, Sertagyn, Zalain] D01AC14		[99592-32-2] $C_{20}H_{15}Cl_3N_2OS$ 437.78 mononitrate: [99592-39-9] $C_{20}H_{15}Cl_3N_2OS$ ·HNO₃ 500.79	ergosterol biosynthesis inhibitor. antifungal and antimycotic. cream 20 mg/g.	synthetic [55, 56]
Fluconazole (UK-49858) [Canesoral, Canifug-Fluco, Diflucan, Flucazol, Fungata] D01AC15; J02AC01 (→ Antiinfectives for Systemic Use, 2. Antimycotics (Antifungals) and Antimycobacterials)		[86386-73-4] $C_{13}H_{12}F_2N_6O$ 306.28	14-α-demethylase inhibitor; biosynthesis inhibitor of fungal ergosterols. antifungal and antimycotic treatment of vaginal, oropharyngeal, and atrophic oral candidiasis. tabl. 100, 150, 200 mg; powder for suspension 350 mg/35 mL; solution 2 mg/mL; cps. 50, 100, 150, 200 mg; susp. 50 mg/5 mL; syrup 50 mg/10 mL; tabl. 50, 100, 150, 200 mg; vial 50 mg/50 mL, 100 mg/50 mL, 200 mg/100 mL, 400 mg.	Pfizer USA/1990 synthetic [57]
Flutrimazole (UR-4056) [Flusporan] D01AC16		[119006-77-8] $C_{22}H_{16}F_2N_2$ 346.38	ergosterol biosynthesis inhibitor. topical antifungal. cream 1%.	Menarini synthetic [58]

(*continued*)

Table 2. (*Continued*)

INN (Synonyms) [Brand names] ATC	Structure (Remarks)	[CAS-No.] Formula MW, g/mol	Target (if known) Medical use Formulation DDD	Originator Approval (country/year) Production method [References]
Eberconazole [Ebermac, Ebernet] D01AC17		[*128326-82-9*] $C_{18}H_{14}Cl_2N_2$ 329.22	ergosterol biosynthesis inhibitor. antifungal for treatment of dermatophytosis, candidiasis, and pityriasis. cream 1%.	Salvat ES/2015 synthetic [59]
Luliconazole [Lulicon, Luzarn, Luzu] D01AC18		[*187164-19-8*] $C_{14}H_9Cl_2N_3S_2$ 354.28	ergosterol biosynthesis inhibitor. antifungal. cream 1%.	Valeant Pharmaceuticals 2013 synthetic [60]
Efinaconazole [Clenafin, Jublia] D01AC19		[*164650-44-6*] $C_{18}H_{22}F_2N_4O$ 348.39	14-α-demethylase-inhibitor. topical antifungal for treatment of onychomycosis. 10% topical solution.	Dow Pharma/Valeant Pharmaceuticals 2014 synthetic [61]
Croconazole (croconazole) [Pilzine] D01AC30		[*77175-51-0*] $C_{18}H_{15}ClN_2O$ 310.78 hydrochloride salt: [*77174-66-4*] $C_{18}H_{15}ClN_2O \cdot HCl$ 347.25	topical antifungal for treatment of candidiasis. cream 1%; gel 1%; sol. 10 mg/g (1%) as hydrochloride.	Shionogi synthetic [62, 63]

2.1.3. D01AE Other Antifungals for Topical Use (Table 3)

Table 3. D01AE Other antifungals for topical use

INN (Synonyms) [Brand names] ATC	Structure (Remarks)	[CAS-No.] Formula MW, g/mol	Target (if known) Medical use Formulation DDD	Originator Approval (country/year) Production method [References]
Bromochlorosalicylanilide [Multifungin, Salifungin] D01AE01		[3679-64-9] $C_{13}H_9BrClNO_2$ 326.57	antifungal.	synthetic [64]
Methylrosanilin (gentian violet) [Gentian violet, Pyoktanin] D01AE02		[548-62-9] $C_{25}H_{30}ClN_3$ 407.99	dye that acts as mitotic poison. antibacterial, antiseptic, anthelmintic, and antimycotic for treatment of thrush. 0.5–2% solution.	synthetic [65]
Tribromometacresol [Micatex, Triphysan, Triphysol] D01AE03		[4619-74-3] $C_7H_5Br_3O$ 344.83	antifungal.	synthetic [66]
Undecylenic acid (10-undecenoic acid) [Gordochom solution, Skinman soft] D01AE04		[112-38-9] $C_{11}H_{20}O_2$ 184.28	antifungal and antimycotic used against fungal skin infections. cream 43 mg; liquid 30 mg; ointment 43 mg, 5%; powder 53 mg; soap 1 g; sol. 0.1 g/100 g.	synthetic [67, 68]
Polynoxylin [Anaflex] D01AE05; A01AB05		[9011-05-6]	formaldehyde releasing antimicrobial polymer. antiseptic.	synthetic [69]
Chlorphenesin [Musil, Mycil] D01AE07		[104-29-0] $C_9H_{11}ClO_3$ 202.65	muscle relaxant and antifungal. emulgel.	synthetic [70]
Ticlatone [Landromil] D01AE08		[70-10-0] C_7H_4ClNOS 185.63	local fungicide. solution.	Sandoz DE synthetic [71]
Sulbentine (dibenzthione) [Fungiplex] D01AE09 wfm		[350-12-9] $C_{17}H_{18}N_2S_2$ 314.48	antifungal. gel 3 g/100 g; ointment 3 g/100 g; sol. 3 g/100 g.	synthetic [72]

(continued)

Table 3. (*Continued*)

INN (Synonyms) [Brand names] ATC	Structure (Remarks)	[CAS-No.] Formula MW, g/mol	Target (if known) Medical use Formulation DDD	Originator Approval (country/year) Production method [References]
Ethyl hydroxybenzoate (ethylparaben, E-214) [Nipagin A] D01AE10		[*120-47-8*] $C_9H_{10}O_3$ 166.18	antifungal and preservative.	synthetic [73]
Haloprogin [Aloprogen, Halotex, Mycanden] D01AE11 wfm		[*777-11-7*] $C_9H_4Cl_3IO$ 361.39	antifungal and antiseptic. cream 1%; ointment 1%; sol. 1%.	synthetic [74]
Salicylic acid [Acnex, Akurza, Avosil, Keralyt, Salacyn, Salvax, Virasal] D01AE12 **WHO essential medicine**		[*69-72-7*] $C_7H_6O_3$ 138.12	salicylate may competitively inhibit prostaglandin formation and oxidation of UDPG; phenolic phyto-hormone. used in many skin care products for treatment of dermatitis, acne, and psoriasis. 2% solution, cream, soap, ointment, and shampoo.	synthetic [75]
Selendisulfide [Selsep, Selsun] D01AE13 **WHO essential medicine**		[*7488-56-4*] SeS_2 143.03	antifungal. treatment of dermatitis and dandruff. 1% and 2.5% solutions.	US/1951 synthetic [76, 77]
Ciclopirox [Batrafen, Loprox, Miclast, Mocomicen, Penlac, Sebiprox] D01AE14; G01AX12		[*29342-05-0*] $C_{12}H_{17}NO_2$ 207.27 olamine: [*41621-49-2*] $C_{12}H_{17}NO_2$ $\cdot C_2H_7NO$ 268.36	antifungal and antimycotic. cream 1%; powder 1% (as olamine); sol. 1% (as ciclopirox); vaginal cream 1% (as olamine). 20 mg T.	Hoechst Aventis DE/1981 synthetic [78–80]
Terbinafine (SF-86327) [Lamisil, Myconormin, Octosan] D01AE15 **WHO essential medicine**	(E)	[*91161-71-6*] $C_{21}H_{25}N$ 291.44 hydrochloride salt: [*78628-80-5*] $C_{21}H_{25}N \cdot HCl$ 327.90	inhibition of squalene epoxidase leads to ergosterol synthesis inhibition. orally and topically active antifungal and antimycotic. cream 10 mg/g; tabl. 125, 250 mg (as hydrochloride). 10 mg T.	Novartis EU/1991 US/1996 synthetic [81–84]

Table 3. *(Continued)*

INN (Synonyms) [Brand names] ATC	Structure (Remarks)	[CAS-No.] Formula MW, g/mol	Target (if known) Medical use Formulation DDD	Originator Approval (country/year) Production method [References]
Amorolfine [Loceryl, Locetan, Pekiron, Penlac] D01AE16		[*78613-35-1*] $C_{21}H_{35}NO$ 317.52 hydrochloride salt: [*78613-38-4*] $C_{21}H_{35}NO \cdot HCl$ 353.98	morpholine antifungal that inhibits delta14-sterol reductase. topical antifungal. cream 0.25%, 0.5%; sol. 5%.	Aventis US/1999 synthetic [85, 86]
Dimazole (diamthazole) [Asterol] D01AE17		[*95-27-2*] $C_{15}H_{23}N_3OS$ 293.44 dihydrochloride salt: [*136-96-9*] $C_{15}H_{23}N_3OS \cdot 2\,HCl$ 366.36	topical antifungal. topical 5%.	Roche synthetic [87]
Tolnaftate [Tinactin, Tinaderm, Tinatox, Tonoftal] D01AE18		[*2398-96-1*] $C_{19}H_{17}NOS$ 307.42	squalene epoxidase inhibitor. antimycotic and fungicide. cream 10 mg/g; ointment 2%; powder 5 mg/g; sol. 10 mg/mL, 2%. 15 mg T.	Nippon Soda synthetic [88, 89]
Tolciclate [Kilmicen] D01AE19		[*50838-36-3*] $C_{20}H_{21}NOS$ 323.45	sterol biosynthesis inhibitor in fungal cells. antifungal and antimycotic.	synthetic [90]
Flucytosine [Ancobon, Ancotil] D01AE21; J02AX01 **WHO essential medicine** (→ Antiinfectives for Systemic Use, 2. Antimycotics (Antifungals) and Antimycobacterials)		[*2022-85-7*] $C_4H_4FN_3O$ 129.09	fungicide and antimycotic. cps. 250, 500 mg; gran. 50%; tabl. 500 mg; vial 2.5 g/250 mL.	Valeant US/2000 synthetic [91–93]
Naftifine (naftifungin) [Exoderil, Naftin] D01AE22		[*65472-88-0*] $C_{21}H_{21}N$ 287.41	inhibitor of squalene epoxidase. antibacterial, antifungal, and anti-inflammatory.	Merz US/1988 synthetic [94, 95]

(continued)

Table 3. (*Continued*)

INN (Synonyms) [Brand names] ATC	Structure (Remarks)	[CAS-No.] Formula MW, g/mol	Target (if known) Medical use Formulation DDD	Originator Approval (country/year) Production method [References]
Butenafine [Lotrimin, Mentax, Volley] D01AE23		[*101828-21-1*] $C_{23}H_{27}N$ 317.48 hydrochloride salt: [*101827-46-7*] $C_{23}H_{27}N \cdot HCl$ 353.94	squalene epoxidase inhibitor. antifungal and antimycotic. cream 1%; sol. 1% (as hydrochloride).	Schering-Plough US/2001 synthetic [96–98]
Dichlorophen [Beaphar, Hyosan, Korium, Yvosan] D01AE24, P02DX02 (→ Antiparasitics (PC))		[*97-23-4*] $C_{13}H_{10}Cl_2O_2$ 269.12	fungicide, anthelmintic, anticestodal, and antimicrobial agent for veterinary use.	synthetic [99]
Bromosalicylic acid isopropylamide [Actol, Brosalpramide] D01AE26		[*7771-03-1*] $C_{10}H_{12}BrNO_2$ 258.12	topical antifungal.	synthetic [100]

3. D02 Emollients and Protectives

3.1. D02A Emollients and Protectives

3.1.1. D02AA Silicon Products (Table 4)

Table 4. D02AA Silicon products

INN (Synonyms) [Brand names] ATC	Structure (Remarks)	[CAS-No.] Formula MW, g/mol	Target (if known) Medical use Formulation DDD	Originator Approval (country/year) Production method [References]
Dimethicone (dimethylpolysiloxane, dimethicone, simethicone) [Ceolat, Gascon, Infacol, Mylicon, Remedy Repair, Vasogen] D02AA01		[*8050-81-5*]	silicon oil that delays the evaporation of water targeting respiratory systems of head lice. used as antacid, antiflatulent, and surfactant in hair and skin care. cream 4 oz.; gran. 10%; lotion 4 oz.; powder 10%; syrup 2%; tabl. 40, 50, 80 mg.	synthetic [101, 102]

3.2. D02B Protectives Against UV-Radiation

3.2.1. D02BA Protectives Against UV-Radiation for Topical Use (Table 5)

Table 5. D02BA Protectives against UV-radiation for topical use

INN (Synonyms) [Brand names] ATC	Structure (Remarks)	[CAS-No.] Formula MW, g/mol	Target (if known) Medical use Formulation DDD	Originator Approval (country/year) Production method [References]
Aminobenzoic acid (PABA, "vitamin B10") [Potaba] D02BA		[150-13-0] $C_7H_7NO_2$ 137.18 potassium salt: [138-84-1]	part of vitamin B9. used against fibrotic skin disorders, e.g., Peyronie's disease and vitiligo. tabl. and caps. 500 mg.	synthetic [103, 104]

3.2.2. D02BB Protectives Against UV-Radiation for Systemic Use (Table 6)

Table 6. D02BB Protectives against UV-radiation for systemic use

INN (Synonyms) [Brand names] ATC	Structure (Remarks)	[CAS-No.] Formula MW, g/mol	Target (if known) Medical use Formulation DDD	Originator Approval (country/year) Production method [References]
Betacarotene (β-carotene) [Absorbene, Aces, Betaselen, Carotaben, Fotoretin, Mirtilen] D02BB01		[7235-40-7] $C_{40}H_{56}$ 536.89	carotenoid and provitamin A. antioxidant, nutrient supplement, and sunscreen; reduction of photosensitivity in patients with EPP and other photosensitivity diseases. cps. 25 mg 0.1 g O.	Roche US/1987 synthetic [105–107]
Afamelanotide (melanotan, NDP-MSH, EPT 1647) [Scenesse] D02BB02		[75921-69-6] $C_{78}H_{111}N_{21}O_{19}$ 1646.8	α-melanocyte-stimulating hormone (α-MSH) analogue for prevention of phototoxicity in adults with erythropoietic protoporphyria (EPP). implant for s.c. administration. 16 mg.	Clinuvel EU/2015 US/2019 synthetic [108–110]

4. D04 Antipruritics, Incl. Antihistamines, Anesthetics, etc.

4.1. D04A Antipruritics, Incl. Antihistamines, Anesthetics, etc.

4.1.1. D04AA Antihistamines for Topical Use (Table 7)

Table 7. D04AA Antihistamines for topical use

INN (Synonyms) [Brand names] ATC	Structure (Remarks)	[CAS-No.] Formula MW, g/mol	Target (if known) Medical use Formulation DDD	Originator Approval (country/year) Production method [References]
Thonzylamine (neohetramine) [Nasopen; Poly Hist] D04AA01; R01AC06; R06AC06 (→ Respiratory Disorders, 2. Nasal Preparations and Decongestants for Topical Use (R01)) (→ Respiratory Disorders, 7. Antihistamines for Systemic Use (R06))		[91-85-0] $C_{16}H_{22}N_4O$ 286.37 hydrochloride: [63-56-9]	histamine H1-receptor antagonist. antihistaminic and anticholinergic. oral liquid 50 mg/15 mL.	synthetic [111]
Mepyramine (pyrilamine) [Anthisan; Neo-Antergan, Nortussine, Triaminic, Viscosan] D04AA02; R06AC01 (→ Respiratory Disorders, 7. Antihistamines for Systemic Use (R06))		[91-84-9] $C_{17}H_{23}N_3O$ 285.39 hydrochloride: [6036-95-9] $C_{17}H_{23}N_3O$ ·HCl 321.85 maleate; [59-33-6] $C_{17}H_{23}N_3O$ ·$C_4H_4O_4$ 401.46	histamine H1-receptor antagonist. antihistaminic and anticholinergic. amp. 15, 25 mg; cream 2%; tabl. 25, 100 mg. 0.2 g O, P.	synthetic [112]
Thenaldine [Sandosten] D04AA03; R06AX03 wfm (→ Respiratory Disorders, 7. Antihistamines for Systemic Use (R06))		[86-12-4] $C_{17}H_{22}N_2S$ 286.44 tartrate: [2784-55-6] $C_{17}H_{22}N_2S$ ·$C_4H_6O_6$ 436.53	histamine H1-receptor antagonist. antihistaminic and anticholinergic. drg. 25 mg (as tartrate).	Sandoz synthetic [113]
Tripelennamine [Azaron; Piristin; Pyribenzamine] D04AA04; R06AC04 (→ Respiratory Disorders, 7. Antihistamines for Systemic Use (R06))		[91-81-6] $C_{16}H_{21}N_3$ 255.54 hydrochloride: [154-69-8] maleate: [57116-36-6] citrate: [6138-56-3]	histamine H1-receptor antagonist. antihistaminic. tablet 50 mg; cream 1% and 2%; syrup. 0.15 g O.	Ciba/Novartis USA/1947 synthetic [114]
Chloropyramine (halopyramine) [Allergosan; Sinopen; Synpen] D04AA09; R06AC03 (→ Respiratory Disorders, 7. Antihistamines for Systemic Use (R06))		[59-32-5] $C_{16}H_{20}ClN_3$ 289.81 hydrochloride: [6170-42-9] $C_{16}H_{20}ClN_3$ ·HCl 326.27	histamine H1-receptor antagonist. antihistaminic and anticholinergic. amp. 20 mg; cream 1%; tabl. 25 mg. 0.15 g O; 0.20 mg P.	synthetic [115, 116]

(continued)

Table 7. (*Continued*)

INN (Synonyms) [Brand names] ATC	Structure (Remarks)	[CAS-No.] Formula MW, g/mol	Target (if known) Medical use Formulation DDD	Originator Approval (country/year) Production method [References]
Promethazine [Atosil; Fargan; Hiberna; Phenergan; Prothazin; Pyrethia; Vegetamin] D04AA10; N05CM22; R06AD02 (→ Hypnotics and Sedatives (N05C)) (→ Respiratory Disorders, 7. Antihistamines for Systemic Use (R06))		[60-87-7] $C_{17}H_{20}N_2S$ 284.43 hydrochloride: [58-33-3] $C_{17}H_{20}N_2S \cdot HCl$ 329.89	histamine H1-receptor antagonist. antihistaminic, neuroleptic, sedative, and antiemetic. amp. 25 mg/1 mL, 56 mg/2 mL; drops 20 mg/mL; f.c. tabl. 25 mg; drg. 25 mg; gran. 10%, powder 10%; suppos. 12.5, 25, 50 mg; syrup 1 mg/mL, 5.65 mg; tabl. 5, 12.5, 25, 50 mg (as hydrochloride). 25 mg O, P, R.	Rhone-Poulenc/ Aventis USA/1949 synthetic [117, 118]
Tolpropamine [Brondilat, Pragman] D04AA12		[5632-44-0] $C_{18}H_{23}N$ 253.39 hydrochloride salt: [3339-11-5] $C_{18}H_{23}N \cdot HCl$ 289.85	antihistaminic, anti-allergic, and anticholinergic. gel 1%.	Hoechst synthetic [119]
Dimethindene [Fenistil; Forhistal; Triten] D04AA13; R06AB03 (→ Respiratory Disorders, 7. Antihistamines for Systemic Use (R06))		[5636-83-9] $C_{20}H_{24}N_2$ 292.43 maleate: [3614-69-2] $C_{20}H_{24}N_2 \cdot C_4H_4O_4$ 408.50	histamine H1-receptor antagonist. antihistaminic, anticholinergic, and antipruritic. amp. 4 mg; drg. 1 mg; drops 1 mg/mL; gel 1 mg/g; s.r. drg. 2.5 mg; s.r. tabl. 2.5 mg; syrup 0.122 mg/mL (as maleate). 4 mg O, P.	Ciba/Novartis synthetic [120, 121]
Clemastine (meclastine) [Tavegil; Tavist] D04AA14; R06AA04 (→ Respiratory Disorders, 7. Antihistamines for Systemic Use (R06))		[15686-51-8] $C_{21}H_{26}ClNO$ 343.90 hydrogen fumarate: [14976-57-9] $C_{21}H_{26}ClNO \cdot C_4H_4O_4$ 459.97	histamine H1-receptor antagonist. antihistaminic, anti-allergic, and anticholinergic. amp. 2 mg/5 mL; dry syrup 0.1%; gel 300 mg/g (as hydrogen fumarate); gran. 0.1%, 1%; powder 0.1%, 1%; syrup 0.01%, 0.5 mg/10 mL; tabl. 1 mg. 2 mg O, P.	Sandoz/Novartis synthetic [122, 123]
Bamipine [Soventol; Taumidrine] D04AA15; R06AX01 (→ Respiratory Disorders, 7. Antihistamines for Systemic Use (R06))		[4945-47-5] $C_{19}H_{24}N_2$ 280.42 hydrochloride: [1229-69-2] $C_{19}H_{24}N_2 \cdot HCl$ 316.88 lactate: [61670-09-5] $C_{19}H_{24}N_2 \cdot C_3H_6O$ 370.49	histamine H1-receptor antagonist. antihistaminic and anticholinergic. cream 20 mg; drg. 20 mg; f.c. tabl. 50 mg; gel 20 mg. 0.1 g O.	Knoll synthetic [124]

Table 7. (*Continued*)

INN (Synonyms) [Brand names] ATC	Structure (Remarks)	[CAS-No.] Formula MW, g/mol	Target (if known) Medical use Formulation DDD	Originator Approval (country/year) Production method [References]
Pheniramine [Avil; Daneral; Naphcon, Neorestamin; Polaramin; Poly-Histin; Rhinosovil; Senodin; Triaminic; Trimeton] D04AA16; R06AB05 (→ Respiratory Disorders, 7. Antihistamines for Systemic Use (R06))		[86-21-5] $C_{16}H_{20}N_2$ 240.35 maleate: [132-20-7] $C_{16}H_{20}N_2$ $\cdot C_4H_4O_4$ 356.42 aminosalicylate: [3269-83-8] $C_{16}H_{20}N_2$ $\cdot C_7H_7NO_3$ 393.49	antihistaminic and anticholinergic. tablets and drg. 75 mg (as maleate). 75 mg O.	synthetic [125–127]
Isothipendyl [Andantol; Apaisyl; Calmogel; Nilergix] D04AA22; R06AD09 (→ Respiratory Disorders, 7. Antihistamines for Systemic Use (R06))		[482-15-5] $C_{16}H_{19}N_3S$ 285.42 hydrochloride: [1225-60-1] $C_{16}H_{19}N_3S\cdot HCl$ 321.88	histamine H1-receptor antagonist. antihistaminic and anti-allergic. drg. 12 mg; gel 0.75% (as hydrochloride); ointment 0.75%, gel.	Degussa/Asta Medica synthetic [128, 129]
Diphenhydramine [Benadryl; Betadorm; Butix; Nautamine; Nytol; Restamin; Sediat; Sedopretten; Simpy Sleep; Vivinox] D04AA32; N05CM20; R06AA02 (→ Hypnotics and Sedatives (N05C)) (→ Respiratory Disorders, 7. Antihistamines for Systemic Use (R06))		[58-73-1] $C_{17}H_{21}NO$ 255.36 hydrochloride: [147-24-0] $C_{17}H_{21}NO\cdot HCl$ 291.82	histamine H1-receptor antagonist antihistaminic, antiemetic, sedative, and anticholinergic. drops 12.5 mg; oil 5 g; ointment 1%; s.r. cps. 30 mg; suppos. 10, 20, 50 mg; syrup 2.67 mg/mL, 12.5 mg/mL; tabl. 25 mg, 50 mg (as hydrochloride). 0.2 g O, R.	Parke Davis 1946 synthetic [130, 131]
Chlorphenoxamine [Clorevan; Phenoxene; Systral] D04AA34; R06AA06 (→ Respiratory Disorders, 7. Antihistamines for Systemic Use (R06))		[77-38-3] $C_{18}H_{22}ClNO$ 303.83 hydrochloride: [562-09-4] $C_{18}H_{22}ClNO$ $\cdot HCl$ 340.29	antihistaminic. cream 15 mg/g; drg. 20 mg, 30 mg (combination); gel 15 mg/g; suppos. 24 mg, 60 mg; tabl. 20 mg. 10 mg O, P.	Asta Medica/Degussa AG D/1960 synthetic [132, 133]
Diphenylpyraline [Agiell; Arbid; Difrin; Hispril; Plokon] D04AA39; R06AA07 (→ Respiratory Disorders, 7. Antihistamines for Systemic Use (R06))		[147-20-6] $C_{19}H_{23}NO$ 281.40 hydrochloride: [132-18-3] $C_{19}H_{23}NO\cdot HCl$ 317.86	antihistaminic. amp. 2 mg/1 mL; gel 15 mg/g. 16 mg O.	synthetic [134, 135]
Dioxopromethazine [Prothanon] D04AA40; R06AD10 (→ Respiratory Disorders, 7. Antihistamines for Systemic Use (R06))		[13754-56-8] 316.42 hydrochloride: [15374-15-9] 352.88	histamine H1-receptor antagonist. antihistaminic. gel.	synthetic [136]

4.1.2. D04AB Anesthetics for Topical Use (Table 8)

Table 8. D04AB Anesthetics for topical use

INN (Synonyms) [Brand names] ATC	Structure (Remarks)	[CAS-No.] Formula MW, g/mol	Target (if known) Medical use Formulation DDD	Originator Approval (country/year) Production method [References]
Lidocaine (lignocaine, xylocaine) [Dynexan; Emia; Licain; Lidoderm Patch; Lidran; Xylocain; Versatis] D04AB01; R02AD02; C01BB01; C05AD01; N01BB02; S01AH07; S02DA01 **WHO essential medicine** (→ Cardiovascular System (C)) (→ Anesthetics (N01)) (→ Respiratory Disorders, 3. Throat Preparations (R02))		[137-58-6] $C_{14}H_{22}N_2O$ 234.34 monohydrochloride: [73-78-9] $C_{14}H_{22}N_2O \cdot HCl$ 270.80	blocking of fast voltage-gated sodium channels. local anesthetic, cough suppressor, and antiarrhythmic. amp. 0.5%, 1%, 2%, 25 mg; gel 2%; inj. 0.5%, 1%, 2%, 3%; inj. 5, 10, 20 mg (as hydrochloride with epinephrine); jelly 2%; ointment 5%; ophth. sol. 4%, plaster 18 mg; sol. 4% (as hydrochloride); spray 8%; syringe 100 mg/5 mL.	Astra 1949 synthetic [137–139]
Cinchocaine (dibucaine) [Deliproct, Nupercainal, Percamin, Ultraproct] D04AB02; C05AD04; N01BB06; S01HA06 (→ Anesthetics (N01))		[85-79-0] $C_{20}H_{29}N_3O_2$ 343.47 hydrochloride salt: [200-498-1] $C_{20}H_{29}N_3O_2 \cdot HCl$ 379.93	local anesthetic. amp. 6 mg/3 mL, 9 mg/3 mL (as hydrochloride); powder 98% and over; rectal ointment 5 mg/100 g; suppos. 1 mg.	synthetic [140–142]
Oxybuprocaine (benoxinate) [Benoxil, Cebesine, Lacrimine, Novesine, Thilorbin] D04AB03 (→ Anesthetics (N01))		[99-43-4] $C_{17}H_{28}N_2O_3$ 308.42 hydrochloride salt: [5987-82-6] $C_{17}H_{28}N_2O_3$ HCl 344.88	local anesthetic. eye drops 2 mg, 4 mg/mL; external sol. 0.3%, 1%; jelly 0.2%; eye drops 0.05%, 0.4%; sol. 10 mg/mL (as hydrochloride).	synthetic [143]

Table 8. (*Continued*)

INN (Synonyms) [Brand names] ATC	Structure (Remarks)	[CAS-No.] Formula MW, g/mol	Target (if known) Medical use Formulation DDD	Originator Approval (country/year) Production method [References]
Benzocaine (ethoforme) [AAA spray; Americaine; Anacaine; Anesthesin; Anbesol; Auralgan; Cepacol; Gengivarium; Gingicaine; Lanacane; Orajel] D04AB04; C05AD03; N01BA05; R02AD01 (→ Anesthetics (N01)) (→ Respiratory Disorders, 3. Throat Preparations (R02))		[*94-09-7*] $C_9H_{11}NO_2$ 165.19	inhibition of voltage-dependent sodium channels. local anesthetic used as topical pain reliever. cream 100 mg; gel 20%, 21.2%; jelly 20%; ointment 5%, 10%, 20%; pills 4, 8, 20 mg; powder 99% and over, 60 mg; sol. 20%, 20.3%; suppos. 100 mg, lozenges.	Dr. R. Ritsert DE/1902 synthetic [144]
Quinisocaine (dimethisoquin) [Haenal akut, Isochinol, Quotane] D04AB05		[*86-80-6*] $C_{17}H_{24}N_2O$ 272.39 hydrochloride salt: [*2773-92-4*] $C_{17}H_{24}N_2O$ HCl 308.85	local anesthetic against pruritus. ointment 0.5 g/100 g (as hydrochloride).	synthetic [145]
Tetracaine (amethocaine) [Cetacaine, Drill, Ophtocain, Tetocaine] D04AB06 (→ Anesthetics (N01))		[*94-24-6*] $C_{15}H_{24}N_2O_2$ 264.37 hydrochloride salt: [*136-47-0*] $C_{15}H_{24}N_2O_2 \cdot HCl$ 300.83	local anesthetic. amp. 20 mg; ear drops 5 mg/g; in combination preparations: eye drops 6 mg/g; sol. 1 g/100 g; spray 5 mg/mL, 2% (as hydrochloride).	synthetic [146]
Pramocaine (pramoxine) [Proctofoam, Sarna] D04AB07		[*140-65-8*] $C_{17}H_{27}NO_3$ 293.4 hydrochloride salt: [*637-58-1*]	topical anesthetic and antipruritic. aerosol and foam.	Abbott US/1953 synthetic [147]

4.1.3. D04AX Other Antipruritics (Table 9)

Table 9. D04AX Other antipruritics

INN (Synonyms) [Brand names] ATC	Structure (Remarks)	[CAS-No.] Formula MW, g/mol	Target (if known) Medical use Formulation DDD	Originator Approval (country/year) Production method [References]
Crotamiton [Crotamitex, Crotan, Euracin, Eurax] D04AX02		[483-63-6] $C_{13}H_{17}NO$ 203.28	antiparasitic that is toxic to the scabies mite. scabicidal and antipruritic agent. cream or lotion for topical treatment.	1949 synthetic [148]
Bufexamac [Anderm, Faktu, Parfenac, Viafen] D04AX03; M01AB17 wfm EU		[2438-72-4] $C_{12}H_{17}NO_3$ 223.27	histone deacetylase (HDAC) inhibitor. nonsteroidal anti-inflammatory (NSAID). cream 5%; ointment 5%; suppos. 250 mg. 0.1 g T.	1970 synthetic [149–151]
Isoprenaline (isoproterenol) [Ingelan; Isomenyl; Isuprel; Many; Medihaler Iso; Proternol; Saventrine] D04AX04; C01CA02; R03AB02; R03CB01 (→ Respiratory Disorders, 4. Inhalants for Obstructive Airway Diseases (R03))		[7683-59-2] $C_{11}H_{17}NO_3$ 211.26 bitartrate: [59-60-9] $C_{11}H_{17}NO_3$ $\cdot C_4H_6O_6$ 361.35 hydrochloride: [51-30-9] $C_{11}H_{17}NO_3 \cdot HCl$ 247.72 sulfate: [299-95-6] $C_{11}H_{17}NO_3$ $\cdot \frac{1}{2} H_2SO_4$ 520.6	β_2-adrenoreceptor agonist. bronchodilator and antiasthmatic. aerosol 0.1 mg/push; aerosol 20 mg/5 mL (comb. with phenylephrine hydrochloride); amp. 0.2, 1, 2 mg (as hydrochloride); inhalation sol. 0.5%, 1%; s.r. cps. 7.5 mg; s.r. tabl. 15 mg; sol. for inhalation 0.25%, 0.5% (as sulfate); tabl. 10, 15, 20 mg; 0.64 mg inhal. aerosol; 20 mg inhal. solution.	Boehringer Ingelheim USA/1947 synthetic [152, 153]
Camphor [Vicks VapoRub] D04AX05; M02AX04		[76-22-2] $C_{10}H_{16}O$ 152.23	terpenoid that activates ion channel receptors creating a sensation of heat. topical ointments and creams.	isolated and distilled from camphor laurel wood [154]

List of Abbreviations

		inj.	injection
		INN	international nonproprietary name
amp.	ampoule(s)	iu	international unit
cps.	capsules	MSH	melanocyte-stimulating hormone
DDD	defined daily dose	MW	molecular mass
drg.	dragee	N	nasal
EPP	erythropoietic protoporphyria	NSAID	nonsteroidal anti-inflammatory drug
f.c.	film-coated	mL	milliliter (cubic centimeter)
gran.	granules	O	oral

P   parenteral
PABA   *p*-aminobenzoic acid
PDMS   polydimethylsiloxane
SD   seborrheic dermatitis
sol.   solution
s.r.   slow release
SSTI   skin and soft tissue infection
suppos.   suppositories
susp.   suspension
T   topical
tabl.   tablets
wfm   withdrawn from market
WHO   World Health Organization

References

1 Eaglestein, W.H. and Corcoran, G. (2011) *Arch. Derrmatol.* **147** (5), 568–572.
2 De Groote, M.A. and Huitt, G. (2006) *Clin. Infect. Dis.* **42**, 1756–1763.
3 Rotta, I., et al. (2012) *Rev. Assoc. Med. Bras.* **58** (3), 308–318.
4 Chowdhry, S. and Gupta, S. (2017) *J. Bacteriol. Mycol.* **4** (1), 1–7.
5 Gupta, A.K., et al. (2014) *J. Eur. Acad. Dermatol. Venerol.* **28** (19), 16–26.
6 van Zuuren, E.J., et al. (2017) *Cochrane Database Syst. Rev.* **2**, CDO12119.
7 Burgess, I.F. (2009) *BMC Pharmacol.* **9**, 3.
8 Meredith, P., et al. (2004) *Photochem. Photobiol.* **79** (2), 211–216.
9 Lecha, M., et al. (2009) *Orphanet J. Rare Dis.* **10**, 4–17.
10 Hadley, M.E. and Dorr, R.T. (2006) *Peptides* **27** (4), 921–930.
11 Greaves, M.W. (2005) *Skin Pharmacol. Physiol.* **18** (5), 220–229.
12 Fischer, M.F., et al. (1996) *Inflamm. Res.* **45**, 564–573.
13 Jankowska-Konsur, A., et al. (2016) *J. Eur. Acad. Dermatol. Venereol.* **30** (1), 67–71.
14 Lee, E.E. and Maibach, H.I. (2001) *Am. J. Clin. Dermatol.* **2**, 27–32.
15 Kavosh, E.R. and Khan, D.A. (2011) *Am. J. Clin. Dermatol.* **12**, 361–376.
16 Elmariah, B., et al. (2011) *Semin. Cutan. Med. Surg.* **30** (2), 118–126.
17 Olin Mathieson, (1957) US 2 786 781 (prior. 1954).
18 Research Corp., (1957) US 2 797 183 (prior. 1952).
19 Königl. Niederl. Gist. & Spiritusfabr., GB 844 289 (appl. 1957; NL-prior. 1956).
20 American Cyanamid, GB 846 933 (appl. 1957; USA-prior. 1956).
21 Mechlinski, W., et al. (1980) *J. Antibiot.* **33** (6), 591–599.
22 Japan Antibiotics Research Assoc., GB 866 425 (appl. 7.4.1959).
23 Tweit, R.C., et al. (1982) *J. Antibiot.* **35**, 997.
24 Spa, DE 2 154 436 (appl. 2.11.1971; GB-prior. 3.11.1970).
25 Nakano, H., et al. (1966) *Tetrahedron Lett.* **7**, 737–740.
26 Fujisawa, US 3 428 648 (18.2.1969; J-prior. 8.4.1965, 12.10.1964).
27 Glaxo, GB 784 618 (appl. 28.3.1955).
28 ICI, US 2 900 304 (18.8.1959; GB-prior. 21.9.1956).
29 Olin Mathieson, (1959) US 2 908 611 (prior. 1954).
30 Bayer, DE 1 617 481 (appl. 15.9.1967).
31 Berg, D., et al. (1984) *Arzneim.-Forsch.* **34** (I), 139.
32 Janssen, US 3 717 655 (20.2.1973; appl. 19.8.1968).
33 Janssen, US 3 839 574 (1.10.1974; prior. 23.7.1969).
34 Thienpont, D., et al. (1975) *Arzneimittel-Forsch.*, **25**(2), 224–230.
35 Janssen, DAS 1 940 388 (appl. 8.8.1969; USA-prior. 19.8.1968).
36 Janssen, US 3 717 655 (20.2.1973; prior. 19.8.1968).
37 Grünenthal, US 2 876 233 (3.3.1959; appl. 29.10.1956).
38 Godefroi, E.F., et al. (1969) *J. Med. Chem.* **12**, 784.
39 Veraldi, S, et al. (2013) *Mycoses.* **56** Suppl 1: 3–15
40 Janssen, US 3 839 574 (1.10.1974; appl. 19.6.1972; USA-prior. 19.8.1968).
41 Merck & Co., US 3 017 415 (16.1.1962; prior. 18.1.1960).
42 Pfizer, DE 2 619 381 (appl. 30.4.1976; GB-prior. 30.4.1975).
43 Pfizer, US 4 062 966 (13.12.1977; appl. 30.4.1976; GB-prior. 30.4.1975).
44 Heeres, J., et al. (1979) *J. Med. Chem.* **22**, 1003–1005.
45 Janssen, DOS 2 804 096 (appl. 31.1.1978; USA-prior. 31.1.1977, 21.11.1977).
46 Janssen, US 4 144 346 (13.3.1979; appl. 21.11.1977; USA-prior. 31.1.1977).
47 Fromtling, R.A. (1988) *Clin. Microbiol. Rev.* **1** (2), 187–217.
48 Bayer, DOS 2 461 406 (appl. 5.12.1975; USA-prior. 24.12.1974).
49 Bayer, US 4 118 487 (3.11.1978; appl. 5.12.1975; prior. 24.12.1974).
50 Siegfried AG, US 4 124 767 (7.11.1978; CH-prior. 24.12.1975).
51 Recordati, DE 2 917 244 (appl. 9.5.1979; I-prior. 18.5.1978).
52 Recordati, US 4 221 803 (9.9.1980; appl. 9.5.1979; I-prior. 18.5.1978).
53 Thiele, K., et al. (1987) *Helv. Chim. Acta* **70**, 441.
54 Siegfried AG, US 4 210 657 (1.7.1980; D-prior. 11.9.1978).
55 Raga, M.M., et al. (1992) *Arzneim.-Forsch.* **42**, 691.
56 Ferrer, EP 151 477 (appl. 2.1.1985; E-prior. 8.6.1984, 2.2.1984, 6.1.1984).
57 Pfizer, US 4 404 216 (13.9.1983; GB-prior. 6.6.1981, 17.10.1981, 4.3.1982).
58 J. Uriach & Cia., EP 352 352 (appl. 31.1.1990; prior. 28.7.1988).
59 Del Palacio, A., et al. (2001) *Mycoses* **44** (5), 173–180.
60 Glenmark Pharms., WO 2 016 092 478 (16.6.2016; appl. 9.12.2015; IN-prior. 12.12.2014).
61 Patel, T., et al. (2013) *Drugs* **73** (17), 1977–1983.
62 Shionogi, US 4 328 348 (4.5.1982; J-prior. 7.6.1979, 7.9.1979).
63 Shionogi, US 4 463 011 (31.7.1984; J-prior. 7.6.1979, 7.9.1978).
64 Weuffen, W., et al. (1966) *Pharmazie* **21** (10), 613–619.
65 Maley, A.M., et al. (2013) *Exp. Dermatol.* **22** (12), 775–780.
66 Zsolnai, T. (1960) *Biochem. Pharmacol.* **5**, 1–19.
67 Krafft, F. (1877) *Ber. Dtsch. Chem. Ges.* **10**, 2034.
68 Perkins, G.A. and Cruz, A.O. (1927) *J. Am. Chem. Soc.* **49**, 1070.
69 Kingston, D. (1960) *J. Clin. Pathol.* **18** (5), 666–667.
70 Burns, D.A. (1986) *Contact Dermatitis* **14** (33), 246.
71 Lundell, E. (1971) *Mykosen* **14** (11), 531–533.
72 DD 20 634 (appl. 17.8.1958)
73 Giordano, F., et al. (1999) *J. Pharm. Sci.* **88**, 1210–1216.
74 Meiji Seika, US 3 322 813 (30.5.1967).
75 Madan, R.K., et al. (2014) *J. Am. Acad. Dermatol.* **70** (4), 788–792.
76 Steudel, R., et al. (1982) *Top. Curr. Chem.* **102**, 177–197.
77 Maddin, F.W., et al. (1993) *J. Am. Acad. Dermatol.* **29** (6), 1008–1012.

78 Hoechst AG, US 3 883 545 (13.5.1975; appl. 16.11.1971; prior. 22.12.1972).
79 Hoechst AG, US 3 972 888 (3.8.1976; D-prior. 25.3.1972).
80 Hoechst AG, DE 1 795 270 (appl. 31.8.1968).
81 Alami, M., et al. (1996) *Tetrahedron Lett.* **37** (1), 57–58.
82 Stütz, A. and Petranyi, G. (1984) *J. Med. Chem.* **27**, 1539.
83 Banyu Pharmaceutical, EP 421 302 (J-prior. 2.10.1989).
84 Sandoz, US 4 755 534 (5.7.1988; CH-prior. 22.8.1979).
85 Hoffmann-La Roche, DE 2 752 135 (appl. 22.11.1976).
86 Hoffmann-La Roche, EP 24 334 (appl. 7.8.1980; CH-prior. 17.8.1979, 29.5.1980).
87 Hoffmann-La Roche, (1951) US 2 578 757 (prior. 1949).
88 Nippon Soda, US 3 334 126 (1.8.1967; J-prior. 21.6.1961, 25.8.1961, 9.4.1962, 13.4.1962).
89 Nippon Soda, GB 967 897 (appl. 31.5.1962; J-prior. 21.6.1961, 25.8.1961, 9.4.1962, 13.4.1962).
90 Ryder, N.S., et al. (1986) *Antimicrob. Agents Chemother.* **29** (5), 858–860.
91 Duschinsky, R., et al. (1957) *J. Am. Chem. Soc.* **79**, 4559.
92 Undheim, K. and Gacek, M. (1969) *Acta Chem. Scand.* **23** (1), 294.
93 Roche, US 3 040 026 (19.6.1962; appl. 3.6.1959).
94 Sandoz, DOS 2 716 943 (appl. 16.4.1977; CH-prior. 28.4.1976).
95 Sandoz, US 4 282 251 (CH-prior. 28.4.1976).
96 Mitsui Toatsu Chem., EP 221.781 (appl. 31.10.1986; J-prior. 1.11.1985).
97 Kaken Pharmaceutical Co., EP 164 697 (appl. 6.6.1985; J-prior. 9.6.1984).
98 Kaken Pharm., US 5 021 458 (4.6.1991; J-prior. 9.6.1984).
99 Dixon, P.A., et al. (1978) *Eur. J. Drug Metab. Pharmakokin.* **3**, 95–98.
100 Weuffen, W., et al. (1966) *Pharmazie* **21**, 477–482.
101 Corning Glass, (1948) US 2 441 098 (appl. 1946).
102 McNeil, US 6 103 260 (15.8.2000; appl. 17.7.1997).
103 Weidner, W., et al. (2005) *Eur. Urol.* **47** (4), 535–536.
104 SK Corp., US 6 222 051 (24.4.2001; appl. 18.10.1999).
105 Isler, O., et al. (1956) *Helv. Chim. Acta* **39**, 249.
106 Isler, O. (1956) *Angew. Chem.* **68**, 547.
107 Roche, DE 855 399 (appl. 26.5.1950).
108 Langendonk, J.G., et al. (2015) *N. Engl. J. Med.* **373**, 48–59.
109 Univ. Arizona, US 4 485 039 (27.11.1984; appl. 11.6.1982).
110 Clinuvel, EP 2 865 422 (29.4.2015; appl. 31.8.2007; AU-prior. 31.8.2006).
111 Armino, J., et al. (1949) *Ind. Med. Surg.* **18** (12), 509–511.
112 Rhône-Poulenc, (1950) US 2 502 151 (F-prior. 1943).
113 Sandoz, (1955) US 2 717 251 (CH-prior. 1951).
114 Ciba, US 2 425 730 (19.8.1947, appl. 9.9.1946).
115 Vaughan, J.R., et al. (1949) *J. Org. Chem.* **14**, 228.

116 American Cyanamid, (1951) US 2 569 314 (appl. 1947).
117 Rhône-Poulenc, (1950) US 2 530 451 (F-prior. 1946).
118 Rhône-Poulenc, (1952) US 2 607 773 (GB-prior. 1949).
119 Hoechst, DE 925 468 (appl. 1941).
120 Ciba, US 2 947 756 (2.8.1960; appl. 5.5.1959; prior. 12.8.1958, 3.11.1958).
121 Ciba, US 2 970 149 (31.1.1961; appl. 3.11.1958).
122 Ebnöther, A. and Weber, H.-P. (1976) *Helv. Chim. Acta* **59**, 2462.
123 Sandoz, GB 942 152 (appl. 14.12.1960; CH-prior. 19.1.1960, 3.8.1960, 27.9.1960).
124 Knoll AG, (1954) US 2 683 714 (D-prior. 1949).
125 Schering Corp., (1951) US 2 567 245 (prior. 1948).
126 Schering Corp., (1954) US 2 676 964 (prior. 1950).
127 Farbw. Hoechst, DE 830 193 (appl. 1948).
128 Degussa, DE 1 001 684 (appl. 1954).
129 Degussa, US 2 974 139 (7.3.1961; D-prior. 2.10.1954).
130 Parke Davis & Co., (1947) US 2 421 714 (prior. 1944).
131 Parke Davis, (1947) US 2 427 878 (appl. 1947).
132 ASTA-Werke, US 2 785 202 (12.3.1957; D-prior. 1952).
133 ASTA-Werke, DE 1 009 193 (appl. 1955).
134 Nopco Chem. Comp., (1949) US 2 479 843 (prior. 1948).
135 Promonta, DE 934 890 (appl. 1951).
136 Borchert, H.H., et al. (1977) *Pharmazie* **32** (8), 489–496.
137 AB Astra, (1948) US 2 441 498 (S-prior. 1943).
138 AB Astra, DE 968 561 (appl. 1944; S-prior. 1943).
139 Brown, C.L.M. and Poole, A. (1957) US 2 797 241 (GB-prior. 1953).
140 Miescher, K. (1932) *Helv. Chim. Acta* **15**, 163.
141 Ciba, DRP 537 104 (appl. 1926).
142 Ciba, (1931) US 1 825 623 (D-prior. 1926).
143 Dr. A. Wander AG, GB 654 484 (appl. 1948; CH-prior. 1947).
144 *Org. Synth.* (ORSYAT) 8, 66 (1928).
145 SmithKline & French, (1952) US 2 612 503 (CDN-prior. 1949).
146 Winthrop, (1932) US 1 889 645 (CH-prior. 1930).
147 Schmidt, J.L., et al. (1953) *Curr. Res. Anesth. Analg.* **32** (6), 418–425.
148 Buffet, M., et al. (2003) *Fundam. Clin. Pharmacol.* **17** (2), 217–235.
149 Buu-Hoï, N.P., et al. (1965) *C. R. Hebd. Seances Acad. Sci.* **261**, 2259.
150 Madan, BE 661 226 (appl. 17.3.1965).
151 Madan, US 3 479 396 (18.11.1969; B-prior. 5.6.1964, 17.3.1965).
152 Boehringer Ingelheim, (1943) US 2 308 232 (D-prior. 1939).
153 Boehringer Ingelheim, DRP 723 278 (appl. 1939).
154 Lincoln, D.E. and Lawrence, B.M. (1984) *Phytochemistry* **23**, 933–934.

Dermatologicals (D), 3. Antipsoriatics (D05), Antibiotics and Chemotherapeutics (D06), and Corticosteroids (D07) for Dermatological Use

ULLMANN'S ENCYCLOPEDIA OF INDUSTRIAL CHEMISTRY

BERNHARD KUTSCHER, Maintal, Germany

1. Introduction

The general topic of "Dermatologicals" is reviewed in this and in two further articles. Please see → Dermatologicals (D), 1. Introduction for a general introduction to the topic of dermatologic drug development [1].

Psoriasis is a common chronic inflammatory skin disorder typically characterized by erythematous papules and plaques. The underlying cause of the disease is unknown. It affects 1–3% of population [2], and an aberrant T-cell-mediated immune response seems crucial in the development of chronic plaque psoriasis.

Most cases are not severe enough to affect general health and are treated in the outpatient setting. However, rare life-threatening presentations can occur that require inpatient management. In any case, psoriasis can have significant effect on patients' quality of life. Numerous topical and systemic therapies are available for treatment of cutaneous manifestations of psoriasis, and the desired outcome differs for individual patients and preferences. Topical treatment may provide symptomatic relief, minimize required dose of expensive systemic medications, and may even be psychologically beneficial. Limited plaque psoriasis responds well to topical steroids and emollients. Emollients, such as mineral oil, are frequently used alone or in combination with salicylic acid or urea to facilitate hydration and removal of hypokeratotic plaques [3]. This

Ullmann's Pharmaceuticals. Edited by Axel Kleemann and Bernhard Kutscher.
© 2022 Wiley-VCH GmbH
Set ISBN: 978-3-527-34252-5/ DOI: 10.1002/14356007.w08_w02

keeps the psoriatic skin soft and moist, minimizing the symptoms of itching and tenderness. Alternatives include the vitamin D analogues, such as calcipotriene and calcitriol, or tar as well as topical retinoids.

Retinoids include natural and synthetic compounds that have similar activity to vitamin A. Vitamin A helps regulate the immune system, and analogues have anti-inflammatory effects. They control the proliferation and differentiation of epithelial tissue. Currently, three generations of synthetic retinoids exist. First-generation agents include tretinoin, isotretinoin, and alitretinoin (→ Antineoplastic and Immunomodulating Agents (L); → Dermatologicals (D), 4. Antiseptics and Disinfectants (D08), Anti-Acne Preparations (D10), and Other Dermatological Preparations (D11)). Second-generation retinoids are etretinate and acitretin, while third-generation retinoids are polyaromatic derivatives including adapalene, tazarotene, and bexarotene. Retinoids act by activating receptors located in the cell nucleus, which leads to the expression of appropriate genes [4].

Dithranol (anthralin) is a synthetic derivative of the natural antipsoriatic agent chrysarobin that reduces DNA synthesis and epidermal replication.

Combinations of potent topical steroids and either calcipotriene, calcitriol, tazarotene, or UV phototherapy are commonly prescribed by dermatologists for some patients. Hence, e.g., calcipotriene and betamethasone as well as tazarotene and ulobetasol (halobetasol) combinations are available. Moderate to severe psoriasis requires phototherapy or systemic therapies such as retinoids, apremilast, or biologic immune-modifying agents.

Psoralens are natural furocoumarins and potent photosensitizers with antiproliferative and immunosuppressive effects in combination with UV light therapy.

Phototherapy has much advanced since the time of the ancient Egyptians when exposure to natural sunlight was used as a form of medical treatment for a variety of skin conditions [5]. Fifty percent of patients with psoriasis have self-treated with sunbeds [6].

1.1. Antibiotics and Chemotherapeutics

Theoretically, topical antibiotics use offers several advantages over systemic administration, including delivery of high concentrations of antimicrobial at the required site of action and a reduction in systemic activity [7]. However, the widespread use of commonly used topical antibiotics has led to an increasing bacterial resistance, limiting the potential efficacy of such agents. This makes the management of superficial skin infections a major challenge.

Topical antibiotics are among the most commonly prescribed antimicrobial agents. In 2015, there were 4.7 million primary care prescriptions for topical antibiotics in the United Kingdom. Topical application allows use and development of agents that may not be able to be used systemically (e.g., neomycin or bacitracin) and may be easier for patients as well as caregivers. Currently used topical antibiotics against skin infections are mupirocin, fusidinic acid, polymyxin B, and retapamulin. For production of antibiotics by fermentation.

Sulfonamides are the first successfully synthesized toxic antimicrobial drugs [8] and can be administered by oral or topical route. The mechanism of sulfonamides antimicrobial action involves competitive inhibition of p-amino benzoic acid (PABA), the compound essential for the synthesis of folic acid and thus for bacterial growth. Topical sulfonamides, such as sulfacetamide sodium or silver sulfadiazine, are used as creams, ointments, or eye solutions against acne, vaginal yeast infections or conjunctivitis [9], and other dermatological disorders [10].

The first antiviral drug, idoxuridine, was approved in 1963 for topical and ophthalmic use [11]. Most of the antiviral drugs in dermatology are nucleoside analogues approved for systemic and topical treatment of herpes simplex virus (HSV), varicella zoster virus (VZV), or cytomegalovirus (CMV) infections [12]. Topical acyclovir and penciclovir are available as ointment as well as cream against HSV and docosanol against cold sore/fever blisters (see → Antiinfectives for Systemic Use, 3. Antivirals). Imiquimod is an imidazoquinoline that has been introduced to dermatologic and genitourinary medicine as a topical immune modifier for treatment of external warts caused by human papillomavirus (HPV). Palliative therapies for HSV also cover local anesthetics such as lidocaine or benzocaine preparations for pain relief [13].

1.2. Corticosteroids (→ Hormone Therapeutics – Systemic Hormonal Preparations, Excluding Reproductive Hormones and Insulins (H))

Glucocorticosteroids afford anti-inflammatory and anti-immune effects and are widely used to relieve inflammatory and pruritic manifestations of various acute and chronic skin diseases. They are administered orally for long periods in autoimmune skin diseases such as pemphigus and pemphigoid. Short-term oral use of corticosteroids may be given for autosensitization dermatitis and psoriasis. However, oral use of steroids may cause a wider variety of side effects than topical use. Since its introduction

in early 1950s, topical hydrocortisone has revolutionized dermatology [14]. Pharmacological action of topical glucosteroids takes place through cytoplasmatic receptors that transport the drug to the cell nucleus, where the complex modifies gene transcription. In the nucleus itself there are also receptors for glucosteroids.

Modifications of the chemical structure have led to enhanced efficacy and also more adverse effects. The topical side effect is primarily the thinning of epidermis and dermis. In general, acute inflammatory eruptions respond well to mild/moderate strength topical steroids of group I (hydrocortisone) or group II (alclometasone or triamcinolone). Chronic, thickened, or hyperkeratotic dermatoses may require potent or very potent steroids of group III (methylprednisolone, mometasone, or betamethasone) or group IV (halcinonide). For a given steroid, ointments are more potent than creams, and for the scalp steroids are often delivered as lotion or gel [15].

2. D05 Antipsoriatics

2.1. D05A Antipsoriatics for Topical Use

2.1.1. D05AC Anthracene Derivatives (Table 1)

Table 1. D05AC Anthracene derivatives

INN (Synonyms) [Brand names] ATC	Structure (Remarks)	[CAS-No.] Formula MW, g/mol	Target (if known) Medical use Formulation DDD	Originator Approval (country/year) Production method [References]
Dithranol (anthralin, cignolin) [Drithocreme, Micanol, Psoralon, Psoriderm, Psoralen] D05AC01		[1143-38-0] $C_{14}H_{10}O_3$ 226.23	anthracene derivative that accumulates in mitochondria and interferes with the energy supply to the cell and has strong reducing properties. antipsoriatic. cream 0.5, 1, 2 mg/g; ointment 0.5%, 1%, 2%, 3%; pencil sticks 0.2 g/10 g, 0.5 g/10 g.	synthetic [16, 17]

2.1.2. D05AD Psoralens for Topical Use (Table 2)

Table 2. D05AD Psoralens for topical use

INN (Synonyms) [Brand names] ATC	Structure (Remarks)	[CAS-No.] Formula MW, g/mol	Target (if known) Medical use Formulation DDD	Originator Approval (country/year) Production method [References]
Trioxsalen (trimethylpsoralen, trioxysalen) D05AD01, D05BA01		[*3902-71-4*] $C_{14}H_{12}O_3$ 228.24	photosensitizer that after photoactivation creates interstrand cross-links in DNA, which can cause programmed cell death. used for treatment of vitiligo and hand eczema. tabl. 5 mg. 10 mg O.	ICN Pharmaceuticals US/1993 synthetic [18]
Methoxsalen (ammoidin, methoxypsoralen, methoxsalen, xanthotoxin) [Meladinine, Oxsoralen, Psoraderm, Uvadex] D05AD02; D05BA02		[*298-81-7*] $C_{12}H_8O_4$ 216.19	radioprotector. ointment 0.3%; lotion 0.3%, 1%; sol. 0.1 g/100 mL, 0.75 g/100 mL; gelatin-coated caps. 10 mg/1 mL; tabl. 10 mg. 10 mg O.	synthetic [19–21]

2.1.3. D05AX Other Antipsoriatics for Topical Use (Table 3)

Table 3. D05AX Other antipsoriatics for topical use

INN (Synonyms) [Brand names] ATC	Structure (Remarks)	[CAS-No.] Formula MW, g/mol	Target (if known) Medical use Formulation DDD	Originator Approval (country/year) Production method [References]
Fumaric acid (BG-12; fumaderm) [Tecfidera] D05AX01, N07XX09		[624-49-7] $C_6H_8O_4$ 144.1	immunomodulator for treatment of psoriasis and multiple sclerosis. hard caps.; 120 and 240 mg with delayed release.	Biogen US/2013 synthetic [22–24]
Calcipotriol (calcipotriene) [Daivobet, Daivonex, Dovonex, Psorcutan] D05AX02		[112965-21-6] $C_{27}H_{40}O_3$ 412.61	topical vitamin D_3 analogue and synthetic derivative of calcitriol. antipsoriatic for treatment of chronic plaque psoriasis and alopecia. ointment 0.005%. 75 µg T.	BMS US/1997 synthetic [25, 26]
Calcitriol (1α,25-dihydroxy-vitamin D_3;1α, 25-dihydroxychole-calciferol; Ro 21-5535, vitamin D) [Calcijex, Osteotriol, Rocaltrol, Silkis] D05AX03, A11CC04		[32222-06-3] $C_{27}H_{44}O_3$ 416.65	calcium regulator, biological active form of vitamin D_3. treatment of osteoporosis and psoriasis. amp. 0.5 µg, 1 µg/mL, 2 µg/mL; cps. 0.25 µg, 0.5 µg; tabl. 0.25 µg, 0.5 µg, ointment. 6 µg T.	Roch US/1978 synthetic [27–29]
Tacalcitol (TV-02) [Apsor, Bonalfa, Curatoderm] D05AX04		[57333-96-7] $C_{27}H_{44}O_3$ 416.65	vitamin D derivative. antipsoriatic. cream 0.0002%; lotion 0.0002%; ointment 4.17 µg/g (as hydrate); ointment 0.0002%, 0.0004%. 4 µg T.	synthetic [30–32]
Tazaroten (AGN-190168) [Avage, Tazorac, Zorac] D05AX05		[118292-40-3] $C_{21}H_{21}NO_2S$ 351.4	retinoid that binds to the retinoic acid receptors (RAR). antipsoriatic, acne therapeutic, retinoid. gel 0.5 mg/g, 1 mg/g, cream and foam.	Allergan synthetic [33, 34]

2.2. D05B Antipsoriatics for Systemic Use

2.2.1. D05BA Psoralens for Systemic Use (Table 4)

Table 4. D05BA Psoralens for systemic use

INN (Synonyms) [Brand names] ATC	Structure (Remarks)	[CAS-No.] Formula MW, g/mol	Target (if known) Medical use Formulation DDD	Originator Approval (country/year) Production method [References]
Bergapten [Heraclin, Majudin] D05BA03		[484-20-8] $C_{12}H_8O_4$ 216.2	natural furanocoumarin that increases the skin tolerability to light. antipsoriatic, photosensitizer, and used for treating leukodermia (vitiligo).	isolated from plants like carrots and citrus family [35]

2.2.2. D05BB Retinoids for Treatment of Psoriasis (Table 5)

Table 5. D05BB Retinoids for treatment of psoriasis

INN (Synonyms) [Brand names] ATC	Structure (Remarks)	[CAS-No.] Formula MW, g/mol	Target (if known) Medical use Formulation DDD	Originator Approval (country/year) Production method [References]
Etretinate [Tigason, Tegison} D05BB01		[54350-48-0] $C_{23}H_{30}O_3$ 354.49	retinoid that binds to RAR. antipsoriatic. cps. 10 mg, 25 mg. 35 mg O.	Roche US/1986 synthetic [36, 37]
Acitretin (Ro-10-1670, etretin) [Neotigason, Soriatane] D05BB02		[55079-83-9] $C_{21}H_{26}O_3$ 326.44	second-generation retinoid. antipsoriatic. cps. 10 mg, 25 mg. 35 mg O.	Roche US/1996 synthetic [38, 39]

3. D06 Antibiotics and Chemotherapeutics for Dermatological Use

3.1. D06A Antibiotics for Topical Use

3.1.1. D06AA Tetracycline and Derivatives (Table 6)

Table 6. D06AA Tetracycline and derivatives

INN (Synonyms) [Brand names] ATC	Structure (Remarks)	[CAS-No.] Formula MW, g/mol	Target (if known) Medical use Formulation DDD	Originator Approval (country/year) Production method [References]
Demeclocycline (demethylchlortetracycline) [Declostatin, Ledermycin, Meciclin] D06AA01; J01AA01 (→ Antiinfectives for Systemic Use, 1. Antibacterials)		[127-33-3] $C_{21}H_{21}ClN_2O_8$ 464.86 hydrochloride salt: [200-592-1] $C_{21}H_{21}ClN_2O_8$ ·HCl 501.32	inhibits protein synthesis of bacteria binding to the 30S ribosomal unit. tetracycline analogue. antibiotic used to treat acne, bronchitis, and Lyme disease. cps. 150 mg; ointment 0.5%; tabl. 150 mg, 300 mg (as hydrochloride).	Corepharma US/1966 from fermentation solutions of a *Streptomyces aureofaciens* mutant. [40, 41]
Chlortetracycline [Aureomycin; Aureocort] D06AA02; A01AB21; J01AA03; S01AA02 (→ Antiinfectives for Systemic Use, 1. Antibacterials)		[57-62-5] $C_{22}H_{23}ClN_2O_8$ 478.89 hydrochloride salt: [64-72-2] $C_{22}H_{23}ClN_2O_8$ ·HCl 515.35	antibiotic. cream 10 mg/g, 30 mg/g, 3%; eye ointment 10 mg/g (1%); ointment 30 mg/10 g (3%); pastes 30 mg; pessaries 100 mg (as hydrochloride).	Lederle US/1950 from fermentation solutions of *Streptomyces aureofaciens* [42, 43]
Oxytetracycline [Posicycline; Sterdex; Terramycin, Urobiotic] D06AA03; G01AA07; J01AA06; S01AA04 (→ Antiinfectives for Systemic Use, 1. Antibacterials)		[79-57-2] $C_{22}H_{24}N_2O_9$ 460.44 hydrochloride salt: [2058-46-0] $C_{22}H_{24}N_2O_9$ ·HCl 496.90	inhibits protein synthesis of bacteria binding to the 30S ribosomal unit. antibiotic. cps. 250 mg, 500 mg; dental insertion 5 mg; ear drops 5 mg (comb. with hydrocortisone); eye drops 5 mg (comb. with hydrocortisone); eye ointment 10 mg/g; eye ointment 5 mg (comb. with polymyxin B); ointment 10 mg/g; ointment 30 mg (comb. with hydrocortisone or polymyxin B); spray 300 mg (comb. with hydrocortisone); vial 5 mL (as hydrochloride).	from fermentation solutions of *Streptomyces rimosus* [44, 45]

(continued)

Table 6. (*Continued*)

INN (Synonyms) [Brand names] ATC	Structure (Remarks)	[CAS-No.] Formula MW, g/mol	Target (if known) Medical use Formulation DDD	Originator Approval (country/year) Production method [References]
Tetracycline [Achromycin, Amphocycline, Colicort, Sumycin, Tetracycline] D06AA04; A01AB13; J01AA07; S01AA09; S02AA08; S03AA02 **WHO essential medicine** (→ Antiinfectives for Systemic Use, 1. Antibacterials)		[60-54-8] $C_{22}H_{24}N_2O_8$ 444.44 hydrochloride salt: [64-75-5] $C_{22}H_{24}N_2O_8 \cdot$ HCl 480.90	inhibits protein synthesis of bacteria binding to the 30S ribosomal unit. antibiotic. cps. 50, 250, 500 mg; cream 30 mg/g; f.c. tabl. 500 mg; ointment 3%; powder 100%; troche 15 mg; vial 500 mg.	Lederle US/1978 from fermentation solutions of *Streptomyces viridifaciens* [46, 47]
Meclocycline (GS-22989) [Anti-acne, Meclosorb] D06AA05; D10AF04		[2013-58-3] $C_{22}H_{21}ClN_2O_8$ 476.9	antibiotic. cream.	synthetic [48]

3.1.2. D06AX Other Antibiotics for Topical Use (Table 7)

Table 7. D06AX Other antibiotics for topical use

INN (Synonyms) [Brand names] ATC	Structure (Remarks)	[CAS-No.] Formula MW, g/mol	Target (if known) Medical use Formulation DDD	Originator Approval (country/year) Production method [References]
Fusidic Acid (ZN-6; ramycin) [Fucidin, Fusicutan, Fucithalmic, Kerlone] D06AX01; D09AA02; S01AA13		[6990-06-3] $C_{31}H_{48}O_6$ 516.72 sodium salt: [751-94-0] $C_{31}H_{47}NaO_6$ 538.70	bacterial protein synthesis inhibitor. antibacterial steroid, antibiotic. tbl. 250 mg; creams and ointments for topical use. 60 mg T.	Leo Pharma US/1984 synthetic or fermentation of *Fusidium coccineum* [49, 50]
Chloramphenicol [Cebenicol, Elase-Chloromycetin, Kemicetine, Pentamycetin, Posifenicol] D06AX02; D10AF03; G01AA05; J01BA01; S01AA01; S02AA01; S03AA08 **WHO essential medicine** (→ Antiinfectives for Systemic Use, 1. Antibacterials)		[56-75-7] $C_{11}H_{12}Cl_2N_2O_5$ 323.13	inhibitor of bacterial protein synthesis. antibiotic. amp. 1 g (as hydrogen succinate sodium salt); cps. 250 mg, 500 mg; ear drops 5 g/100 mL, 50 mg/g; eye drops 5 mg, 10 mg; ointment 1% (10 mg/g), 2%; powder 90% and over; sol. 0.5%, 5%; tabl. 50 mg, 250 mg; vaginal tabl. 100 mg.	Parke-Davis US/1950 synthetic [51, 52]
Neomycin [Myciguent; Mycifradin; Neo-Rx; Neo-Tab; Neosporin] D06AX04; R02AB01; A01AB08; B05CA09; G01AA14; J01CB05; R01AX08; S01AA03; S03AA01 (→ Antiinfectives for Systemic Use, 1. Antibacterials) (→ Respiratory Disorders, 3. Throat Preparations (R02))		[1404-04-2] $C_{23}H_{46}N_6O_{13}$ 614.64 sulfate: [1405-10-3]	aminoglycoside antibiotic (mixture of isomers) blocking protein biosynthesis of bacteria while binding to the 30S subunit of the ribosome. antibacterial. creams, ointment 0.5% and eye drops.	Upjohn USA/1951 fermentation [53]

(continued)

Table 7. (*Continued*)

INN (Synonyms) [Brand names] ATC	Structure (Remarks)	[CAS-No.] Formula MW, g/mol	Target (if known) Medical use Formulation DDD	Originator Approval (country/year) Production method [References]
Bacitracin [Baciguent; Nebacetin; Neobac; Bacicoline; Cicatrin; Neosporin; Polysporin] D06AX05 D06AX05, R02AB04 (→ Respiratory Disorders, 3. Throat Preparations (R02))	Bacitracin A	[1405-87-4] $C_{66}H_{103}N_{17}O_{16}S$ 1422.72	polypeptide antibiotic (mixture of related cyclopeptides) that interferes with bactoprenol, a membrane carrier molecule that transports building blocks of the bacterial cell wall. antibacterial. topical use; amp. 50.000 iu; vial 5.000 iu; nasal ointment 300 iu; ointment 300 iu, 500 iu; powder 300 iu; troche 250 iu.	Upjohn USA/1948 fermentation [54–57]
Gentamicin (gentamycin) [Cidomycin, Garamycin, Gentacin, Gentalene, Gentacort, Voltamycin] D06AX07; J01GB03; S01AA11; S03AA06 **WHO essential medicine** (→ Antiinfectives for Systemic Use, 1. Antibacterials)		[1403-66-3] sulfate salt: [1405-41-0]	inhibitor of bacterial protein synthesis binding to 30S ribosomal subunit. aminoglycoside antibiotic. amp. 10 mg/1 mL, 40 mg/1 mL, 80 mg, 120 mg, 160 mg; cream 0.1%; eye drops 0.3%; cream 1 mg (in comb. with betamethasone); lotion 1 mg (in comb. with betamethasone); ointment 3 mg/g (as sulfate); ointment 1 mg (in comb. with betamethasone). 2.5 mg T.	Schering-Plough US/1964 from fermentation solutions of *Micromonospora purpurea*; *Micromonospora echinospora* [58, 59]

Tyrothricin [Dorithricin; Lemocin; Limexx; Solmucain; Tyroqualin; Tyrosur] D06AX08; R02AB02; S01AA05 (→ Respiratory Disorders, 3. Throat Preparations (R02))	Cyclo[D-Phe-L-Pro-X-Y-L-Asn-L-Gln-Z-L-Val-L-Orn-L-Leu]	[1404-88-2] $C_{65}H_{85}N_{11}O_{13}$ 1228.44	antimicrobial polypeptide (mixture of tyrocidine and gramicidin) that irreversibly damages cell membranes of microorganism. antibacterial, antiviral. local application with spray, solution, tablets, and lozenges.	fermentation [60]
Mupirocin (pseudomonic acid) [Bactroban, Mupiderm, Turixin] D06AX09; R01AX06 **WHO essential medicine** (→ Respiratory Disorders, 2. Nasal Preparations and Decongestants for Topical Use (R01))		[12650-69-0] $C_{26}H_{44}O_{9}$ 500.63	inhibitor of bacterial protein synthesis binding to the isoleucyl t-RNA synthetase. topical antibiotic. cream 20 mg/g; ointment 2% (as calcium salt). 40 mg T.	Clay-Park Labs US/2002 fermentation of *Pseudomonas fluorescens* NCIB 10586 [61–63]
Virginiamycin (pristinamycin) [Stafac] D06AX10	Mixture of cyclic polypeptide and macrolide	[11006-76-1]	streptogramin antibiotic. combination and mixture of virginiamycin M1 and virginiamycin S1. veterinary use.	isolated from *Streptomyces virginiae* [64]

(continued)

Table 7. (Continued)

INN (Synonyms) [Brand names] ATC	Structure (Remarks)	[CAS-No.] Formula MW, g/mol	Target (if known) Medical use Formulation DDD	Originator Approval (country/year) Production method [References]
Rifaximin (L-105) [Normix, Rifacol, Xifaxan] D06AX11; A07AA11	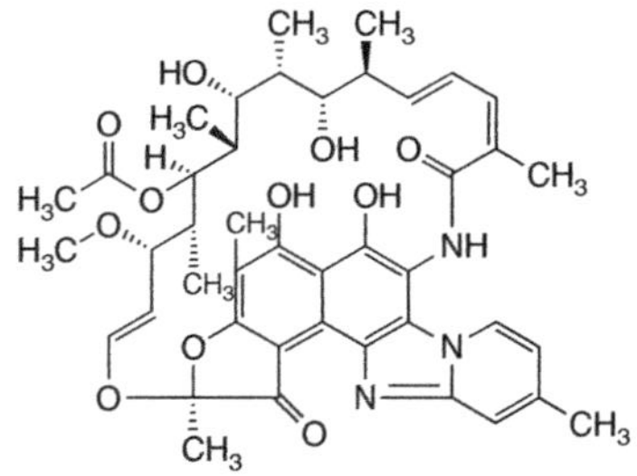	[80621-81-4] $C_{43}H_{51}N_3O_{11}$ 785.89	interferes with bacterial RNA polymerase. antibiotic. ointment 5%; susp. 2%; tabl. 200 mg.	Salix Pharmaceuticals US/2004 semisynthetic [65–67]
Amikacin [Amiklin, Arikayce, Biklin, Likacin, Mikavir, Pierami] D06AX12; J01GB06; S01AA21 **WHO essential medicine** (→ Antiinfectives for Systemic Use. 1. Antibacterials)	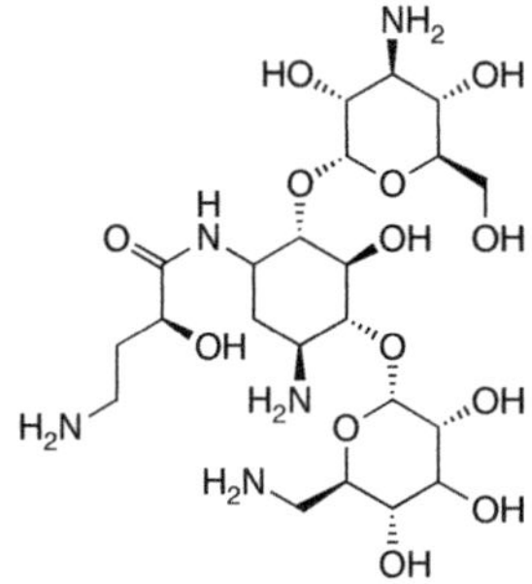	[37517-28-5] $C_{22}H_{43}N_5O_{13}$ 585.61 sulfate salt: [39831-55-5] $C_{22}H_{43}N_5O_{13} \cdot H_2SO_4$ 781.76	inhibits bacterial protein synthesis binding to ribosomal 30S subunit. aminoglycoside antibiotic. amp. 100 mg/1 mL, 200 mg/2 mL; cream 2.5%, 5%; eye drops 0.3%, 0.5%; gel 5.5; vial 100 mg/2 mL, 250 mg/2 mL, 500 mg/2 mL, oral inhalation suspension 590 mg/8.4 mL.	Insmed US/1976 US/2018 as liposomal inhalation suspension synthetic [68, 69]
Retapamulin (SB-275833) [Altabax, Altargo] D06AX13	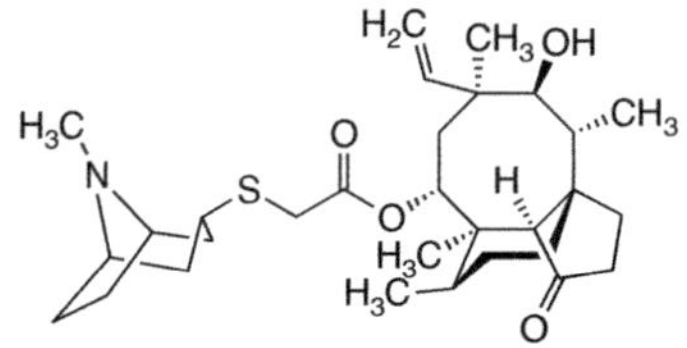	[224452-66-8] $C_{30}H_{47}NO_4S$ 517.78	bacterial protein synthesis inhibitor. antibiotic for treatment of impetigo. topical ointment 1%. 20 mg T.	GSK US/2007 first isolated from *Pleurotus mutilus*. semisynthetic [70, 71]

3.2. D06B Chemotherapeutics for Topical Use

3.2.1. D06BA Sulfonamides (Table 8)

Table 8. D06BA Sulfonamides

INN (Synonyms) [Brand names] ATC	Structure (Remarks)	[CAS-No.] Formula MW, g/mol	Target (if known) Medical use Formulation DDD	Originator Approval (country/year) Production method [References]
Silver sulfadiazine [Flamazine, Silvadene, Silvazine, Sulfose; Theradia; Terfonyl] D06BA01; J01EC01 **WHO essential medicine**		[68-35-9] $C_{10}H_{10}N_4O_2S$ 250.28 silver salt: [22199-08-2] $C_{10}H_9AgN_4O_2S$ 357.14	competitive antagonist of *p*-amino benzoic acid (PABA), the compound essential for the synthesis of folic acid and thus for bacterial growth. chemotherapeutic, sulfonamide antibiotic; used to treat or prevent infections on areas of burned skin. cream 1%; ointment 5%; tabl. 500 mg (as silver salt).	King Pharmaceuticals US/1973 synthetic [72]
Sulfathiazole [Neothalidine, Sulfathalidine, Thalazole] D06BA02; A07AB02 wfm		[85-73-4] $C_{17}H_{13}N_3O_5S_2$ 403.44	competitive antagonist of *p*-amino benzoic acid (PABA), the compound essential for the synthesis of folic acid and thus for bacterial growth. chemotherapeutic, topical antimicrobial. tabl. 500 mg, vaginal cream.	GW Labs US/1983 synthetic [73, 74]
Mafenide [Sulfamylon] D06BA03; G01AE01		[138-39-6] $C_7H_{10}N_2O_2S$ 186.24 acetate salt: [13009-99-9] $C_7H_{10}N_2O_2S$ $\cdot C_2H_4O_2$ 246.29	competitive antagonist of *p*-amino benzoic acid (PABA), the compound essential for the synthesis of folic acid and thus for bacterial growth. chemotherapeutic against severe burns. cream 8.5%, 11.2 g/100 g (as acetate); eye drops 2.5 mg/g, 5%.	US/1948 synthetic [75, 76]
Sulfamethizole (sulfamethizole) [Rufol, Urolucosil] D06BA04; B05CA04; J01EB02; S01AB01		[144-82-1] $C_9H_{10}N_4O_2S_2$ 270.34	competitive antagonist of *p*-amino benzoic acid (PABA), the compound essential for the synthesis of folic acid and thus for bacterial growth. chemotherapeutic, antibiotic. cps. 250 mg in comb. with oxytetracycline·HCl (250 mg) and phenazopyridine·HCl (50 mg); drg. 350 mg in comb. with sulfaethidole.	synthetic [77, 78]

(continued)

Table 8. (*Continued*)

INN (Synonyms) [Brand names] ATC	Structure (Remarks)	[CAS-No.] Formula MW, g/mol	Target (if known) Medical use Formulation DDD	Originator Approval (country/year) Production method [References]
Sulfanilamide [AVC, Elixir, Tablamide] D06BA05; J01EB06		[63-74-1] $C_6H_8N_2O_2S$ 172.21	competitive antagonist of *p*-amino benzoic acid (PABA), the compound essential for the synthesis of folic acid and thus for bacterial growth. chemotherapeutic. vaginal ointment 15%; vaginal suppos. 1.05 g.	synthetic [79]
Sulfamerazine (methylsulfadiazine) [Berlocombine, Polagin] D01BA06; J01ED07		[127-79-7] $C_{11}H_{12}N_4O_2S$ 264.31 sodium salt: [127-58-2] $C_{11}H_{11}N_4NaO_2S$ 286.29	competitive antagonist of *p*-amino benzoic acid (PABA), the compound essential for the synthesis of folic acid and thus for bacterial growth. chemotherapeutic, antibiotic. susp. 60 mg/5 mL; tabl. 120 mg (as sodium salt).	synthetic [80]
Sulfisomidine (sulfasomidine) [Elcosine, Elkosin, Sulfisomidine] D01BA07		[515-64-0] $C_{12}H_{14}N_4O_2S$ 278.34 sodium salt: [2462-17-1] $C_{12}H_{13}N_4NaO_2S$ ·H_2O 318.33	competitive antagonist of *p*-amino benzoic acid (PABA), the compound essential for the synthesis of folic acid and thus for bacterial growth. chemotherapeutic. eye drops 114.4 mg; ointment 5%; ophthalmic ointment 100 mg.	Ciba Geigy US/1976 synthetic [81]

3.2.2. D06BB Antivirals (Table 9)

Table 9. D06BB Antivirals

INN (Synonyms) [Brand names] ATC	Structure (Remarks)	[CAS-No.] Formula MW, g/mol	Target (if known) Medical use Formulation DDD	Originator Approval (country/year) Production method [References]
Idoxuridine [Dendrid, Herpid, Herplex, Idustatin, IDU] D06BB01; 05AB02; S01AD01 (→ Antiinfectives for Systemic Use, 3. Antivirals)		[54-42-2] $C_9H_{11}IN_2O_5$ 354.10	inhibits viral DNA synthesis. antiviral against herpes simplex. eye drops 0.1%; eye ointment 0.25%; ointment 0.2%, 0.5%; sol. 5%, 10%, 40%.	Alcon US/1963 synthetic [82, 83]
Tromantadine [Viru-Merz, Viruserol] D06BB02; J05AC03 (→ Antiinfectives for Systemic Use, 3. Antivirals)		[53783-83-8] $C_{16}H_{28}N_2O_2$ 280.41 hydrochloride salt: [41544-24-5] $C_{16}H_{28}N_2O_2$ ·HCl 316.87	antiviral. cream 10 mg, 100 mg/10 g; gel 10 mg, 100 mg/10 g; ointment 1% (as monohydrochloride).	Merz synthetic [84, 85]
Aciclovir (acyclovir, acycloguanosine) [Acyvir, Remex, Zovirax] D06BB03; J05AB01; S01AD03 **WHO essential medicine** (→ Antiinfectives for Systemic Use, 3. Antivirals)		[59277-89-3] $C_8H_{11}N_5O_3$ 225.21 sodium salt: [69657-51-8] $C_8H_{10}N_5NaO_3$ 247.19	viral DNA replication inhibitor active against herpes simplex virus (HSV) and varicella zoster virus (VZV). antiviral, antiherpes virus agent; ophthalmologic use. amp. 125, 250 mg; cps. 200 mg; cream 5%; eye ointment 3%; gran. 40%; jelly oral 200, 800 mg; susp. 8%; syrup 8%; tabl. 200, 400, 800 mg; vial 250, 500 mg. 25 mg T.	Glaxo USA/1985 synthetic [86–88]
Podophyllotoxin (podophyllinic acid lactone, PPT) [Esberiven, Podofilox, Wartec, Warticon] D06BB04		[518-28-5] $C_{22}H_{22}O_8$ 414.41	inhibits replication of cellular and viral DNA. antimitotic, dermatological agent, precursor for antineoplastics etoposide and teniposide. 0.5% sol. 3.0 mL and 3.5 mL; 0.15% cream 5 g.	Valeant synthetic [89–91]

(continued)

Table 9. (*Continued*)

INN (Synonyms) [Brand names] ATC	Structure (Remarks)	[CAS-No.] Formula MW, g/mol	Target (if known) Medical use Formulation DDD	Originator Approval (country/year) Production method [References]
Inosine [Catechol, Correctol, Isoprinosine] D06BB05; G01AX02; S01XA10		[*58-63-9*] $C_{10}H_{12}N_4O_5$ 263.28	nucleoside that also binds to GABA A receptor. immuno- and feed stimulant. eye drops and solution.	synthetic [92]
Penciclovir [Denavir] D06BB06 (→ Antiinfectives for Systemic Use, 3. Antivirals)		[*39809-25-1*] $C_{10}H_{15}N_5O_3$ 253.26	guanosine nucleotide analogue that inhibits viral DNA polymerase. antiviral prodrug; used to treat cold sore after herpes simplex virus (HSV) infections. cream 10 mg/g and 1%. 6 mg T.	Perrigo 1996 synthetic [93]
Lysozyme (muramidase) [Acdeam, Lysopaine, Neuzyme] D06BB07		[*9000-63-2*]	antimicrobial enzyme that acts as glycoside hydrolase and thus compromises the bacterial cell-wall stability causing lysis of the bacteria. tabl. and solution	first isolated from hen egg white [94]
Ibacitabine [Cuterpes, Herpes-Gel] D06BB08		[*611-53-0*] $C_9H_{12}IN_3O_4$ 353.12	antiviral agent. topical 1% gel.	Chauvin, Bausch & Lomb FR synthetic [95–97]
Edoxudine [Virostat Cream] D06BB09		[*15176-29-1*] $C_{11}H_{16}N_2O_5$ 256.55	thymidine nucleoside analogue. antiviral drug against herpes simplex. cream 3%.	McNeil US/1992 synthetic [98]

Table 9. (*Continued*)

INN (Synonyms) [Brand names] ATC	Structure (Remarks)	[CAS-No.] Formula MW, g/mol	Target (if known) Medical use Formulation DDD	Originator Approval (country/year) Production method [References]
Imiquimod [Aldara, Beselna] D06BB10		[99011-02-6] $C_{14}H_{16}N_4$ 240.31 hydrochloride salt: [99011-78-6] $C_{14}H_{16}N_4 \cdot HCl$ 276.77	immune modulator that upregulates the immune response at the infection site activating monocytes/macrophages and dendritic cells. antiviral, immunomodulator, interferon alfa inducer. used to treat genital warts caused by human papillomavirus (HPV). cream 1–5%.	3M USA/1997 synthetic [99–101]
Docosanol (behenyl alcohol) [Abreva, Erazaban] D06BB11		[661-19-8] $C_{22}H_{46}O$ 326.61	inhibition of viral entry into the cell. antiviral against herpes simplex virus (HSV), CMV, and cold sores. cream 1%.	Avanir Pharms USA/2000 synthetic or derived from plant extracts [102]
Foscarnet [Foscavir, Triapten] D06BB13; J05AD01 (→ Antiinfectives for Systemic Use, 3. Antivirals)		[63585-09-1] CNa_3O_5P 191.95 hexahydrate: [34156-56-4] $CNa_3O_5P \cdot 6\,H_2O$ 300.04	viral DNA polymerase inhibitor. antiviral. cream 2 g/100 g; vial 6000 mg/250 mL (hexahydrate). 12 mg T.	AstraZeneca US/2012 synthetic [103–105]
Vidarabine [Arasena, Silberan, Vira-A, Vira-MP] D06BB15; J05AB03; S01AD06 (→ Antiinfectives for Systemic Use, 3. Antivirals)		[5536-17-4] $C_{10}H_{13}N_5O_4$ 267.25	viral DNA polymerase inhibitor. antiviral. cream 3%; ointment 3%; vial 300 mg.	Monarch Pharmaceuticals US/1976 synthetic [106, 107]

3.2.3. D06BX Other Chemotherapeutics (Table 10)

Table 10. D06BX Other chemotherapeutics

INN (Synonyms) [Brand names] ATC	Structure (Remarks)	[CAS-No.] Formula MW, g/mol	Target (if known) Medical use Formulation DDD	Originator Approval (country/year) Production method [References]
Metronidazole [Flagyl, Metrogel, Metrosa] D06BX01; A01AB17; G01AF01; J01XD01; P01AB01 **WHO essential medicine**		[*443-48-1*] $C_6H_9N_3O_3$ 171.16	inhibits nucleic acid synthesis by disrupting DNA in microbial cells. Chemotherapeutic, antiprotozoal against trichomonas, anti-infective. cps. 250, 375 mg; f.c. tabl. 250, 400 mg; suppos. 100 mg (vaginal); tabl. 250, 400 mg; vaginal tabl. 100, 250 mg; vial 5 g/1000 mL. 15 mg T.	Rhone-Poulenc/GD Searle FR/1960 US/1963 synthetic [108]
Ingenol mebutate (ingenol-3-angelate) [Picato] D06BX02		[*75567-37-2*] $C_{25}H_{34}O_6$ 430.53	induces cell death. topical treatment of actinic keratosis. gel 0.015% and 0.05%.	Leo Pharma USA/2012 synthetic [109, 110]

4. D07 Corticosteroids, Dermatological Preparations

4.1. D07A Corticosteroids, Plain

4.1.1. D07AA Corticosteroids, Weak (Group I, Table 11)

Table 11. D07AA Corticosteroids, weak (group I)

INN (Synonyms) [Brand names] ATC	Structure (Remarks)	[CAS-No.] Formula MW, g/mol	Target (if known) Medical use Formulation DDD	Originator Approval (country/year) Production method [References]
Methylprednisolone [Depo-Medrol, Medrol, Neo-Medrol, Urbason] D07AA01; D10AA02; H02AB04 **WHO essential medicine** (→ Hormone Therapeutics – Systemic Hormonal Preparations, Excluding Reproductive Hormones and Insulins (H))		[83-43-2] $C_{22}H_{30}O_5$ 374.48 acetate: [53-36-1] $C_{24}H_{32}O_6$ 416.51 succinate: [2921-57-5] $C_{26}H_{34}O_8$ 474.55	glucocorticoids cross cell membrane and bind with high affinity to cytoplasmatic receptors. anti-inflammatory. amp. 20, 30, 40, 60, 80 mg; cream 0.1%; ointment 0.1%; tabl. 2, 4, 6, 24, 60 mg.	Pharmacia Upjohn US/1959 synthetic [111, 112]
Hydrocortisone [Cortifoam, Cortril, Hydrocortone, Gentisone, Mitocortyl, Systral Hydrocort] D07AA02; A01AC03; A07EA02; C05AA01; D07XA01; H02AB09; S01BA02; S01CB03; S02BA01 **WHO essential medicine** (→ Hormone Therapeutics – Systemic Hormonal Preparations, Excluding Reproductive Hormones and Insulins (H))		[50-23-7] $C_{21}H_{30}O_5$ 362.47	glucocorticoids cross cell membrane and bind with high affinity to cytoplasmatic receptors, natural hormone cortisol. glucocorticoid, anti-inflammatory. cream 0.5%; lotion 0.5%; ointment 1%, 2.5%; tabl. 10 mg.	Upjohn US/1941 synthetic [113, 114]
Prednisolone [Aersolin D; Blephamide; Decortin; Deliproct; Dontisolon; Farnerate; Farnezone; Hydrocortancyl; Lidomex; Linola; Solupred; Pred Forte; Predsol] D07AA01; A07EA01; C05AA04; D07XA02; H02AB06; R01AD02; R01AD52; S01BA04; S01CB02; S02BA03; S03BA02 **WHO essential medicine** (→ Hormone Therapeutics – Systemic Hormonal Preparations, Excluding Reproductive Hormones and Insulins (H)) (→ Respiratory Disorders, 2. Nasal Preparations and Decongestants for Topical Use (R01))		[50-24-8] $C_{21}H_{28}O_5$ 360.45 acetate: [52-21-1] $C_{23}H_{30}O_6$ 402.49	corticosteroid hormone receptor agonist; glucocorticoid. antiasthmatic, anti-inflammatory. amp. 10 mg/mL, 25 mg/mL, 50 mg/mL (as acetate); cream 0.5%; eye drops 1.2 mg/mL, 10 mg/mL (as acetate); eye ointment 0.25%; ointment 0.5%, 100 mg/g; 1%, powder 97–102%; suppos. 100 mg (as acetate); syrup 15 mg/5 mL; tabl. 1, 5, 20 mg.	Schering Corp. USA/1955 synthetic [115–117]

4.1.2. D07AB Corticosteroids, Moderately Potent (Group II, Table 12)

Table 12. D07AB Corticosteroids, moderately potent (group II)

INN (Synonyms) [Brand names] ATC	Structure (Remarks)	[CAS-No.] Formula MW, g/mol	Target (if known) Medical use Formulation DDD	Originator Approval (country/year) Production method [References]
Clobetasone [Emovate, Eumovate, Kindavate, Trimovate] D07AB01; S01BA09		[54063-32-0] $C_{22}H_{26}ClFO_4$ 408.90 butyrate: [25122-57-0] $C_{26}H_{32}ClFO_5$ 478.99	glucocorticoids cross cell membrane and bind with high affinity to cytoplasmatic receptors. topical corticoid. eye drops 0.1%; cream 0.05%; lotion 0.05%; ointment 0.05%.	Glaxo DE/1980 synthetic [118]
Hydrocortisone butyrate [Alfason, Laticort, Locoid, Pandel] D07AB02		[13609-67-1] $C_{25}H_{36}O_6$ 432.56	glucocorticoids cross cell membrane and bind with high affinity to cytoplasmatic receptors. topical glucocorticoid. cream 0.1%; emulsion 1 mg/g; lotion 1 mg/g; ointment 0.1%.	synthetic [119, 120]
Flumetasone [Locasalen, Locorten] D07AB03; D07BB01; D07CB05; D07XB01; S02CA02		[2135-17-3] $C_{22}H_{28}F_2O_5$ 410.46 pivalate: [2002-29-1] $C_{27}H_{36}F_2O_6$ 494.58	glucocorticoids cross cell membrane and bind with high affinity to cytoplasmatic receptors. topical glucocorticoid, anti-inflammatory. sol. 0.02 g/100 g (0.02%); cream 0.02%; lotion 0.02%; ointment 0.02%.	Novartis 1964 synthetic [121, 122]
Fluocortine [Lenen, Vaspit] D07AB04		[33124-50-4] $C_{22}H_{27}FO_5$ 390.45 butyl ester: [41767-29-7] $C_{26}H_{35}FO_5$ 446.56	glucocorticoid. cream 7.5 mg/g; ointment 7.5 mg/g; powder 100 mg/4 g.	synthetic [123, 124]

Table 12. (*Continued*)

INN (Synonyms) [Brand names] ATC	Structure (Remarks)	[CAS-No.] Formula MW, g/mol	Target (if known) Medical use Formulation DDD	Originator Approval (country/year) Production method [References]
Fluperolone acetate [Alacortil, Methral] D07AB05; H02AB wfm	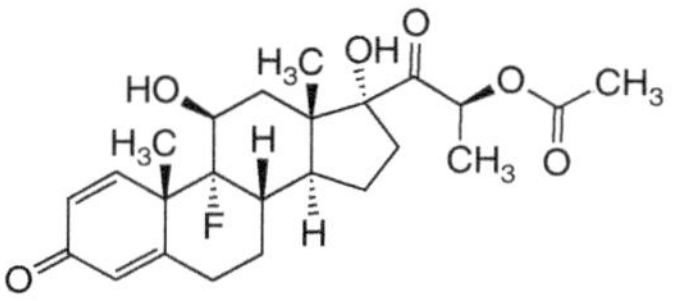	[2119-75-7] $C_{24}H_{31}FO_6$ 434.50	glucocorticoid, anti-inflammatory.	Pfizer synthetic [125]
Fluorometholone [Flucon, FML, Gentacort, Odomel, Oxylone] D07AB06; C05AA06; D07XB04; D10AA01; S01BA07; S01CB05	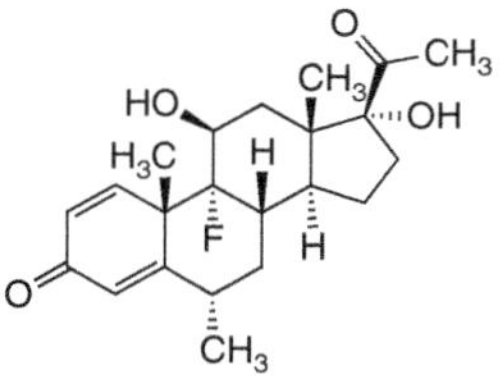	[426-13-1] $C_{22}H_{29}FO_4$ 376.47	glucosteroid. against conjunctivitis. eye drops 0.02%, 0.05%, 0.1%.	synthetic [126]
Fluprednidene acetate [Decoderm, Sali-Decoderm] D07AB07; D07CB02; D07XB03	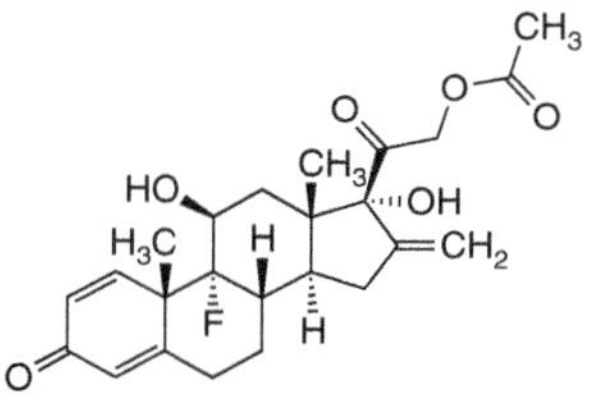	[1255-35-2] $C_{24}H_{29}FO_6$ 432.49	topical glucosteroid. cream 1 mg/g; ointment 0.05 g/100 g, 1 mg/g; sol. 0.025 g/100 g, 0.15 g/100 g, 1 mg/mL.	synthetic [127, 128]
Desonide (prednacinolone) [Locapred, Locatrop, Sterades, Tridesilon] D07AB08; S01BA11	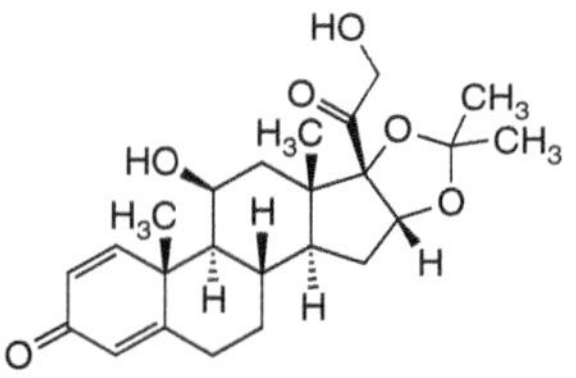	[638-94-8] $C_{24}H_{32}O_6$ 416.51	topical glucocorticoid. cream 1 mg/g; gel 0.5 mg/g; foam 0.5 mg/g; lotion 0.5 mg/g.	Dow Pharma US/2006 synthetic [129, 130]
Triamcinolone [Delphicort, Kenacort, Kenalog, Ledercort] D07AB09; A01AC01; D07AB09; D07BB03; D07CB01; D07XB02; H02AB08; S01BA05; S02CA04 (→ Hormone Therapeutics – Systemic Hormonal Preparations, Excluding Reproductive Hormones and Insulins (H)) (→ Respiratory Disorders, 2. Nasal Preparations and Decongestants for Topical Use (R01))	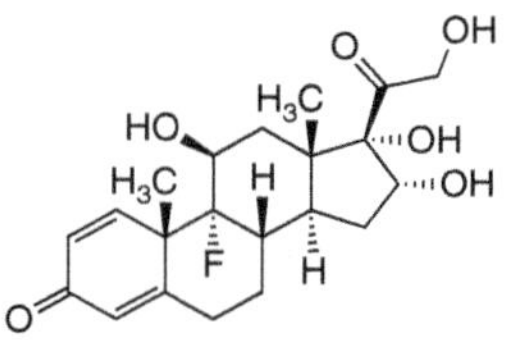	[124-94-7] $C_{21}H_{27}FO_6$ 394.44	glucosteroid. tabl. 2, 4, 8, 16 mg. 2 mg T.	Apothecon US/1965 synthetic [131–133]

(*continued*)

Table 12. (*Continued*)

INN (Synonyms) [Brand names] ATC	Structure (Remarks)	[CAS-No.] Formula MW, g/mol	Target (if known) Medical use Formulation DDD	Originator Approval (country/year) Production method [References]
Alclometasone [Aclosone, Almeta, Delonal, Legederm] D07AB10; S01BA10		[*66734-13-2*] $C_{22}H_{29}ClO_5$ 408.9	topical glucocorticoid. cream and ointment 0.5 mg/1 g; ointment 0.1%.	Essex Pharma DE/1980 synthetic [134, 135]
Hydrocortisone buteprate (hydrocortisone 17-butyrate 21-proprionate) [Pandel] D07AB11		[*72590-77-3*] $C_{28}H_{40}O_7$ 488.13	glucosteroid; cortisol ester cream 1 mg/1 g.	Fougera Pharma USA/1997 synthetic [136]
Dexamethasone [Decadron, Dextenza, Dexycu, Eurason, Maxidex, Otomize, Sofradex] D07AB19; A01AC02; C05AA09; D07AB19; D07XB05; D10AA03; H02AB02; R01AD03; S01BA01; S01CB01; S02BA06; S03BA01 **WHO essential medicine** (→ Hormone Therapeutics – Systemic Hormonal Preparations, Excluding Reproductive Hormones and Insulins (H)) (→ Respiratory Disorders, 2. Nasal Preparations and Decongestants for Topical Use (R01))		[*50-02-2*] $C_{22}H_{29}FO_5$ 392.47	glucocorticoid. aerosol 0.075 mg/per pump; amp. 5 mg/mL; amp. 0.4%/5 mL, 0.45%/5 mL (as acetate); amp. 2.5 mg (as palmitate); amp. 0.2%, 0.3%, 0.33% (as metasulfobenzoate sodium); cream 0.1%; cream 0.1% (as propionate); dental ointment 0.1%; dry syrup 0.1%; ear drop 0.1% (as metasulfobenzoate sodium); elixir 0.01%; eye drops 1 0.1%; eye drops 0.02%, 0.05%, 0.1% (as metasulfobenzoate sodium); f.c. tabl. 0.5, 0.75, 1.5 mg; lotion 0.1%; lotion 0.1% (as propionate); nasal drops 0.1% (as metasulfobenzoate sodium); ointment 0.1%; ointment 1 mg (glymesason); ointment 0.1% (as propionate); ophth. ointment 0.05%, 0.1%; sol. 0.03%; suppos. 2.2 mg; tabl 0.5 mg; vial 0.2%, 0.3%, 0.33% (as metasulfobenzoate sodium); intraocular suspension 9%	MSD 1961 Icon Bioscience US/2018 as intraocular suspension synthetic [137–139]

Table 12. (*Continued*)

INN (Synonyms) [Brand names] ATC	Structure (Remarks)	[CAS-No.] Formula MW, g/mol	Target (if known) Medical use Formulation DDD	Originator Approval (country/year) Production method [References]
Clocortolone [Cloderm, Cilder] D07AB21	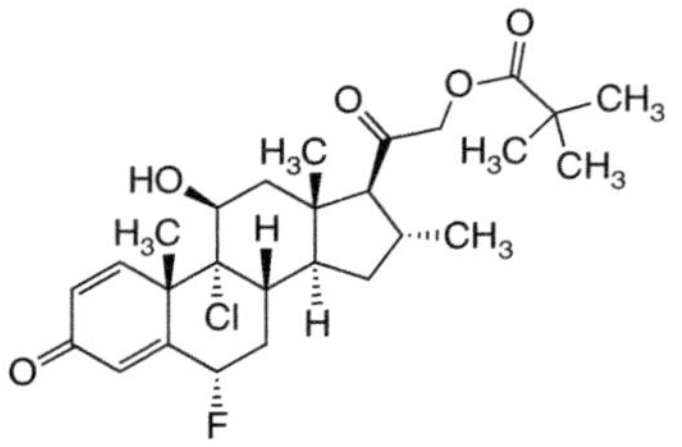	[*4828-27-7*] $C_{22}H_{28}ClFO_4$ 410.91 pivalate: [*34097-16-0*] $C_{27}H_{36}ClFO_5$ 495.03	glucocorticoid. cream 1 mg/g.	Epi Health US/1977 synthetic [140]

4.1.3. D07AC Corticosteroids, Potent (Group III, Table 13)

Table 13. D07AC Corticosteroids, potent (group III)

INN (Synonyms) [Brand names] ATC	Structure (Remarks)	[CAS-No.] Formula MW, g/mol	Target (if known) Medical use Formulation DDD	Originator Approval (country/year) Production method [References]
Betamethasone (NSC-39470, Sch-4831) [Beben; Bebesil; Beclospray; Bentelan, Betametasol; Betnelan; Betnesol; Betneval; Betnovate; Buccobet; Celestamine; Celestene; Celestoderm; Celestone; Diprolene; Diprosalic; Diprosone; Ecoval; Rinderon] D07AC01; A07EA04; C05AA05; D07XC01; H02AB01; R01AD06; R03BA04; S01BA06; S01CB04; S03CA06 **WHO essential medicine** (→ Hormone Therapeutics – Systemic Hormonal Preparations, Excluding Reproductive Hormones and Insulins (H)) (→ Respiratory Disorders, 2. Nasal Preparations and Decongestants for Topical Use (R01))		[378-44-9] $C_{22}H_{29}FO_5$ 392.47	corticosteroid hormone receptor agonist. glucocorticoid; antiasthmatic, anti-inflammatory, antirheumatic. powder 0.1%; suppos. 0.5, 1 mg; syrup 0.01%, 0.6 mg/5 mL; tabl. 0.5, 0.6, 1 mg. 2 mg T.	Schering-Plough USA/1961 synthetic [141, 142]
Fluclorolone acetonide (flucloronide) [Topilar] D07AC02 wfm		[3693-39-8] $C_{24}H_{29}Cl_2FO_5$ 487.40	topical glucosteroid. cream 0.025%, 0.25%.	Syntex synthetic [143]
Desoximetasone [Flubason, Topicort, Topisolon] D07AC03; D07X02		[382-67-2] $C_{22}H_{29}FO_4$ 376.47	topical glucosteroid. cream 0.25%, 0.05%; lotion 0.25%; ointment 0.35%.	Sanofi/Tavo Pharm US/1985 synthetic [144, 145]

Table 13. (*Continued*)

INN (Synonyms) [Brand names] ATC	Structure (Remarks)	[CAS-No.] Formula MW, g/mol	Target (if known) Medical use Formulation DDD	Originator Approval (country/year) Production method [References]
Fluocinolone acetonide [Flucort, Jelin, Localyn, Synalar] D07AC04, C05AA10; D07AC04		[*67-73-2*] $C_{24}H_{30}F_2O_6$ 452.49	glucosteroid, anti-inflammatory. cream 0.025%; external sol. 0.01%; ointment 0.025%; spray 0.007%; tape 8 µg/cm^2 topical sol. 0.01%. 0.3 mg T.	Syntex 1961 synthetic [146, 147]
Fluocortolone [Ultralan, Ultralanum] D07AC05; C05AA08; D07BC03; H02AB03; S01CA04 (→ Hormone Therapeutics – Systemic Hormonal Preparations, Excluding Reproductive Hormones and Insulins (H))		[*152-97-6*] $C_{22}H_{29}FO_4$ 376.47	glucosteroid. cream 2.5 mg/g; lotion 2.5 mg/g; ointment 2.5 mg/g; tabl. 5, 20, 50 mg.	Bayer/Schering synthetic [148, 149]
Difluocortolone [Nerisona, Temetex, Texmeten] D07AC06		[*2607-06-9*] $C_{22}H_{28}F_2O_4$ 394.46 valerate: [*59198-70-8*] $C_{27}H_{36}F_2O_5$ 478.58	glucocorticoid. cream 0.1%; ointment 0.1%; sol. 0.1%.	Schering/GSK CA/1983 synthetic [150, 151]
Fludroxycortide (flurandrenolide) [Cordran, Drenison] D07AC07		[*1524-88-5*] $C_{24}H_{33}FO_6$ 436.52	glucocorticoid, anti-inflammatory. lotion 0.05% (15, 60 mL); plaster 4 µg; tape 4 µg/cm^2.	synthetic [152]
Fluocinonide [Jelliproct, Lidecomb, Topsym, Topsyn, Vanos Cream] D07AC08; C05AA11		[*356-12-7*] $C_{26}H_{32}F_2O_7$ 494.53	glucocorticoid, anti-inflammatory. cream 0.05%; gel 0.05%; lotion 0.05%; ointment 0.05%, sol. 0.5 mg/g; spray 0.0143%.	Medicis US/1980 synthetic [153, 154]

(*continued*)

Table 13. (*Continued*)

INN (Synonyms) [Brand names] ATC	Structure (Remarks)	[CAS-No.] Formula MW, g/mol	Target (if known) Medical use Formulation DDD	Originator Approval (country/year) Production method [References]
Budesonide [Aircort; Bidien; Budecort Novolizer; Budenofalk; Budes; Budiair; Easyhaler; Entocort; Miflonide; Miflonil; Novopulmon; Novopulmon Novolizer; Novolizer Budesonide; Preferid; Pulmax; Pulmaxan Turbohaler; Pulmicort; Rafton; Rhinocort; Symbicort; Symbicort Turbohaler; Uceris] D07AC09; A07EA06; H02AB16; R01AD05; R03AB; R03BA02 **WHO essential medicine** (→ Respiratory Disorders, 2. Nasal Preparations and Decongestants for Topical Use (R01))		[*51333-22-3*] $C_{25}H_{34}O_6$ 430.54	corticosteroid receptor agonist. glucocorticoid; antiasthmatic, anti-inflammatory. aerosol 0.2 mg/puff; cream 0.025%, 0.05%; nasal aerosol 0.05 mg/puff; ointment 0.025 mg, 0.05%; pump spray 0.05 mg/puff; susp. 0.5 mg/2 mL, 1 mg/2 mL.	AstraZeneca 1981 synthetic [155, 156]
Diflorasone [Apexicon, Dermaflor, Diacort, Diflal, Florone] D07AC10		[*2557-49-5*] $C_{22}H_{28}F_2O_5$ 410.46 diacetate: [*33564-31-7*] $C_{26}H_{32}F_2O_7$ 494.53	topical glucocorticoid. cream 0.05%; lotion 0.05%; ointment 0.05%.	Galderma DE/1982 synthetic [157]
Amcinonide (Triamcinolone acetate cyclopentanoide) [Amciderm, Cyclocort, Penticort, Visderm] D07AC11		[*51022-69-6*] $C_{28}H_{35}FO_7$ 502.58	modulator of transcription factors regulating phospholipase A2 activity. topical glucosteroid. cream 0.1%; lotion 0.1%; ointment 0.1%. 1.5 mg T.	Fujisawa USA/1979 synthetic [158]
Halometasone [Sicorten] D07AC12		[*50629-82-8*] $C_{22}H_{27}ClF_2O_5$ 444.90	topical corticosteroid (glucocorticoid), anti-inflammatory. cream 0.5 mg/g (0.05%); ointment 0.5 mg/g (0.05%).	synthetic [159]

Table 13. (*Continued*)

INN (Synonyms) [Brand names] ATC	Structure (Remarks)	[CAS-No.] Formula MW, g/mol	Target (if known) Medical use Formulation DDD	Originator Approval (country/year) Production method [References]
Mometasone [Altasone; Asmanex; Asmanex Twisthaler; Ecural; Elocom; Elocon; Nasaflex; Nasonex; Rinelon; Sensicort; Uniclar] D07AC13; D07XC03; R01AD09; R03BA07 (→ Respiratory Disorders, 2. Nasal Preparations and Decongestants for Topical Use (R01))		[105102-22-5] $C_{22}H_{28}Cl_2O_4$ 427.37 furoate: [83919-23-7] $C_{27}H_{30}Cl_2O_6$ 521.44	corticosteroid receptor agonist. glucocorticoid; antiasthmatic; anti-inflammatory. cream 0.1%; lotion 0.1%; ointment 0.1% (1 mg/g); powder inhaler. 1 mg T.	Schering Corp./MSD USA/1991 synthetic [160–162]
Methylprednisolone aceponate [Advantan] D07AC14		[86401-95-8] $C_{27}H_{36}O_7$ 472.57	glucocorticosteroid derivate of methylprednisolone. 1 mg T.	synthetic [163]
Beclometasone [Aerobec; Aldecin; Beclomet Easyhaler; Beclovent; Beconase; Becotide; Bronchocort; Filair; Inuvais; Junik Autohaler; Miflasone; Nexxair; Prolair Autohaler; Propaderm; Rhinocort; Vancenase; Vanceril; Ventolair Autohaler] D07AC15; A07EA07; R01AD01; R03BA01 **WHO essential medicine** (→ Respiratory Disorders, 2. Nasal Preparations and Decongestants for Topical Use (R01))		[4419-39-0] $C_{22}H_{29}ClO_5$ 408.92 dipropionate: [5534-09-8] $C_{28}H_{37}ClO_7$ 521.05	corticosteroid hormone receptor agonist. glucocorticoid; antiasthmatic, antiallergic, anti-inflammatory. cream 0.025%; dose aerosol (0.05 µg, 0.25 µg/puff); nasal aerosol 50 µg/puff; nasal and oral inhaler 50 µg; nasal spray (0.05 µg/puff); ointment 0.025%; oral aerosol 50 µg/puff; powder inhaler.	MSD/GSK US/1987 synthetic [164, 165]

(*continued*)

Table 13. (*Continued*)

INN (Synonyms) [Brand names] ATC	Structure (Remarks)	[CAS-No.] Formula MW, g/mol	Target (if known) Medical use Formulation DDD	Originator Approval (country/year) Production method [References]
Hydrocortisone aceponate [Efficort] D07AC16		[74050-20-7] $C_{26}H_{36}O_7$ 460.50	cream and spray.	synthetic [166]
Fluticasone [Adoair; Advair; Aliflus Diskus; Atemur; Cutivate; Dymista; Flixoderm; Flixonase; Flixotide; Flonase; Flovent; Flunase; Fluspiral; Flutide; Flutivate; Seretide; Seretide Diskus; Viani] D07AC17; R01AD08; R03BA05 (→ Respiratory Disorders, 2. Nasal Preparations and Decongestants for Topical Use (R01))		[90566-53-3] $C_{22}H_{27}F_3O_4S$ 444.51 propionate: [80474-14-2] $C_{25}H_{31}F_3O_5S$ 500.58	corticosteroid receptor agonist. glucocorticoid, antiasthmatic, anti-inflammatory. aerosol for inh. 44, 110, 220 µg; cream 0.05%; ointment 0.005%; nasal spray 0.05%, 0.51 mg/1 mL; powder inhaler 50, 100, 200 µg.	GlaxoSmithKline/ Pharma Derm USA/1990 synthetic [167, 168]
Prednicarbate (HOE777) [Dermatop, Prednitop] D07AC18		[73771-04-7] $C_{27}H_{36}O_8$ 488.58	topical glucocorticoid, steroidal anti-inflammatory. cream 2.5 mg/1 g; ointment 2.5 mg/1 g; sol. (in aqueous ethanol, 20%) 2.5 mg/1 g. 2.5 mg T.	Valeant US/1993 synthetic [169, 170]
Difluprednate [Durezol, Epitopic, Myser] D07AC19		[23674-86-4] $C_{27}H_{34}F_2O_7$ 508.56	topical glucosteroid. cream 0.05%; gel 0.05%; lotion 0.05%; ointment 0.05%; ophthalmic emulsion 0.05%.	Sirion Therapeutics US/2008 ophthalmic emulsion synthetic [171, 172]
Ulobetasol propionate (halobetasol) [Lexette, Ultravate] D07AC21		[66852-54-8] $C_{25}H_{31}ClF_2O_5$ 484.97	induction of phospholipase A2. inhibitory of proteins, topical glucosteroid. cream 0.05%; ointment 0.05%, foam.	Bausch Health CA/1993 synthetic [173]

4.1.4. D07AD Corticosteroids, Very Potent (Group IV, Table 14)

Table 14. D07AD Corticosteroids, very potent (group IV)

INN (Synonyms) [Brand names] ATC	Structure (Remarks)	[CAS-No.] Formula MW, g/mol	Target (if known) Medical use Formulation DDD	Originator Approval (country/year) Production method [References]
Clobetasol [Clobesol, Clobex, Dermoxin, Dermoval, Temovate] D07AD01		[25122-41-2] $C_{22}H_{28}ClFO_4$ 410.91 propionate: [25122-46-7] $C_{25}H_{32}ClFO_5$ 466.98	topical glucosteroid. cream 0.05%; gel 0.05%; lotion 0.05%; ointment 0.05%; sol. 0.5 mg/g (0.05%). 0.5 mg T.	GlaxoSmithKline DE/1976 synthetic [174, 175]
Halcinonide [Adcortin, Halciderm, Halog, Sawastin] D07AD02		[3093-35-4] $C_{24}H_{32}ClFO_5$ 454.97	topical glucosteroid. cream 0.1%; ointment 0.1%; sol. 0.1%.	Westwood Squibb US/1977 synthetic [176, 177]

List of Abbreviations

amp.	ampule(s)
CMV	cytomegalovirus
cps.	capsules
COX	cyclooxygenase
DDD	defined daily dose
drg.	dragee
eff.	effervescent
f.c.	film coated
gran.	granules
HPV	human papilloma virus
HSV	herpes simplex virus
i.m.	intramuscular
inj.	injection
INN	international nonproprietary name
MW	molecular mass
N	nasal
NSAID	nonsteroidal anti-inflammatory drug
mL	milliliter (cubic centimeter)
O	oral
P	parenteral
PABA	*p*-aminobenzoic acid
RAR	retinoic acid receptor
sol.	solution
s.r.	slow release
suppos.	suppositories
susp.	suspension
T	topical
tabl.	tablets
U	unit
VZV	varicella zoster virus
wfm	withdrawn from market
WHO	World Health Organization

References

1 Eaglestein, W.H. and Corcoran, G. (2011) *Arch. Derrmatol.* **147** (5), 568–572.

2 Aslam, A. and Griffiths, C. (2014) *Clin. Med.* **14** (1), 47–53.

3 Rendon, A. and Schäkel, K. (2019) *Int. J. Mol. Sci.* **20** (6), 1475.

4 Khalil, S., et al. (2017) *J. Dermatol. Treat.* **28** (8), 1–56.

5 Nolan, B.V., et al. (2010) *Dermatol. Online J.* **16** (2), 1.

6 Turner, R.F., et al. (1998) *Br. Med. J.* **317** (7155), 412.

7 Williamson, D.A., et al. (2017) *Clin. Microbiol. Rev.* **30** (3), 827.

8 Van Meter, K.C. and Hubert, R.J. (2016) *Microbiology for the Healthcare Professional*, Elsevier, China, pp. 214–232.

9 Tacic, A., et al. (2017) *Adv. Technol.* **6** (19), 58–71.

10 Moschella, S.L. (1977) *Cutis* **19** (5), 603–605.

11 De Clercq, E. and Li, G. (2016) *Clin. Microbiol. Rev.* **29** (3), 695–747.

12 Abdel-Haq, N., et al. (2006) *Indian J. Pediatr.* **73** (4), 313–321.

13 Stoopler, E.T., et al. (2013) *CDA J.* **41** (4), 259–265.

14 Lee, M. and Marks, R. (1998) *Aust. Prescr.* **21**, 9–11.

15 Kleyn, E.C., et al. (2019) *J. Dermatol. Treat.* **30** (4), 311–319.

16 Zahn, K. and Koch, H. (1938) *Ber. Dtsch. Chem. Ges.* **71**, 172.

17 Bayer, DRP 296 091 (appl. 1915).

18 Coeverden, A.M., et al. (2004) *Arch. Dermatol.* **140** (12), 1463–1466.

19 US-Secret. of Agriculture, (1959) US 2 889 337 (appl. 1956).

20 T. C. Elder, US 4 129 576 (12.12.1978; prior. 12.4.1976, 24.6.1976).

21 T. C. Elder, US 4 129 575 (12.12.1978; prior. 12.4.1976).

22 Fumapharm AG, US 6 509 376 (21.1.2003; appl. 10.5.2000; DE-prior. 19.11.1998).

23 Fumapharm AG, US 7 320 999 (22.1.2008; appl. 17.7.2002; DE-prior. 19.11.1998).

24 Biogen Idec, US 2014 0 200 363 (17.7.2014; appl. 8.6.2012; USA-prior. 8.6.2011).

25 Calverley, M.J. (1987) *Tetrahedron* **43**, 4609.

26 Leo, EP 227 826 (appl. 14.7.1986; GB-prior. 2.8.1985).

27 Zhu, G.-D. and Okamura, W.H. (1995) *Chem. Rev.* **95**, 1877–1952.

28 Labler, L. (2012) Vitamins, 3. Vitamin D, in *Ullmann's Encyclopedia of Industrial Chemistry*, Wiley-VCH Verlag, pp. 185–196.

29 Wisconsin Alumni Research Found, US 3 565 924 (23.02.1971; appl. 01.07.1968).

30 *Synform (SNFMDF)* **5** (1), 1–8 (1987).

31 Morisaki, M., et al. (1975) *J. Chem. Soc., Perkin Trans. 1* 1421–1424.

32 Teijin, DE 2 526 981 (18.6.1975; J-prior. 18.6.1974).

33 Allergan Inc., EP 284 261 (5.1.1994; appl. 28.9.1988; USA-prior. 13.3.1987).

34 Allergan Inc., US 5 089 509 (18.2.1992; USA-prior. 20.3.1989).

35 Honigsman (1979) *Br. J. Dermatol.* **101**, 13–16.

36 Mayer, H., et al. (1978) *Experientia* **34**, 1105.

37 Roche, US 4 105 681 (8.8.1978; prior. 22.3.1974; 1.8.1975; 13.8.1976).

38 Hoffmann-La Roche, DE 2 414 619 (17.10.1974; appl. 26.3.1974; CH-prior. 30.3.1973).

39 Hoffmann-La Roche, US 4 105 681 (8.8.1978; appl. 13.8.1976; USA-prior. 22.3.1974).

40 McCormick, J.R.D., et al. (1957) *J. Am. Chem. Soc.* **79**, 4561.

41 American Cyanamid, US 2 878 289 (17.3.1959; prior. 28.5.1956).

42 Duggar, B.M. (1948) *Ann. N. Y. Acad. Sci.* **51**, 175.

43 American Cyanamid, (1949) US 2 482 055 (prior. 1948).

44 Finlay, A.C., et al. (1950) *Science* **111**, 85.

45 Pfizer, (1950) US 2 516 080 (prior. 1949).

46 Conover, L.H., et al. (1963) *J. Am. Chem. Soc.* **75**, 4622.

47 Conover, L. H. (1955) US 2 699 054 (prior. 1953).

48 Chopra, I., et al. (2001) *Microbiol. Mol. Biol. Rev.* **65** (2), 232–260.

49 Godtfredsen, W.O. and Vangedal, S. (1962) *Tetrahedron* **18**, 1029.

50 Lovens Kemiske, (1966) US 3 230 240 (GB-prior.1962).

51 Rebstock, M.C., et al. (1949) *J. Am. Chem. Soc.* **71**, 2458–2468.

52 Parke Davis, (1949) US 2 483 871 (appl. 1948).

53 Rutgers Res. Fund, US 2 799 620 (16.7.1957; appl. 29.6.1956).

54 US-Secret. of War, (1950) US 2 498 165 (appl. 1946).

55 Ben Venue Labs., (1949) US 2 457 887 (appl. 1947).

56 Commercial Solvents Corp., (1952) US 2 609 324 (appl. 1949).

57 Merck & Co., (1959) US 2 915 432 (appl. 1955).

58 Weinstein, M.J., et al. (1963) *Antimicrob. Agents Chemother.* **161**, 138.

59 Schering Corp., US 3 091 572 (28.5.1963; prior. 16.7.1962).

60 Penick, S.B. US 2 482 832 (27.9.1949, appl. 21.11.1945).

61 Fuller, A.T., et al. (1971) *Nature* **234**, 416.

62 Beecham, DE 2 227 739 (GB-prior. 12.6.1971).

63 Beecham, US 3 977 943 (31.8.1976; appl. 7.7.1975; prior. 27.3.1974).

64 J. Carlson et al., US 20 010 018 417 (30.8.2001; appl. 10.4.2001; USA.prior. 1.7.1996).

65 Marchi, E., et al. (1985) *J. Med. Chem.* **28**, 960.

66 Alfa Farm., DE 3 120 460 (appl. 22.5.1981; I-prior. 22.5.1980).

67 Alfa Farm., US 4 341 785 (27.7.1982; I-prior. 22.5.1980).

68 Kawaguchi, H., et al. (1972) *J. Antibiot.* **25**, 695.

69 Bristol Myers, GB 1 401 221 (appl. 13.7.1972; USA-prior. 13.7.1971).

70 Hirokawa, Y., et al. (2008) *J. Med. Chem.* **51**, 1991–1994.

71 SmithKline Beecham, WO 9 921 855 (6.5.1995; appl. 27.10.1998; GB-prior. 29.10.1997).

72 Sharp & Dohme, (1946) US 2 407 966.

73 Sharp & Dohme, (1943) US 2 324 013 (prior. 1941).

74 Sharp & Dohme, (1943) US 2 324 015 (prior. 1941).

75 I. G. Farben, DRP 726 386 (appl. 1939).

76 Winthrop, (1942) US 2 288 531 (D-prior. 1939).

77 American Cyanamid, (1944) US 2 358 031 (prior. 1940).

78 Lundbeck & Co., (1948) US 2 447 702 (DK-prior. 1942).

79 Mietzsch, Klarer, (1938) US 2 132 178.

80 Sharp & Dohme, (1946) US 2 407 966 (appl. 1940).

81 Geigy, (1944) US 2 351 333 (CH-prior. 1940).

82 Chang, P.K. and Welch, A.D. (1963) *J. Med. Chem.* **6**, 428.

83 Roussel-Uclaf, FR 1 336 866 (6.9.1963; appl. 27.7.1962).

84 Peteri, D. and Sterner, W. (1973) *Arzneim.-Forsch.* **23**, 577.

85 Merz & Co., DOS 1 941 218 (appl. 13.8.1969).

86 Schaeffer, H.J., et al. (1978) *Nature* **272**, 583.

87 Matsumoto, H., et al. (1988) *Chem. Pharm. Bull.* **36**, 1153.

88 Wellcome, US 4 199 574 (22.4.1980; GB-prior. 2.9.1974).

89 Hartwell, J.L. and Schrecker, A.W. (1958) *Fortschr. Chem. Org. Naturst.* **15**, 83.

90 Andrews, R.C., et al. (1988) *J. Am. Chem. Soc.* **110**, 7854–7858.

91 Berkowitz, D.B., et al. (2000) *J. Org. Chem.* **65**, 847–860.

92 Kyowa Hakko, US 3 616 212 (26.10.1971; appl. 9.9.1969; JP-prior. 27.9.1967).

93 Beecham, US 5 075 445 (24.12.1991; appl. 12.8.1987; GB-prior. 18.8.1983).

94 Chandan, R.C., et al. (1964) *Nature* **204** (4953), 76–77.

95 Kumar, V., et al. (2009) *Synthesis* **23**, 3957–3962.

96 M. Hartmann, H. Venner, DD 69 813 (20.11.1969; appl. 2.8.1968).

97 L'Oreal, EP 0 755 675 (29.1.1997; appl. 5.7.1996; F-prior. 28.7.1995).

98 Bergstroem, D., et al. (1978) *J. Am. Chem. Soc.* **100** (26), 8106–8112.

99 Minnesota Mining and Manufacturing Company, WO 9 748 704 (appl. 22.10.1996; USA-prior. 21.6.1996).

100 Riker Lab., EP 145 340 (appl. 16.11.1984; USA-prior. 18.11.1983).

101 Riker Lab., US 5 238 944 (24.8.1993; appl. 3.3.1992; USA-prior. 15.12.1988, 30.11.1989).

102 Katz, D.H., et al. (1991) *Proc. Natl. Acad. Sci. U. S. A.* **88** (23), 10825–10829.

103 Astra, US 4 339 445 (13.7.1982; appl. 21.12.1978; S-prior. 1.7.1976).

104 Nylen, P. (1924) *Ber. Dtsch. Chem. Ges.* **57b**, 1023.

105 Esp. Latinas Med. Universales, ES 541 567 (appl. 26.3.1985).

106 Parke Davis, GB 1 159 290 (appl. 29.12.1967; USA-prior. 30.12.1966, 29.9.1967).

107 Parke Davis, US 3 703 507 (21.11.1972; prior. 26.9.1969).

108 Rhône-Poulenc, US 2 944 061 (5.7.1960; F-prior. 20.9.1975).

109 Hecker, E. (1968) *Cancer Res.* **28**, 2338.

110 Peplin Biotech, US 6 432 452 (13.08.2002; AU-prior. 19.08.1997).

111 Speero, G.G., et al. (1956) *J. Am. Chem. Soc.* **78**, 6213. **79**, 1515 (1957).

112 Upjohn, US 2 897 218 (28.7.1959; appl. 23.11.1956; prior. 23.4.1956).
113 Upjohn, (1953) US 2 649 401 (appl. 1950).
114 Pfizer, (1953) US 2 658 023 (appl. 1952).
115 Schering Corp., (1959) US 2 897 216 (prior. 1952).
116 Wettstein, A., et al. (1956) *Helv. Chim. Acta* **39**, 734.
117 Nobile, A., et al. (1955) *J. Am. Chem. Soc.* **77**, 4184.
118 Glaxo, US 3 721 687 (20.3.1973; GB-prior. 4.4.1968).
119 Schering AG, DAS 2 644 556 (DAS 2 441 284 appl. 16.9.1974).
120 Taisho, JP 52 010 489 (appl. 15.7.1975).
121 Djerassi, C., et al. (1959) *J. Am. Chem. Soc.* **81**, 3156. **82**, 2318 (1961).
122 Syntex, (1954) US 2 671 752 (appl. 1951).
123 Laurent, H., et al. (1977) *Arzneim.-Forsch.* **27** (II), 2187.
124 Schering AG, DOS 2 150 268 (appl. 4.10.1971).
125 Agnello, E.J., et al. (1963) *J. Org. Chem.* **28**, 1531.
126 Upjohn, US 2 867 638 (6.1.1959; appl. 17.5.1967; prior. 10.9.1956).
127 Irmscher, K., et al. (1968) *Arzneim.-Forsch.* **18**, 7.
128 Merck Patent GmbH, GB 1 230 671 (5.5.1971; appl. 10.7.1969).
129 Bernstein, S., et al. (1959) *J. Am. Chem. Soc.* **81**, s4573.
130 Squibb, US 3 536 586 (27.10.1970; prior. 25.1.1968).
131 Bernstein, S., et al. (1959) *J. Am. Chem. Soc.* **81**, 1689.
132 American Cyanamid, US 2 789 118 (16.4.1957; prior. 30.3.1956).
133 Squibb, US 3 536 586 (27.10.1970; prior. 25.1.1958).
134 Shue, H.-J. and Green, M.J. (1980) *J. Med. Chem.* **23**, 430.
135 Schering Corp., US 4 124 707 (7.11.1978; prior. 12.12.1976, 7.11.1977).
136 Sears, H.W., et al. (1997) *Clin. Ther.* **19** (4), 710–719.
137 Oliveto, E.P., et al. (1958) *J. Am. Chem. Soc.* **80**, 4431.
138 Merck & Co., DE 1 113 690 (appl. 22.2.1958; USA-prior. 27.2.1957).
139 Olin Mathieson, (1958) US 2 852 511 (prior. 1953).
140 Schering AG, NL-appl. 6 412 708 (appl. 2.11.1964; D-prior. 9.11.1963).
141 Julian, P.L., et al. (1955) *J. Am. Chem. Soc.* **77**, 4601.
142 Schering Corp., US 3 164 618 (5.1.1965; prior. 23.7.1957, 8.5.1958).
143 Syntex, US 3 201 391 (17.8.1965; Mex-prior. 18.2.1959, 20.10.1959).
144 Joly, R., et al. (1974) *Arzneim.-Forsch.* **24**, 1.
145 Roussel-Uclaf, US 3 099 654 (30.6.1963; F-prior. 17.8.1960).
146 Djerassi, C., et al. (1960) *J. Am. Chem. Soc.* **82**, 3399.
147 Syntex, US 3 014 938 (26.12.1961, appl. 23.8.1960; Mex-prior. 7.9.1959).
148 Domenico, A., et al. (1965) *Arzneim.-Forsch.* **15**, 46.
149 Schering AG, DE 1 135 899 (appl. 20.5.1960).
150 Kieslich, K., et al. (1976) *Arzneim.-Forsch.* **26**, 1462.
151 Schering, DE 1 211 194 (27.7.1963) continuation of DE 1 169 444.
152 Syntex, US 3 014 938 (26.12.1961; appl. 23.8.1960; Mex-prior. 7.9.1959).
153 Olin Mathieson, GB 916 996 (appl. 21.7.1959; USA-prior. 6.8.1958).
154 Syntex, US 3 124 571 (10.3.1964; appl. 19.5.1960; Mex-prior. 26.1.1960).
155 Bofors, US 3 929 768 (30.12.1975; appl. 14.5.1973; S-prior. 19.5.1972).
156 Falk Pharma, US 5 932 249 (3.8.1999; appl. 15.7.1996, DE-prior. 23.9.1993).
157 Upjohn, US 3 980 778 (14.9.1976; appl. 20.5.1975; prior. 9.3.1972).
158 American Cyanamid, GB 1 442 925 (USA-prior. 17.8.1973).
159 Ciba-Geigy, US 3 652 554 (28.3.1972; appl. 15.11.1968; CH-prior. 17.11.1967).
160 Shapiro, E.L., et al. (1987) *J. Med. Chem.* **30**, 1581.
161 Schering Corp., EP 57 401 (appl. 25.1.1982; USA-prior. 2.2.1981).
162 Schering Corp., US 4 472 393 (18.9.1984; appl. 29.7.1982; prior. 2.2.1981).
163 Bieber, T., et al. (2007) *Allergy* **62** (29), 184–189.
164 Merck & Co., GB 912 378 (appl. 3.6.1959; USA-prior. 19.6.1958).
165 Glaxo, US 3 312 591 (4.4.1967; GB-prior. 10.5.1963, 28.1.1964).
166 Baghel, V., et al. (2010) *Indian. J. Dermatol.* **76** (5), 591.
167 Phillipps, G.H., et al. (1994) *J. Med. Chem.* **37**, 3717.
168 Glaxo, US 4 335 121 (15.6.1982; appl. 13.2.1981; GB-prior. 15.2.1980).
169 Stache, U., et al. (1985) *Arzneim.-Forsch.* **35** (II), 1753.
170 Hoechst, EP 742 (appl. 27.7.1978; D-prior. 4.8.1977).
171 Gardi, R., et al. (1972) *J. Med. Chem.* **15**, 556.
172 Warner-Lambert, US 3 780 177 (18.12.1973; I-prior. 6.6.1967).
173 Ciba-Geigy, CH 631 185 (appl. 1.1.1978).
174 Glaxo, DE 1 902 340 (appl. 17.1.1969; GB-prior. 19.1.1968).
175 Glaxo, US 3 721 687 (20.3.1973; GB-prior. 4.4.1968).
176 Bernstein, S., et al. (1962) *J. Org. Chem.* **27**, 690.
177 Squibb, DE 2 355 710 (appl. 7.11.1973; USA-prior. 24.11.1972).

Dermatologicals (D), 4. Antiseptics and Disinfectants (D08), Anti-Acne Preparations (D10), and Other Dermatological Preparations (D11)

BERNHARD KUTSCHER, Maintal, Germany

1. Introduction

The general topic of "Dermatologicals" is reviewed in this and two further articles. Please see → Dermatologicals (D), 1. Introduction for a general introduction to the topic of dermatologic drug development [1]. For an overview, please see → Dermatologicals (D), 1. Introduction.

Increasing bacterial resistance limiting the potential efficacy of antibiotics makes the management of superficial skin infections a major challenge [2]. Antiseptics should be distinguished from antibiotics that destroy microorganisms inside the body and from disinfectants that destroy microorganisms found on nonliving objects. However, antiseptics are often referred to as skin disinfectants.

1.1. Antiseptics

Antiseptics with a broader spectrum of antimicrobial activity are alternatives to antibiotics for the prevention and treatment of superficial skin infections with a much lower risk of bacterial resistance selection [3]. Antiseptics are chemical agents applied to the skin to reduce the microbial count and risk of surgical site infections. The use of antiseptics dates back to the mid-1800s when IGNAZ SEMMELWEIS noted a dramatic decrease in periportal sepsis events with appropriate handwashing techniques, later reaffirmed with LISTER's use of carbolic acid (phenol) showing improved infection control [4].

Ullmann's Pharmaceuticals. Edited by Axel Kleemann and Bernhard Kutscher.
© 2022 Wiley-VCH GmbH
Set ISBN: 978-3-527-34252-5/ DOI: 10.1002/14356007.w08_w03

Generally, antiseptics can be classified according to their chemical structure into alcohols (ethyl and isopropyl alcohol), quaternary ammonium compounds (such as benzalkonium chloride, cetylpyridinium chloride, or cetrimide), diguanides (such as chlorhexidine), antibacterial dyes (such as proflavine, euflavine, or gentian violet), peroxides (hydrogen or benzoyl peroxide), inorganic iodine compounds (povidone–iodine), phenol derivatives (phenol, triclosan, hexachlorophene, or chlorocresol), or quinoline derivatives (hydroxyquinolone, dequalium chloride, or chlorquinaldol).

Each antiseptic possesses a specific mechanism of action and a spectrum of microbes targeted. Mechanistically, they are considered in two major classes, small molecules, such as iodine, penetrating bacterial membrane channels and large molecules, such as chlorhexidine, as positively charged bisguanidine binding to the negatively charged cell walls of bacteria causing a disruption in microbial cell membranes. Povidone–iodine, polyhexanide, chlorhexidine, and octenidine are the most widely used antiseptics. Povidone–iodine is a complex of povidone, hydrogen iodide, and elemental iodine that can be considered as a first-choice antiseptic due to its broad microbiocidal activity.

1.2. Anti-Acne Preparations

Acne is a common dermatological disorder characterized by noninflammatory comedones and inflammatory papules and cysts [5]. Acne medications work by reducing sebum oil production, speeding up skin cell turnover, fighting bacterial skin infections, and reducing inflammation in order to help prevent scarring. Comedolytic agents such as retinoids modify the altered epithelial replication process [6].

Retinoids and retinoid-like drugs are derived from vitamin A and include tretinoin, adapalene, and tazarotene in formulations such as creams, gels, and lotions. Antibiotics and antibacterial agents are used for killing excess skin bacteria and reducing redness. In the first few months of treatment, retinoids and antibiotics are used in combination. The only

retinoid used orally to treat severe forms of acne vulgaris and rosacea is isotretinoin. The topical antibiotics such as meclocycline, clindamycin, or erythromycin are often combined with benzoyl peroxide in order to reduce the likelihood of developing antibiotic resistance. Benzoyl peroxide as oxidizing agent is commonly used for topical acne treatment due to its comedolytic and antibacterial properties.

Salicylic acid and the naturally occurring azelaic acid are used in creams alone or in combination with erythromycin for acne treatment. Sulfur preparations containing bithionol, tioxolone, mesulfen, zinc pyrithion, or sulfonamides are also useful in the topical control of acne vulgaris.

1.3. Other Dermatological Preparations

New treatment opportunities of cutaneous pathologies were provided by macrolides inhibiting the phosphatase calcineurin [7] which under normal circumstances induces the transcription of interleukin-2. The most known and used nonantibiotic calcineurin inhibitors are tacrolimus and pimecrolimus [8]. In addition, these drugs inhibit lymphokine production and interleukin release, which leads to a reduced function of effector T-cells [9]. Due to the significant anti-inflammatory and immunomodulatory activity both are used to treat atopic dermatitis and psoriasis as well as vitiligo and rosacea. Unlike topical corticosteroids, these agents do not cause skin thinning.

The US Food and Drug Administration (FDA) has approved two drugs to treat male pattern baldness and alopecia: Minoxidil as liquid or foam and finasteride as tablet. Finasteride as synthetic 4-azasteroid and 5-α-reductase inhibitor [10] works by blocking dihydrotestosterone, which is responsible for damaging hair follicles and causing baldness, whereas minoxidil as vasodilator and opener of potassium channels provides more oxygen, blood, and nutrients to the hair follicles.

2. D08 Antiseptics and Disinfectants

2.1. D08AA Acridine Derivatives (Table 1)

Table 1. D08AA Acridine derivatives

INN (Synonyms) [Brand names] ATC	Structure (Remarks)	[CAS-No.] Formula MW, g/mol	Target (if known) Medical use Formulation DDD	Originator Approval (country/year) Production method [References]
Ethacridine (acrinol, aethacridin) [Acrinol, Dentinox, Metifex, Rivanol] D08AA01; B05CA08		[442-16-0] $C_{15}H_{15}N_3O$ 253.31 lactate salt: [1837-57-6] $C_{15}H_{15}N_3O$ $\cdot C_3H_6O_3$ 343.38	inhibits bacterial protein synthesis binding to DNA and RNA. wound antiseptic and intestinal disinfectant. drg. 200 mg; eye drops 1 mg/g (as free base); tabl. 25 mg; ointment 2 mg/g; powder 99% and over; sol. 0.1%, 0.2%, 0.5%; tabl. 0.1 g.	synthetic [11]
Aminoacridine (aminacrine) [Hepro, Iglu, Medijel, Septogel] D08AA02		[90-45-9] $C_{13}H_{10}N_2$ 194.23 hydrochloride salt: [134-50-9]	topical antiseptic and fluorescent dye. gel.	synthetic [12]
Euflavine (acriflavine) [Burncure, Neutroflavine] D08AA03		[86-40-8] $C_{14}H_{14}ClN_3$ 259.8	antiseptic, antimicrobial, and disinfectant. used against burns. cream 0.1% and foam spray.	synthetic [13]

2.2. D08AC Biguanides and Amidines (Table 2)

Table 2. D08AC Biguanides and amidines

INN (Synonyms) [Brand names] ATC	Structure (Remarks)	[CAS-No.] Formula MW, g/mol	Target (if known) Medical use Formulation DDD	Originator Approval (country/year) Production method [References]
Dibrompropamidine [Brolene, Brulidine, Phenergan] D08AC01; S01AX14		[496-00-4] $C_{17}H_{18}Br_2N_4O_2$ 470.17 diisethionate salt: [614-87-9] $C_{17}H_{18}Br_2N_4O_2$ $\cdot 2\,C_2H_6O_4S$ 722.43	antiseptic, chemotherapeutic. eye drops 0.1%; eye ointment 0.15%.	May & Baker synthetic [14]
Chlorhexidine [Betasept, Chloraprep, Corsodyl; Hibitane; Hydrex; Meridol; Peridex; Plurexid; Unisept] D08AC02; R02AA05; A01AB03; B05CA02; D09AA12; S01AX09; S02AA09; S03AA04 **WHO essential medicine** (→ Respiratory Disorders, 3. Throat Preparations (R02))		[55-56-1] $C_{22}H_{30}Cl_2N_{10}$ 505.46 dihydrochloride: [3697-425] $C_{22}H_{30}Cl_2N_{10}$ $\cdot 2\,HCl$ 578.38 digluconate: [18472-51-0] $C_{22}H_{30}Cl_2N_{10}$ $\cdot 2\,C_6H_{12}O_7$ 897.77	binding to bacterial cell wall. antiseptic, antibacterial, and disinfectant. gel 1 g/100 g; powder 1 g/100 g; sol. 0.1 g/100 g. 0.2 g/100 g, 1 g/50 mL (as digluconate); troche 5 mg.	1952 synthetic [15]

Propamidine
[Brolene]
D08AC03

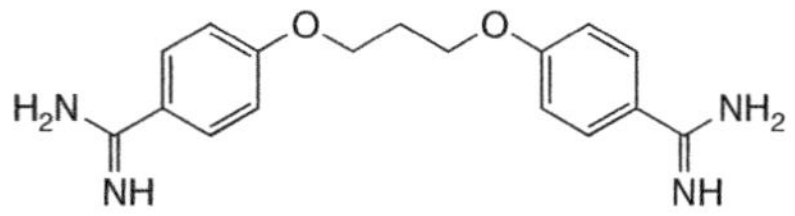

[104-32-5]
$C_{17}H_{20}N_4O_2$
312.32
isethionate salt:
[140-63-6]

disinfectant and antiseptic.
treatment of *Acanthamoeba* infections.
eye drops.

synthetic
[16]

Hexamidine
[Desomedine; Eucerin;
Hexaseptine; Hexomedine]
D08AC04; R01AX07

(→ Respiratory Disorders, 2.
Nasal Preparations and
Decongestants for Topical Use
(R01))

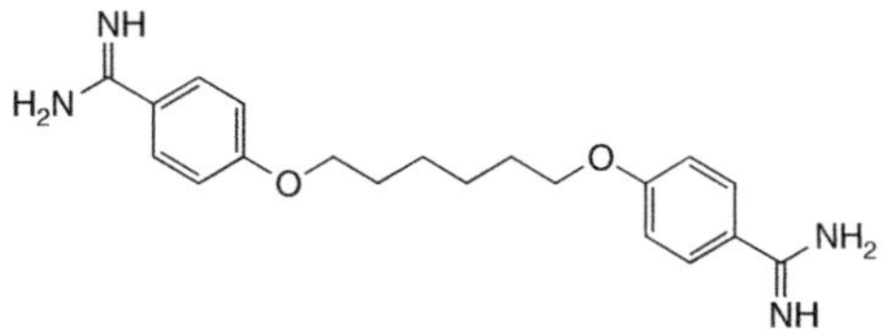

[3811-75-4]
$C_{20}H_{26}N_4O_2$
354.44
diisethionate
salt:
[659-40-5]
dihydrochloride:
[50357-46-5]

antiseptic: disinfectant, preservative.
solution 0.1% and 0.15%; spray
30 mg/mL.

synthetic
[17]

Polyhexanide
(PHMB)
[Lavasept, Omnicide, Prontosan,
Serasept]
D08AC05

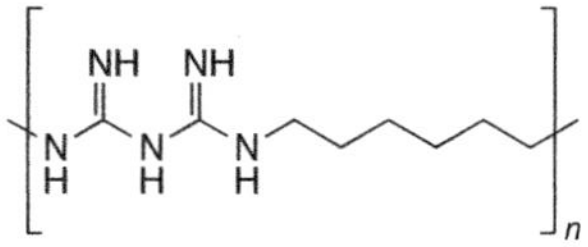

[28757-47-3]
$(C_8H_{17}N_5)_n$
hydrochloride
salt:
[32289-58-0]

polymer used as disinfectant and
antiseptic.
eye drops and solution.

synthetic
[18]

2.3. D08AE Phenol and Derivatives (Table 3)

Table 3. D08AE Phenol and derivatives

INN (Synonyms) [Brand names] ATC	Structure (Remarks)	[CAS-No.] Formula MW, g/mol	Target (if known) Medical use Formulation DDD	Originator Approval (country/year) Production method [References]
Hexachlorophene [Aknefug, Dermalex, Phisohex, Vestene] D08AE01 wfm in most countries		[70-30-4] $C_{13}H_6Cl_6O_2$ 406.91	topical anti-infective, disinfectant, parasiticide. cream 0.5 g/100 g; emulsion 0.5 g/100 g; lotion 0.5 g/100 g.	synthetic [19]
Policresulen [Albothyl, Polilen] D08AE02, G01AX03		[101418-00-2] $C_9H_8O_4S)_n$	topical hemostatic and antiseptic. spray, suppository, and gel 45−55%.	synthetic [20]
Phenol [Chloraseptic; Carmex] R02AA19; D05BB05; D08AE03; N01BX03		[108-95-2] C_6H_6O 91.11	antiseptic, disinfectant. oral spray and rinse.	synthetic [21]
Triclosan (cloxifenol) [Clearasil, Irgasan, Rutisept, Sicorten] D08AE04; D09AA04		[3380-34-5] $C_{12}H_7Cl_3O_2$ 289.55	antimicrobial, antiseptic, and disinfectant. cream 10 mg/g; sol. 0.1 g/100 g.	synthetic [22]
Chloroxylenol (parachlorometaxylenol) [Cortic, Foille, Gehwol, Rinstead, Zeasorb, Zoto] D08AE05		[88-04-0] C_8H_9ClO 156.61	antiseptic and disinfectant. cream 0.33 g/100 g; powder 0.33 g/100 g (combination); sol. 1 g/100 g (combination).	synthetic [23]
Biphenylol (2-phenylphenol) [Dowicide, Fungal, Nipacide, Preventol] D08AE06		[90-43-7] $C_{12}H_{10}O$ 170.2	biocide, fungicide, preservative, and disinfectant. spray and deodorant.	synthetic [24]

2.4. D08AF Nitrofuran Derivatives (Table 4)

Table 4. D08AF Nitrofuran derivatives

INN (Synonyms) [Brand names] ATC	Structure (Remarks)	[CAS-No.] Formula MW, g/mol	Target (if known) Medical use Formulation DDD	Originator Approval (country/year) Production method [References]
Nitrofural (nitrofurazone) [Furacin, Furacillin, Otofuran] D08AF01; B05CA03; D09AA03; P01CC02; S01AX04; S02AA02 (→ Antiparasitics (PC))		[59-87-0] $C_6H_6N_4O_4$ 198.14	antiseptic and topical antibacterial. cream 0.2 g/100 g; ointment 0.2 g/100 g; sol. 0.2 g/100 g.	synthetic [25, 26]

2.5. D08AG Iodine Products (Table 5)

Table 5. D08AG Iodine products

INN (Synonyms) [Brand names] ATC	Structure (Remarks)	[CAS-No.] Formula MW, g/mol	Target (if known) Medical use Formulation DDD	Originator Approval (country/year) Production method [References]
Povidone−Iodine (PVP-I, Iodopovidone) [Betadine; Betaisodona; Isodine; Pyodine; Wokadine] D08AG02; R02AA15; D09AA09; D11AC06; G01AX11; S01AX18 **WHO essential medicine** (→ Respiratory Disorders, 3. Throat Preparations (R02))		[25655-41-8] $(C_6H_9NO)_n \cdot xI$	chemical complex of polyvinylpyrrolidone (povidone), hydrogen iodide, and iodine that directly causes protein denaturation and precipitation of bacteria. antiseptic, antibacterial, and disinfectant. solution 2.5–10%, gels, ointment, eye drops.	1955 synthetic [27, 28]
Diiodohydroxypropane [Iothion, Jothion] D08AG04		[534-08-7] $C_3H_6I_2O$ 311.3	antiseptic and disinfectant.	synthetic [29]

2.6. D08AH Quinoline Derivatives (Table 6)

Table 6. D08AH Quinoline derivatives

INN (Synonyms) [Brand names] ATC	Structure (Remarks)	[CAS-No.] Formula MW, g/mol	Target (if known) Medical use Formulation DDD	Originator Approval (country/year) Production method [References]
Dequalinium [Aperdan; Evazol; Dequadin; Fluomizin; Golandin; Pulmisan] D08AH01; R02AA02; G01AC05 (→ Respiratory Disorders, 3. Throat Preparations (R02))		[522-51-0] $C_{30}H_{40}Cl_2N_4$ 527.58	action via increasing cell permeability with subsequent loss of enzyme activity. antiseptic, bactericidal, disinfectant, antifungal, and antiparasitic. troche 0.25 mg; different tabl., creams, sol. and gels, lozenges.	Glaxo CA/1958 synthetic [30, 31]
Chlorquinaldol [Anginazol; Lacoid; Nerisone] D08AH02; R02AA11 G01AC03; P01AA04 (→ Respiratory Disorders, 3. Throat Preparations (R02))		[72-80-0] $C_{10}H_7Cl_2NO$ 228.08	antimicrobial. cream 10 mg, 130 mg; lozenges.	1954 synthetic [32, 33]
Oxyquinoline (quinolin-8-ol) [Bioquin; Chinosol; Fennosan; NeoChinosol] D08AH03; R02AA14 (→ Respiratory Disorders, 3. Throat Preparations (R02))		[148-24-3] C_9H_7NO 145.2 sulfate: [207386-91-2]	transcription inhibitor. antiseptic, antibacterial, antifungal; disinfectant; pesticide. tablet 0.5 g/1.0 g, ointment, cream; lozenges, liquid.	synthetic [34]
Cloxiquine [Dermofungin] D08AH10		[130-16-5] C_9H_5ClNO 179.6	disinfectant, antituberculosis agent. caps. 250 mg.	synthetic [35]
Clioquinol (iodochlorhydroxyquin) [Aristoform, Vioform] D08AH30; D09AA10; P01AA02 (→ Antiparasitics (PC))		[130-26-7] C_9H_5ClINO 305.5	inhibitor of DNA replication. antifungal and antiprotozoal. cream and 3% ointment.	synthetic [36]

2.7. D08AJ Quaternary Ammonium Compounds (Table 7)

Table 7. D08AJ Quaternary ammonium compounds

INN (Synonyms) [Brand names] ATC	Structure (Remarks)	[CAS-No.] Formula MW, g/mol	Target (if known) Medical use Formulation DDD	Originator Approval (country/year) Production method [References]
Benzalkonium [Alfa; Bactine; Bradosol; Dorithricin; Laudamonium; Biseptine; Osamane; Zephirane] D08AJ01; R02AA16; D09AA11 (→ Respiratory Disorders, 3. Throat Preparations (R02))		[8001-54-5]	dissociation of cellular membrane lipid bilayer that compromises cellular permeability and causes bactericidal effect, mixture of C8 to C18 chains. antiseptic, surfactant, biocide, preservative, disinfectant. nail lacquer 1 oz.; sol. 0.01%,0.02%, 0.025%, 0.05%, 0.1%, 0.2%, 10%, 44%, 1 oz., drops, spray; liquid 78 mg/60 mL	synthetic [37, 38]
Cetrimonium [Lemocin; Nostril; Savlon; Xylonor] D08AJ02; R02AA17 (→ Respiratory Disorders, 3. Throat Preparations (R02))		[57-09-0] $C_{19}H_{42}BrN$ 364.46 hydroxide: [505-86-2] $C_{19}H_{43}NO$ 301.56	antiseptic, surfactant. sol. 117 mg/100 g; tabl. 4 mg; cream 0.5%, gel.	synthetic [39]
Cetylpyridinium (CPC) [Calgel; Cepacol; Dobendan; Lysopaine; Neo Cepacol; Neocoricidin; Pyrisept] D08AJ03; R02AA06 B05CA01; D09AA07 (→ Respiratory Disorders, 3. Throat Preparations (R02))		[123-03-5] $C_{21}H_{38}ClN$ 340.00 monohydrate: [6004-24-6]	rapid bactericidal effect due to binding to bacterial cell wall. antiseptic and antibacterial. eff. tabl. 1.5 mg, 3 mg; lozenge 1.4 mg; sol. 5 mg/10 mL (0.01%, 0.05%); tabl. 2 mg; troche 2 mg.	synthetic [40]
Cetrimide [Cetavlon] D08AJ04	mixture of different quaternary ammonium salts including cetrimonium bromide	[1119-97-7]	antiseptic. 1−3% solution	ICI synthetic [41]
Benzoxonium chloride [Absonal] D08AJ05		[19379-90-9] $C_{23}H_{42}ClNO_2$ 400.0	antiseptic and disinfectant.	synthetic [42]
Didecyldimethylammonium chloride (DDAC) [Amosept, Farmasept, Quaternium-12, Septidil] D08AJ06		[7173-51-5] $C_{22}H_{48}ClN$ 362.0	disruption of intramolecular interactions and lipid bilayers. antiseptic and disinfectant.	synthetic [43]

(continued)

Table 7. (*Continued*)

INN (Synonyms) [Brand names] ATC	Structure (Remarks)	[CAS-No.] Formula MW, g/mol	Target (if known) Medical use Formulation DDD	Originator Approval (country/year) Production method [References]
Benzethonium chloride [Antiseptol; Disylin, Emko; Hyamine; Neostelin; Salaline] D08AJ08; R02AA09 (→ Respiratory Disorders, 3. Throat Preparations (R02))		[*121-54-0*] $C_{27}H_{42}ClNO_2$ 448.09	interaction with and subsequent leakage of microbial cytoplasmatic membranes. antimicrobial, antiseptic, antibacterial, disinfectant, preservative, and surfactant. sol. 0.01%, 0.02%, 0.025%, 0.05%, 0.1%, 10%.	synthetic [44, 45]
Decametho-xine D08AJ10		[*38146-42-8*] $C_{38}H_{74}Cl_2N_2O$ 693.9	antiseptic and disinfectant.	synthetic [46]
Octenidine [Octeniderm, Octenisept; Octenilin] D08AJ57; G01AX66; R02AA21 (→ Respiratory Disorders, 3. Throat Preparations (R02))		[*71251-02-0*] $C_{36}H_{62}N_4$ 550.90 dihydro-chloride: [*70775-75-6*] $C_{36}H_{64}Cl_2N_4$ 623.84	antiseptic, surfactant, and disinfectant. gel and solution 0.1–2%.	synthetic [47]

3. D10 Anti-Acne Preparations

3.1. D10A Anti-Acne Preparations for Topical Use

3.1.1. D10AB Preparations Containing Sulfur (Table 8)

Table 8. D10AB Preparations containing sulfur

INN (Synonyms) [Brand names] ATC	Structure (Remarks)	[CAS-No.] Formula MW, g/mol	Target (if known) Medical use Formulation DDD	Originator Approval (country/year) Production method [References]
Bithionol [Actamer, Bitin] D10AB01; P02BX01 (→ Antiparasitics (PC))	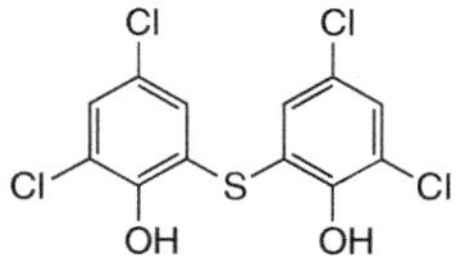	[97-18-7] $C_{12}H_6Cl_4O_2S$ 356.06	inhibitor of adenylyl cyclase. anthelmintic and antiparasitic. soaps and cosmetics.	synthetic [48]
Tioxolone [Acnosan, Loscon, Wasacne] D10AB03	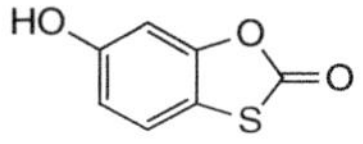	[4991-65-5] $C_7H_4O_3S$ 168.17	carbonic anhydrase inhibitor. antiseborrheic and anti-acne. sol. 200 mg/100 g in comb. with 100 mg benzoxonium chloride.	synthetic [49]
Mesulfen (thianthol) [Citemul, Mitigal] D10AB05; P03AA03 (→ Antiparasitics (PC))	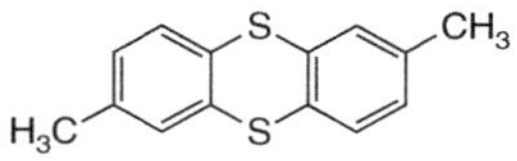	[135-58-0] $C_{14}H_{12}S_2$ 244.38	anti-acne preparation; topical scabicide, antipruritic. ointment 5−25%	synthetic [50, 51]

3.1.2. D10AD Retinoids for Topical Use in Acne (Table 9)

Table 9. D10AD Retinoids for topical use in acne

INN (Synonyms) [Brand names] ATC	Structure (Remarks)	[CAS-No.] Formula MW, g/mol	Target (if known) Medical use Formulation DDD	Originator Approval (country/year) Production method [References]
Tretinoin (retinoic acid, vitamin A acid) [Airol, Curacne, Tretinoin Kefran, Retin-A Micro, Vesanoid] D10AD01; L01XX14 **WHO essential medicine** (→ Antineoplastic and Immunomodulating Agents (L))		[*302-79-4*] $C_{20}H_{28}O_2$ 300.44	retinoid. acne therapeutic, keratolytic. cream 25 mg/100 g, 50 mg/100 g, 100 mg/100 g; cps. 10 mg; gel 25 mg/100 g, 50 mg/100 g; sol. 50 mg/100 mL, 100 mg/100 mL.	1962 synthetic [52–54]
Retinol (axerophthol, vitamin A) [Aklief, Arovit; Arovit Roche; Avibon; Carencyl; Chocola; Differin, Oculotect mono; Panviatan; Solan; Visutein; Vitafluid; Vitagel; Vitamin A Dulcis] D10AD02; A11CA01; R01AX02; S01XA02 **WHO essential medicine** (→ Respiratory Disorders, 2. Nasal Preparations and Decongestants for Topical Use (R01))		[*68-26-8*] $C_{20}H_{30}O$ 286.46] acetate: [*127-47-9*] $C_{22}H_{32}O_2$ 328.50 propionate: [*7069-42-3*] $C_{23}H_{34}O_2$ 342.52 palmitate: [*79-81-2*] $C_{36}H_{60}O_2$ 524.87	epithelial protective vitamin; retinoid. anti-acne. amp. 33.333 mg/1 mL; cps. 2500 iu, 30×10^3 iu, 50×10^3 iu; drg. 10×10^3 iu (as acetate); drops 1000 iu/mL, 10×10^3 iu/g, 40×10^3 iu/mL; emulsion 30×10^3 iu/mL, 300×10^3 iu/g; gel 1000 iu/mL; ointment 250 iu/g, 10×10^3 iu/g; sol. 2%; tabl. 20×10^3 iu.	Roche synthetic [55, 56]

Table 9. *(Continued)*

INN (Synonyms) [Brand names] ATC	Structure (Remarks)	[CAS-No.] Formula MW, g/mol	Target (if known) Medical use Formulation DDD	Originator Approval (country/year) Production method [References]
Adapalene [Differin, Dipalen, Epiduo] D10AD03; D10AD53	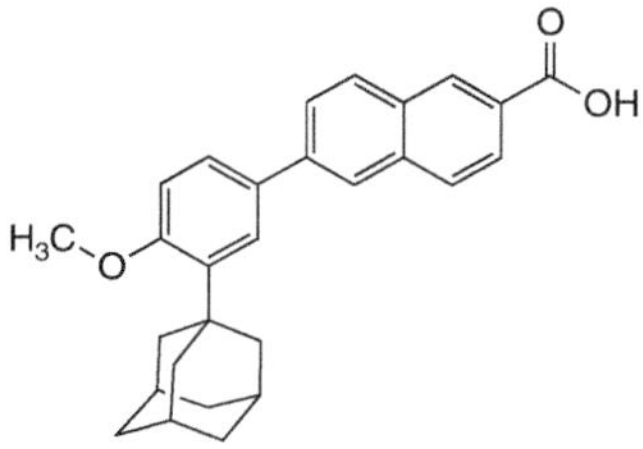	[106685-40-9] $C_{28}H_{28}O_3$ 412.52	third-generation retinoid, inhibits keratinocyte differentiation. anti-acne. cream and gel 0.1%. 1 mg T.	Galderma US/1996 synthetic [57]
Isotretinoin [Accutane, Roaccutane] D10AD04; D10BA01	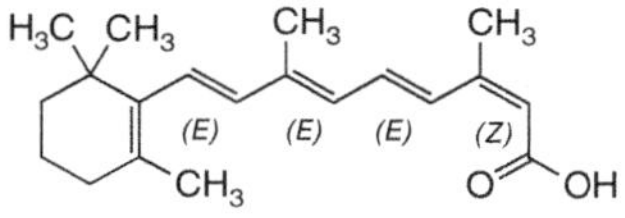	[4759-48-2] $C_{20}H_{28}O_2$ 300.44 sodium salt: [13497-05-7] $C_{20}H_{27}NaO_2$ 322.42	13-*cis*-retinoic acid that induces apoptosis. keratolytic, acne therapeutic, and anticancer drug. cps. 2.5, 5, 10, 20 mg; gel 0.05%.	Roche USA/1982 synthetic [58–60]
Motretinide (RO-11-1430) [Tasmaderm] D10AD05	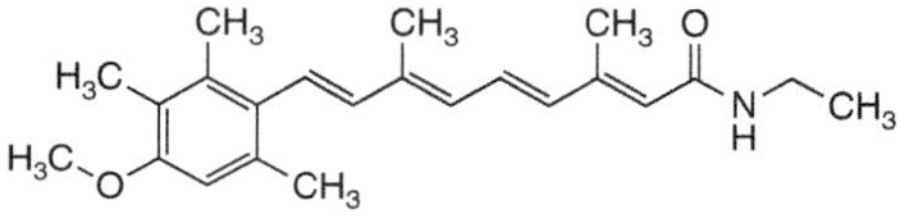	[56281-36-8] $C_{23}H_{31}NO_2$ 353.5	aromatic analog of retinoic acid. Anti-acne agent for topical use. 0.1% cream.	Gebro synthetic [61]

3.1.3. D10AE Peroxides (Table 10)

Table 10. D10AE Peroxides

INN (Synonyms) [Brand names] ATC	Structure (Remarks)	[CAS-No.] Formula MW, g/mol	Target (if known) Medical use Formulation DDD	Originator Approval (country/year) Production method [References]
Benzoyl peroxide [Benzac, Benzaknen, Clearasil, PanOxyl] D10AE01 **WHO essential medicines**	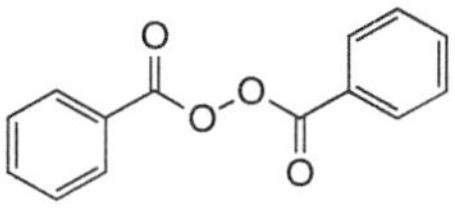	[94-36-0] $C_{14}H_{10}O_4$ 243.2	inhibits growth of *Cutibacterium acnes*. bactericidal and sebostatic; treatment of mild to moderate acne. cream 2.5–10%, gel and liquids. 0.1 g T.	1930 synthetic [62]

3.1.4. D10AF Antiinfectives for Treatment of Acne (Table 11)

Table 11. D10AF Antiinfectives for treatment of acne

INN (Synonyms) [Brand names] ATC	Structure (Remarks)	[CAS-No.] Formula MW, g/mol	Target (if known) Medical use Formulation DDD	Originator Approval (country/year) Production method [References]
Clindamycin [Cleocin, Dalacin, Sobelin, Turimycin] D10AF01; G01AA10; J01FF01 **WHO essential medicine** (→ Antiinfectives for Systemic Use, 1. Antibacterials)		[*18323-44-9*] C$_{18}$H$_{33}$ClN$_2$O$_5$S 424.99 hydrochloride salt: [*21462-39-5*] C$_{18}$H$_{33}$ClN$_2$O$_5$S ·HCl 461.45	antibiotic. topical anti-acne agent. amp. 300 mg/2 mL, 600 mg/4 mL, 900 mg/6 mL; cps. 75, 150, 300 mg; gel 10 mg/g; sol. 10 mg/mL (as hydrochloride or phosphate); vaginal cream 20 mg.	Pharmacia & Upjohn USA/1970 semisynthetic [63, 64]
Erythromycin [Eritrocina, Ery-tab, Erythrocin, Eryacne, Erymax, Llosone] D10AF02; D10AF52; J01FA01; S01AA17 **WHO essential medicine** (→ Antiinfectives for Systemic Use, 1. Antibacterials)		[*114-07-8*] C$_{37}$H$_{67}$NO$_{13}$ 733.94	macrolide antibiotic decreasing bacterial protein production binding to the 50S subunit of the bacterial rRNA complex. anti-acne. cps. 250 mg; eye ointment 0.5%; f.c. tabl. 250 mg, 500 mg; gel 0.5 mg/100 g, 1 g/100 g, 2 g/100 g (2%), 4 g/100 g; sol. 0.2 g/10 g, 1.68 g/100 mL; ointment 1%; spray 20 mg/mL; s.r. tabl. 250, 333, 500 mg; suppos. 250 mg (as free base); tabl. 100 mg, 200 mg.	Eli Lilly 1952 from fermentation solutions of *Streptomyces erythreus* [65, 66]
Chloramphenicol [Chloromycetin, Elase, Mycetin, Posifenicol] D10AF03; G01AA05; J01BA01; S01AA01; S02AA01; S03AA08 **WHO essential medicine** (→ Antiinfectives for Systemic Use, 1. Antibacterials)		[*56-75-7*] C$_{11}$H$_{12}$Cl$_2$N$_2$O$_5$ 323.13	inhibitor of bacterial protein synthesis. antibiotic, anti-acne. amp. 1 g, cps. 250 mg, 500 mg; ear drops 5 g/100 mL, 50 mg/g; eye drops 5 mg, 10 mg; ointment 1% (10 mg/g), 2%; powder 90% and over; sol. 0.5%, 5%; tabl. 50 mg, 250 mg; vaginal tabl. 100 mg.	Parke-Davis US/1950 synthetic [67, 68]

Table 11. (*Continued*)

INN (Synonyms) [Brand names] ATC	Structure (Remarks)	[CAS-No.] Formula MW, g/mol	Target (if known) Medical use Formulation DDD	Originator Approval (country/year) Production method [References]
Nadifloxacin (OPC-7251, jinofloxacin) [Acuatim, Nadixa] D10AF05		[*124858-35-1*] $C_{19}H_{21}FN_2O_4$ 360.39	gyrase inhibitor. antibiotic, topical use against acne. cream, 1 g contains 10 mg nadifloxacin.	Otsuka Pfleger DE/2004 synthetic [69–71]
Sulfacetamide (*N'*-acetylsulfanilamide, sulfacetamide) [Antebor, Blephamide, Klaron lotion] D10AF06; S01AB04		[*144-80-9*] $C_8H_{10}N_2O_3S$ 214.25 sodium salt: [*127-56-0*] $C_8H_9N_2NaO_3S$ 236.23	competitive inhibitor of bacterial *p*-aminobenzoic acid (PABA). chemotherapeutic (eye infection), sulfonamide antibiotic. eye drops 100 mg/mL; ophthalmic ointment 2 mg/g, 100 mg/g (as sodium salt); pessaries 143.75 mg. 20 mg T.	synthetic [72]
Chlortetracycline [Aureomycine; Aureocort] D10AF07; D06AA02; A01AB21; J01AA03; S01AA02 (→ Antiinfectives for Systemic Use, 1. Antibacterials)		[*57-62-5*] $C_{22}H_{23}ClN_2O_8$ 478.89 hydrochloride salt: [*64-72-2*] $C_{22}H_{23}ClN_2O_8 \cdot HCl$ 515.35	antibiotic. cream 10 mg/g, 30 mg/g, 3%; eye ointment 10 mg/g (1%); ointment 30 mg/10 g (3%); pastes 30 mg; pessaries 100 mg (as hydrochloride).	Lederle US/1950 from fermentation solutions of *Streptomyces aureofaciens* [73, 74]
Tetracycline [Achromycin, Amphocycline, Colicort, Sumycin] D10AF08; D06AA04; A01AB13; J01AA07; S01AA09; S02AA08; S03AA02 (→ Antiinfectives for Systemic Use, 1. Antibacterials)		[*60-54-8*] $C_{22}H_{24}N_2O_8$ 444.44 hydrochloride salt: [*64-75-5*] $C_{22}H_{24}N_2O_8 \cdot HCl$ 480.90	inhibits protein synthesis by bacteria binding to the 30S ribosomal unit. antibiotic. cps. 50, 250, 500 mg; cream 30 mg/g; f.c. tabl. 500 mg; ointment 3%; powder 100%; troche 15 mg; vial 500 mg.	Lederle US/1955 from fermentation solutions of *Streptomyces viridifaciens* [75, 76]

4. D11 Other Dermatological Preparations

4.1. D11AH Agents for Dermatitis, Excluding Corticosteroids (Table 12)

Table 12. D11AH Agents for dermatitis, excluding corticosteroids

INN (Synonyms) [Brand names] ATC	Structure (Remarks)	[CAS-No.] Formula MW, g/mol	Target (if known) Medical use Formulation DDD	Originator Approval (country/year) Production method [References]
Tacrolimus (FK-506, FR-900506, fujimycin, L-679934) [Advagraf, Prograf, Protopic] D11AH01; L04AA05 (→ Antineoplastic and Immunomodulating Agents (L))		[104987-11-3] $C_{44}H_{69}NO_{12}$ 804.03	calcineurin inhibitor with subsequent interleukin-2 and T-cell suppression. immunosuppressive amp. 5 mg/1 mL; cps. 0.5 mg, 1 mg, 5 mg; gran. 0.2%; ointment 0.1%	Fujisawa/Astellas US/1994 first isolated from fermentation broth of *Streptomyces tsukubaensis* No. 9993 synthetic [77–79]
Pimecrolimus (SDZ-ASM 981, ASM 981) [Elidel] D11AH02		[137071-32-0] $C_{43}H_{68}ClNO_{11}$ 810.47	calcineurin inhibitor anti-allergic, antipruritic, antipsoriatic, and immunosuppressive cream 1%; 15 g, 30 g, 100 g tubes 20 mg T	Novartis USA/2002 synthetic [80, 81]

Table 12. (*Continued*)

INN (Synonyms) [Brand names] ATC	Structure (Remarks)	[CAS-No.] Formula MW, g/mol Formulation DDD	Target (if known) Medical use Formulation DDD	Originator Approval (country/year) Production method [References]
Cromoglicic acid (acidum cromoglicicum, cromoglycate) [Aarane; Allergocrom; Allergospasmin; Colimune; Intal; Lomipren; Opticrom; Vividrin; Lomudal; Multicron; Nalcron; Opticron; Nalcrom; Cromantal; Intal Inhaler] D11AH03; R01AC01; R03BC01; S01GX01; A07EB01 (→ Respiratory Disorders, 2. Nasal Preparations and Decongestants for Topical Use (R01))	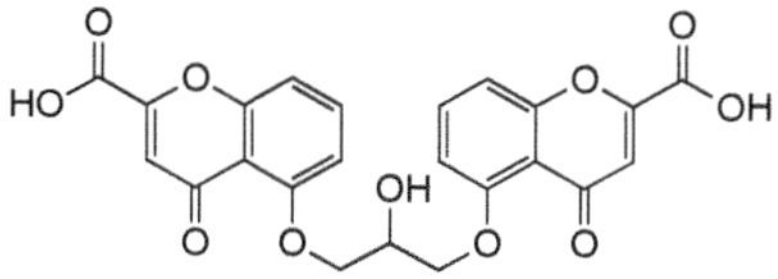	[*16110-51-3*] $C_{23}H_{16}O_{11}$ 468.37 disodium salt: [*15826-37-6*] $C_{23}H_{14}Na_2O_{11}$ 512.33	mast cell stabilizer preventing release of histamine and other mediators of type I allergic reactions anti-allergic and antiasthmatic aerosol 1 mg/0.05 mL, 20 mg/mL; cps. 20 mg, 100 mg; eye drops 20 mg/mL; gran. 10%, 100 mg, 200 mg; nasal drops 20 mg/mL; nasal spray 2.8 mg/0.14 mL, 20 mg/mL; ophthalmic drops 10 mg/0.5 mL, 20 mg/mL (as disodium salt); sol. 20 mg/mL	Fisons synthetic [82]
Alitretinoin (ACRT1057, AGN 192013, LGD 1057, NSC 659772) [Panretin] D11AH04; D11AX19; L01XX22 (→ Antineoplastic and Immunomodulating Agents (L))	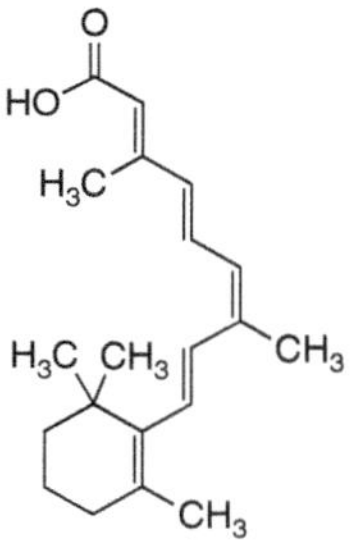	[*5300-03-8*] $C_{20}H_{28}O_2$ 300.44	9-*cis*-retinoic acid; retinoid. topical treatment of cutaneous lesions, antineoplastic. gel 0.1% (w/w). 20 mg O.	Ligand USA/1999 synthetic [83, 84]

4.2. D11AX Other Dermatologicals (Table 13)

Table 13. D11AX Other dermatologicals

INN (Synonyms) [Brand names] ATC	Structure (Remarks)	[CAS-No.] Formula MW, g/mol	Target (if known) Medical use Formulation DDD	Originator Approval (country/year) Production method [References]
Minoxidil [Alostil, Belohair, Loniten, Lonolox, Rogaine] D11AX01; C02DC01 (→ Cardiovascular System (C))		[38304-91-5] $C_9H_{15}N_5O$ 209.25	potassium channel opener. treatment of male hair loss; antihypertensive, vasodilator. topical gel 2%; topical sol. 2%; foam, tabl. 2.5 mg, 10 mg.	Pharmacia & Upjohn USA/1979 synthetic [85–87]
Gamolenic acid (GLA, evening primrose oil, Nachtkerzensamenöl) [Efamast, Epogam] D11AX02		[506-26-3] $C_{18}H_{30}O_2$ 278.44	leukotriene inhibitor. treatment of eczema. cps. 40 mg, 80 mg, 466–536 mg, 932–1073 mg extract of evening primrose seeds. 0.4 g T.	Beiersdorf DE/1990 from fermentation of *Mortierella* or mucor or extraction and isolation from other natural sources (seeds of black currant, evening primrose, borage) [88–90]
Mequinol (4-hydroxyanisole) [Solage] D11AX06		[150-76-5] $C_7H_8O_2$ 124.13	used for skin depigmentation. 2% ethanolic solution in comb. with tretinoin.	synthetic [91]
Tiratricol [Triacana] D11AX08		[51-21-1] $C_{14}H_9I_3O_4$ 621.9	thyroid hormone analogue.	synthetic [92]
Oxaceprol (aceprolinum, N-acetyl-4-hydroxy-L-proline) [AHP 200; Jonctum] D11AX09; M01AX24 (→ Drugs for the Musculo-Skeletal System (M))		[33996-33-7] $C_7H_{11}NO_4$ 173.17	connective tissue therapeutic, antirheumatic. f.c. tabl. 200 mg.	synthetic [93, 94]

Table 13. (*Continued*)

INN (Synonyms) [Brand names] ATC	Structure (Remarks)	[CAS-No.] Formula MW, g/mol	Target (if known) Medical use Formulation DDD	Originator Approval (country/year) Production method [References]
Finasteride [Propecia, Proscar] D11AX10; G04CA01		[98319-26-7] $C_{23}H_{36}N_2O_2$ 372.55	α-reductase inhibitor. treatment of benign prostatic hypertrophy (BPH) and male hair loss. f.c. tabl. 5 mg. 1 mg O.	Merck & Co. USA/1992 synthetic [95, 96]
Hydroquinone [Eldoquin, Lustra, Melanex] D11AX11		[123-31-9] $C_6H_6O_2$ 110.1	skin whitening and depigmentation. antioxidant. cream 2% and 4%.	synthetic [97]
Zinc Pyrithione [Pyrithione Zinc] D11AX12		[13463-41-7] $C_{10}H_8N_2O_2S_2Zn$ 317.7	fungistatic and bacteriostatic. used for seborrheic dermatitis. antidandruff shampoo.	synthetic [98]
Monobenzone [Benoquin, Benoquik, Depigman] D11AX13		[103-16-2] $C_{13}H_{12}O_2$ 200.24	depigmentant melanin inhibitor against hyperpigmentation of skin and vitiligo. cream 20%.	synthetic [99]
Eflornithine (DFMO, RMI-71782) [Ornidyl, Vaniqa] D11AX14; P01CX03 **WHO essential medicine** (→ Antiparasitics (PC))		[70052-12-9] $C_6H_{12}F_2N_2O_2$ 182.17 hydrochloride salt: [68278-23-9] $C_6H_{12}F_2N_2O_2 \cdot HCl$ 218.63	hirsutism treatment inhibitor of ornithine decarboxylase. vial 200 mg/mL (20 g as hydrochloride hydrate), cream 13.9%.	BMS US/2000 synthetic [100–102]
Diclofenac [Arthrotec; Artotec; Novapirina; Optalidon; Voltaren; Voltarol] D11AX18; M01AB05; M02AA15; S01BC03 (→ Drugs for the Musculo-Skeletal System (M))		[15307-86-5] $C_{14}H_{11}Cl_2NO_2$ 296.15 monosodium salt: [15307-79-6] $C_{14}H_{10}Cl_2NNaO_2$ 318.14	COX inhibitor. anti-inflammatory, antirheumatic, and antipyretic. amp. 75 mg; cps. and drg. 25, 50, 100, 140 mg; eye drops 1 mg, 0.3 mg, 5 mg/5 mL, 0.1%; gel 11.6 mg, 1%; inj. sol. 75 mg/3 mL; ointment 50 mg/1 g; s.r. cps. 37.5 mg; suppos. 12.5, 25, 50, 100 mg; tabl. 25, 50, 75 mg. 0.1 g O, P, R.	Ciba Geigy EU 1973/USA 1988 synthetic [103, 104]

(*continued*)

Table 13. (*Continued*)

INN (Synonyms) [Brand names] ATC	Structure (Remarks)	[CAS-No.] Formula MW, g/mol	Target (if known) Medical use Formulation DDD	Originator Approval (country/year) Production method [References]
Brimonidine (UK-14304, UK-14304-08, AGN-190342LF) [Alphagan, Combigan, Lumify, Simbrinza] D11AX21, S01EA05		[*59803-98-4*] $C_{11}H_{10}BrN_5$ 292.14 tartrate salt: [*70359-46-5*] $C_{11}H_{10}BrN_5$ $\cdot C_4H_6O_6$ 442.23	α_2-adrenoceptor antagonist, α-blocker. antihypertensive against ocular-hypertension and rosacea. eye drops 0.2%, ophthalmic solution 0.025% and gel 0.33%. 1.65 mg T.	Allergan 1996 US/2005 synthetic [105–107]
Ivermectin [Ivomec, Sklice, Soolantra, Stromectol] D11AX22; P02CF01 **WHO essential medicine** (→ Antiparasitics (PC))	R = C_2H_5 ivermectin B_{1a} R = CH_3 ivermectin B_{1b}	[*70288-86-7*] $C_{48}H_{74}O_{14}$ 875.1	binds to glutamate-gated chloride channels in invertebrate nerve cells followed by paralysis and death. treatment of parasite infestations such as head lice, scabies, and river blindness. cream, lotion, and tabl. 10 mg T.	MSD USA/1981 semisynthetic, derived from *Streptomyces avermitills* [108]
Estradiol (oestradiol) [Climara, Estraderm, Estrasorb, Menostar, Nuvella, Vagimex] D11AX26; G03CA03 (→ Antineoplastic and Immunomodulating Agents (L))		[*50-28-2*] $C_{18}H_{24}O_2$ 272.39	estrogen that binds as agonist to estrogen and progesterone receptors in the skin. hormone therapy. gel 0.5 mg/g, 1 mg/g; tabl. 2 mg, 4 mg; plaster 0.72, 2, 2.17 4.33 mg; transdermal plaster 0.75, 1.5, 2, 3, 4, 8 mg; vaginal tabl. 0.025 mg.	synthetic [109, 110]

List of Abbreviations

amp.	ampule(s)		INN	international nonproprietary name
			MW	molecular mass
cps.	capsules		N	nasal
COX	cyclooxygenase		mL	milliliter (cubic centimeter)
DDD	defined daily dose		O	oral
drg.	dragee		P	parenteral
eff.	effervescent		PABA	*p*-aminobenzoic acid
f.c.	film-coated		sol.	solution
gran.	granules		s.r.	slow release
inj.	injection		suppos.	suppositories
iu	international units		susp.	suspension
			T	topical

tabl.   tablets
wfm   withdrawn from market
WHO   World Health Organization

References

1 Eaglestein, W.H. and Corcoran, G. (2011) *Arch. Dermatol.* **147** (5), 568–572.
2 Kramer, A., et al. (2018) *Skin Pharmacol. Physiol.* **31** (1), 28–58.
3 Lachapelle, J.-M., et al. (2013) *Clin. Pract.* **10** (5), 579–592.
4 Bednareck, R.S. and Ramsey, M.L. (2019) *Transposition Flaps*, StatPearls.
5 Knop, J., et al. (1982) *Arch. Dermatol. Res.* **274**, 267–275.
6 Krycyk-Poprawa, A., et al. (2020) *Pharmaceutics* **12**, 10.
7 Carr, W.W. (2013) *Pediatr. Drugs* **15** (4), 303–310.
8 Rodriguez-Cerdeira, C., et al. (2012) *Mediat. Inflammat.*, ID563709 doi: 10.1155/2012/563709.
9 Czarnecka-Operacz, M. and Jenerowitz, D. (2012) *J. Ger. Soc. Dermatol.* **10** (3), 167–173.
10 Tian, G., et al. (1994) *Biochemistry* **33** (8), 2291–2296.
11 Hoechst, (1922) DRP 360 421.
12 Bunch, E.A. (1987) *J. Assoc. Off. Anal. Chem.* **70** (3), 560–565.
13 Gandrup, P., et al. (1982) *J. Urol.* **14** (1), 60–61.
14 May & Baker, GB 598 911 (appl. 1945).
15 ICI, GB 705 838 (appl. 1951; valid from 1952).
16 Perrine, D., et al. (1995) *Antimicrob. Agents Chemother.* **39** (2), 339–342.
17 May & Baker Ltd., (1939) UK Patent 507 565.
18 Ansorg, R., et al. (2003) *Arzneimittel-Forschung* **53** (5), 368–371.
19 Bush B.T. (1941) US 2 250 480 (appl. 1939).
20 Salem, H., et al. (2018) *J. Adv. Biomed. Pharm. Sci.* **1**, 85–89.
21 Lister, J., (1867) Antiseptic principle.
22 Geigy, US 3 506 720 (14.4.1970; CH-prior. 22.2.1963).
23 Wiggins Cocker, W. (1944) US 2 350 677 (GB-prior. 1939).
24 Perold, G.W. (1975) *J. South Afr. Chem. Inst.* **28**, 300–303.
25 Eaton Labs, (1947) US 2 416 234 (prior. 1945).
26 Norwich Pharma, US 2 927 110 (1.3.1960; appl. 23.1.1958).
27 Purdue Frederick, US 4 113 857 (12.9.1978, appl. 19.5.1977).
28 Fleischer, W., et al. (1997) *Dermatology* **195** (2), 3–9.
29 Mazzei, M., et al. (1985) *Il Farmaco* **40** (8), 266–272.
30 Allen & Hanburys, GB 745 956 (appl. 1953).
31 Flow Inc., US 4 946 849 (7.8.1990, appl. 10.10.1989).
32 Bourquin, J.-P., et al. (1962) *Arch. Pharm. Ber. Dtsch. Pharm. Ges. (APBDAJ)* **295**, 383.
33 Geigy, (1946) US 2 411 670 (CH-prior. 1942).
34 Sadik, F. (1970) *J. Am. Pharm. Assoc.* 18–24.
35 Zhang, W., et al. (2019) *Cell Death Dis.* **10**, 404.
36 Rohde, W., et al. (1976) *Antimicrob. Agents Chemother.* **10** (2), 243–240.
37 Guyer, et al. (1937) *Helv. Chim. Acta (HCACAV)* **20**, 1462.
38 Ralston, A.W., et al. (1947) *J. Am. Chem. Soc. (JACSAT)* **69**, 2095.
39 Shelton, R.S., et al. (1946) *J. Am. Chem. Soc. (JACSAT)* **68**, 753.
40 Pitten, F.A., et al. (2001) *Arzneimittel-Forschung* **51** (7), 588–595.
41 Ferrer-Luque, C.M., et al. (2014) *Int. J. Oral Sci.* **6** (19), 46–49.
42 Weibel, M.A., et al. (1987) *Arzneimittel-Forschung* **37** (4), 467–471.
43 Anderson, S.E., et al. (2016) *J. Immunotoxicol.* **13** (4), 557–566.
44 Rohm & Haas, (1938) US 2 115 250 (appl. 1936).

45 Rohm & Haas, (1939) US 2 170 111 (appl. 1936).
46 Natalia Derkach, WO 2 017 003 403 (5.1.2017; appl. 2.7.2016).
47 Sedlock, D., et al. (1987) *Antimicrob. Agents Chemother.* **28**, 786–790.
48 Kleinbeulting, S., et al. (2016) *J. Biol. Chem.* **291** (18), 9776–9784.
49 Winthrop, (1943) US 2 332 418 (D-prior. 1938).
50 Barber, H.J. and Smiles, S. (1928) *J. Chem. Soc. (JCSOA9)* 1141.
51 Bayer, DE 365 169 (appl. 1919).
52 Pommer, H. (1960, 1977) *Angew. Chem. (ANCEAD)* **72**, **89**, 811, 437.
53 König, H., et al. (1974) *Arzneim.-Forsch. (ARZNAD)* **24**, 1184.
54 BASF, US 3 006 939 (31.10.1961; D-prior. 17.1.1957).
55 Freyschlag, H., et al. (1965) *Angew. Chem. (ANCEAD)* **77**, 277.
56 Roche, (1947) DE 839 495 (appl. 1949; CH-prior. 1945).
57 Thiboutot, D.M., et al. (2007) *J. Am. Acad. Dermatol.* **57** (5), 791–799.
58 Garbers, C.F., et al. (1968) *J. Chem. Soc. C (JSOOAX)* 1982.
59 Pattenden, G. and Weedon, B.C.L. (1968) *J. Chem. Soc. (JSOOAX)* 1984.
60 Hoffmann-La Roche, US 4 556 518 (3.12.1985; appl. 16.7.1984; USA-prior. 10.12.1982).
61 Zouboulis, C.C. (2003) *Akt. Dermatol.* **29** (2), 49–57.
62 Simonart, T. (2012) *Am. J. Clin. Dermatol.* **13** (6), 357–364.
63 Upjohn, US 3 418 414 (24.12.1968; appl. 31.8.1966).
64 Upjohn, US 3 475 407 (28.10.1969; appl. 22.12.1967).
65 Lilly, (1953) US 2 653 899 (prior. 1952).
66 Abbott, (1958) US 2 823 203 (appl. 1954).
67 Rebstock, M.C., et al. (1949) *J. Am. Chem. Soc. (JACSAT)* **71**, 2458–2468.
68 Parke Davis, (1949) US 2 483 871 (appl. 1948).
69 Ishikawa, H., et al. (1989) *Chem. Pharm. Bull. (CPBTAL)* **37** (8), 2103–2108.
70 Hashimoto, K., et al. (1996) *Chem. Pharm. Bull. (CPBTAL)* **44** (4), 642–645.
71 Otsuka, US 4 399 134 (16.8.1983; appl. 10.11.1981; J-prior. 10.11.1980).
72 Schering Corp., (1946) US 2 411 495 (D-prior. 1938).
73 Duggar, B.M. (1948) *Ann. N. Y. Acad. Sci. (ANYAA9)* **51**, 175.
74 American Cyanamid, (1949) US 2 482 055 (prior. 1948).
75 Conover, L.H., et al. (1963) *J. Am. Chem. Soc. (JACSAT)* **75**, 4622.
76 Conover, L.H. (1955) US 2 699 054 (prior. 1953).
77 Fujisawa Pharmaceutical, EP 184 162 (appl. 11.6.1986; GB-prior. 5.2.1985, 1.4.1985).
78 Ireland, R., et al. (1996) *J. Org. Chem. (JOCEAH)* **61**, 6856.
79 Fujisawa Pharmaceutical, EP 378 318 (appl. 18.7.1990; USA-prior. 11.1.1989, 30.6.1989).
80 Sandoz, DE 3 937 336 (appl. 7.11.1990; D-prior. 9.11.1989).
81 Novartis, US 6 423 722 (23.7.2002; GB-prior. 30.6.1997; USA-prior. 16.12.1999).
82 Fisons, US 3 419 578 (31.12.1968; GB-prior. 25.3.1965, 9.12.1965).
83 Wada, A., et al. (1997) *J. Org. Chem. (JOCEAH)* **62**, 4343–4348.
84 Ligand Pharm., WO 9 532 946 (appl. 24.5.1995; USA-prior. 27.5.1994).
85 McCall, J.M., et al. (1975) *J. Org. Chem. (JOCEAH)* **40**, 3304.
86 Upjohn, DE 1 620 649 (prior. 28.10.1966).
87 Upjohn, US 3 382 247 (7.5.1968; appl. 1.11.1965).
88 Hansson, L., et al. (1989) *Appl. Microbiol. Biotechnol. (AMBIDG)* **31**, 223.
89 Agency of Ind. Sciences and Techn., JP 59 130 191 (appl. 12.1.1983).

90 Efamol, EP 153 134 (appl. 12.2.1985; GB-prior. 21.2.1984).

91 Galderma, US 493 350 (28.2.2007; appl. 13.5.2005; FR-prior. 14.5.2004).

92 Brenta, G., et al. (2003) *J. Clin. Endocrinol. Metab.* **88** (11), 5287–5292.

93 P. and B. Coirre, GB 1 246 141 (appl. 30.8.1968; F-prior. 14.9.1967).

94 Franco-Chimie, S.A.R.L. DAS 1 795 327 (appl. 13.9.1968; F-prior. 14.9.1967).

95 Bhattacharya, A., et al. (1988) *J. Am. Chem. Soc. (JACSAT)* **110**, 3318.

96 Merck & Co., US 4 760 071 (26.7.1988; appl. 21.11.1985; USA-prior. 27.2.1984).

97 Lonza A.G., US 3 355 503 (28.11.1967; appl. 13.8.1964; CH-prior. 15.8.1963).

98 Faergemann, J. (2000) *J. Clin. Dermatol.* **1**, 75–80.

99 Schiff, H. and Pellizzari, G. (1883) *Justus Liebigs Ann. Chem. (JLACBF)* **221**, 365.

100 Metcalf, B.W., et al. (1978) *J. Am. Chem. Soc. (JACSAT)* **100**, 2551.

101 Merrell-Toraude, US 4 413 141 (1.11.1983; appl. 17.9.1982; prior. 11.7.1977, 2.7.1979).

102 Merrell-Toraude, US 4 330 559 (18.5.1982; appl. 3.2.1981; prior. 11.7.1977, 10.4.1979).

103 Geigy, US 3 558 690 (26.1.1971; CH-prior. 8.4.1965, 25.2.1966).

104 Ciba-Geigy, DAS 1 543 639 (appl. 7.4.1966; CH-prior. 8.4.1965).

105 Pfizer, US 3 890 319 (17.6.1975; GB-prior. 29.2.1972, 8.11.1972).

106 Pfizer, US 4 029 792 (14.6.1977; GB-prior. 29.2.1972, 8.11.1972).

107 Allergan, WO 9 510 280 (appl. 19.9.1994; USA-prior. 13.10.1993).

108 Merck & CO., US 4 963 667 (16.10.1990; appl. 19.12.1989; USA-prior.10.11.1987).

109 Inhoffen, H.H. and Zühlsdorff, G. (1941) *Ber. Dtsch. Chem. Ges. (BDCGAS)* **74**, 1911.

110 Schering Corp., (1937) US 2 096 744 (D-prior. 1932).

Genitourinary System and Sex Hormones (G)

BERNHARD KUTSCHER, Maintal, Germany

ULLMANN'S ENCYCLOPEDIA OF INDUSTRIAL CHEMISTRY

Ullmann's Pharmaceuticals. Edited by Axel Kleemann and Bernhard Kutscher.
© 2022 Wiley-VCH GmbH
Set ISBN: 978-3-527-34252-5/ DOI: 10.1002/14356007.w12_w01

1. Introduction: G Genitourinary System and Sex Hormones

The following sections review the different drugs of the genitourinary system and sex hormones. They are arranged according to the Anatomical Therapeutical Chemical (ATC) Classification System — a system of alphanumeric codes developed by the World Health Organization for the classification of drugs and other medical products. Here, the active substances are divided into different groups according to the organ or system on which they act and their therapeutic, pharmacological, and chemical properties. Each drug is given an ATC classification number, specifically for the genitourinary system starting with a G. As new drugs are numbered consecutively in the ATC classification system, novel approved chemical entities (NCEs) can be easily added. Biopharmaceuticals or biologics, e.g., antibodies, polypeptides, and proteins, are not included (except gonadotropins, G03GA). Also, homeopathic, plant-derived, or inorganic preparations are not enclosed in the tables. Focus is on the so-called synthetic "small molecules."

In the tables of the particular chapters, the individual drug substances are described by their INN (international nonproprietary name), synonyms, brand or trade names, ATC number, structure, CAS registry number, formula, molecular weight, mode of action, medical use, formulation, DDD (defined daily dose), originator, country and/or date of approval, production method, and references. The references for each drug typically include the basic patents and publications in which the relevant synthetic

or biosynthetic processes for production are described.

Genitourinary tract agents are medicines which are used to treat conditions of the reproductive organs and excretory system or urinary tract [1]. Urinary refers to the system responsible for removal of nitrogenous waste products (urine) from the bloodstream [2]. The drugs encompass a heterogeneous range of disorders from uterine relaxants and stimulants, urinary antispasmodics and antiinfectives, and hormonal therapy to impotence agents. This article covers antiinfectives (G01), gynecologicals (G02), sex hormones (G03), and urologicals (G04) but excludes drugs against genitourinary cancers (→ Antineoplastic and Immunomodulating Agents (L)).

1.1. Gynecological Antiinfectives and Antiseptics (G01)

This group comprises preparations mainly for local use. Vaginal infections are one of the most common gynecological problems. The vaginal microbiota is constituted by a variety of microorganisms, with lactobacilli being the most prevalent. The normal microbiota plays an important function in vaginal health, preventing colonization and infection by pathogens [3]. However, this population can be disturbed, occurring to a loss of beneficial bacteria due to hormone deficiency, contraceptives, or sexual intercourse, and thus, the vaginal tract can be infected by diverse pathogens. Various antimicrobial agents including antibacterials, antifungals, antiparasital, and antiviral drugs are used to treat diseases such as vaginitis, bacterial vaginosis (BV), vulvovaginal candidiasis (VVC), and urinary tract infections [4]. Depending on the antiinfective agent (→ Antiinfectives for Systemic Use, 1. Antibacterials; → Antiinfectives for Systemic Use, 3. Antivirals; → Antiinfectives for Systemic Use, 2. Antimycotics (Antifungals) and Antimycobacterials), different dosage forms have been developed comprising vaginal douches, gels, creams, ointments, foams, suppositories, tablets, ovules, rings, and vaginal films. Vaginal drug delivery offers several advantages over systemic oral administration, including delivery of high concentrations of antimicrobial at the required site of action, easy self-application, and a reduction in systemic activity or side effects. Nystatin, amphotericin B, candicidin, clindamycin, and neomycin are antibiotics active against candidiasis. Organic acids such as lactic acid, ascorbic acid (vitamin C), and acetic acid are antimicrobials used to treat BV by increasing vaginal acidification [5]. Sulfonamides are the first successfully synthesized toxic antimicrobial drugs [6] and can be administered by oral or topical route. The mechanism of sulfonamides antimicrobial action involves competitive inhibition of p-amino benzoic acid (PABA), the compound essential for the synthesis of folic acid and thus for bacterial growth. Topical sulfonamides such as sulfatolamide and sulfanilamide are used as creams or ointments against vaginal yeast infections.

Antifungal drugs are characterized by a variety of chemical structures and a broad range of mechanism of action. The gynecological intervention typically consists in one of the main classes of antifungals: either azole derivatives applied orally and topically (such as ketoconazole, fluconazole, butoconazole, econazole, or clotrimazole) or polyene antifungal (nystatin and amphotericin B) drugs [7]. Azoles chemically belong to either imidazoles or triazoles classes and exert their antifungal effect by blocking the synthesis of ergosterol, an important component of the fungal cell membrane. Ciclopirox as a pyridone is structurally distinct to other antifungals and inhibits transmembrane transport of ions and substrates essential for cell metabolism.

Antiseptics with a broader spectrum of antimicrobial activity are alternatives to antibiotics for the prevention and treatment of infections with a much lower risk of bacterial resistance selection [8]. Antiseptics are chemical agents applied to the skin to reduce the microbial count and risk of surgical site infections. According to their chemical structure, antiseptics can be classified into alcohols, quaternary ammonium compounds, diguanides (chlorhexidine), antibacterial dyes (proflavine, euflavine, or gentian violet), inorganic iodine compounds (povidone–iodine), phenol derivatives, or quinolone derivatives (dequalinium chloride or chlorquinaldol). Each antiseptic possesses a specific mechanism of action and a spectrum of microbes targeted. Mechanistically they are considered in two major classes, small molecules such as iodine penetrating bacterial membrane channels and large molecules such as chlorhexidine, as positively charged bisguanidine binding to the negatively charged cell walls of bacteria causing a disruption in microbial cell membranes. Metronidazole and clindamycin as well as lactic acid are the antimicrobials used to treat BV. Gynecological application of antiseptics and disinfectants in treatment of BV were also studied [9]. Antifungals such as itraconazole, econazole, clotrimazole, and fluconazole as well as nystatin are administered to treat VVC. Various antibiotics (chloramphenicol and neomycin), antifungal agents (clotrimazole, tinidazole, and nystatin), and antiseptics (povidone–iodine and chlorquinaldol) can be used against vaginitis [10].

Vaginal rings with the antiviral dapivirine for risk reduction and prevention of HIV infection were developed for resource-poor countries.

1.2. Other Gynecologicals (G02)

Uterotonics are drugs used to modify uterine contractions. These include oxytocic drugs that stimulate contractions both by induction of labor and control of postpartum hemorrhage [11]. The ergot alkaloids ergometrine as well as methylergometrine and the natural peptide hormone oxytocin differ in their actions on the uterus. While oxytocin in moderate doses produces slow generalized contractions with full relaxation in between, ergometrine produces faster contractions.

Prostaglandins (PGs) help to control functions in the body such as blood pressure and muscle contractions. The natural PGs dinoprost (PGF2alpha) and dinoprostone (PGE2) are used in obstetric practice to soften and shorten the cervix for preinduction of cervical ripening and management of postpartum uterine atony [12]. Carboprost as a synthetic prostaglandin analogue is used to treat severe bleeding after childbirth (postpartum) and to produce an abortion by causing uterine contractions [13]. Misoprostol is a synthetic prostaglandin E1 analogue that is used for medication abortion, medical management of miscarriage, induction of labor, and cervical ripening [14]. Prostaglandins such as dinoprost, dinoprostone, gemeprost, or misoprostol can be applied as vaginal tablets, gels, or pessaries for labor induction or against bleeding [15].

Nonoxynol-9 is a spermicidal agent and surfactant used as topical (vaginal) contraceptive worldwide. Tocolytic drugs are anticontraction medications that are used to suppress premature labor and immature birth. Tocolytic therapy is provided when delivery would result in premature birth, postponing delivery for several days. Agents used to delay premature uterine activity are β2-adrenergic receptor agonists such as ritodrine, fenoterol, or buphenine [16], as well as the oxytocin receptor antagonist atosiban [17].

The principal role of prolactin in mammals is the regulation of lactation. Prolactin is a peptide hormone that is mainly synthesized and secreted by lactotroph cells in the anterior pituitary gland [18]. Prolactin production by the pituitary is mainly under the inhibitory control of dopamine acting via the dopamine 2 receptor (D2R) expressed in lactotroph cells. Thus, dopamine D2 receptor agonists such as bromocriptine, lisuride, cabergoline, quinagolide, metergoline, and terguride are used as therapeutic inhibitors of prolactin secretion in patients with hyperprolactinemia [19]. Hyperprolactinemia is associated with gonadal dysfunction, including infertility, and reduced libido, as well as long-term complications such as osteoporosis.

Nonsteroidal antiinflammatory drugs (NSAIDs) for vaginal administration comprise ibuprofen, naproxen, benzydamine, and flunoxaprofen. In addition to specific antimicrobial treatment, antiinflammatory drugs applied systemic or topical via vaginal douche could be beneficial in reducing symptoms and speeding up recovery from vulvovaginitis [20]. Flibanserin is a medication approved for the treatment of premenopausal women with hypoactive sexual desire disorder (HSDD). By modulating serotonin and dopamine activity in the brain, flibanserin may improve the balance between these neurotransmitter systems in the regulation of sexual response [21]. Flibanserin was originally developed as antidepressant before being repurposed to treat HSDD [22].

1.3. Sex Hormones and Modulators of the Genital System (G03)

Sex hormones, also known as gonadosteroids and gonadal steroids, are steroid hormones that interact with vertebrate steroid hormone receptors [23]. The sex steroids include androgens (male sex hormones) and estrogens and progestogens (both considered female sex hormones). Natural sex steroids are produced by ovaries or testes or by adrenal glands. Birth control or contraception is a method or device used to prevent pregnancy. Contraceptive drugs describe hormone-based methods available in the form of pills, injections, implants, gels, vaginal rings, or patches and have also established means of emergency contraception. Use of contraceptive drugs can reduce risk on women's health due to early childbearing, unsafe abortion, and infant mortality rate [24].

Introduced in 1960, the oral contraceptive "pill" is a medical innovation that has dramatically transformed generations [25]. Animal experiments in the late 1930s demonstrated that high-dose progesterone could arrest ovulation. Progesterone is a natural female steroid hormone that given orally is rapidly deactivated. However, structural modifications such as eliminations at C-19 led to synthetic progestogens with oral stability and progesterone-like effects. The chemist PAUL DJERASSI synthesized the progestogens (also called progestins) norethisterone and norethynodrel from an extract of the Mexican wild yam root [26], and the concept of ovulation inhibition in women became reality. In early clinical studies with norethynodrel the ovulation was inhibited, but significant intramenstrual vaginal bleeding was also observed.

Subsequently, it was discovered that earlier batches of norethynodrel with better menstrual cycle control were contaminated with the estrogen mestranol. This observation led to the controlled inclusion of estrogens into future combined oral contraceptives (COCs) [27]. Ever since, hormonal contraception in which synthetic hormones, either progestogens alone or in combination with estrogen, prevent ovulation is the innovative contribution made by the pill. The first marketed pill, Enovid, contained 9.85 mg of the progestin norethynodrel and 150 µg of the estrogen mestranol [28]. Modern COCs contain lower hormone doses, e.g., 0.1–3.0 mg of modern progestins such as levonorgestrel and 15–50 µg of estrogens such as estradiol or the most widely used ethinylestradiol. Low-dose estrogen COCs are called "micropills." Progestogen-only pills (POPs) were developed in the early 1970s and are called "minipill" with reduced risk of cardiovascular side effects. The progestins used today such as desogestrel, drospirenone, dienogest, gestodene, norgestimate, or nomegestrol either alone or in combination are much more focused on their progesterone-like drug effect and also address further safety concerns [29].

The pill cleared the way for the introduction of an expanded range of hormone-based contraceptives. The first progestogen-only injectable long-term delivery forms provided a two or three months protection against pregnancy with, e.g., norethindrone enanthate and medroxyprogesterone acetate, respectively [30]. Contraceptive biodegradable implants, patches, and hormonal intrauterine devices, as well as cervical rings, are available with progestogen-only or combined formulations [31]. Emergency contraception is also called the "morning-after pill" or postcoital contraception and can prevent up to 95% of pregnancies when taken within 5 days after intercourse. Emergency contraceptive pills (ECPs) are the antiprogestins or progesterone receptor modulators ulipristal acetate and mifepristone (only approved in China), levonorgestrel-only, or COCs consisting of ethinyl estradiol plus levonorgestrel [32]. COCs seems to be the most popular form of reversible contraception in Europe and the United States due to the benefits of efficacy, control, and low failure rate. Hormonal contraception has been used by more than 100 million women worldwide, has the widest geographic distribution of any method [33], and the market is steadily increasing.

Androgens are produced by the body in greater amounts in males. However, androgens are also present in females in small amounts. Estrogens are necessary for normal sexual development of the female and for regulation of the menstrual cycle during the childbearing years. The ovaries and adrenal glands begin to produce less of these hormones after menopause, and climacteric syndrome occurs. Estrogens alone [34] and in combinations with androgens are prescribed to make up for this lower production of hormones and may relieve signs of menopause, such as hot flashes, unusual sweating, chills, vaginal dryness, or dizziness [35]. Early estrogen and ethinyl estradiol formulations including a patch were used for menopausal hormone therapy (MHT). In 1941, the nonsteroidal estrogen diethylstilbestrol (DES) was approved for treatment of menopausal symptoms [36]. However, due to increased risk for endometrial cancer and/or cardiovascular side effects, the estrogen therapy products got a label restriction. Subsequent evidence, that the addition of a progestin could prevent estrogen-induced endometrial changes, led to various new formulations of estrogens, progestin, and estrogen–progestin combinations for treating vasomotor symptoms and vaginal atrophy associated with menopause [37]. Estrogens such as estradiol, estrone, or dienestrol can be administered orally, transdermally, intramuscularly, intranasally, subcutaneously, or locally as creams or gels with transdermal route being preferred for MHT. Tibolon is a progestogen with selective tissue estrogenic activity and the therapy of choice for women with a history of endometriosis or unwanted side effects with conventional MHT [38]. Prasterone is a natural anabolic androgen sometimes used in MHT and against vaginal atrophy alone or with estrogen. In addition, estrogen–androgen (methyltestosterone or testosterone) combinations are used in this indication.

Gonadotropins are peptide hormones that stimulate gonadal functions such as sex steroid hormone production or gametogenesis in testes and ovaries. Major gonadotropins are glycoproteins produced primarily by the adenohypophysis (pituitary gonadotropins) and placenta (chorionic gonadotropins). According to their specific actions they are named follicle-stimulating hormone (FSH) and luteinizing hormone (LH). The biological activity of gonadotropins suggested that they might be useful for the treatment of infertility [39]. The first commercially available gonadotropin product was a human chorionic gonadotropin (hCG) extract, followed by animal pituitary gonadotropin extracts. These extracts were effective, leading to ovarian stimulation using animal gonadotropins followed by ovulation triggering using hCG. However, ovarian response to animal gonadotropins was only short in duration and prompted the development of human pituitary gonadotropins as well as urinary human menopausal gonadotropin (hMG). In 1962, the first pregnancy in a hypogonadotropic patient following ovulation induction with hMG and final oocyte maturation with hCG was reported and became standard protocol [40]. Ongoing issues with gonadotropins derived from urine donations were overcome by the development of recombinant gonadotropin products. The first recombinant human FSH molecules received approval in 1995/1996 followed by recombinant versions of LH and hCG. Currently available products, including rFSH manufactured using a human cell line and a long-acting FSH preparation, can be injected subcutaneously with ready-to-use pen injection devices [41]. Controlled ovarian stimulation with exogenous gonadotropin administration is an important component of assisted reproductive technology (ART) [42].

Selective estrogen receptor modulators (SERMs) such as cyclofenil or clomifene are small synthetic molecules orally used as gonadotropin stimulants or ovulation inducers and in MHT [43]. Epimestrol as synthetic steroidal estrogen [44] is also used as a component for ovulation induction in combination with luteinizing hormone-releasing hormone (LHRH) (→ Hormone Therapeutics – Systemic Hormonal Preparations, Excluding Reproductive Hormones and Insulins (H)). Further SERMs such as raloxifene, bazedoxifene, lasofoxifen, or ospemifen are used to prevent or treat osteoporosis in postmenopausal women [45]. Antiandrogenes, also known as androgen antagonists or testosterone blockers, are drugs that prevent androgens from mediating their biological effects in the body. Cyproterone is an antiandrogen and progestin medication used in the treatment of early puberty, prostate cancer, and birth control. Danazol and gestrinone are synthetic steroids with antigonadotropic and antiestrogenic activities that act as anterior pituitary suppressants used in the treatment of endometriosis.

1.4. Urologicals (G04)

Lower urinary tract symptoms, in particular storage disorders such as urinary incontinence (UI) as well as bladder underactivity, are major health-related problems that increase with age. UI is 1.5 times more common in women than men, affecting approximately 45% and 77% adult men and women in industrialized countries with predicted pharmacological treatment cost around USD $4.2 billion annually [46]. The most common types of UI in women are stress (involuntary loss of urine with physical exertion or sneezing or coughing), urgency, and mixed incontinence [47]. Antimuscarinics (anticholinergics such as oxybutynin, tolterodine, propiverine, darifenacin, solifenacin, or trospium) remain widely used treatment options available as oral immediate as well as extended release and transdermal formulations. Oxybutynin is the most prescribed antimuscarinic drug against UI. For muscarinic receptors, see the chapter "Spasmolytics" in Ullmann's Encyclopedia, Online Edition [351]. Other drugs, such as the β3-agonist mirabegron, duloxetine as serotonin/noradrenaline reuptake inhibitor, or phosphodiesterase 5 (PDE5) inhibitors, relax the smooth muscle via dual mechanism of action.

Overactive bladder (OAB) syndrome is a chronic medical condition which has tremendous impact on the quality of life in both men and women. This syndrome was shown to be prevalent in European and American populations by 12−17%, and significant budgets are allocated for its medical management [48, 49]. The costs of OAB in the USA alone have been estimated at about USD 66 billion per year [2]. The state-of-the-art pharmacological treatment is the use of antimuscarinic agents such as oxybutynin.

Erectile dysfunction (ED) is one of the most common chronic diseases affecting men, and its prevalence increases with aging. It is also the most frequently diagnosed sexual dysfunction in the older male population [50]. ED in aging males is the result of various factors which exert negative effects on multiple levels in erectile biology [51] such as compromised vascular supply, modulated α-adrenergic receptors in aging arteries, and decreased smooth muscles. Since 1998, phosphodiesterase-5 (PDE-5) inhibitors such as sildenafil, tadalafil, vardenafil, avanafil, and udenafil are commonly used for on-demand or chronic treatment of ED [52]. Sildenafil was the first oral treatment for ED and became a great commercial success under the brand Viagra. With ending patent protection and most of the drugs being generic, the ED market is in decline. The intracavernous application of vasoactive substances such as the prostaglandin E1 alprostadil and papaverine is recommended for second-line treatment in patients who fail to respond to PDE-5 inhibitors. Yohimbine is a peripheral and central α2-blocking agent derived from the bark of an evergreen tree. Other urologicals used for treatment of urinary infections, incontinence, and urine flow or as analgesics and antiseptics are acetohydroxamic acid, phenazopyridine, chlorhexidine, phenyl salicylate, and tolvaptan. Dapoxetine and duloxetine are also applied to treat premature ejaculation. Benign prostatic hyperplasia (BPH) is a common disorder in men with an incidence that increases with age. BPH often requires therapy when patients begin to experience urinary tract symptoms that affect quality of life, such as increased frequency of urination, nocturia, or weak urinary stream. Current management strategies involve lifestyle modifications, pharmacotherapy, phytotherapy, and surgical interventions [53]. Prostatic enlargement with increased age is correlated with smooth muscle hyperplasia and lower urinary tract symptoms (LUTSs). Thus, α-blockers (such as silodosin, tamsulosin, alfuzosin, terazosin, prazosin, or doxazosin), 5-α-reductase inhibitors (dutasteride and azasteroid finasteride), muscarinic receptor antagonists (trospium), and PDE-5 inhibitors (such as tadalafil) are oral treatment options. Fixed-dose combinations are also available for BPH treatment [54]. Other drugs such as the antibiotic mepartricin, the peptide fexapotide, and the phytosteroid β-sitosterol are used in BPH.

2. G01 Gynecological Antiinfectives and Antiseptics

2.1. G01A Antiinfectives and Antiseptics; Excluding Combinations with Corticosteroids

2.1.1. G01AA Antibiotics (Table 1)

Table 1. G01AA Antibiotics

INN (Synonyms) [Brand names] ATC	Structure (Remarks)	[CAS-No.] Formula MW, g/mol	Target (if known) Medical use Formulation DDD	Originator Approval (country/year) Production method [References]
Nystatin [Mycolog, Nystan, Nystatin] G01AA01; A07AA02; D01AA01 (→ Dermatologicals (D), 2. Antifungals (D01), Emollients and Protectives (D02), and Antipruritics (D04) for Dermatological Use)		[1400-61-9] $C_{47}H_{75}NO_{17}$ 926.1	fungicidal antibiotic, antimycotic. cream 100 000 iu/g; drg. 500 000 iu; f.c. tabl. 500 000 iu/g; gel 100 000 iu/g; ointment 100 000 iu; pessaries 100 000 iu; susp. 100 000 iu/mL; tabl. 500 000 iu. 0.1×10^6 iu V.	BMS US/1998 from fermentation solutions of *Streptomyces noursei* [55, 56]
Natamycin (pimaricin) [Natafucin, Pimafucin] G01AA02; A01AB10; A07AA03; D01AA02; S01AA10 (→ Dermatologicals (D), 2. Antifungals (D01), Emollients and Protectives (D02), and Antipruritics (D04) for Dermatological Use)		[7681-93-8] $C_{33}H_{47}NO_{13}$ 665.73	fungicidal antibiotic, antimycotic. cream 2 g/100 g; drg. 100 mg; eye drops 5%; eye ointment 1%; ointment 10 mg/g, 2%; tabl. 10 mg. 25 mg V.	Gist/American Cyanamide from fermentation solutions of *Streptomyces natalensis* [57, 58]
Amphotericin B [Abelcet, Ambisome, Fungillin, Fungizone, Mysteclin] G01AA03; A01AB04; A07AA07; D01AA10; J02AA01 (→ Dermatologicals (D), 2. Antifungals (D01), Emollients and Protectives (D02), and Antipruritics (D04) for Dermatological Use)		[1397-89-3] $C_{47}H_{73}NO_{17}$ 924.09	antimycotic, antibiotic. caramels 10 mg; cream 30 mg/g; ointment 30 mg/1 g; powder 50 mg; susp. 100 mg, 500 mg; syrup 10%; tabl. 10 and 100 mg; liposome-encapsulated amphotericin B in a complex with dimyristoyl phosphatidylcholine and dimyristoyl phosphatidylglycerol, vial 20 mL; vial 50 mg. 0.2 g V.	Gilead 1999 fermentation from *Streptomyces nodosus* [59]

Candicidin

[Vanobid]

G01AA04

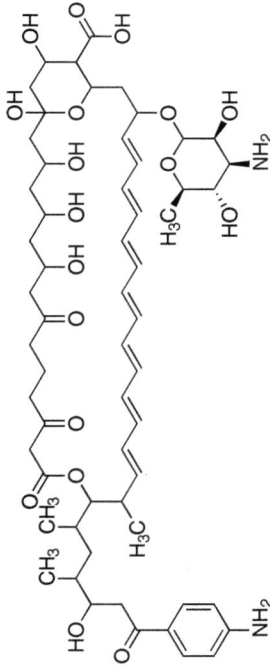

[1403-17-4]
for mixture of
candicidin A–D
[39372-30-0]
$C_{59}H_{84}N_2O_{18}$
1109.32
for candicidin D

macrolide antibiotic and
fungicide.
intravaginal administered for
treatment of vulvovaginal
candidiasis.
6 mg V.

isolated from
Streptomyces griseus
[60]

Chloramphenicol

[Chloromycetin, Elase,
Mycetin, Posifenicol]

G01AA05; D10AF03;
J01BA01; S01AA01;
S02AA01; S03AA08

WHO essential medicine

(→ Dermatologicals (D). 4.
Antiseptics and Disinfectants
(D08), Anti-Acne
Preparations (D10), and
Other Dermatological
Preparations (D11);
→ Antiinfectives for
Systemic Use, 1.
Antibacterials)

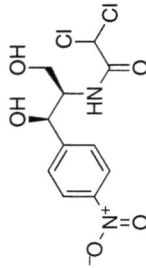

[56-75-7]
$C_{11}H_{12}Cl_2N_2O_5$
323.13

inhibitor of bacterial protein
synthesis.
antibiotic, antiacne.
amp. 1 g. cps. 250 mg,
500 mg; ear drops
5 g/100 mL, 50 mg/g; eye
drops 5 mg, 10 mg; ointment
1% (10 mg/g), 2%; powder
90% and over; sol. 0.5%, 5%;
tabl. 50 and 250 mg; vaginal
tabl. 100 mg.

Parke-Davis
US/1950
synthetic
[61, 62]

(continued)

Table 1. (*Continued*)

INN (Synonyms) [Brand names] ATC	Structure (Remarks)	[CAS-No.] Formula MW, g/mol	Target (if known) Medical use Formulation DDD	Originator Approval (country/year) Production method [References]
Hachimycin (trichomycin) [Tricomycin] G01AA06; D01AA03; JAA02 (→ Dermatologicals (D), 2. Antifungals (D01), Emollients and Protectives (D02), and Antipruritics (D04) for Dermatological Use)	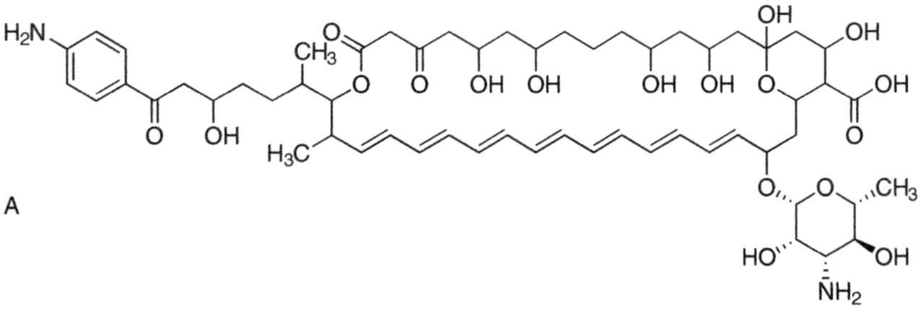 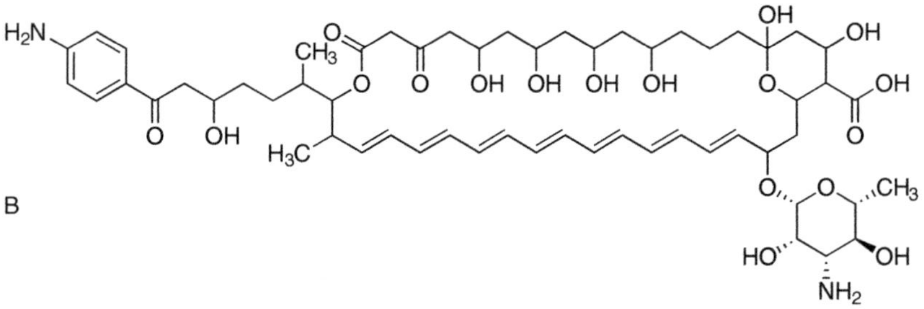	[1394-02-1] $C_{116}H_{168}N_4O_{36}$ 2194.58	polyene macrolide antibiotic. antifungal agent.	derived from *Streptomyces* [63]

Oxytetracycline

[Posicycline; Sterdex; Terramycin, Urobiotic]

G01AA07; D06AA03; J01AA06; S01AA04

(→ Dermatologicals (D). 3. Antipsoriatics (D05), Antibiotics and Chemotherapeutics (D06), and Corticosteroids (D07) for Dermatological Use;

→ Antiinfectives for Systemic Use, 1. Antibacterials)

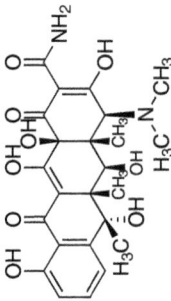

[79-57-2]
$C_{22}H_{24}N_2O_9$
460.44
hydrochloride salt:
[2058-46-0]
$C_{22}H_{24}N_2O_9 \cdot HCl$
496.90

inhibits protein synthesis of bacteria binding to the 30S ribosomal unit. antibiotic.

cps. 250 mg, 500 mg; dental insertion 5 mg; ear drops 5 mg (comb. with hydrocortisone); eye drops 5 mg (comb. with hydrocortisone); eye ointment 10 mg/g; eye ointment 5 mg (comb. with polymyxin B); ointment 10 mg/g; ointment 30 mg (comb. with hydrocortisone or polymyxin B); spray 300 mg (comb. with hydrocortisone); vial 5 mL (as hydrochloride).

Pfizer
US/1950
from fermentation solutions of *Streptomyces rimosus*.
[64, 65]

Carfecillin

[Uricillina, Urpcarf]

G01AA08

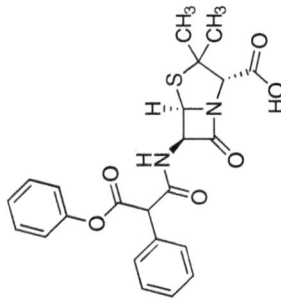

[27025-49-6]
$C_{23}H_{22}N_2O_6S$
454.50

β-lactam antibiotic. prodrug.
tabl. 500 mg.

synthetic
[66, 67]

(continued)

Table 1. (*Continued*)

INN (Synonyms) [Brand names] ATC	Structure (Remarks)	[CAS-No.] Formula MW, g/mol	Target (if known) Medical use Formulation DDD	Originator Approval (country/year) Production method [References]
Mepartricin (methylpartricin) [Tricandil] G01AA09; A01AB16; D01AA06; (→ Dermatologicals (D), 2. Antifungals (D01), Emollients and Protectives (D02), and Antipruritics (D04) for Dermatological Use)	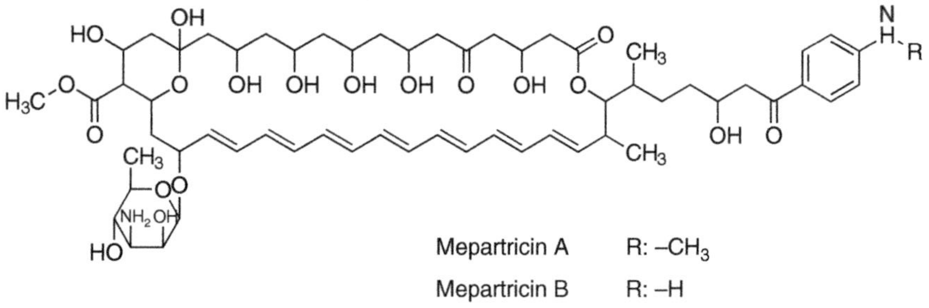 Mepartricin A R: –CH$_3$ Mepartricin B R: –H	[11121-32-7] C$_{60}$H$_{88}$N$_2$O$_{19}$ 1141.36	polyene antibiotic. treatment of candidal and trichomonal gynecological infections, treatment of benign prostatic hypertrophy. tabl. 50 000 iu/g, 40 mg; vaginal cream 5000 iu/g; vaginal tabl. 25 000 iu.	Spa 1975 fermentation of *Streptomyces aureofaciens* [68, 69]
Clindamycin [Cleocin, Dalacin, Sobelin, Turimicin] G01AA10; D10AF01; J01FF01 **WHO essential medicine** (→ Dermatologicals (D), 4. Antiseptics and Disinfectants (D08), Anti-Acne Preparations (D10), and Other Dermatological Preparations (D11); → Antiinfectives for Systemic Use, 1. Antibacterials	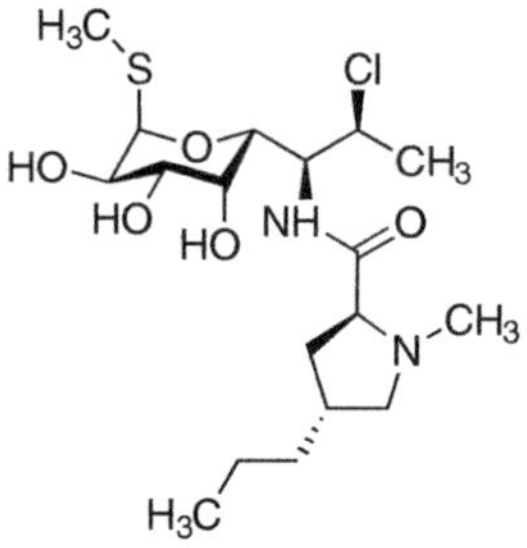	[18323-44-9] C$_{18}$H$_{33}$ClN$_2$O$_5$S 424.99 hydrochloride salt: [21462-39-5] C$_{18}$H$_{33}$ClN$_2$O$_5$S · HCl 461.45	antibiotic. gyno antiinfective and topical anti-acne agent. amp. 300 mg/2 mL, 600 mg/4 mL, 900 mg/6 mL; cps. 75 mg, 150 mg, 300 mg; gel 10 mg/g; sol. 10 mg/mL (as hydrochloride or phosphate); vaginal cream 20 mg. 0.1 g V.	Pharmacia & Upjohn US/1970 semisynthetic [70, 71]

Pentamycin

(fungichromin)

[Femi-Fect, Pruri-Ex, RespiFect]

G01AA11

[6834-98-6]
C$_{35}$H$_{58}$O$_{12}$
670.83

macrolide antibiotic.
antifungal.
treatment of vaginal infection (vaginitis).
tabl., 3 mg vaginal pessary; dry powder inhalation.

isolated from *Streptomyces pentaticus* [72]

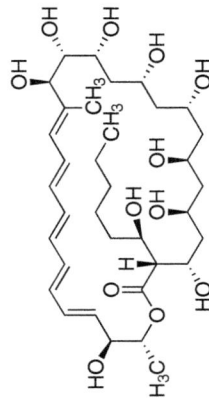

Neomycin

[Myciguent; Mycifradin; Neo-rx; Neo-Tab; Neosporin]

G01AA14; A01AB08; B05CA09; J01CB05; R01AX08; R02AB01; S01AA03; S03AA01

(→ Antiinfectives for Systemic Use, 1. Antibacterials;
→ Respiratory Disorders. 3. Throat Preparations (R02))

[1404-04-2]
C$_{23}$H$_{46}$N$_6$O$_{13}$
614.64
sulfate:
[1405-10-3]

aminoglycoside antibiotic (mixture of isomers) blocking protein biosynthesis of bacteria while binding to the 30S subunit of the ribosome.
antibacterial.
creams, ointment 0.5%, and eye drops.

Upjohn
US/1951
fermentation
[73]

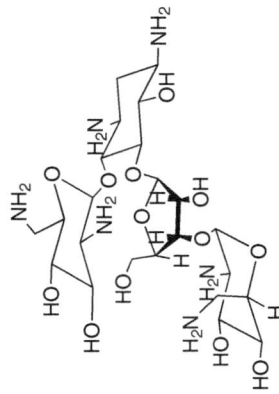

2.1.2. G01AB Arsenic Compounds (Table 2)

Table 2. G01AB Arsenic compounds

INN (Synonyms) [Brand names] ATC	Structure (Remarks)	[CAS-No.] Formula MW, g/mol	Target (if known) Medical use Formulation DDD	Originator Approval (country/year) Production method [References]
Acetarsol (Acetarsone) [Gynoplix, Polygynax, Sanogyl, Stovarsol] G01AB01; A07AX02; P01CD (→ Antiparasitics (PC))		[97-44-9] $C_8H_{10}AsNO_5$ 275.09 sodium salt: [5892-48-8] $C_8H_9AsNNaO_5$ 297.07	antiprotozoal. collutorium (mouth wash) 0.5 mg/100 g. 0.5 g V.	synthetic [74, 75]

2.1.3. G01AC Quinoline Derivatives (Table 3)

Table 3. G01AC Quinoline derivatives

INN (Synonyms) [Brand names] ATC	Structure (Remarks)	[CAS-No.] Formula MW, g/mol	Target (if known) Medical use Formulation DDD	Originator Approval (country/year) Production method [References]
Diiodohydroxyquinoline (diiodohydroxyquin, iodoquinol) [Diiodoquin, Entero-sediv, Ioquin, Vytone] G01AC01		[83-73-8] $C_9H_5I_2NO$ 396.95	intestinal antiseptic, antiamebic. cream 1% (comb. with hydrocortisone); tabl. 210 and 650 mg. 0.2 g V.	synthetic [76]
Clioquinol (iodochlorhydroxyquin) [Aristoform, Vioform] G01AC02: D08AH30: D09AA10 (→ Dermatologicals (D), 4. Antiseptics and Disinfectants (D08), Anti-Acne Preparations (D10), and Other Dermatological Preparations (D11))		[130-26-7] C_9H_5ClINO 305.5	inhibitor of DNA replication. antifungal and antiprotozoal. cream and 3% ointment.	synthetic [77]
Chlorquinaldol [Anginazol; Lacoid; Nerisone] G01AC03: D08AH02: R02AA11: P01AA04 (→ Dermatologicals (D), 4. Antiseptics and Disinfectants (D08), Anti-Acne Preparations (D10), and Other Dermatological Preparations (D11); → Respiratory Disorders, 3. Throat Preparations (R02); (→ Antiparasitics (PC))		[72-80-0] $C_{10}H_7Cl_2NO$ 228.08	antimicrobial. cream 10 mg, 130 mg: lozenges. 0.2 g V.	1954 synthetic [78, 79]

(continued)

Table 3. (*Continued*)

INN (Synonyms) [Brand names] ATC	Structure (Remarks)	[CAS-No.] Formula MW, g/mol	Target (if known) Medical use Formulation DDD	Originator Approval (country/year) Production method [References]
Dequalinium [Aperdan; Evazol; Dequadin; Fluomizin; Golandin; Pulmisan] G01AC05: D08AH01; R02AA02 (→ Dermatologicals (D), 4. Antiseptics and Disinfectants (D08), Anti-Acne Preparations (D10), and Other Dermatological Preparations (D11); → Respiratory Disorders, 3. Throat Preparations (R02))		[522-51-0] $C_{30}H_{40}Cl_2N_4$ 527.58	action via increasing cell permeability with subsequent loss of enzyme activity. antiseptic, bactericidal, disinfectant, antifungal, antiparasitic. troche 0.25 mg; different tabl., creams, sol. and gels, and lozenges.	Glaxo CA/1958 synthetic [80, 81]
Broxyquinoline [Entercine, Fenilor, Intestopan, Sandoin] G01AC06: A07AX01; P01AA01 (→ Antiparasitics (PC))		[521-74-4] $C_9H_5Br_2NO$ 302.95	intestinal antiseptic. ointment 1.5%. 0.1 g V.	synthetic [82, 83]

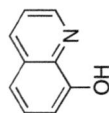

Oxyquinoline
(quinolin-8-ol)
[Bioquin, Chinosol,
Fennosan,
NeoChinosol,
Uveline]

G01AC30;
D08AH03;
R02AA14

(→ Dermatologicals
(D), 4. Antiseptics
and Disinfectants
(D08), Anti-Acne
Preparations (D10),
and Other
Dermatological
Preparations (D11);
→ Respiratory
Disorders, 3. Throat
Preparations (R02))

[148-24-3]
C_9H_7NO
145.2
sulfate:
[207386-91-2]

transcription
inhibitor.
antiseptic,
antibacterial,
antifungal;
disinfectant;
pesticide.
tabl. 0.5 g/1.0 g,
ointment, cream,
lozenges, liquid.

synthetic
[84]

2.1.4. G01AD Organic Acids (Table 4)

Table 4. G01AD Organic acids

INN (Synonyms) [Brand names] ATC	Structure (Remarks)	[CAS-No.] Formula MW, g/mol	Target (if known) Medical use Formulation DDD	Originator Approval (country/year) Production method [References]
Lactic acid (AHA) [Lactinol, Lactovagan] G01AD01		[50-21-5] $C_3H_6O_3$ 90.1	endogenous agonist for hydroxycarboxylic receptor 1. used in topical preparations and cosmetics to adjust acidity and for its keratolytic properties. component of Ringer's or dialysis solutions and biodegradable polymer precursor. lotion 1 g/10 mL, cream 100 mg/g or injection concentrate.	Boehringer Ingelheim 1895 by bacterial fermentation from corn or beets and chemical synthesis [85, 86]
Acetic acid [Borofair] G01AD02 **WHO essential medicine**		[64-19-7] $C_2H_4O_2$ 60.05	antiinfective and antiseptic. antibiotic for the ear canal (otitis), skin infections, and as irrigant solution. solution 20.65 mg/1 mL and drops 2 mg/1 mL	produced by oxidation of ethanol, fermentation, or distillation from wood [87]
Ascorbic acid (ascorbate, vitamin C) [Acerotab, Ascor, Redoxon] G01AD03 **WHO essential medicine**		[50-81-7] $C_6H_8O_6$ 176.12	antioxidant and essential nutrient involved in tissue repair and production of neurotransmitters. antimicrobial; dietary supplement used to prevent and treat scurvy. caps., tabl., injection solution. 0.25 g V.	produced by fermentation from glucose and sorbose or by chemical synthesis [88]

2.1.5. G01AE Sulfonamides (Table 5)

Table 5. G01AE Sulfonamides

INN (Synonyms) [Brand names] ATC	Structure (Remarks)	[CAS-No.] Formula MW, g/mol	Target (if known) Medical use Formulation DDD	Originator Approval (country/year) Production method [References]
Sulfatolamide [Marbaletten, Sulfomyl] G01AE01		[1161-88-2] $C_7H_{10}N_2O_2S$ $\cdot C_7H_9N_3O_2S_2$ 417.54	antibacterial. combination of sulfathiourea and mafenide used in gynecology. cream 8.5%	synthetic [89, 90]

2.1.6. G01AF Imidazole Derivatives (Table 6)

Table 6. G01AF Imidazole derivatives

INN (Synonyms) [Brand names] ATC	Structure (Remarks)	[CAS-No.] Formula MW, g/mol	Target (if known) Medical use Formulation DDD	Originator Approval (country/year) Production method [References]
Metronidazole [Flagyl, Metrogel, Plagyl] G01AF01; D06BX01; A01AB17; J01XD01; P01AB01 **WHO essential medicine** (→ Dermatologicals (D), 3. Antipsoriatics (D05), Antibiotics and Chemotherapeutics (D06), and Corticosteroids (D07) for Dermatological Use; → Antiinfectives for Systemic Use, 1. Antibacterials; → Antiparasitics (PC))		[443-48-1] $C_6H_9N_3O_3$ 171.16	inhibits nucleic acid synthesis by disrupting DNA in microbial cells. chemotherapeutic (trichomonas), antiinfective. cps. 250 mg, 375 mg; f.c. tabl. 250 and 400 mg; suppos. 100 mg (vaginal); tabl. 250 and 400 mg; vaginal tabl. 100 and 250 mg; vial 5 g/1 000 mL. 0.5 g V.	Rhone-Poulenc/GD Searle FR/1960 US/1963 synthetic [91]
Clotrimazole [Canesten, Empecid, Lotrimin] G01AF02; A01AB18; D01AC01 (→ Dermatologicals (D), 2. Antifungals (D01), Emollients and Protectives (D02), and Antipruritics (D04) for Dermatological Use)		[23593-75-1] $C_{22}H_{17}ClN_2$ 344.85	blocking the synthesis of ergosterol, an important component of the fungal cell membrane. antifungal, antimycotic. ointment 1%; pessaries 100 mg, 200 mg, powder 10 mg/g (1%); 500 mg; sol. 1%, 10 mg; spray 10 mg/mL (1%); topical cream 1%, 10 mg; troche 10 mg; vaginal cream 10 mg (2%), 20 mg/mL (10%), 100 mg; vaginal tabl. 100, 200, and 500 mg. 0.1 g V.	Bayer/Schering 1973/1976 synthetic [92, 93]
Miconazole [Daktar, Gyno-Daktar, Daktarin, Fungoid] G01AF04; A01AB09; A07AC01; D01AC02; J02AB01; S02AA13 (→ Dermatologicals (D), 2. Antifungals (D01), Emollients and Protectives (D02), and Antipruritics (D04) for Dermatological Use; → Antiinfectives for Systemic Use, 2. Antimycotics (Antifungals) and Antimycobacterials)		[22916-47-8] $C_{18}H_{14}Cl_4N_2O$ 416.14 mononitrate: [22832-87-7] $C_{18}H_{14}Cl_4N_2O$ · HNO_3 479.15	ergosterol biosynthesis inhibitor. topical antifungal, antimycotic. amp. 200 mg/20 mL; cream 1%, 2 g/100 g, 20 mg/g; lotion 1%; ointment 1%; oral gel 2%; powder 2 g/100 g, 20 mg/g (as mononitrate); sol. 20 mg/mL; suppos. 100 mg; tabl. 250 mg (as free base); vaginal cream 20 mg/g; vial 400 mg/40 mL. 0.1 g V.	Janssen 1974 synthetic [94, 95]

(continued)

Table 6. (*Continued*)

INN (Synonyms) [Brand names] ATC	Structure (Remarks)	[CAS-No.] Formula MW, g/mol	Target (if known) Medical use Formulation DDD	Originator Approval (country/year) Production method [References]
Econazole [Ecorex, Epipevisone, Gyno-Pevaryl, Palavale] G01AF05; D01AC03 (→ Dermatologicals (D), 2. Antifungals (D01), Emollients and Protectives (D02), and Antipruritics (D04) for Dermatological Use)		[27220-47-9] $C_{18}H_{15}Cl_3N_2O$ 381.69 mononitrate: [24169-02-6] $C_{18}H_{15}Cl_3N_2O$ · HNO_3 444.70	ergosterol biosynthesis inhibitor. antifungal, antimycotic. cream 1%; lotion 1 g/100 g; pastes 10 mg; powder 1 g/100 g; sol. 1%; spray 1 g/100 g (as nitrate); suppos. 50 mg. 0.1 g V.	Janssen 1974 synthetic [96–98]
Ornidazole [Xynor] G01AF06; J01RA05; J01RA09; J01RA12; J01XD03; P01AB03 (→ Antiparasitics (PC))		[16773-42-5] $C_7H_{10}ClN_3O_3$ 219.63	antibiotic. used against protozoal infections.	synthetic [99]
Isoconazole [Adestan, Travogen] G01AF07; D01AC05 (→ Dermatologicals (D), 2. Antifungals (D01), Emollients and Protectives (D02), and Antipruritics (D04) for Dermatological Use)		[27523-40-6] $C_{18}H_{14}Cl_4N_2O$ 416.14 mononitrate: [24168-96-5] $C_{18}H_{14}Cl_4N_2O$ · HNO_3 479.15	ergosterol biosynthesis inhibitor. antifungal, antimycotic. cream 1%; pessaries 100, 300, and 600 mg (as nitrate); spray 10 mg/mL; vaginal tabl. 100 and 300 mg. 0.6 g V.	Janssen EU/1979 synthetic [100–102]
Tioconazole [Mykontral, Trosyd, Trosyl] G01AF08; D01AC07 (→ Dermatologicals (D), 2. Antifungals (D01), Emollients and Protectives (D02), and Antipruritics (D04) for Dermatological Use)		[65899-73-2] $C_{16}H_{13}Cl_3N_2OS$ 387.72	ergosterol biosynthesis inhibitor. antimycotic, topical antifungal. cream 10 mg/g; lotion 10 mg/g; ointment vaginal 6.5%; powder 1 g/100 g; spray 1 g/100 g. 0.3 g V.	Pfizer/Combe US/1998 synthetic [103, 104]
Ketoconazole [Ketoderm, Nizoral, Terzolin, Triatop] G01AF11; D01AC08; J02AB02 (→ Dermatologicals (D), 2. Antifungals (D01), Emollients and Protectives (D02), and Antipruritics (D04) for Dermatological Use; → Antiinfectives for Systemic Use, 2. Antimycotics (Antifungals) and Antimycobacterials)		[65277-42-1] $C_{26}H_{28}Cl_2N_4O_4$ 531.44	ergosterol biosynthesis inhibitor. antimycotic. cream 2%; shampoo 2%; sol. 20 mg/mL; susp. 100 mg; tabl. 200 mg. 0.4 g V.	Janssen/J&J US/1996 synthetic [105–107]

Table 6. (*Continued*)

INN (Synonyms) [Brand names] ATC	Structure (Remarks)	[CAS-No.] Formula MW, g/mol	Target (if known) Medical use Formulation DDD	Originator Approval (country/year) Production method [References]
Fenticonazole [Falvin, Fenizolan, Lomexin, Terlomexin] G01AF12; D01AC12 (→ Dermatologicals (D), 2. Antifungals (D01), Emollients and Protectives (D02), and Antipruritics (D04) for Dermatological Use)	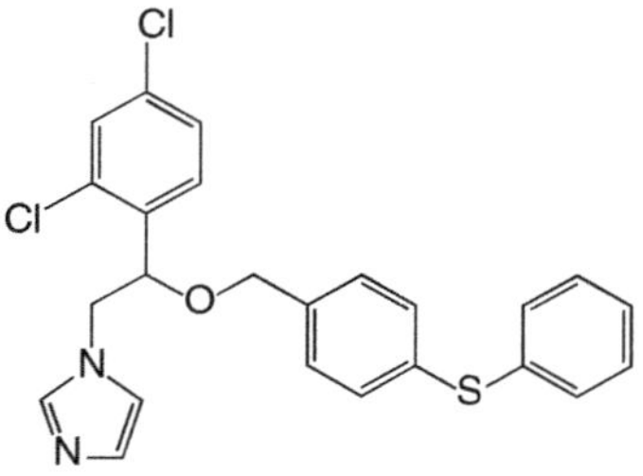	[72479-26-6] $C_{24}H_{20}Cl_2N_2OS$ 455.41 mononitrate: [73151-29-8] $C_{24}H_{20}Cl_2N_2OS$ · HNO_3 518.42	ergosterol biosynthesis inhibitor. antifungal, antimycotic. cream 2%, gel 2%; vaginal ovules 200 mg. 0.1 g V.	Recordati 1986 synthetic [108, 109]
Azanidazole [Premium, Triclose] G01AF13; P01AB04 (→ Antiparasitics (PC))	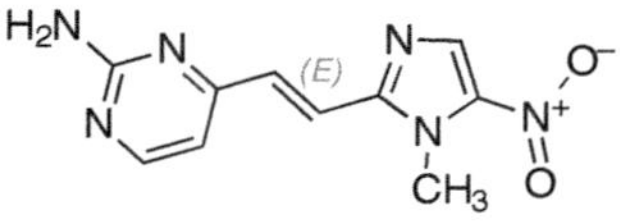	[53409-75-9] $C_{10}H_{10}N_6O_2$ 246.26	antiinfective; antiprotozoal used in gynecology.	synthetic [110]
Propenidazole G01AF14	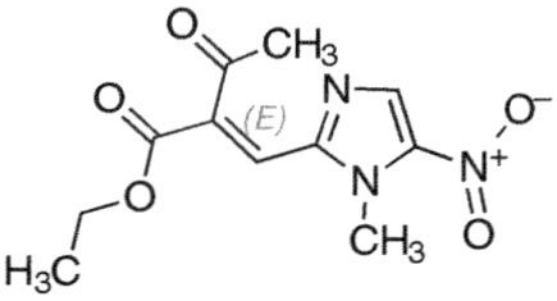	[76448-31-2] $C_{11}H_{13}N_3O_5$ 267.24	antiinfective; antiprotozoal used in gynecology.	synthetic [111]
Butoconazole [Femstat, Gynazole, Gynomyk] G01AF15	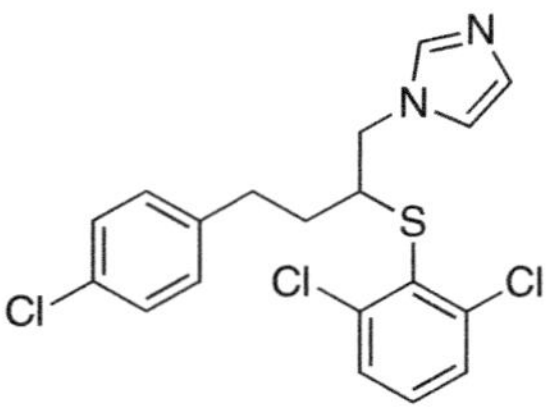	[64872-76-0] $C_{19}H_{17}Cl_3N_2S$ 411.78 mononitrate: [64872-77-1] $C_{19}H_{17}Cl_3N_2S$ · HNO_3 474.80	topical antifungal used in gynecology. vaginal cream 2%. 0.1 g V.	synthetic [112, 113]
Omoconazole [Fungisan, Fongamil, Fongarex] G01AF16; D01AC13 (→ Dermatologicals (D), 2. Antifungals (D01), Emollients and Protectives (D02), and Antipruritics (D04) for Dermatological Use)	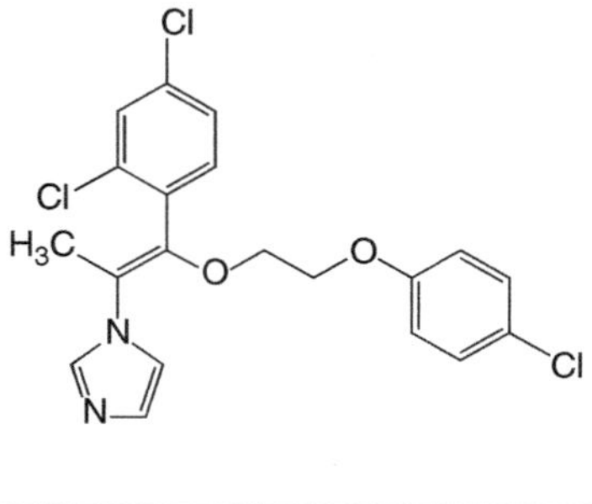	[74512-12-2] $C_{20}H_{17}Cl_3N_2O_2$ 423.73 mononitrate: [83621-06-1] $C_{20}H_{17}Cl_3N_2O_2$ · HNO_3 486.74	ergosterol biosynthesis inhibitor. topical antifungal, antimycotic. cream 10 mg/1 g	synthetic [114, 115]

(*continued*)

Table 6. (*Continued*)

INN (Synonyms) [Brand names] ATC	Structure (Remarks)	[CAS-No.] Formula MW, g/mol	Target (if known) Medical use Formulation DDD	Originator Approval (country/year) Production method [References]
Oxiconazole [Fonx, Myfungar, Okinazole, Oxistat] G01AF17; D01AC11 (→ Dermatologicals (D), 2. Antifungals (D01), Emollients and Protectives (D02), and Antipruritics (D04) for Dermatological Use)		[64211-45-6] $C_{18}H_{13}Cl_4N_3O$ 429.13 mononitrate: [64211-46-7] $C_{18}H_{13}Cl_4N_3O \cdot HNO_3$ 492.15	ergosterol biosynthesis inhibitor. antifungal, antimycotic. cream 1%; pessaries 600 mg (as mononitrate); powder 10 mg/g; sol. 1%; vaginal tabl. 100 and 600 mg.	synthetic [116]
Flutrimazole (UR-4056) [Flusporan] G01AF18; D01AC16 (→ Dermatologicals (D), 2. Antifungals (D01), Emollients and Protectives (D02), and Antipruritics (D04) for Dermatological Use)		[119006-77-8] $C_{22}H_{16}F_2N_2$ 346.38	ergosterol biosynthesis inhibitor. topical antifungal. cream 1%.	Menarini synthetic [117]
Sertaconazole [Monazol, Sertaderm, Sertagyn, Zalain] G01AF19; D01AC14 (→ Dermatologicals (D), 2. Antifungals (D01), Emollients and Protectives (D02), and Antipruritics (D04) for Dermatological Use)		[99592-32-2] $C_{20}H_{15}Cl_3N_2OS$ 437.78 mononitrate: [99592-39-9] $C_{20}H_{15}Cl_3N_2OS \cdot HNO_3$ 500.79	ergosterol biosynthesis inhibitor. antifungal, antimycotic. cream 20 mg/g.	synthetic [118, 119]

2.1.7. G01AG Triazole Derivatives (Table 7)

Table 7. G01AG Triazole derivatives

INN (Synonyms) [Brand names] ATC	Structure (Remarks)	[CAS-No.] Formula MW, g/mol	Target (if known) Medical use Formulation DDD	Originator Approval (country/year) Production method [References]
Terconazol [Terazol] G01AG01		[67915-31-5] $C_{26}H_{31}Cl_2N_5O_3$ 532.47	binds to heme iron in cytochrome P450 enzyme lanosterol of fungi. antifungal for treatment of vaginal thrush. intravaginal cream 0.4%, lotion, and 80 mg suppository. 80 mg V.	Janssen 1983 synthetic [120]

2.1.8. G01AX Other Antiinfectives and Antiseptics (Table 8)

Table 8. G01AX Other antiinfectives and antiseptics

INN (Synonyms) [Brand names] ATC	Structure (Remarks)	[CAS-No.] Formula MW, g/mol	Target (if known) Medical use Formulation DDD	Originator Approval (country/year) Production method [References]
Clodantoin (chlordantoin) [Sporostacin] G01AX01		[5588-20-5] $C_{11}H_{17}Cl_3N_2O_2S$ 347.69	antifungal used in gynecology. 0.1 g V.	Ortho 1960 synthetic [121]
Inosine [Catachol, Correctol, Isoprinosine] G01AX02; D06BB05; S01XA10		[58-63-9] $C_{10}H_{12}N_4O_5$ 263.28	nucleoside that also binds to GABA A receptor. immuno- and feed stimulant. eye drops and solution.	synthetic [122]
Policresulen [Albothyl, Faktu, Polilen] G01AX03		[101418-00-2] $(C_9H_8O_4S)_n$	topical hemostatic and antiseptic used for gynecological infections. spray, suppository, and gel 45–55%. 90 mg V.	synthetic [123]
Nifuratel [Inimur, Macmiror] G01AX05	(E)	[4936-47-4] $C_{10}H_{11}N_3O_5S$ 285.28	chemotherapeutic (trichomonas), antibiotic, antifungal, antiprotozoal. drg. 200 mg; ointment 100 mg/g; vaginal pessaries 250 mg. 0.6 g O, V.	synthetic [124]
Furazolidone [Dependal, Furoxan, Furoxon, Nifuran Ovula] G01AX06		[67-45-8] $C_8H_7N_3O_5$ 225.16	monoamine oxidase inhibitor and binds to bacterial DNA. topical antiinfective, topical antiprotozoal, and chemotherapeutic (trichomonas). liquid 50 mg/15 mL; tabl. 100 mg.	Roberts/GSK synthetic [125, 126]

Table 8. (*Continued*)

INN (Synonyms) [Brand names] ATC	Structure (Remarks)	[CAS-No.] Formula MW, g/mol	Target (if known) Medical use Formulation DDD	Originator Approval (country/year) Production method [References]
Methylrosanilin (gentian violet) [Gentian violet, Pyoktanin] G01AX09; D01AE02 (→ Dermatologicals (D), 2. Antifungals (D01), Emollients and Protectives (D02), and Antipruritics (D04) for Dermatological Use)		[*548-62-9*] $C_{25}H_{30}ClN_3$ 407.99	dye that acts as mitotic poison. antibacterial, antiseptic, anthelmintic, and antimycotic for treatment of thrush. 0.5−2% solution.	synthetic [127]
Povidone−iodine (PVP-I, iodopovidone) [Betadine; Betaisodona; Isodine; Pyodine; Wokadine] G01AX11; D08AG02; D09AA09; D11AC06; R02AA15; S01AX18 **WHO essential medicine** (→ Dermatologicals (D), 4. Antiseptics and Disinfectants (D08), Anti-Acne Preparations (D10), and Other Dermatological Preparations (D11); → Respiratory Disorders, 3. Throat Preparations (R02))		[*25655-41-8*] $(C_6H_9NO)_n \cdot xI$	chemical complex of polyvinylpyrrolidone (povidone), hydrogen iodide, and iodine that directly causes protein denaturation and precipitation of bacteria. antiseptic, antibacterial, and disinfectant. solution 2.5−10%, gels, ointment, eye drops. 0.2 g V.	1955 synthetic [128, 129]
Ciclopirox [Batrafen, Loprox, Miclast, Mocomicen, Penlac, Sebiprox] G01AX12; D01AE14 (→ Dermatologicals (D), 2. Antifungals (D01), Emollients and Protectives (D02), and Antipruritics (D04) for Dermatological Use)		[*29342-05-0*] $C_{12}H_{17}NO_2$ 207.27 olamine: [*41621-49-2*] $C_{12}H_{17}NO_2$ $\cdot C_2H_7NO$ 268.36	antifungal and antimycotic. cream 1%; powder 1% (as olamine); sol. 1% (as ciclopirox); vaginal cream 1% (as olamine). 50 mg V.	Hoechst Aventis DE/1981 synthetic [130−132]

(*continued*)

Table 8. (*Continued*)

INN (Synonyms) [Brand names] ATC	Structure (Remarks)	[CAS-No.] Formula MW, g/mol	Target (if known) Medical use Formulation DDD	Originator Approval (country/year) Production method [References]
Protiofate [Atrimycon] G01AX13		[58416-00-5] $C_{12}H_{16}O_6S$ 288.32	topical fungicide for use in gynecology.	synthetic [133]
Hexetidine [Hexoral; Collu-Hextril; Colludol; Hextril; Oraldene; Oraseptic] G01AX16; A01AB12; R02AA28; (→ Respiratory Disorders, 3. Throat Preparations (R02))		[141-94-6] $C_{21}H_4N_3$ 339.61	antiseptic. sol. 100 mg/100 mL, 200 mg/100 mL; spray 0.1 g/100 g; vaginal tabl. 10 mg.	synthetic [134, 135]
Dapivirine (TMC-120) G01AX17		[244767-67-7] $C_{20}H_{19}N_5$ 329.40	nonnucleotide reverse transcriptase inhibitor (NNRTI). microbiocide; antiviral for HIV prevention for women with intravaginal ring. 25 mg per silicon polymer ring.	Janssen/Tibotec synthetic [136]
Chlorphenesin [Musil, Mycil] G01AX21; D01AE07 (→ Dermatologicals (D), 2. Antifungals (D01), Emollients and Protectives (D02), and Antipruritics (D04) for Dermatological Use)		[104-29-0] $C_9H_{11}ClO_3$ 202.65	muscle relaxant, antifungal. emulgel.	synthetic [137]
Monalazone [Malun, Spergisin, Speton] G01AX26		[106145-03-3] $C_7H_6ClNO_4S$ 235.64 disodium salt: [61477-95-0]	antiseptic, spermicidal contraceptive, and vaginal disinfectant. tabl. 10 mg in combination with estradiol.	synthetic [138]

Table 8. (*Continued*)

INN (Synonyms) [Brand names] ATC	Structure (Remarks)	[CAS-No.] Formula MW, g/mol	Target (if known) Medical use Formulation DDD	Originator Approval (country/year) Production method [References]
Octenidin [Octeniderm, Octenisept; Octenillin] G01AX66; D08AJ57; R02AA21 (→ Dermatologicals (D), 4. Antiseptics and Disinfectants (D08), Anti-Acne Preparations (D10), and Other Dermatological Preparations (D11); → Respiratory Disorders, 3. Throat Preparations (R02))	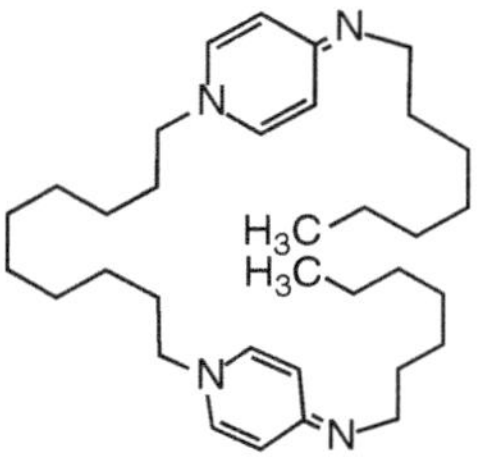	[*71251-02-0*] $C_{36}H_{62}N_4$ 550.90 dihydrochloride: [*70775-75-6*] $C_{36}H_{64}Cl_2N_4$ 623.84	antiseptic, surfactant, and disinfectant. gel and solution 0.1−2%.	synthetic [139]
Guaiazulene (guajazulene) [Azulon, Cicatryl, Thrombocid] G01AX75; S01AX01	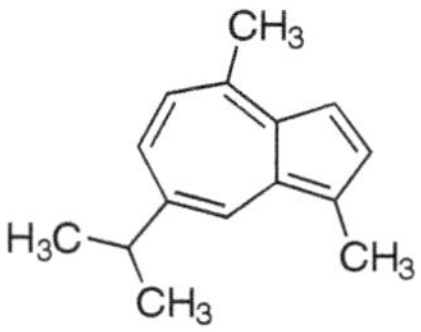	[*489-84-9*] $C_{15}H_{18}$ 198.31	antiinflammatory. cream; drg. 20 mg; eye drops 0.02%; gran. 0.4%, 1%; ointment 0.033% (ethanolic chamomile extract); powder 0.4%; tabl 2 mg.	synthetic [140]

3. G02 Other Gynecologicals

3.1. G02A Uterotonics

3.1.1. G02AB Ergot Alkaloids (Table 9)

Table 9. G02AB Ergot alkaloids

INN (Synonyms) [Brand names] ATC	Structure (Remarks)	[CAS-No.] Formula MW, g/mol	Target (if known) Medical use Formulation DDD	Originator Approval (country/year) Production method [References]
Methylergometrine (methylergonovine) [Ergotrate, Ergotyl, Methergin, Partan] G02AB01		[113-42-8] $C_{20}H_{25}N_3O_2$ 339.44 maleate salt: [57432-61-8] $C_{20}H_{25}N_3O_2 \cdot C_4H_4O_4$ 455.51 tartrate salt: [6209-37-6] $C_{20}H_{25}N_3O_2 \cdot \frac{1}{2} C_4H_6O_6$ 828.96	5-HT2A-mGlu2-receptor partial agonist acting directly on smooth muscle of the uterus; ergot alkaloid. uterotonic, oxytocic; used for uterus contraction and prevention of bleeding after vaginal childbirth. amp. 0.2 mg/1 mL; drg. 0.125 mg; drops 0.25 mg/mL; sol. 0.24 mg/100 mL; tabl. 0.125 mg (as maleate). 0.2 mg O, P.	Sandoz/Novartis US/1946 synthetic [141]
Ergometrine (Ergobasine, Ergonovine) [Ergotrate, Ergonovine, Secalysat, Syntometrine] G02AB03 **WHO essential medicine**		[60-79-7] $C_{19}H_{23}N_3O_2$ 325.41 maleate salt: [129-51-1] $C_{19}H_{23}N_3O_2 \cdot C_4H_4O_4$ 441.48 tartrate salt: [129-50-0] $C_{19}H_{23}N_3O_2 \cdot \frac{1}{2} C_4H_6O_6$ 800.91	5-HT2 (serotonin) and dopaminergic receptor agonist acting directly on uterus smooth muscle; ergot alkaloid. oxytocic for treatment of postpartum hemorrhage. amp. 0.2 mg/1 mL; sol. 50 mg/100 mL (as maleate). 0.2 mg O, P.	Sandoz/E. Lilly synthetic [142, 143]

3.1.2. G02AC Ergot Alkaloids and Oxytocin Including Analogues, in Combination (Table 10)

Table 10. G02AC Ergot alkaloids and oxytocin including analogues, in combination

INN (Synonyms) [Brand names] ATC	Structure (Remarks)	[CAS-No.] Formula MW, g/mol	Target (if known) Medical use Formulation DDD	Originator Approval (country/year) Production method [References]
Oxytocin [Bakumokon, Oxytocin] G02AC01, H01BB02 (→ Hormone Therapeutics – Systemic Hormonal Preparations, Excluding Reproductive Hormones and Insulins (H))		*[50-56-6]* $C_{43}H_{66}N_{12}O_{12}S_2$ 1007.14	hormone and neuropeptide produced in the hypothalamus and released to posterior pituitary; agonist to oxytocin receptor in brain. inj. 10 units/mL and combinations.	synthetic [144]

3.1.3. G02AD Prostaglandins (Table 11)

Table 11. G02AD Prostaglandins

INN (Synonyms) [Brand names] ATC	Structure (Remarks)	[CAS-No.] Formula MW, g/mol	Target (if known) Medical use Formulation DDD	Originator Approval (country/year) Production method [References]
Dinoprost (prostaglandin F$_{2\alpha}$) [Glandin, Glandinon, Prosmon, Prostin, Prostamodin] G02AD01		[551-11-1] C$_{20}$H$_{34}$O$_5$ 354.49 tromethamine salt: [38562-01-5] C$_{20}$H$_{34}$O$_5$ · C$_4$H$_{11}$NO$_3$ 475.62	natural PG that binds to prostaglandin F receptor and increases oxytocin levels in uterus. oxytocic, abortifacient. amp. 50 µg/1 mL; 1 mg/1 mL; 2 mg/2 mL; 5 mg/mL. 25 mg P.	Upjohn synthetic [145–147]
Dinoprostone (prostaglandin E$_2$) [Cervidil, Minprostin, Prepidil, Prostin, Propess] G02AD02 **WHO essential medicine**		[363-24-6] C$_{20}$H$_{32}$O$_5$ 352.47	naturally occurring prostaglandin binding to prostaglandin receptor. oxytocic, abortifacient, and labor induction. amp. 0.5 g/0.5 mL, 0.75 mg/0.75 mL; rectangular tabl. 0.5 mg; syringe with gel 0.5 mg; tabl. 0.5 mg; vaginal gel 0.5 mg/3 g, 1 mg/3 g, 2 mg/3 g; vaginal tabl. 3 mg. 0.5 mg O, V.	Upjohn/Pharmacia US/1977 synthetic [148–150]
Gemeprost (DSC-37681) [Cergem, Preglandin] G02AD03		[64318-79-2] C$_{23}$H$_{38}$O$_5$ 394.55	prostaglandin E1 analogue and antiprogestogen that binds as agonist to prostaglandin E2/E3 receptors causing myometrical contractions. used for obstetric bleeding with pessary and in combination with mifepristone for termination of pregnancy. vaginal suppositories and pessary. 1 mg V.	synthetic [151, 152]
Carboprost [Hemabate, Tham] G02AD04		[35700-23-3] C$_{21}$H$_{36}$O$_5$ 368.51 tromethamine salt: [58551-69-2] C$_{25}$H$_{47}$NO$_8$ 489.64	prostaglandin PGF2 analogue that binds to prostaglandin E2 receptor causing myometrial contractions and induction of labor. oxytocic used for prevention of postpartum hemorrhage and abortifacient agent. amp. 250 µg/mL. 2.5 mg P.	Pharmacia & Upjohn US/1979 synthetic [153, 154]

Table 11. (*Continued*)

INN (Synonyms) [Brand names] ATC	Structure (Remarks)	[CAS-No.] Formula MW, g/mol	Target (if known) Medical use Formulation DDD	Originator Approval (country/year) Production method [References]
Sulprostone (CP34089) [Nalador] G02AD05		[60325-46-4] $C_{23}H_{31}NO_7S$ 465.56	prostaglandin E2 analogue that activates prostaglandin E3 receptor inducing medical abortion. oxytocic and abortifacient agent. amp. 0.5 mg. 0.5 mg P.	synthetic [155]
Misoprostol [Artotec, Cytotec, Misodelle, Misodex, Gymiso, Misofenac] G02AD06; A02BB01 **WHO essential medicine** (→ Drugs for Acid-Related Disorders (A02))		[59122-46-2] $C_{22}H_{38}O_5$ 382.54	prostaglandin E1 analogue that stimulates E1 receptors in stomach and smooth muscle cells. peptic ulcer therapeutic and agent used in miscarriages. f.c. tabl. 0.2 mg (comb. with diclofenac sodium); tabl. 100 and 200 µg. 0.2 mg O, V.	G.D. Searle/Pfizer US/1986 synthetic [156, 157]

3.2. G02B Contraceptives for Topical Use

3.2.1. G02BB Intravaginal Contraceptives (Table 12)

Table 12. G02BB Intravaginal contraceptives

INN (Synonyms) [Brand names] ATC	Structure (Remarks)	[CAS-No.] Formula MW, g/mol	Target (if known) Medical use Formulation DDD	Originator Approval (country/year) Production method [References]
Nonoxinol 9 (nonoxynol 9) [Patentex, Patentex Oval, Semicid] G02BB09		[26027-38-3] $[C_2H_4O] \times C_{15}H_{24}O$	nonionic surfactant that attacks the acrosomal membrane of sperms. spermatocide. foam 12.5%; suppos. 0.075 g; vaginal gel 2 g/100 g.	synthetic [158]

3.3. G02C Other Gynecologicals

3.3.1. G02CA Sympathomimetics, Labor Repressants (Table 13)

Table 13. G02CA Sympathomimetics, labor repressants

INN (Synonyms) [Brand names] ATC	Structure (Remarks)	[CAS-No.] Formula MW, g/mol	Target (if known) Medical use Formulation DDD	Originator Approval (country/year) Production method [References]
Ritodrine [Fremove, Miolene, Pre-Par, Utemerin, Utezol, Yutopar] G02CA01		[26652-09-5] $C_{17}H_{21}NO_3$ 287.36 hydrochloride salt: [23239-51-2] $C_{17}H_{21}NO_3$ · HCl 323.82	β2-receptor agonist. uterus relaxant, tocolytic. amp. 50 mg/5 mL; s.r. cps. 40 mg; tabl. 5 and 10 mg (as hydrochloride). 40 mg O, P.	Bristol Labs US/1984 synthetic [159]
Buphenine (nylidrine) [Adrin, Apoleptal, Ophtadil, Opino] G02CA02		[447-41-6] $C_{19}H_{25}NO_2$ 299.41 hydrochloride salt: [849-55-8] $C_{19}H_{25}NO_2$ · HCl 335.88	β2-adrenergic receptor agonist. vasodilator, sympathomimetic. amp. 5 mg; drops 4 mg; tabl. 6 mg. 30 mg P.	synthetic [160, 161]
Fenoterol [Berodual; Berodual Aerosol; Berotec Dosier Aerosol; Brochodual; Dosberotec; Duovent; Iprafen; Partusisten G02CA03, R02CC03; R03AC04; (→ Respiratory Disorders, 4. Inhalants for Obstructive Airway Diseases (R03))		[13392-18-2] $C_{17}H_{21}NO_4$ 303.36 hydrobromide: [1944-2-3] $C_{17}H_{21}NO_4$ · HBr 384.27	β2-adrenergic receptor agonist. bronchodilator, antiasthmatic, and tocolytic. aerosol 0.05 mg/puff in comb, 0.2%; amp. 0.025 mg/mL, 0.5 mg/10 mL; cps. 200 μg; dry syrup 0.25%; 0.5% sol. for inhalation 0.5 mg/mL in comb., 1 mg/mL; syrup 0.05%; tabl. 2.5 and 5 mg. 0.6 mg inhal. aerosol/powder; 4 mg inhal. solution.	Boehringer Ingelheim US/1977 synthetic [162, 163]

3.3.2. G02CB Prolactine Inhibitors (Table 14)

Table 14. G02CB Prolactine inhibitors

INN (Synonyms) [Brand names] ATC	Structure (Remarks)	[CAS-No.] Formula MW, g/mol	Target (if known) Medical use Formulation DDD	Originator Approval (country/year) Production method [References]
Bromocriptine (2-bromoergocriptine) [Bromo-Kin, Kirim, Kirim-gyn, Parlodel, Serocryptin] G02CB01; N04B01 (→ Anti-Parkinson Drugs (N04))		[25614-03-3] $C_{32}H_{40}BrN_5O_5$ 654.61 mesylate salt: [22260-51-1] $C_{32}H_{40}BrN_5O_5$ $\cdot CH_4O_3S$ 750.71	dopamine D_2-agonist and prolactin inhibitor. antiparkinsonian and treatment of agromegaly. cps. 5 mg, 10 mg; tabl. 2.5 mg (as mesylate). 5 mg O, P.	Sandoz 1975 synthetic [164]
Lisuride [Arolac, Dipergon, Dopergin, Proclatam, Revanil] G02CB02; N02CA07 (→ Anti-Parkinson Drugs (N04))		[18016-80-3] $C_{20}H_{26}N_4O$ 338.45 maleate salt: [9875-60-6]	ergot derivative that acts as dopamin $D_2/D_3/D_4$ receptor agonist and serotonin 5-HT2B receptor antagonist. prolactin inhibitor, antiparkinsonian. 0.6 mg O.	Bayer synthetic [165]
Cabergoline [Caberseril, Cabaser, Dostinex] G02CB03; N04BC06 (→ Anti-Parkinson Drugs (N04))		[81409-90-7] $C_{26}H_{37}N_5O_2$ 451.62	dopamine D_2-antagonist prolactin inhibitor, prevention or suppression of puerperal lactation. tabl. 0.25, 0.5, 1, 2, and 4 mg. 0.6 mg O.	Pharmacia/Farmitalia 1993 synthetic [166–168]
Quinagolide (CV-205502, SDZ-205502) [Norprolac] G02CB04		[87056-78-8] $C_{20}H_{33}N_3O_3S$ 395.57 hydrochloride salt: [94424-50-7] $C_{20}H_{33}N_3O_3S$ $\cdot$ HCl 432.03	nonergot-derived dopamine D_2-antagonist. antiparkinsonian, prolactin secretion inhibitor. tabl. 0.025, 0.050, 0.075, and 0.150 mg. 75 µg O.	Ferring 2004 synthetic [169, 170]

(continued)

Table 14. (*Continued*)

INN (Synonyms) [Brand names] ATC	Structure (Remarks)	[CAS-No.] Formula MW, g/mol	Target (if known) Medical use Formulation DDD	Originator Approval (country/year) Production method [References]
Metergoline [Liserdol] G02CB05		[*17692-51-2*] $C_{25}H_{29}N_3O_2$ 403.52	ergot derivative that binds to dopamin and serotonin receptors as antagonist. prolactin regulation and antianxiety treatment. tabl.	synthetic [171]
Terguride (*trans*-dihydrolisuride) [Teluron] G02CB06		[*37686-84-3*] $C_{20}H_{28}N_4O$ 340.46	ergot derivative that binds to dopamin and serotonin receptors as antagonist. tabl.	Pfizer synthetic [172]

3.3.3. G02CC Antiinflammatory Products for Vaginal Administration (Table 15)

Table 15. G02CC Antiinflammatory products for vaginal administration

INN (Synonyms) [Brand names] ATC	Structure (Remarks)	[CAS-No.] Formula MW, g/mol	Target (if known) Medical use Formulation DDD	Originator Approval (country/year) Production method [References]
Ibuprofen [Advel; Advil; Aktren; Brufen; Dolobene; Dolormin; Ibu Bejuron; Nurefast; Nurofen; Vicoprofen] G02CC01; M01AE01; M02AA13; R02AX02 **WHO essential medicine** (→ Drugs for the Musculo-Skeletal System (M); → Respiratory Disorders, 3. Throat Preparations (R02))		[*15687-27-1*] $C_{13}H_{18}O_2$ 206.29 lysine salt: [*57469-77-9*]	nonsteroidal antiinflammatory drug (NSAID), analgesic, and antirheumatic. amp. 400 mg; drg. 200 mg, 400 mg; eff. gran. 200 mg; f.c. tabl. 200, 400, and 600 mg; gran. 20%, 40%, 50%; s.r. tabl. 800 mg; suppos. 50, 100, and 600 mg; syrup 100 mg/5 mL; tabl. 100 mg, 200 mg.	Boots UK/1969 synthetic [173–175]
Naproxen [Aleve; Anaprox; Mobilat; Naixan, Naprosyn; Naprosyne; Prevacid; Synflex] G02CC02; M01AE02; M02AA12 (→ Drugs for the Musculo-Skeletal System (M))		[*22204-53-1*] $C_{14}H_{14}O_3$ 230.26 sodium salt: [*26159-34-2*] $C_{14}H_{13}NaO_3$ 252.25	COX-inhibitor. nonsteroidal antiinflammatory drug (NSAID), antirheumatic, and analgesic. cps. 300 mg; f.c. tabl. 250, 500, and 1000 mg; suppos. 250 and 500 mg; susp. 125 mg/5 mL, 250 mg/5 mL; tabl. 250, 375, 500, and 750 mg (as acid); f.c. tabl. 550 mg; tabl. 100, 275, 550, 375, and 500 mg (as sodium salt).	Syntex/Roche US/1976 synthetic [176–178]
Benzydamine [Afloben; Diflam; Opalgyne; Tantum; Verax; Xentafid] G02CC03; A01AD02; M01AX07; M02AA05; M02AX (→ Stomatological Preparations (A01); → Drugs for the Musculo-Skeletal System (M))		[*642-72-8*] $C_{19}H_{23}N_3O$ 309.41 hydrochloride salt: [*132-69-4*] $C_{19}H_{23}N_3O$ · HCl 345.87	NSAID. analgesic, antipyretic, antiinflammatory, and antimicrobial. amp. 25 mg; cps. 50 mg; cream 30 mg; drg. 50 mg; drops 50 mg; liquid 1.5 mg; powder 500 mg (as hydrochloride).	Angelini synthetic [179]
Flunoxaprofen [Priaxim] G02CC04; M01AE15 (→ Drugs for the Musculo-Skeletal System (M))		[*66934-18-7*] $C_{16}H_{12}FNO_3$ 285.27 lysine salt: [*124816-13-3*] $C_{16}H_{12}FNO_3$ · $C_6H_{14}N_2O_2$ 431.46	COX-2 inhibitor. nonsteroidal antiinflammatory and antirheumatic. gel 5%; tabl. 50, 100, and 200 mg.	synthetic [180, 181]

3.3.4. G02CX Other Gynecologicals (Table 16)

Table 16. G02CX Other gynecologicals

INN (Synonyms) [Brand names] ATC	Structure (Remarks)	[CAS-No.] Formula MW, g/mol	Target (if known) Medical use Formulation DDD	Originator Approval (country/year) Production method [References]
Atosiban (antocin II, CAP449, F314, ORF22164, RWJ-22164) [Tractocile] G02CX01		[90779-69-4] $C_{43}H_{67}N_{11}O_{12}S_2$ 994.21	inhibitor of the peptide hormones oxytocin and vasopressin. treatment of preterm labor oxytocin antagonist, tocolytic. inf. concentrate 7.5 mg/mL; inj. vial 7.5 mg/mL. 175 mg P.	Ferring EU/2000 synthetic [182, 183]
Flibanserin (BIMT-17) [Addyi, Ectris] G02CX02		[67933-07-5] $C_{20}H_{21}F_3N_4O$ 390.41 hydrochloride salt: [147359-76-0] $C_{20}H_{21}F_3N_4O \cdot HCl$ 426.87	serotonin 5-HTA1 receptor agonist and 5-HTA2 receptor antagonist. treatment of female hypoactive sexual desire disorder (HSDD). tbl. 100 mg.	Sprout Pharmaceuticals US/2015 synthetic [184, 185]

4. G03 Sex Hormones and Modulators of the Genital System

4.1. G03A Hormonal Contraceptives for Systemic Use

4.1.1. G03AA Progestogens and Estrogens, Fixed Combinations (Table 17)

Table 17. G03AA Progestogens and estrogens, fixed combinations

INN (Synonyms) [Brand names] ATC	Structure (Remarks)	[CAS-No.] Formula MW, g/mol	Target (if known) Medical use Formulation DDD	Originator Approval (country/year) Production method [References]
Etynodiol diacetate (äthynodioldiacetat, ethynodiol diacetate, etynodiol diacetate) [Demulen, Edulen, Femulen, Ovulen] G03AA01		[297-76-7] $C_{24}H_{32}O_4$ 384.52	as progestin agonist of the progesterone receptor. progestogen (in combination with estrogen as oral contraceptive). tabl. 0.05 and 1 mg.	Searle/Pharmacia/ Pfizer US/1965 synthetic [186, 187]
Quingestanol acetate (W-4540) [Demovis, Pilomin, Unovis] G03AA02; G03AC04 wfm		[3000-39-3] $C_{27}H_{36}O_3$ 408.57	as progestin agonist of the progesterone receptor. prodrug of norethisterone; postcoital contraceptive. tabl.	IT/1972 synthetic [188]
Lynestrenol (lynelol) [Exluton, Ministat, Lyndiol, Orgametril] G03AA03		[52-76-6] $C_{20}H_{28}O$ 284.36	as progestin agonist of the progesterone receptor. prodrug of norethisterone; contraceptive and used for treatment of menstrual disorders. tabl.	Organon NL/1962 synthetic [189]
Megestrol acetate (MGA) [Megace, Ovaban] G03AA04; G03AC05		[595-33-5] $C_{24}H_{32}O_4$ 384.51	as progestin agonist of the progesterone receptor. appetite stimulant, antineoplastic, and contraceptive. tabl., 5, 20, and 40 mg; oral suspensions and injectable suspensions.	Syntex US/1963 synthetic [190]
Norethisterone (norethindrone) [Alone, Aygestin, Camila, Micronor, Nonyl, Norlutin, Primolut] G03AA05; G03AC01		[68-22-4] $C_{20}H_{26}O_2$ 298.42	as progestin agonist of the progesterone receptor. contraceptive and used for treatment of menstrual disorders. tabl. 1, 5, and 10 mg	Syntex US/1957 synthetic [191]

(continued)

Table 17. (*Continued*)

INN (Synonyms) [Brand names] ATC	Structure (Remarks)	[CAS-No.] Formula MW, g/mol	Target (if known) Medical use Formulation DDD	Originator Approval (country/year) Production method [References]
Norgestrel (DL-norgestrel) [Cyclo-Progynova, Eugynon, Ovral, Ovrette, Neogest, Stediril, Schering PC4, Tetragynon] G03AA06; G03FA10; G03FB01	(racemic mixture)	[6533-00-2] $C_{21}H_{28}O_2$ 312.45	progestogen and agonist of the progesterone receptor. drg. 0.5 mg in comb. with ethinylestradiol; tabl. 0.5 mg.	Schering AG DE/1966 synthetic [192, 193]
Levonorgestrel ((−)-norgestrel) [Adepal, Climara, Levogynon, Levonelle, Microgynon, Mirena, Neogynon, Norplant, Postinor, Trigynon] G03AA07 **WHO essential medicine**		[797-63-7] $C_{21}H_{28}O_2$ 312.45	as progestogen (gestagen) agonist of the progesterone receptor. enantiopure form of norgestrel; contraceptive and emergency birth control. tabl. 0.05, 0.075, and 0.125 mg; drg. 0.03, 0.1, 0.15, and 0.25 mg; pessaries 52 mg and implants.	Schering AG 1970 synthetic [194–196]
Medroxyprogesterone acetate (MPA) [Amen, Climapax, Cycrin, Depo-Provera, Farlutal, Prempro, Provera] G03AA08; G03AC06; G03DA02; L02AB02 **WHO essential medicine**		[71-58-9] $C_{24}H_{34}O_4$ 386.53	as progestogen (gestagen) agonist of the progesterone receptor. contraceptive. amp. 500 mg, 1 g; susp. 500 mg; susp. 150 mg/mL, 500 mg/mL; tabl. 2.5, 5, 10, 100, 200, 250, 400, and 500 mg.	Upjohn US/1959 synthetic [197, 198]
Desogestrel [Biviol, Cerazette, Desogen, Marvalon, Oviol, Varnoline] G03AA09, G03AC09		[54024-22-5] $C_{22}H_{30}O$ 310.48	progestogen, oral contraceptive (in combination with ethinylestradiol). tabl. 150 µg (in combination with ethinylestradiol)	Organon NL/1981 synthetic [199, 200]

Table 17. (*Continued*)

INN (Synonyms) [Brand names] ATC	Structure (Remarks)	[CAS-No.] Formula MW, g/mol	Target (if known) Medical use Formulation DDD	Originator Approval (country/year) Production method [References]
Gestodene (SHB 331) [Avadane, Fedra, Femovan, Harmonet, Milvane, Minesse, Minulet, Triminulet] G03AA10; G03AB06	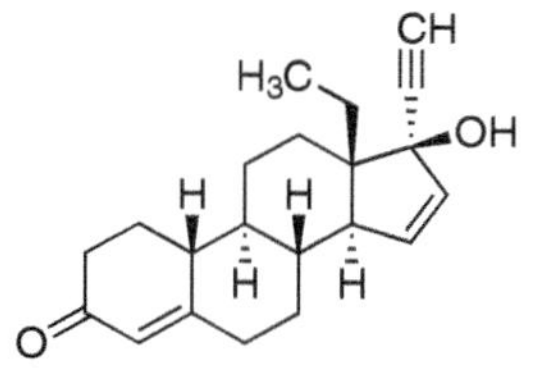	[*60282-87-3*] $C_{21}H_{26}O_2$ 310.44	as progestogen (gestagen) agonist of the progesterone receptor. progestogen (gestagen) and oral contraceptive. drg. and tabl. 75 µg in combination with 30 µg ethinylestradiol.	Bayer DE/1987 synthetic [201, 202]
Norgestimate [Cilest, Ortho-Tri-Cyclen, Pramino, Previfem, Tricilest, Triciclen] G03AA11; G03AB08	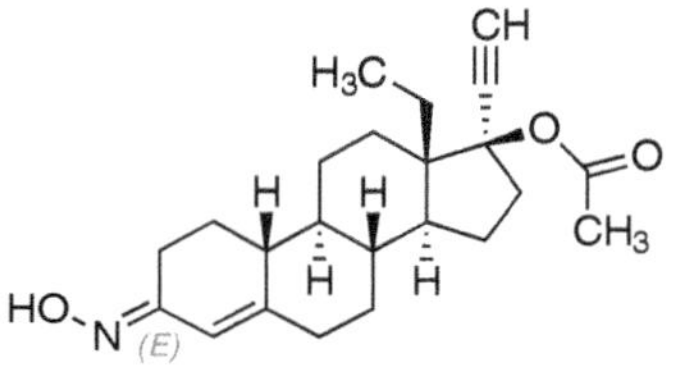 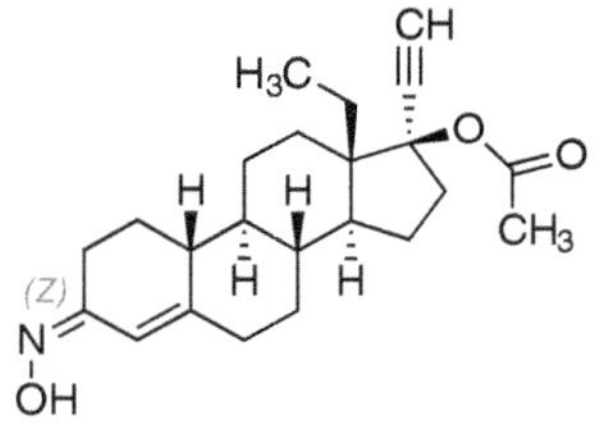	[*35189-28-7*] $C_{23}H_{31}NO_3$ 369.51	as progestogen (gestagen) agonist of the progesterone receptor. progestogen (gestagen) and oral contraceptive. tabl. 0.25 mg in comb. with ethinylestradiol.	Janssen/Ortho EU/1986 US/1999 synthetic [203]
Drospirenone (1,2-dihydro-spirorenone, ZK 30595) [Angeliq, Jasminella, Petibelle, Yasmin, Yaz] G03AA12; G03FA17	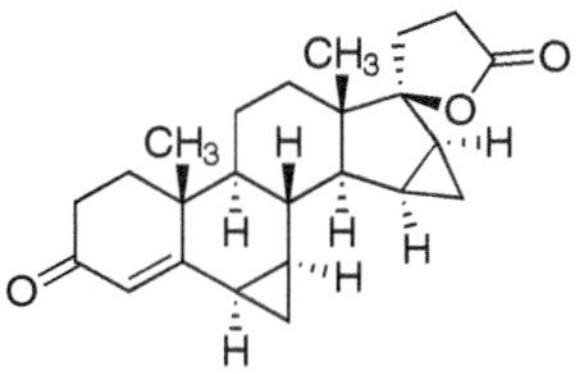	[*67392-87-4*] $C_{24}H_{30}O_3$ 366.50	as progestin agonist of the progesterone receptor; antiandrogen and gestagen. used as oral contraceptive and treatment of premenstrual dysphoric disorder. tabl. 3 mg in comb. with ethinylestradiol.	Schering AG US/2005 synthetic [204, 205]

(continued)

Table 17. (*Continued*)

INN (Synonyms) [Brand names] ATC	Structure (Remarks)	[CAS-No.] Formula MW, g/mol	Target (if known) Medical use Formulation DDD	Originator Approval (country/year) Production method [References]
Norelgestromin (levonorgestrel 3-oxime, 18-methylnore-thindrone oxime, norplant 3-oxime, RWJ 10553, 17-deacetyln-orgestimate) [Evra, Ortho Evra, Xulane] G03AA13		[*53016-31-2*] $C_{21}H_{29}NO_2$ 327.47	as progestin agonist of the progesterone receptor. contraceptive, progestogen (gestagen), and prodrug of levonorgestrel. plaster 6 mg (in comb. with ethinylestradiol).	Janssen synthetic [206, 207]
Nomegestrol acetate (NOMAC) [Lutenyl, Naemis, Zoely] G03AA14; G03DB04		[*58652-20-3*] $C_{23}H_{30}O_4$ 370.49	as progestin agonist of the progesterone receptor. synthetic progestogen for treatment of gynecological disturbances. tabl. 3.75 and 5 mg.	Theramex/MSD EU/1986 synthetic [208, 209]
Chlormadinone acetate [Belara, Chlormadinon, Luteran, Neo-Eunomon, Prostal] G03AA15; G03DB06		[*302-22-7*] $C_{23}H_{29}ClO_4$ 404.93	progestogen and antiandrogen. tabl. 2, 5, and 25 mg; s.r. tabl. 50 mg.	synthetic [210, 211]
Dienogest (STS-557) [Climodien, Lafamme, Natazia, Valette] G03AA16; G03FA15		[*65928-58-7*] $C_{20}H_{25}NO_2$ 311.43	as progestin agonist of the progesterone receptor. progestogen and oral contraceptive. f.c. tabl. 1, 2, and 3 mg.	Jenapharm EU/1995 synthetic [212, 213]

4.1.2. G03AC Progestogens (Table 18)

Table 18. G03AC Progestogens

INN (Synonyms) [Brand names] ATC	Structure (Remarks)	[CAS-No.] Formula MW, g/mol	Target (if known) Medical use Formulation DDD	Originator Approval (country/year) Production method [References]
Norgestrienone [Miniplanor, Ogyline, Planor] G03AC07 wfm		[*848-21-5*] $C_{20}H_{22}O_2$ 294.39	as progestin agonist of the progesterone receptor. progestogen, contraceptive. tabl. 0.35 mg.	Roussel synthetic [214, 215]
Etonogestrel (ORG-3236) [Circlet, Implanon, Nuvoring] G03AC08		[*54048-0-1*] $C_{22}H_{28}O_2$ 324.46	as progestin agonist of the progesterone receptor. contraceptive. implant s.c. 68 mg/1 month	Organon UK/1998; US/2001 synthetic [216]

4.1.3. G03AD Emergency Contraceptives (Table 19)

Table 19. G03AD Emergency contraceptives

INN (Synonyms) [Brand names] ATC	Structure (Remarks)	[CAS-No.] Formula MW, g/mol	Target (if known) Medical use Formulation DDD	Originator Approval (country/year) Production method [References]
Ulipristal acetate (CDP-2914) [Ella, Ella-One, Esmya, Fibristal] G03AD02 **WHO essential medicine**		[*126784-99-4*] $C_{30}H_{37}NO_4$ 475.62	selective progesterone modulator. used for emergency birth control, postcoital contraception, and uterine fibroids. tabl. 5 and 30 mg. 30 mg O.	HRA Pharma/ Watson Pharmaceuticals EU/2009; US/2010 synthetic [217]

4.2. G03B Androgens

4.2.1. G03BA 3-Oxoandrosten (4) Derivatives (Table 20)

Table 20. G03BA 3-Oxoandrosten (4) derivatives

INN (Synonyms) [Brand names] ATC	Structure (Remarks)	[CAS-No.] Formula MW, g/mol	Target (if known) Medical use Formulation DDD	Originator Approval (country/year) Production method [References]
Fluoxymesterone [Halotestin, Halotin, Ultandren] G03BA01 **controlled substance**		[76-43-7] $C_{20}H_{29}FO_3$ 336.45	androgen binding to androgen receptors and promoting protein synthesis. anabolic steroid medication used to treat low testosterone levels in men, delayed puberty in boys, and breast cancer in women. tabl. 1, 2, 2.5, 5, and 10 mg. 5 mg O.	Upjohn US/1957 synthetic [218, 219]
Methyltestosterone [Android, Enarmon, Estratest, Hormovistan, Metandren, Oreton methyl, Testovis] G03BA02; G03EK01 **controlled substance**		[58-18-4] $C_{20}H_{30}O_2$ 302.46	while binding to androgen receptors promotes protein synthesis. androgen used to treat low testosterone levels in men and delayed puberty. cps. 5, 10, and 25 mg; tabl. 10 and 25 mg. 25 mg O.	Ciba/Novartis 1936 synthetic [220, 221]
Testosterone [Androtop, Andropatch, Bothermon, Mydrotest, Pantestone, Restandol, Testopatch, Tlando, Virormone] G03BA03 **WHO essential medicine**		[58-22-0] $C_{19}H_{28}O_2$ 288.43	while binding to androgen receptors promotes protein synthesis. androgen, primary male sex hormone and anabolic used for hormone replacement therapy. amp. 4.76 mg/1 mL; cps. 40 mg (as undecanoate); plaster 2.5 mg/37 cm^2, 4 mg/40 cm^2, 5 mg/44 cm^2, and 6 mg/60 cm^2. 0.12 g O, R; 18 mg P; 50 mg TD-gel.	Lipocine US/2019 synthetic [222]

4.2.2. G03BB 5-Androstanon (3) Derivatives (Table 21)

Table 21. G03BB 5-Androstanon (3) derivatives

INN (Synonyms) [Brand names] ATC	Structure (Remarks)	[CAS-No.] Formula MW, g/mol	Target (if known) Medical use Formulation DDD	Originator Approval (country/year) Production method [References]
Mesterolone [Proviron, Vistimon] G03BB01 **controlled substance**		[*1424-00-6*] $C_{20}H_{32}O_2$ 304.47	agonist of androgen receptor; androgen and anabolic steroid. treatment of male infertility and low testosterone levels. tabl. 25 and 50 mg. 50 mg O.	Schering AG 1967 synthetic [223]
Androstanolone (stanolone, DHT) [Anabolex, Andractim, Neodrol] G03BB02 **controlled substance**		[*521-18-6*] $C_{19}H_{30}O_2$ 290.45	dihydrotestosterone (DHT); androgen and anabolic steroid. used to treat low testosterone levels in men. amp. 2% and 5%; gel 2.5%; tabl. 5 and 25 mg.	Schering AG US/1953 synthetic [224]

4.3. G03C Estrogens

4.3.1. G03CA Natural and Semisynthetic Estrogens, Plain (Table 22)

Table 22. G03CA Natural and semisynthetic estrogens, plain

INN (Synonyms) [Brand names] ATC	Structure (Remarks)	[CAS-No.] Formula MW, g/mol	Target (if known) Medical use Formulation DDD	Originator Approval (country/year) Production method [References]
Ethinylestradiol (ethinyloestradiol) [Adepal, Alesse, Brevicon, Estinyl, Femoden, Levlen, Nordette, Nuvaring, Ovral, Triphasil, Triquilar] G03CA01; L02AA03 (→ Antineoplastic and Immunomodulating Agents (L))		[*57-63-6*] $C_{20}H_{24}O_2$ 296.41	agonist to estrogen receptor. estrogen in combination with progestogen as oral contraceptive and against gynecological disorders. tabl. 0.02, 0.025, 0.05, and 0.5 mg; drg. 1 mg. 25 µg O.	Schering AG US/1943 synthetic [225, 226]

(continued)

Table 22. (*Continued*)

INN (Synonyms) [Brand names] ATC	Structure (Remarks)	[CAS-No.] Formula MW, g/mol	Target (if known) Medical use Formulation DDD	Originator Approval (country/year) Production method [References]
Estradiol (oestradiol) [Climara, Climagest, Cutanum, Estraderm, Estring, Menostar, Nuvella, Progynova, Qlaria, Vagifem] G03CA03 (→ Antineoplastic and Immunomodulating Agents (L))		[*50-28-2*] $C_{18}H_{24}O_2$ 272.39	as natural estrogen steroid hormone agonist to estrogen receptor. estrogen used, e.g., for menopausal hormone therapy. gel 0.5 and 1 mg/g; tabl. 2 and 4 mg; plaster 0.72, 2, 2.17, and 4.33 mg; transdermal plaster 0.75, 1.5, 2, 3, 4, and 8 mg; vaginal tabl. 0.025 mg. 0.3 mg N, 2 mg O, 1 mg P, 50 µg patch, and 7.5 µg V.	Schering AG synthetic [227, 228]
Estriol [Estriol-Ovulum, Hormonin, Ortho-Gynest, Ovestin, Synapause, Trofogin] G03AC04		[*50-27-1*] $C_{18}H_{24}O_3$ 288.39	endogenous estrogen and agonist to estrogen receptor. estrogen used, e.g., for menopausal hormone therapy. cream 0.5 mg/g; drg. 1 mg; f.c. tabl. 2 mg; ovula 0.03 and 0.5 mg; tabl. 1 and 2 mg. 2 mg O, P; 0.2 mg V.	synthetic [229]
Chlorotrianisene (CTA) [Diuril, Merbentul, Tace] G03CA06 wfm		[*569-57-3*] $C_{23}H_{21}ClO_3$ 380.87	nonsteroidal estrogen binding as agonist to estrogen receptor. used to treat menopausal symptoms. cps. 12, 24, and 72 mg. 24 mg O.	Merrell US/1952 synthetic [230]
Estrone (oestron) [Colpormon, Ogen, Prempak] G03CA07; G03CC04 (→ Antineoplastic and Immunomodulating Agents (L)		[*53-16-7*] $C_{18}H_{22}O_2$ 270.37	endogenous estrogen and agonist to estrogen receptor. estrogen used, e.g., for menopausal hormone therapy. 1.4 mg in comb.; amp. 20 mg; drg. 0.625, 1.25, and 2.5 mg; tabl. with 1 mg pregnenolone, 1 mg androstenedione, 0.5 mg androstenediol, 0.1 mg testosterone, 5 µg estrone, and 7.5 mg dried thyroid; vial 20 mg. 1 mg O.	first discovered in 1929 synthetic [231, 232]

Table 22. (*Continued*)

INN (Synonyms) [Brand names] ATC	Structure (Remarks)	[CAS-No.] Formula MW, g/mol	Target (if known) Medical use Formulation DDD	Originator Approval (country/year) Production method [References]
Promestriene [Colposeptine, Colpotrophine, Delipoderm] G03CA09		[*39219-28-8*] $C_{22}H_{32}O_2$ 328.50	synthetic steroidal estrogen. cream 1%; vaginal cps. 10 mg.	Theramex synthetic [233]
Mestranol [Enovid, Metrulene, Necon, Norinyl-1, Ovaras] G03CA10		[*72-33-3*] $C_{21}H_{26}O_2$ 310.44	agonist to estrogen receptor. estrogen for treatment of menstrual disorders and contraceptive. drg. 0.05 and 0.08 mg; tabl. 0.02, 0.05, and 0.08 mg.	G.D. Searle US/1957 synthetic [234, 235]

4.3.2. G03CB Synthetic Estrogens, Plain (Table 23)

Table 23. G03CB Synthetic estrogens, plain

INN (Synonyms) [Brand names] ATC	Structure (Remarks)	[CAS-No.] Formula MW, g/mol	Target (if known) Medical use Formulation DDD	Originator Approval (country/year) Production method [References]
Dienestrol (dienoestrol) [Dienol, Dinovex, Sebohormal, Ortho Dienestrol, Sexadien] G03CB01; G03CC02		[*84-17-3*] $C_{18}H_{18}O_2$ 266.34	synthetic nonsteroidal estrogen of stilbestrol group. used for treatment of prostate cancer and vaginal atrophy. cream 0.01%; tabl. 5 and 25 mg. 2.5 mg O; 0.2 mg V.	Schering US/1947 synthetic [236, 237]
Diethylstilbestrol [Cyren A, Distilbene, Menopax, Tylosterone] G03CB02; G03CC05; L02AA01 wfm		[*56-53-1*] $C_{18}H_{20}O_2$ 268.36	synthetic nonsteroidal estrogen of stilbestrol group. formerly used in estrogenic hormone therapy and prostate cancer, listed as a known carcinogen. tabl. 1 and 5 mg. 0.2 mg O; 0.2 mg V.	US/1941 synthetic [238, 239]
Methallenestril (methallenoestril, methallenoestrol) [Vallestril] G03CB03; G03CC03 wfm		[*517-18-0*] $C_{18}H_{22}O_3$ 286.37	synthetic nonsteroidal estrogen. treatment of menstrual issues. tabl. 3 mg. 9 mg O.	G.D. Searle US/1950 synthetic [240, 241]

4.3.3. G03CX Other Estrogens (Table 24)

Table 24. G03CX Other estrogens

INN (Synonyms) [Brand names] ATC	Structure (Remarks)	[CAS-No.] Formula MW, g/mol	Target (if known) Medical use Formulation DDD	Originator Approval (country/year) Production method [References]
Tibolone [Livial, Liviella, Tinox] G03CX01		[5630-53-5] $C_{21}H_{28}O_2$ 312.45	synthetic steroid and agonist to various receptors. anabolic steroid, immunomodulating steroid, and treatment of postmenopausal vasomotor symptoms. tabl. 2.5 mg. 2.5 mg O.	Organon NL/1988 synthetic [242]

4.4. G03D Progestogens

4.4.1. G03DA Pregnen (4) Derivatives (Table 25)

Table 25. G03DA Pregnen (4) derivatives

INN (Synonyms) [Brand names] ATC	Structure (Remarks)	[CAS-No.] Formula MW, g/mol	Target (if known) Medical use Formulation DDD	Originator Approval (country/year) Production method [References]
Gestonorone caproate (gestronol) [Depostat, Primostat] G03DA01		[1253-28-7] $C_{26}H_{38}O_4$ 414.58	progestin. treatment of enlarged prostate and anticancer. inj. amp. 200 mg/2 mL. 30 mg P.	Schering AG 1968 synthetic [243, 244]
Hydroxyproge-sterone caproate [Delautin, Lentogest, Makena, Proge, Proluton Depot] G03DA03		[630-56-8] $C_{27}H_{40}O_4$ 428.61	endogenous proge-stogen steroid hormone, agonist of the progesterone receptor. depot progestogen (depot gestagen) for prevention of preterm birth in pregnant women and contraceptive in comb. amp. 250 mg/mL. 10 mg P.	Schering US/1955 synthetic [245]
Progesterone [Crinone, Lutnum, Progestasert, Progestol, Prometium] G03DA04		[57-83-0] $C_{21}H_{30}O_2$ 314.47	endogenous progestogen. menopausal hormone therapy. amp. 10, 25, and 50 mg/1 mL; amp. 20 mg in comb. with estradiol benzoate; cps. 100 mg; gel 1%, 4%, and 8%. 0.3 g O; 5 mg P; 0.2 g R; 90 mg V.	1934 first extracted from urine/synthetic [246, 247]

4.4.2. G03DB Pregnadiene Derivatives (Table 26)

Table 26. G03DB Pregnadiene derivatives

INN (Synonyms) [Brand names] ATC	Structure (Remarks)	[CAS-No.] Formula MW, g/mol	Target (if known) Medical use Formulation DDD	Originator Approval (country/year) Production method [References]
Dydrogesterone [Climaston, Duphaston] G03DB01		[152-62-5] $C_{21}H_{28}O_2$ 312.45	progestogen. menopausal hormone therapy. tabl. 5 and 10 mg. 10 mg O.	Duphar/Solvay 1961 synthetic [248, 249]
Medrogestone [Colprone, Etogyn, Presomen] G03DB03		[977-79-7] $C_{23}H_{32}O_2$ 340.51	progestogen. used for menopausal hormone therapy and anticancer. tabl. 5 and 25 mg. 5 mg O.	Ayerst 1969 synthetic [250, 251]
Nomegestrol acetate (NOMAC) [Lutenyl, Naemis, Zoely] G03DB04		[58652-20-3] $C_{23}H_{30}O_4$ 370.49	progestin. synthetic progestogen for treatment of gynecological disturbances; contraceptive. tabl. 5 mg.	Theramex EU/1986 synthetic [252]
Demegestone [Lutionex] G03DB05		[10116-22-0] $C_{21}H_{28}O_2$ 312.45	progestogen. tabl. 500 mg.	Roussel synthetic [253, 254]
Promegestone [Surgestone] G03DB06		[34184-77-5] $C_{22}H_{30}O_2$ 326.48	progestogen. tabl. 0.125, 0.25, and 0.5 mg.	Cassene synthetic [255]

4.4.3. G03DC Estren Derivatives (Table 27)

Table 27. G03DC Estren derivatives

INN (Synonyms) [Brand names] ATC	Structure (Remarks)	[CAS-No.] Formula MW, g/mol	Target (if known) Medical use Formulation DDD	Originator Approval (country/year) Production method [References]
Allylestrenol [Arandal, Elmolan, Gestanon, Perselin, Turinal] G03DC01		[432-60-0] $C_{21}H_{32}O$ 300.49	progestogen. treatment of BPH and premature labor. tabl. 5 and 25 mg. 10 mg O.	Organon 1961 synthetic [256]
Ethisterone [Menstrogen, Progestoral, Trosinone] G03DC04 wfm		[434-03-7] $C_{21}H_{28}O_2$ 312.45	progestogen. cps. 50, 100, and 250 mg; tabl. 25 mg. 5 mg O.	Schering AG 1939 synthetic [257]
Methylestrenolone (normethandrone) [Methalutin, Lutenin, Orgasteron] G03DC31		[514-61-4] $C_{19}H_{28}O_2$ 288.43	progestogen. tabl. 1 mg. 5 mg O.	Syntex 1957 synthetic [258]

4.5. G03E Androgens and Female Sex Hormones in Combination (Table 28)

Table 28. G03E Androgens and female sex hormones in combination

INN (Synonyms) [Brand names] ATC	Structure (Remarks)	[CAS-No.] Formula MW, g/mol	Target (if known) Medical use Formulation DDD	Originator Approval (country/year) Production method [References]
Prasterone enanthate (DHEA-E) [Binodian, Gynodian Depot] G03EA03; A14AA07		[53-43-0] $C_{19}H_{28}O_2$ 288.42	androgen. menopausal hormone therapy. tabl. 10 mg, amp. 200 mg in comb. with estradiol valerate.	Schering AG 1975 synthetic [259]

4.6. G03G Gonadotropins and Other Ovulation Stimulants

4.6.1. G03GA Gonadotropins (Table 29)

Table 29. G03GA Gonadotropins

INN (Synonyms) [Brand names] ATC	Structure (Remarks)	[CAS-No.] Formula MW, g/mol	Target (if known) Medical use Formulation DDD	Originator Approval (country/year) Production method [References]
Chorionic gonadotropin (hCG) [Ovidrel, Novarel, Pubergen, Pregnyl] G03GA01	glycoprotein composed of 237 amino acids in two subunits >Alpha Chain APDVQDCPECTLQENPFFSQPGAPILQCMGCCFSRAYPTPLRSKKTMLVQKNVTSESTCC VAKSYNRVTVMGGFKVENHTACHCSTCYYHKS >Beta Chain SKEPLRPRCRPINATLAVEKEGCPVCITVNTTICAGYCPTMTRVLQGVLPALPQVVCNYR DVRFESIRLPGCPRGVNPVVSYAVALSCQCALCRRSTTDCGGPKDHPLTCDDPRFQDSSS SKAPPPSLPSPSRLPGPSDTPILPQ	[9002-61-3]	gonad-stimulating hormone produced by placenta that interacts with LHCG receptor. induction of ovulation and fertility. powder for solution, 1000 units/vial. 250 U P.	Ayerst 1951 extracted from urine of pregnant women or produced by recombinant DNA technology [260]
Human menopausal gonadotropin (menotropin, hMG) [Fertinex, Humegon, Menopur, Repronex] G03GA02	>Alpha Chain (LH) APDVQDCPECTLQENPFFSQPGAPILQCMGCCFSRAYPTPLRSKKTMLVQKNVTSESTCC VAKSYNRVTVMGGFKVENHTACHCSTCYYHKS >Beta Chain (LH) SREPLRPWCHPINAILAVEKEGCPVCITVNTTICAGYCPTMMRVLQAVLPPLPQVVCTYR DVRFESIRLPGCPRGVDPVVSFPVALSCRCGPCRRSTSDCGGPKDHPLTCDHPQLSGLLFL >Alpha Chain (FSH) APDVQDCPECTLQENPFFSQPGAPILQCMGCCFSRAYPTPLRSKKTMLVQKNVTSESTCC VAKSYNRVTVMGGFKVENHTACHCSTCYYHKS >Beta Chain (FSH) NSCELTNITIAIEKEECRFCISINTTWCAGYCYTRDLVYKDPARPKIQKTCTFKELVYET VRVPGCAHHADSLYTYPVATQCHCGKCDSDSTDCTVRGLGPSYCSFGEMKE	[61489-71-2]	hormone medication containing the peptide hormones follicle-stimulating hormone (FSH) and luteinizing hormone (LH). treatment of fertility disturbances. powder for solution 75 units/vial. 75 U P.	Ferring US/2003 extracted from urine of postmenopausal women [261]
Serum gonadotropin (eCG, PMSG) [Folligon. Prospecbio] G03GA03	glycoprotein hormone consisting of 247 amino acids		gonadotropic hormone; pregnant mare's serum gonadotropin (PMSG) veterinary use for ovulation induction and artificial insemination. lyophilisate. 750 U P.	isolated from pregnant mare's urine [262]
Urofollitropin [Bravelle, Fertinex, Fostimon] G03GA04	>Alpha chain APDVQDCPECTLQENPFFSQPGAPILQCMGCCFSRAYPTPLRSKKTMLVQKNVTSESTCC VAKSYNRVTVMGGFKVENHTACHCSTCYYHKS >Beta chain NSCELTNITIAIEKEECRFCISINTTWCAGYCYTRDLVYKDPARPKIQKTCTFKELVYET VRVPGCAHHADSLYTYPVATQCHCGKCDSDSTDCTVRGLGPSYCSFGEMKE	[146479-72-3]	purified form of follicle-stimulating hormone (FSH). used in combination with hCG to assist to ovulation and fertility. powder for solution 75 units/vial. 75 U P.	extraction from human urine and purified [263]

(continued)

Table 29. (Continued)

INN (Synonyms) [Brand names] ATC	Structure (Remarks)	[CAS-No.] Formula MW, g/mol	Target (if known) Medical use Formulation DDD	Originator Approval (country/year) Production method [References]
Follitropin alfa (r-FSH alfa) [Bemfola, Gonal-f, Gonal-F RFF Redi-ject, Ovaleap] G03GA05	glycoprotein hormone consisting of 203 amino acids >Alpha chain APDVQDCPECTLQENPFFSQPGAPILQCMGCCFSRAYPTPLRSKKTMLVQKNVTSESTCC VAKSYNRVTVMGGFKVENHTACHCSTCYYHKS >Beta chain NSCELTNITIAIEKEECRFCISINTTWCAGYCYTRDLVYKDPARPKIQKTCTFKELVYET VRVPGCAHHADSLYTYPVATQCHCGKCDSDSTDCTVRGLGPSYCSFGEMKE	[146479-72-3]	hormone identical to FSH. female ovulation and fertility induction. vial with lyophilized powder for injection, syringe and pen injector. 75 U P.	Serono US/2004 recombinant DNA production in CHO cells [264]
Follitropin beta (r-FSH beta) [Follistim, Puregon] G03GA06	glycoprotein hormone consisting of 203 amino acids identical to follitropin alpha	[146479-72-3]	hormone identical to FSH. female ovulation and fertility induction; fertility stimulation. vial with lyophilized powder for injection. 75 U P.	Organon US/2004 recombinant DNA production in CHO cells [265]
Lutropin alfa (r-hLH) [Luveris, Pergoveris] G03GA07	glycoprotein hormone consisting of 213 amino acids >Alpha Chain APDVQDCPECTLQENPFFSQPGAPILQCMGCCFSRAYPTPLRSKKTMLVQKNVTSESTCC VAKSYNRVTVMGGFKVENHTACHCSTCYYHKS >Beta Chain (LH) SREPLRPWCHPINAILAVEKEGCPVCITVNTTICAGYCPTMMRVLQAVLPPLPQVVCTYR DVRFESIRLPGCPRGVDPVVSFPVALSCRCGPCRRSTSDCCGPKDHPLTCDHPQL.SGLLFL	[152923-57-4]	luteinizing hormone (LH). induction of ovulation. powder for solution. 75 U P.	EMD Serono 2002 recombinant DNA production yeast [266]
Choriogonado-tropin alfa (r-hCG) [Ovidrel, Ovitrelle, Profasi] G03GA08	glycoprotein hormone consisting of 237 amino acids >Alpha chain APDVQDCPECTLQENPFFSQPGAPILQCMGCCFSRAYPTPLRSKKTMLVQKNVTSESTCC VAKSYNRVTVMGGFKVENHTACHCSTCYYHKS >Beta chain SKEPLRPRCRPINATLAVEKEGCPVCITVNTTICAGYCPTMTRVLQGVLPALPQVVCNYR DVRFESIRLPGCPRGVNPVVSYAVALSCQCALCRRSTTDCGGPKDHPLTCDDPRFQDSSS SKAPPPSLPSPSRLPGPSDTPILPQ	[195962-23-3]	gonadotropin. treatment of female infertility. 0.25 mg P.	recombinant DNA production [267]
Corifollitropin alfa (r-hFSH, MK-8962) [Elonva] G03GA09	the agent comprises an α-subunit, which is identical to that of FSH, and a β-subunit, which is produced by the fusion of the C-terminal peptide from the β-subunit of chorionic gonadotropin to the β-subunit of FSH		long-acting gonadotropin similar to FSH. treatment of female infertility with ovulation stimulation. inj. solution, 100 and 150 µg. 0.15 mg P.	MSD US/2010 recombinant DNA production [268]
Follitropin delta (r-hFSH, FE 999049) [Rekovelle] G03GA10	>Alpha chain APDVQDCPECTLQENPFFSQPGAPILQCMGCCFSRAYPTPLRSKKTMLVQKNVTSESTCC VAKSYNRVTVMGGFKVENHTACHCSTCYYHKS >Beta chain NSCELTNITIAIEKEECRFCISINTTWCAGYCYTRDLVYKDPARPKIQKTCTFKELVYET VRVPGCAHHADSLYTYPVATQCHCGKCDSDSTDCTVRGLGPSYCSFGEMKE	[146479-72-3]	hormone identical to FSH. treatment of female infertility with ovulation stimulation and IVF. inj. solution 12, 36, and 72 µg. 12 µg P.	Ferring EU/2016 recombinant DNA production in human cell line PER.C6 [269]

4.6.2. G03GB Ovulation Stimulants, Synthetic (Table 30)

Table 30. G03GB Ovulation stimulants, synthetic

INN (Synonyms) [Brand names] ATC	Structure (Remarks)	[CAS-No.] Formula MW, g/mol	Target (if known) Medical use Formulation DDD	Originator Approval (country/year) Production method [References]
Cyclofenil [Ondogyne, Fertodur, Ondonit, Neoclym, Sexovid] G03GB01		[2624-43-3] $C_{23}H_{24}O_4$ 364.44	selective estrogen receptor modulator (SERM). gonadotropin stimulant (against infertility). tabl. 100, 200, and 400 mg. 0.14 g O.	1970 synthetic [270]
Clomifene [Clomid, Pergotime, Serophene], Serofene] G03GB02 **WHO essential medicine**		[911-45-5] $C_{26}H_{28}ClNO$ 405.97 citrate salt: [50-41-9] $C_{26}H_{28}ClNO$ $\cdot C_6H_8O_7$ 598.09	selective estrogen receptor modulator (SERM). synthetic gonadotropin stimulant, antiestrogen. treatment of infertility. cps. 50 mg; tabl. 50 mg (as citrate). 9 mg O.	Merrell US/1967 synthetic [271]
Epimestrol [Alene, Stimovul] G03GB03		[7004-98-0] $C_{19}H_{26}O_3$ 302.41	estrogen. ovulation stimulant, pituitary activator. tabl. 5 mg. 10 mg O.	Organon synthetic [272]

4.7. G03H Antiandrogens

4.7.1. G03HA Antiandrogens, Plain (Table 31)

Table 31. G03HA Antiandrogens, plain

INN (Synonyms) [Brand names] ATC	Structure (Remarks)	[CAS-No.] Formula MW, g/mol	Target (if known) Medical use Formulation DDD	Originator Approval (country/year) Production method [References]
Cyproterone acetate (CPA) [Androcur, Climen, Cyprostat, Diane, Lumalia, Virilit] G03H01		[427-51-0] $C_{24}H_{29}ClO_4$ 374.91	antiandrogen. hormone therapy and anti-acne. amp. 300 mg/3 mL; tabl. 10 and 50 mg. 0.1 g O, P; 25 mg P depot.	Schering AG EU/1973 synthetic [273]

4.8. G03X Other Sex Hormones and Modulators of the Genital System

4.8.1. G03XA Antigonadotropins and Similar Agents (Table 32)

Table 32. G03XA Antigonadotropins and similar agents

INN (Synonyms) [Brand names] ATC	Structure (Remarks)	[CAS-No.] Formula MW, g/mol	Target (if known) Medical use Formulation DDD	Originator Approval (country/year) Production method [References]
Danazol [Bonzol, Danol, Dainazol, Danocrine] G03XA01		[17230-88-5] $C_{22}H_{27}NO_2$ 337.46	antigonadotropin and antiestrogenic. treatment of endometriosis and fibrocystic breast disease. cps. 50, 100, and 200 mg; tabl. 100 and 200 mg. 0.6 g O.	Sterling Winthrop US/1971 synthetic [274, 275]
Gestrinone [Dimetriose, Dimetrose, Nemestrane] G03XA02		[16320-04-0] $C_{21}H_{24}O_2$ 308.42	orally active progestogen. treatment of endometriosis; antigonadotropin. cps. 2.5 mg. 0.7 mg O.	Roussel-Uclaf 1986 synthetic [276]

4.8.2. G03XB Progesterone Receptor Modulators (Table 33)

Table 33. G03XB Progesterone receptor modulators

INN (Synonyms) [Brand names] ATC	Structure (Remarks)	[CAS-No.] Formula MW, g/mol	Target (if known) Medical use Formulation DDD	Originator Approval (country/year) Production method [References]
Mifepristone (Ru-486) [Mifegyne] G03XB01 **WHO essential medicine**		[84371-65-3] $C_{29}H_{35}NO_2$ 429.60	orally active progesterone and glucocorticoid receptor antagonist. antiprogestogen; abortifacient and contraceptive. tabl. 200 mg. 0.2 g O.	Roussel-Uclaf FR/1987 synthetic [277, 278]

4.8.3. G03XC Selective Estrogen Receptor Modulators (Table 34)

Table 34. G03XC Selective estrogen receptor modulators

INN (Synonyms) [Brand names] ATC	Structure (Remarks)	[CAS-No.] Formula MW, g/mol	Target (if known) Medical use Formulation DDD	Originator Approval (country/year) Production method [References]
Raloxifene hydrochloride (LY-156758, keoxifene) [Evista, Optruma] G03XC01		[82640-04-8] $C_{28}H_{27}NO_4S$ · HCl 510.05	selective estrogen receptor modulator (SERM), mixed agonist and antagonist to estrogen receptor. antiestrogen, prevention of osteoporosis. f.c. tabl. 60 mg (as hydrochloride); tabl. 60 mg. 60 mg O.	E. Lilly US/1998 synthetic [279, 280]
Bazedoxifene [Conbrinza, Duavee, Viviant] G03XC02		[198481-32-2] $C_{30}H_{34}N_2O_3$ 470.6	selective estrogen receptor modulator (SERM), mixed agonist and antagonist to estrogen receptor. treatment of postmenopausal osteoporosis. tabl. 20 mg in combination with conjugated estrogen. 20 mg O.	Pfizer EU/2009; US/2013 synthetic [281]
Lasofoxifene [Fablyn, Oporia] G03XC03		[180916-16-9] $C_{28}H_{31}NO_2$ 413.55	nonsteroidal selective estrogen receptor modulator (SERM), mixed agonist and antagonist to estrogen receptor. treatment of postmenopausal osteoporosis and vaginal atrophy. tabl. 0.5 mg. 0.5 mg O.	Pfizer EU/2009 synthetic [282]
Ormeloxifene (centchroman) [Centron, Saheli, Sevista] G03XC04		[78994-24-8] $C_{30}H_{35}NO_3$ 457.61	nonsteroidal selective estrogen receptor modulator (SERM), mixed agonist and antagonist to estrogen receptor. contraceptive. tabl.	Torrent Pharma IN/1991 synthetic [283]

(continued)

Table 34. (*Continued*)

INN (Synonyms) [Brand names] ATC	Structure (Remarks)	[CAS-No.] Formula MW, g/mol	Target (if known) Medical use Formulation DDD	Originator Approval (country/year) Production method [References]
Ospemifene [Osphena, Senshio] G03XC05		[*128607-22-7*] $C_{24}H_{23}ClO_2$ 378.89	mixed agonist and antagonist to estrogen receptor; selective estrogen receptor modulator (SERM). treatment of dyspareunia. tabl. 60 mg. 60 mg O.	Shionogi US/2013 synthetic [284]

4.8.4. G03XX Other Sex Hormones and Modulators of the Genital System (Table 35)

Table 35. G03XX Other sex hormones and modulators of the genital system

INN (Synonyms) [Brand names] ATC	Structure (Remarks)	[CAS-No.] Formula MW, g/mol	Target (if known) Medical use Formulation DDD	Originator Approval (country/year) Production method [References]
Prasterone (DHEA) [Diandrone, Gynodian, Intrarosa, Levospa, Mylis] G03XX01 **controlled substance**		[*53-43-0*] $C_{19}H_{28}O_2$ 288.43	natural steroid hormone; anabolic, androgen. used to treat vaginal atrophy and as hormone therapy. amp. 200 mg/mL; suppos. 600 mg; vial 100 and 200 mg; intravaginal gel.	US/2016 as gel synthetic [285, 286]

5. G04 Urologicals

5.1. G04B Urologicals

5.1.1. G04BA Acidifiers (Table 36)

Table 36. G04BA Acidifiers

INN (Synonyms) [Brand names] ATC	Structure (Remarks)	[CAS-No.] Formula MW, g/mol	Target (if known) Medical use Formulation DDD	Originator Approval (country/year) Production method [References]
L-Methionine [Clinimix, Pramin, Travasol] G04BA04; QA05BA90; V03AB26		[*59-51-8*] $C_5H_{11}NO_2S$ 149.21	essential amino acid for protein biosynthesis. nutraceutical; component of parenteral nutrition; increases urinary excretion; supplement against copper poisoning. inj. solution. 2.25 g O.	synthetic [287]

5.1.2. G04BD Drugs for Urinary Frequency and Incontinence (Table 37)

Table 37. G04BD Drugs for urinary frequency and incontinence

INN (Synonyms) [Brand names] ATC	Structure (Remarks)	[CAS-No.] Formula MW, g/mol	Target (if known) Medical use Formulation DDD	Originator Approval (country/year) Production method [References]
Emepronium bromide [Catiprin] G04BD01		[3614-30-0] $C_{20}H_{28}BrN$ 362.4	anticholinergic used in urology as antispasmodic. 0.5 g O; 75 mg P.	Pharmacia & Upjohn synthetic [288]
Flavoxate [Bladderon, Genurin, Spasuret, Urispas] G04BD02		[15301-69-6] $C_{24}H_{25}NO_4$ 391.47 hydrochloride salt: [3717-88-2] $C_{24}H_{25}NO_4 \cdot HCl$ 427.93	antispasmodic. f.c. tabl. 200 mg; gran. 20%; tabl. 100 and 200 mg (as hydrochloride). 0.8 g O.	Recordati synthetic [289]
Meladrazine (CIBA 13155) [Lisidonil] G04BD03		[13957-36-3] $C_{11}H_{23}N_7$ 253.47	anticholinergic used in urology as antispasmodic. 0.45 g O.	synthetic [290]
Oxibutynin [Cystrin, Dridase, Ditropan, Lyrinel, Pollakisu] G04BD04		[5633-20-5] $C_{22}H_{31}NO_3$ 357.49	anticholinergic and antispasmodic. syrup 2.5 mg/5 mL; tabl. 1, 2, 2.5, 3, and 5 mg (as hydrochloride). 15 mg O; 3.9 mg P.	Alza/Ortho/Janssen US/1975 synthetic [291]
Terodiline [Mictrol, Micdurin, Midurin] G04BD05		[15793-40-5] $C_{20}H_{27}N$ 281.44 hydrochloride salt: [7082-21-5] $C_{20}H_{27}N \cdot HCl$ 317.90	calcium channel blocker and antagonist. antianginal, antihypertensive, anticholinergic, and treatment of urinary frequency and incontinence. f.c. tabl. 12.5 and 25 mg (as hydrochloride). 50 mg O.	synthetic [292]
Propiverine [Detrunorm, Mictonorm, Mictonetten] G04BD06		[60569-19-9] $C_{23}H_{29}NO_3$ 367.49 hydrochloride salt: [54556-98-8] $C_{23}H_{29}NO_3 \cdot HCl$ 403.95	anticholinergic. treatment of incontinence. drg. 5 and 15 mg (as hydrochloride). 30 mg O; 20 mg O children.	Apogepha synthetic [293, 294]

(continued)

Table 37. (*Continued*)

INN (Synonyms) [Brand names] ATC	Structure (Remarks)	[CAS-No.] Formula MW, g/mol	Target (if known) Medical use Formulation DDD	Originator Approval (country/year) Production method [References]
Tolterodine (Kabi 2234, PNU-200583) [Detrol, Detrustil, Detrusitol] G04BD07		[124937-51-5] $C_{22}H_{31}NO$ 325.50 tartrate salt: [124937-52-6] $C_{22}H_{31}NO$ $\cdot C_4H_6O_6$ 475.58	muscarinic M_3-receptor antagonist. urinary incontinence. f.c. tabl. 1 mg, 2 mg (as maleate). 4 mg O.	Pharmacia DE/1998 synthetic [295]
Solifenacin (YM-53705) [Vesikur, Vesicare] G04BD08		[242478-37-1] $C_{23}H_{26}N_2O_2$ 362.47 succinate salt; [242478-38-2] $C_{27}H_{32}N_2O_6$ 480.56	muscarinic M_3-receptor antagonist. treatment of urinary incontinence. f.c. tabl. 5 mg, 10 mg (as succinate). 4 mg O.	Astellas US/2004 synthetic [296]
Trospium chloride [Regurin, Sanctura] G04BD09		[10405-02-4] $C_{25}H_{30}ClNO_3$ 427.96	muscarinic receptor antagonist; anticholinergic. treatment of overactive bladder. tabl. 20 mg. 40 mg O.	Madaus AG EU/1999; US/2004 synthetic [297]
Darifenacin (UK-88525) [Emselex, Enablex] G04BD10		[133099-04-4] $C_{28}H_{30}N_2O_2$ 426.56 hydrobromide salt: [133099-07-7] $C_{28}H_{30}N_2O_2$ $\cdot$ BrH 507.47	muscarinic M_3 antagonist. urinary incontinence. tabl. 7.5 mg, 15 mg (as hydrobromide). 7.5 mg O.	Novartis US/2004 synthetic [298]
Fesoterodine [Toviaz] G04BD11		[286930-02-7] $C_{26}H_{37}NO_3$ 411.28	muscarinic receptor antagonist; antimuscarinic, prodrug. treatment of overactive bladder. f.c. tabl. 4 and 8 mg. 4 mg O.	Schwarz Pharma AG EU/2007 synthetic [299]
Mirabegron (YM-178) [Betamis, Betmiga, Myrbetriq] G04BD12		[223673-61-8] $C_{21}H_{24}N_4O_2S$ 396.51	activation of β_3 adrenergic receptor in the detrusor muscle in the bladder. treatment of overactive bladder. tabl. 25 and 50 mg. 50 mg O.	Astellas synthetic [300]

Table 37. (*Continued*)

INN (Synonyms) [Brand names] ATC	Structure (Remarks)	[CAS-No.] Formula MW, g/mol	Target (if known) Medical use Formulation DDD	Originator Approval (country/year) Production method [References]
Desfesoterodine [Tovedeso] G04BD13		[*207679-81-0*] $C_{22}H_{31}NO_2$ 341.5 succinate salt: [*1512864-59-3*]	muscarinic receptor antagonist. antimuscarinic, active metabolite of fesoterodine. treatment of overactive bladder. tabl. 3.5 and 7 mg. 3.5 mg O.	Teva EU/2018 synthetic [301]
Atropine (DL-hyoscyamine) [Atquick, Diarsed, Donnatal, Lomotil] G04BD15; A03BA01; S01FA01		[*51-55-8*] $C_{17}H_{23}NO_3$ 289.38	anticholinergic, mydriatic, and antispasmodic. amp. 0.5 mg/1 mL; amp. for inj. 100 mg; eye drops 10 mg, 1%; eye ointment 0.1%; inj. sol. 0.25, 0.5, 1, and 2 mg; powder 98% and over; syringe 0.5 mg/1 mL; tabl. 0.5 mg.	by extraction of solanacean drugs, especially *Atropa belladonna*, *Hyoscyamus niger* or other species. [302, 303]
Dicycloverine (dicyclomine) [Bentyl, Diarrest, Merbentyl] G04BD19		[*77-19-0*] $C_{19}H_{35}NO_2$ 309.49 hydrochloride salt: [*67-92-5*] $C_{19}H_{35}NO_2 \cdot HCl$ 345.96	antispasmodic and anticholinergic. cps. 10 mg; gran. 0%; powder 10%; tabl. 10 mg.	Merrell synthetic [304, 305]
Phenoxybenzamine [Dibenzyline, Dibenzyran] G04BD20		[*59-96-1*] $C_{18}H_{22}ClNO$ 303.83 hydrochloride salt: [*63-92-3*] $C_{18}H_{22}ClNO \cdot HCl$ 340.29	α-blocker. vasodilator, antihypertensive. cps. 1, 5, and 10 mg (as hydrochloride).	synthetic [306]
Ethaverine [Etadil, Ethaquin, Predem] G04BD66		[*486-47-5*] $C_{24}H_{29}NO_4$ 395.50 hydrochloride salt: [*985-13-7*]	antispasmodic. suppos. 30 mg in comb. (as hydrochloride).	synthetic [307]

5.1.3. G04BE Drugs Used in Erectile Dysfunction (Table 38)

Table 38. G04BE Drugs used in erectile dysfunction

INN (Synonyms) [Brand names] ATC	Structure (Remarks)	[CAS-No.] Formula MW, g/mol	Target (if known) Medical use Formulation DDD	Originator Approval (country/year) Production method [References]
Alprostadil [Edex, Caverject, Muse, Vitaros] G04BE01, C01EA01 **WHO essential medicine** (→ Cardiovascular System (C))		[745-65-3] $C_{20}H_{34}O_5$ 354.48	naturally occurring prostaglandin E1; vasodilator. treatment of erectile dysfunction. inj. powder 5.4 µg/mL and urethral suppositories; cream 200/300 µg per 100 mg. 20 µg P; 0.25 mg urethal.	Pharmacia & Upjohn US/1995 synthetic [308]
Papaverine [Aspace, Nyxanthan, Oxadilene, Vasclerran] G04BE02; A03AD01; C04AX38 (→ Cardiovascular System (C))		[58-74-2] $C_{20}H_{21}NO_4$ 339.39 hydrochloride salt: [61-25-6] $C_{20}H_{21}NO_4$ · HCl 375.85	phosphodiesterase (PDE) inhibitor; antispasmodic, vasodilator. treatment of spasm and erectile dysfunction. amp. 40 mg/1 mL, 60 mg/2 mL; multiple-dose vial 30 mg/mL; powder 10%, 98.5% and over.	synthetic [309]
Sildenafil (UK-92480) [Revatio, Viagra] G04BE03; C02KX06 (→ Cardiovascular System (C))		[139755-83-2] $C_{22}H_{30}N_6O_4S$ 474.59 citrate salt: [171599-83-0] $C_{22}H_{30}N_6O_4S$ · $C_6H_8O_7$ 666.71	PDE 5-inhibitor. therapeutic for erectile dysfunction and pulmonary arterial hypertension. f.c. tabl. 25, 50, and 100 mg (as citrate). 50 mg O.	Pfizer US/1998; EU/1999 synthetic [310, 311]
Yohimbine [Yocon] G04BE04		[146-48-5] $C_{21}H_{26}N_2O_3$ 354.44	α2-adrenergic receptor antagonist. treatment of erectile dysfunction, aphrodisiac. tabl. 2 and 5.4 mg. 15 mg O.	isolated from bark of *Pausinystalia johimbe* [312]
Moxisylyte (thymoxamine) [Carlytene, Opilion] G04BE06		[54-32-0] $C_{16}H_{25}NO_3$ 279.38	α1-adrenergic receptor antagonist. treatment of erectile dysfunction. tabl.	Fujirebo/Archimedes Pharma EU/1987 synthetic [313]

Table 38. (*Continued*)

INN (Synonyms) [Brand names] ATC	Structure (Remarks)	[CAS-No.] Formula MW, g/mol	Target (if known) Medical use Formulation DDD	Originator Approval (country/year) Production method [References]
Apomorphine [Apokyn, Ixense, Movapo, Sontane, Uprima] G04BE07; N04BC07 (→ Anti-Parkinson Drugs (N04))		[*58-00-4*] $C_{17}H_{17}NO_2$ 267.23	nonselective dopamine agonist. treatment of erectile dysfunction, addiction, and anxiety. sc. inj. 10 mg/1 mL and 30 mg/3 mL. 2 mg SL.	synthetic [314]
Tadalafil (IC-151) [Adcirca, Cialis] G04BE08; C02KX07 (→ Cardiovascular System (C))		[*171596-29-5*] $C_{22}H_{19}N_3O_4$ 389.41	phosphodiesterase 5 (PDE 5) inhibitor. treatment of erectile dysfunction. tabl. 10 mg, 20 mg. 10 mg O.	E. Lilly/Icos US/2003 synthetic [315, 316]
Vardenafil (bay-38-9456, nuviva) [Levitra, Staxyn, Vivanza] G04BE09		[*224785-90-4*] $C_{23}H_{32}N_6O_4S$ 488.61 dihydrochloride salt: [*224789-15-5*] $C_{23}H_{32}N_6O_4S$ · 2 HCl 561.54	phosphodiesterase 5 (PDE 5) inhibitor. treatment of erectile dysfunction. tabl. 5, 10, and 20 mg. 10 mg O.	Bayer US/2003 synthetic [317, 318]
Avanafil [Spedra, Stendra] G04BE10		[*330784-47-9*] $C_{23}H_{26}ClN_7O_3$ 483.91	phosphodiesterase 5 (PDE 5) inhibitor. treatment of erectile dysfunction. tabl. 50, 100, and 200 mg. 0.1 g O.	Tanabe/Menarini US/2012; EU/2013 synthetic [319]
Udenafil [Zydena] G04BE11		[*268203-93-6*] $C_{25}H_{36}N_6O_4S$ 516.66	phosphodiesterase 5 (PDE 5) inhibitor. treatment of erectile dysfunction. tabl. 100 and 200 mg.	Dong-A Pharmaceuticals SK/2005 synthetic [320]

5.1.4. G04BX Other Urologicals (Table 39)

Table 39. G04BX Other urologicals

INN (Synonyms) [Brand names] ATC	Structure (Remarks)	[CAS-No.] Formula MW, g/mol	Target (if known) Medical use Formulation DDD	Originator Approval (country/year) Production method [References]
Acetohydroxamic acid (AHA) [Lithostat] G04BX3		[546-88-3] $C_2H_5NO_2$ 75.0	urease inhibitor. used against urinary tract infections and prevention of stones.	US/1983 synthetic [321]
Phenazopyridine [Azocline, Pyridium] G04BX06		[94-78-0] $C_{11}H_{11}N_5$ 213.24 hydrochloride salt: [136-40-3] $C_{11}H_{11}N_5 \cdot HCl$ 249.71	inhibition of sodium channels. chemotherapeutic, antiseptic, and urinary tract analgesic. cps. 50 mg (as hydrochloride) in comb.; f.c. tabl. 50 mg; tabl. 100 mg, 200 mg (as hydrochloride). 0.6 g O.	Roche synthetic [322, 323]
Phenyl salicylate (salol) [Azuphen, Darcalma, Watkins Settelz] G04BX012		[118-55-8] $C_{13}H_{10}O_3$ 212.44	prodrug of salicylic acid. antiseptic and analgesic; food supplement caps., tabl., and liquids. 2 g O.	synthetic [324]
Dimethyl sulfoxide (DMSO) [Dermasorb; Dolobene; Dolocur; Domoso; Hyadur; Kernsol; Sclerosol; Topsym] G04BX13; M02AX03 (→ Drugs for the Musculo-Skeletal System (M)		[67-68-5] C_2H_6OS 78.13	skin penetration enhancer. topical analgesic, antiinflammatory, and antioxidant. gels and ointments.	synthetic [325]
Dapoxetine (LY-210448) [Priligy, Westoxetin] G04BX14		[119356-77-3] $C_{21}H_{23}NO$ 305.42 hydrochloride salt: [129938-20-1] $C_{21}H_{23}NO \cdot HCl$ 341.88	serotonin reuptake inhibitor. treatment of premature ejaculation. tabl. 30 and 60 mg. 30 mg O.	E. Lilly/J&J EU/2009 synthetic [326, 327]
Tiopronin (mercamidum) [Acadione; Captimer; Thiola] G04BX16; R05CB12 (→ Respiratory Disorders, 3. Throat Preparations (R02))		[1953-02-2] $C_5H_9NO_3S$ 163.20 monosodium salt: [2015-20-0] $C_5H_8NNaO_3S$ 185.18	mucolytic, hepatoprotectant, and detoxicant. amp. 100 mg, 250 mg; drg. 100 and 250 mg; gran. 150 and 350 mg; tabl. 100 mg.	synthetic [328, 329]

Table 39. (*Continued*)

INN (Synonyms) [Brand names] ATC	Structure (Remarks)	[CAS-No.] Formula MW, g/mol	Target (if known) Medical use Formulation DDD	Originator Approval (country/year) Production method [References]
Chondroitin sulfate (dermatan sulfate) [Remaxazon, Theraflex] G04BX17; B01AX04; M01AX25 (→ Antithrom- botics (B01) and Antihemorrhagics (B02))	(glycosaminoglycan composed of alternating *N*-acetylgalactosamine and glucuronic acid sulfated in variable positions)	[24967-93-9] sodium salt: [39455-18-0]	used for treating osteoarthritis. caps. and tabl. 500 mg; patch.	Bayer US/2015 extraction from cartilaginous cow and pig tissues [330]
Duloxetine (LY-264453, LY-248686) [Cymbalta, Xeristar, Yentreve] G04BX18		[116539-59-4] $C_{18}H_{19}NOS$ 297.42 oxalate salt: [116817-77-7] $C_{18}H_{19}NOS \cdot C_2H_2O_4$ 355.46	serotonin–nor- epinephrine reuptake inhibitor. antidepressant, treatment of stress urinary incontinence and premature ejaculation. cps. 20, 30, 40, and 60 mg. 80 mg O.	E. Lilly/Boehringer Ingelheim EU/2004; US/2004 synthetic [331, 332]
Chlorhexidine [Betasept, Chloraprep, Corsodyl; Hibitane; Hydrex; Meridol; Peridex; Plurexid; Unisept] G04BX19; A01AB03; B05CA02; D08AC02; D09AA12; R02AA05 (→ Stomatological Preparations (A01); → Respiratory Disorders, 3. Throat Preparations (R02))		[55-56-1] $C_{22}H_{30}Cl_2N_{10}$ 505.46 dihydro- chloride: [3697-425] $C_{22}H_{30}Cl_2N_{10}$ $\cdot 2$ HCl 578.38 digluconate: [18472-51-0] $C_{22}H_{30}Cl_2N_{10}$ $\cdot 2\,C_6H_{12}O_7$ 897.77	binding to bacterial cell wall. antiseptic, antibacterial, and disinfectant. gel 1 g/100 g; powder 1 g/100 g; sol. 0.1 g/100 g, 0.2 g/100 g, 1 g/50 mL (as digluconate); troche 5 mg.	1952 synthetic [333]
Tolvaptan (OPC-41061) [Jynarque, Samsca, Tolvat] G04BX21; C03XA01 (→ Cardiovascular System (C))		[150683-30-0] $C_{26}H_{25}ClN_2O_3$ 448.95	vasopressin receptor antagonist. diuretic, hyponatremia treatment. tabl. 15 and 30 mg. 0.12 g O.	Otsuka US/2009 synthetic [334]

5.2. G04C Drugs Used in Benign Prostatic Hypertrophy

5.2.1. G04CA α-Adrenoreceptor Antagonists (Table 40)

Table 40. G04CA α-Adrenoreceptor antagonists

INN (Synonyms) [Brand names] ATC	Structure (Remarks)	[CAS-No.] Formula MW, g/mol	Target (if known) Medical use Formulation DDD	Originator Approval (country/year) Production method [References]
Alfluzosin [Alfunar, Mittoval, Urion, Uroxatral, Xatral] G04CA01		[81403-80-7] $C_{19}H_{27}N_5O_4$ 389.46 hydrochloride salt: [81403-68-1] $C_{19}H_{27}N_5O_4$ · HCl 425.92	α_1-adrenoreceptor antagonist, α-blocker. treatment of benign prostatic hypertrophy (BPH); antihypertensive. film tabl. 2.5 mg; retard tabl. 10 mg (hydrochloride). 7.5 mg O.	Synthelabo/Sanofi EU/1988; US/2003 synthetic [335, 336]
Tamsulosin ((−)-LY 253352, LY 253351, (−)-YM 12617, (R)-(−)-YM 12617) [Alna Ocas, Flomax, Harnal, Josir, Mecir, Omix, Omnic] G04CA02		[06133-20-4] $C_{20}H_{28}N_2O_5S$ 408.52 hydrochloride salt: [106463-17-6] $C_{20}H_{28}N_2O_5S$ · HCl 444.98	α-adrenoreceptor antagonist, α-blocker. antihypertensive, treatment of BPH. cps. 0.1, 0.2, and 0.4 mg. 0.4 mg O.	Boehringer Ingelheim US/1997 synthetic [337, 338]
Terazosin [Dysalfa, Heitrin, Hytracin, Hytrin, Unoprost, Vasomet] G04CA03		[63590-64-7] $C_{19}H_{25}N_5O_4$ 387.44 hydrochloride salt: [63074-08-8] $C_{19}H_{25}N_5O_4$ · HCl 423.90	α-adrenoreceptor antagonist, α-blocker. antihypertensive, treatment for enlarged prostate. cps. 1, 2, 5, and 10 mg; tabl. 0.25, 0.5, 1, 2, 5, and 10 mg. 5 mg O.	Abbott US/1987 synthetic [339, 340]
Silodosin (KMD-3213) [Repaflo, Silodyx, Urorec] G04CA04		[160970-54-7] $C_{25}H_{32}F_3N_3O_4$ 495.54	α-adrenoreceptor antagonist. prostate antihypertrophic. cps. 4 and 8 mg. 8 mg O.	Watson/Recordati US/2008; EU/2010 synthetic [341, 342]
Doxazosin [Cardular, Cardura, Diblocin, Zoxan] G04CA05; C02CA04 (→ Cardiovascular System (C))		[74191-85-8] $C_{23}H_{25}N_5O_5$ 451.48 hydrochloride salt: [70918-01-3] $C_{23}H_{25}N_5O_5$ · HCl 487.94 mesylate salt: [77883-43-3]	α_1-adrenoreceptor antagonist and α-blocker. antihypertensive. tabl. 0.5, 1, 2, and 4 mg. 6 mg O.	Pfizer/Astra Zeneca EU/1989; US/1990 synthetic [343]

5.2.2. G04CB Testosterone-5-α Reductase Inhibitors (Table 41)

Table 41. G04CB Testosterone-5-α reductase inhibitors

INN (Synonyms) [Brand names] ATC	Structure (Remarks)	[CAS-No.] Formula MW, g/mol	Target (if known) Medical use Formulation DDD	Originator Approval (country/year) Production method [References]
Finasteride [Propecia, Proscar] G04CB01; D11AX10; G04CA01 (→ Dermatologicals (D), 4. Antiseptics and Disinfectants (D08), Anti-Acne Preparations (D10), and Other Dermatological Preparations (D11))		[98319-26-7] $C_{23}H_{36}N_2O_2$ 372.55	5α-reductase inhibitor treatment of benign prostatic hypertrophy (BPH) and male hair loss. f.c. tabl. 5 mg 5 mg O	Merck & Co. US/1992 synthetic [344, 345]
Dutasteride (GG 745, GI-198745, GI 198745X) [Avodart] G04CB02		[164656-23-9] $C_{27}H_{30}F_6N_2O_2$ 528.54	5α-reductase inhibitor. treatment of benign prostatic hypertrophy (BPH). cps. 0.5 mg. 0.5 mg O.	GlaxoSmithKline DE/2003 synthetic [346–348]

5.2.3. G04CX Other Drugs Used in Benign Prostatic Hypertrophy (Table 42)

Table 42. G04CX Other drugs used in benign prostatic hypertrophy

INN (Synonyms) [Brand names] ATC	Structure (Remarks)	[CAS-No.] Formula MW, g/mol	Target (if known) Medical use Formulation DDD	Originator Approval (country/year) Production method [References]
Mepartricin (methylpartricin) [Tricandil] G04CX03; A01AB16; D01AA06; G01AA09 (→ Dermatologicals (D), 2. Antifungals (D01). Emollients and Protectives (D02), and Antipruritics (D04) for Dermatological Use)	Mepartricin A R: –CH₃ Mepartricin B R: –H	[11121-32-7] $C_{60}H_{88}N_2O_{19}$ 1141.36	polyene antibiotic. treatment of candidal and trichomonal gynecological infections and of benign prostatic hypertrophy. tabl. 50000 iu/g, 40 mg; vaginal cream 5000 iu/g; vaginal tabl. 25 000 iu.	Spa 1975 fermentation of *Streptomyces aureofaciens* [68, 69]
Fexapotide (NX-1207) G04CX04	H-Ile-Asp-Gln-Gln-Val-Leu-Ser-Arg-Ile-Lys-Leu-Glu-Ile-Lys-Arg-Cys-Leu-OH	[492447-54-8] 2055.5	pro-apoptotic peptide. under development for BPH treatment. inj.	Nymox US/2018 synthetic [349]
Beta-Sitosterol (β-sitosterin, α-phytosterol) [Azuprostat, Cytellin, Flemun, Sito-Lande] G04CX05		[83-46-5] $C_{29}H_{50}O$ 414.72	phytosterol. prostate adenoma therapeutic (benign prostate hypertrophy, BPH), antihypercholes-terolemic. cps. 10 and 65 mg; gran. 1.76 g/2 g; tabl. 75 mg, 100 mg. 60 mg O.	synthetic [350]

List of Abbreviations

amp.	ampule(s)
ART	assisted reproductive technology
BPH	benign prostate hyperplasia
BV	bacterial vaginosis
cps.	capsules
CHO	Chinese hamster ovary
COC	combination oral contraceptive
COX	cyclooxygenase
DDD	defined daily dose
DES	diethylstilbestrol
drg.	dragee
ECP	emergency contraceptive pill
ED	erectile dysfunction
eff.	effervescent
f.c.	film-coated
FSH	follicle-stimulating hormone
gran.	granules
hCG	human chorionic gonadotropin
hMG	human menopausal gonadotropin
HSDD	hypoactive sexual desire disorder
i.m.	intramuscular
inj.	injection
INN	international nonproprietary name
LHRH	luteinizing hormone-releasing hormone
LUTS	lower urinary tract symptom
MHT	menopausal hormone therapy
MW	molecular mass
N	nasal
NSAID	nonsteroidal antiinflammatory drug
mL	milliliter (cubic centimeter)
O	oral
OAB	overactive bladder
P	parenteral
PABA	*p*-aminobenzoic acid
PDE-5	phosphodiesterase-5
PG	prostaglandin
POP	progestin-only pill
SERM	selective estrogen receptor modulator
sol.	solution
s.r.	slow release
suppos.	suppositories
susp.	suspension
T	topical
tabl.	tablets
U	unit
UI	urinary incontinence
V	vaginal
VVC	vulvovaginal candidiasis
wfm	withdrawn from market
WHO	World Health Organization

References

1 Atala, A. and Amin, M. (1991) *Drugs Aging* **1**, 176–193.
2 Michel Br, M.C. (2011) *J. Clin. Pharmacol.* **72** (2), 183–185.
3 Farage, M.A., et al. (2010) *Infect. Dis. Res. Treat.* **3**, 1–15.
4 Borges, S., et al. (2015) *Front. Clin. Drug Res.: Anti-Infect.* **2**, 3–19.
5 Krasnopolsky, V.N., et al. (2013) *J. Clin. Med. Res.* **5** (4), 309–315.
6 Van Meter, K.C. and Hubert, R.J. (2016) *Microbiology for the Healthcare Professional*, Elsevier, China, pp. 214–232.
7 Rotta, I., et al. (2012) *Rev. Assoc. Med. Bras.* **58** (3), 308–318.
8 Lachapelle, J.-M., et al. (2013) *Clin. Pract.* **10** (5), 579–592.
9 Verstraeten, H., et al. (2012) *BMC Infect. Dis.* **12**, 148.
10 Neut, C., et al. (2015) *Open J. Obstetr. Gynecol.* **5**, 173–180.
11 Vallera, C., et al. (2017) *Anesthesiol. Clin.* **35** (2), 207–219.
12 Kent, N.E. (1999) *J. SOGC* **21** (3), 224–228.
13 Sunil Kumar, K.S., et al. (2016) *J. Obstet. Gynecol. India* **66** (Suppl. 1), 229–234.
14 Allen, R., et al. (2009) *Rev. Obstet. Gynecol.* **2** (3), 159–168.
15 Alfirevic, Z., et al. (2015) *BMJ* **350**, h217.
16 Caughey, A.B., et al. (2001) *Semin. Perinatol.* **25** (4), 248–255.
17 Lamont, R.F. (2003) *BJOG* **110** (Suppl. 20), 108–112.
18 Bernard, V., et al. (2019) *Nat. Rev. Endocrinol.* **15**, 356–365.
19 Ferraris, J., et al. (2013) *Neuroendocrinology* **98**, 171–179.
20 Milani, M. and Iacobelli, P. (2012) *Int. Scholarly Res. Notices* ID673131.
21 Pfaus, J.G. (2009) *J. Sex Med.* **6** (6), 1506–1533.
22 Invermizzi, R.W., et al. (2003) *Br. J. Pharmacol.* **139** (7), 1281–1288.
23 Guerriro, J. (2009) *Ann. N.Y. Acad. Sci.* **1163**, 154–168.
24 Tsui, A.O., et al. (2010) *Epidemiol. Rev.* **32** (1), 152–174.
25 Liao, P.V. and Dollin, J. (2012) *Can. Fam. Phys.* **2** (58), e757–e760.
26 Djerassi, C., et al. (1954) *J. Am. Chem. Soc.* **76**, 4092.
27 Kay, C.R. (1980) *J. Royal Col. Gen. Prasct.* **30**, 8.
28 Siegel, E. (2012) *AM. J. Public Health* **102** (8), 1462–1475.
29 Schindler, A.E., et al. (2003) *Maturitas* **46** (Suppl. 1), 7–16.
30 Mitchell, D.R. Jr. ed. (1983) Long-acting Steroid Contraception, in *Advances in Human Fertility and Reproductive Endocrinology*, Raven Press, New York, NY.
31 Zatuchni, G.J., et al. eds. (1984) *Long-acting Contraceptive Delivery Systems*, Harper and Row, Philadelphia, PA.
32 Genzell-Danielson, K. (2010) *Contraception* **82**, 404–409.
33 Casado-Espada, N.M., et al. (2019) *J. Clin. Med* **8**, 908.
34 Simon, J.A. (2002) *Fertil. Steril.* **77** (Suppl. 4), 77–82.
35 Simon, J., et al. (1999) *Menopause* **6** (2), 138–146.
36 Veurink, M., et al. (2005) *Pharm. World Sci.* **27** (3), 139–143.
37 Stefanick, M.L. (2005) *Am. J. Med.* **118** (12B), 64S–73S.
38 Fait, L. (2019) *Drugs in Context* **8**, 212551.
39 Lunenfeld, B., et al. (2019) *Front. Endocrinol.* **10**, 429.
40 Schertz, W., et al. (2018) *Exp. Opin. Drug Deliv.* **15**, 435–442.
41 Schertz, W., et al. (2018) *Exp. Opin. Drug Deliv.* **15**, 435–442.
42 Andersen, C.Y. and Ezcurra, D. (2014) *Reprod. Biol. Endocrinol.* **12**, 128.
43 Dickey, R.P., et al. (1996) *Human Reprod. Update* **2** (6), 483–506.
44 Maia, H., et al. (1980) *Int. J. Gynecol. Obstet.* **17** (5), 431–433.
45 Riggs, B.L. and Hartmann, L.C. (2003) *N. Engl. J. Med.* **348**, 618–629.

46 McDonell, B., et al. (2017) *F1000Research* **6**, 2146.
47 Balk, E.M., et al. (2019) *Ann. Int. Med.* **170**, 465–470.
48 Leron, E., et al. (2017) *Curr. Urol.* **11**, 117.
49 Stewart, W.F. (2003) *World J. Urol.* **20**, 327–336.
50 Gareri, P., et al. (2014) *Int. J. Endocrinol.* ID 87670.
51 Albersen, M., et al. (2013) *Gerontology* **58** (1), 3–14.
52 Bruzziches, R., et al. (2013) *Exp. Opin. Pharmacother.* **14** (10), 1333–1344.
53 Wu, Y., et al. (2016) *US Pharm.* **41** (8), 36–40.
54 Dahm, P., et al. (2017) *Eur. Urol.* **71** (4), 570–581.
55 Olin Mathieson, (1957) US 2 786 781 (prior. 1954).
56 Research Corp., (1957) US 2 797 183 (prior. 1952).
57 Königl. Niederl. Gist- & Spiritusfabr., GB 844 289 (appl. 1957; NL-prior. 1956).
58 American Cyanamid, GB 846 933 (appl. 1957; USA-prior. 1956).
59 Olin Mathieson, (1959) US 2 908 611 (prior. 1954).
60 Rutgers, US 2 992 162 (11.7.1961; USA-prior. 9.9.1952).
61 Rebstock, M.C., et al. (1949) *J. Am. Chem. Soc. (JACSAT)* **71**, 2458–2468.
62 Parke Davis, (1949) US 2 483 871 (appl. 1948).
63 Mechlinski, W., et al. (1980) *J. Antibiot.* **33** (6), 591–599.
64 Finlay, A.C., et al. (1950) *Science (Washington, D.C.) (SCIEAS)* **111**, 85.
65 Pfizer, (1950) US 2 516 080 (prior. 1949).
66 Beecham, US 3 853 849 (10.12.1974; prior. 2.11.1967 and 29.5.1969).
67 Beecham, US 3 881 013 (29.4.1975; GB-prior. 5.11.1966 and 27.1.1967).
68 Tweit, R.C., et al. (1982) *J. Antibiot. (JANTAJ)* **35**, 997.
69 Spa, DE 2 154 436 (appl. 2.11.1971; GB-prior. 3.11.1970).
70 Upjohn, US 3 418 414 (24.12.1968; appl. 31.8.1966).
71 Upjohn, US 3 475 407 (28.10.1969; appl. 22.12.1967).
72 Frey Tizzi, P., et al. (2010) *Chemotherapy* **56** (3), 190–196.
73 Rutgers Res. Fund, US 2 799 620 (16.7.1956; appl. 29.6.1956).
74 Raiziss, G.W. and Gavron, J.L. (1921) *J. Am. Chem. Soc. (JACSAT)* **43**, 583.
75 Bart, H., DRP 250 264 (appl. 1910).
76 Passek, F. (1925) DRP 411 050.
77 Rohde, W., et al. (1976) *Antimicrob. Agents Chemother.* **10** (2), 243–240.
78 Bourquin, J.-P., et al. (1962) *Arch. Pharm. Ber. Dtsch. Pharm. Ges. (APBDAJ)* **295**, 383.
79 Geigy, (1946) US 2 411 670 (CH-prior. 1942).
80 Allen & Hanburys, GB 745 956 (appl. 1953).
81 Flow Inc., US 4 946 849 (7.8.1990, appl. 10.10.1989).
82 Zinnei, F. (1958) *Arch. Pharm. Ber. Dtsch. Pharm. Ges. (APBDAJ)* **291**, 493.
83 Chem. Fabrik Kalk, DOS 2 515 476 (appl. 9.4.1975).
84 Sadik, F. (1970) *J. Am. Pharm. Assoc.* 18–24.
85 Shuklov, I.A., et al. (2016) *Adv. Synth. Catal.* **358** (24), 3910–3931.
86 Andersch, B., et al. US 8 871 244 (28.10.2014; appl. 28.12.2010; EP-prior. 4.6.2008).
87 Nagoba, B.S., et al. (2013) *J. Infect. Public Health* **6** (6), 410–415.
88 Stacey, M., et al. (1978) *Adv. Carbohydr. Chem. Biochem.* **35**, 1–29.
89 Schenley Ind., (1954) US 2 696 454 (CH-prior. 1949).
90 Bayer, DE 836 350 (appl. 1944).
91 Rhône-Poulenc, US 2 944 061 (5.7.1960; F-prior. 20.9.1975).
92 Bayer, DE 1 617 481 (appl. 15.9.1967).
93 Berg, D., et al. (1984) *Arzneim.-Forsch. (ARZNAD)* **34** (I), 139.
94 Janssen, US 3 717 655 (20.2.1973; appl. 19.8.1968).
95 Janssen, US 3 839 574 (1.10.1974; prior. 23.7.1969).
96 Thienpont, D., et al. (1975) *Arzneimittel. Forschung* **25** (2), 224–230.
97 Veraldi, S., et al. (2013) *Mycoses* **56** (Suppl. 1), 3–15.
98 Janssen, US 3 717 655 (20.2.1973; prior. 19.8.1968).
99 Rutgeerts, P., et al. (2005) *Gastroenterology* **128** (4), 856–861.
100 Godefroi, E.F., et al. (1969) *J. Med. Chem. (JMCMAR)* **12**, 784.
101 Janssen, DOS 1 940 388 (appl. 8.8.1969; USA-prior. 19.8.1968).
102 Janssen, US 3 839 574 (1.10.1974; appl. 19.6.1972; USA-prior. 19.8.1968).
103 Pfizer, DE 2 619 381 (appl. 30.4.1976; GB-prior. 30.4.1975).
104 Pfizer, US 4 062 966 (13.12.1977; appl. 30.4.1976; GB-prior. 30.4.1975).
105 Heeres, J., et al. (1979) *J. Med. Chem. (JMCMAR)* **22**, 1003–1005.
106 Janssen, DOS 2 804 096 (appl. 31.1.1978; USA-prior. 31.1.1977, 21.11.1977).
107 Janssen, US 4 144 346 (13.3.1979; appl. 21.11.1977; USA-prior. 31.1.1977).
108 Recordati, DE 2 917 244 (appl. 9.5.1979; I-prior. 18.5.1978).
109 Recordati, US 4 221 803 (9.9.1980; appl. 9.5.1979; I-prior. 18.5.1978).
110 Marchionni, M., et al. (1981) *Clin. Exp. Obstet. Gynecol.* **8** (1), 18–20.
111 Gaba, M., et al. (2015) *Med. Chem. Res.* **25**, 173–210.
112 Walker, K.A.M., et al. (1978) *J. Med. Chem. (JMCMAR)* **21**, 840.
113 Syntex, US 4 078 071 (USA-prior. 28.7.1975).
114 Thiele, K., et al. (1987) *Helv. Chim. Acta (HCACAV)* **70**, 441.
115 Siegfried AG, US 4 210 657 (1.7.1980; D-prior. 11.9.1978).
116 Siegfried AG, US 4 124 767 (7.11.1978; CH-prior. 24.12.1975).
117 J. Uriach & Cia., EP 352 352 (appl. 31.1.1990; prior. 28.7.1988).
118 Ferrer, EP 151 477 (appl. 2.1.1985; E-prior. 8.6.1984, 2.2.1984, 6.1.1984).
119 Raga, M.M., et al. (1992) *Arzneim.-Forsch. (ARZNAD)* **42**, 691.
120 Heeres, J., et al. (1983) *J. Med. Chem.* **26** (4), 611–613.
121 Kupferberg, A.B., et al. (1961) *Antibiot. Chemother.* **11** (2), 73–78.
122 Kyowa Hakko, US 3 616 212 (26.10.1971; appl. 9.9.1969; JP-prior. 27.9.1967).
123 Salem, H., et al. (2018) *J. Adv. Biomed. Pharm. Sci.* **1**, 85–89.
124 Polichimica SAP, BE 635 608 (appl. 30.7.1963; I-prior. 1.8.1962).
125 Norwich Pharm. Co., (1956), US 2 759 931 (prior. 1953).
126 Norwich Pharm. Co., US 2 927 110 (1.3.1960; prior. 23.1.1958).
127 Maley, A.M., et al. (2013) *Exp. Dermatol.* **22** (12), 775–780.
128 Purdue Frederick, US 4 113 857 (12.9.1978, appl. 19.5.1977).
129 Fleischer, W., et al. (1997) *Dermatology* **195** (2), 3–9.
130 Hoechst AG, US 3 883 545 (13.5.1975; appl. 16.11.1971; prior. 22.12.1972).
131 Hoechst AG, US 3 972 888 (3.8.1976; D-prior. 25.3.1972).
132 Hoechst AG, DE 1 795 270 (appl. 31.8.1968).
133 Riviera, L., et al. (1984) *Chemioterapia* **3** (2), 116–118.
134 Senkus, M. (1946) *J. Am. Chem. Soc. (JACSAT)* **68**, 1611.
135 Commercial Solvents, (1947) US 2 415 047 (appl. 1945).
136 Glaubius, R., et al. (2019) *J. Int. AIDS Soc.* **22** (5), e25282.
137 Burns, D.A. (1986) *Contact Dermat.* **14** (33), 246.
138 Schneider Adv. Technol., US 8 003 823 (23.8.2011, appl. 28.10.2009).
139 Sedlock, D., et al. (1987) *Antimicrob. Agents Chemother.* **28**, 786–790.
140 Joos, B., CH 314 487 (appl. 1953).
141 Sandoz, (1941) US 2 265 207 (CH-prior. 1939).

142 Stoll, A. and Hofmann, A. (1943) *Helv. Chim. Acta (HCACAV)* **26**, 956.

143 Sandoz, (1937) US 2 090 430 (CH-prior. 1936).

144 Lee, H.J., et al. (2009) *Progr. Neurobiol.* **88** (2), 127–151.

145 Corey, E.J., et al. (1969) *J. Am. Chem. Soc. (JACSAT)* **91**, 5675.

146 Corey, E.J., et al. (1970) *J. Am. Chem. Soc. (JACSAT)* **92**, 397.

147 Upjohn, DOS 2 145 125 (9.9.1971; USA-prior. 11.9.1970, 2.7.1971).

148 Schneider, W.P., et al. (1973) *J. Chem. Soc., Chem. Commun. (JCCCAT)* 254.

149 Heather, J.B., et al. (1973) *Tetrahedron Lett. (TELEAY)* 2313.

150 Upjohn, US 3 948 981 (6.4.1976; prior. 18.12.1974, 3.10.1973, 2.7.1971, 11.9.1970).

151 Bartley, J., et al. (2001) *Human Reprod.* **16** (10), 2098–2102.

152 Ono Pharms., US 4 052 512 (4.7.1977; appl. 5.10.1975).

153 Yankee, E.W., et al. (1974) *JACS* **96** (18), 5865–5876.

154 Shanghai Techwell, US 20 130 190 404 (25.7.2013; appl. 9.4.2013; CN-prior. 21.7.2010).

155 Moreno, J.J. (2017) *Eur. J. Pharmacol.* **796**, 7–19.

156 Collins, P.W., et al. (1977) *J. Med. Chem. (JMCMAR)* **20**, 1152.

157 Searle, US 3 965 143 (22.6.1976; appl. 26.3.1974).

158 GAF, (1940) US 2 313 477.

159 Philips, US 3 410 944 (12.11.1968; NL-prior. 27.2.1964).

160 Külz, F. and Schöpf, C. (1953) US 2 661 373 (prior. 1953).

161 Troponwerke, (1948) DE 815 043.

162 Boehringer Ingelheim, DE 1 286 047 (appl. 30.11.1962).

163 Boehringer Ingelheim, US 3 341 593 (12.9.1967; D-prior. 30.11.1962).

164 Sandoz, US 3 752 814 (14.8.1973; CH-prior. 31.5.1968).

165 Hofmann, C., et al. (2006) *Clin. Neuropharmacol.* **29** (2), 80–86.

166 Candiani, Cabri, W., Zarini, F. and Bedeschi, A. (1995) *Synlett (SYNLES)* (6), 605.

167 Farmitalia Carlo Erba S.p.A., GB 2 074 566 (appl. 31.3.1981; GB-prior. 3.4.1980).

168 Farmitalia Carlo Erba S.p.A., US 4 526 892 (2.7.1985; USA-prior. 3.3.1981).

169 Sandoz, EP 77 754 (appl. 27.4.1983; CH-prior. 16.10.1982, 25.6.1982).

170 Sandoz, US 4 565 818 (appl. 21.1.1986; CH-prior. 16.10.1981, 25.6.1982).

171 Miller, A.J., et al. (1992) *Eur. J. Pharmacol.* **227** (1), 99–102.

172 Janssen, W., et al. (2015) *BioMed. Res. Int.* 438403.

173 Boots, DE 1 443 429 (appl. 26.1.1962; GB-prior. 2.2.1961).

174 Boots, GB 971 700 (appl. 2.2.1961).

175 Boots, US 3 228 831 (11.1.1966; GB-prior. 2.2.1961).

176 Harrington, P.J. and Lodewijk, E. (1997) *Org. Proc. Res. Dev. (OPRDFK)* **1**, 72.

177 Harrison, J.T., et al. (1970) *J. Med. Chem. (JMCMAR)* **13**, 203.

178 Syntex, US 3 896 157 (22.7.1975; prior. 13.1.1967, 7.12.1967, 4.11.1971).

179 Angelini Francesco, FR 1 382 855 (appl. 21.2.1964; I-prior. 9.8.1963).

180 Dunwell, D.W., et al. (1957) *J. Med. Chem. (JMCMAR)* **18**, 53.

181 Ravizza, DE 2 931 255 (appl. 1.8.1979; I-prior. 4.8.1978).

182 Manning, M., et al. (1995) *Int. J. Pept. Protein Res. (IJPPC3)* **46**, 244–252.

183 Ferring, WO 9 501 368 (appl. 22.6.1994; USA-prior. 29.6.1993).

184 Boehringer Ing., US 5 576 318 (19.11.1996; IT-prior. 30.07.1991).

185 Boehringer Ing., US 7 151 103 (19.12.2006; EP-prior. 20.10.2001).

186 Searle, US 3 176 013 (30.3.1965; appl. 25.7.1963).

187 Gedeon Richter, DE 1 668 604 (appl. 7.9.1967; H-prior. 7.9.1969).

188 Reynaud, J.P., et al. (1986) *J. Steroid Biochem.* **25** (5B), 811–833.

189 Odlind, V., et al. (1979) *Clin. Endocrinol.* **10** (1), 29–38.

190 Ringold, H., et al. (1959) *JACS* **81** (14), 3712–3716.

191 Djerassi, C., et al. (1959) *J. Am. Chem. Soc.* **81**, 436.

192 Buzby, G.C., et al. (1966) *J. Med. Chem. (JMCMAR)* **9**, 782.

193 Smith, H., GB 1 041 279 (appl. 19.10.1961).

194 Rufer, C., et al. (1967) *Justus Liebigs Ann. Chem. (JLACBF)* **702** (141).

195 Smith, H., et al. (1964) *J. Chem. Soc. (JCSOA9)* 4472.

196 Hoffmann-La Roche, DOS 1 806 410 (appl. 31.10.1968; USA-prior. 2.11.1967).

197 Ellis, B., et al. (1957) *J. Chem. Soc. (JCSOA9)* 4092.

198 Upjohn, US 3 147 290 (1.9.1964; appl. 17.5.1961; prior. 23.11.1956).

199 van den Broek, A.S., et al. (1975) *Recl. Trav. Chim. Pays-Bas (RTCPA3)* **94**, 35.

200 Akzona, US 3 927 046 (16.12.1975; appl. 3.12.1973; NL-prior. 9.12.1972).

201 Hofmeister, H., et al. (1986) *Arzneim.-Forsch. (ARZNAD)* **36**, 781.

202 Schering AG, US 4 621 079 (4.11.1986; appl. 21.12.1984; D-prior. 22.12.1983).

203 Ortho, GB 1 123 104 (appl. 2.9.1966; USA-prior. 22.10.1965).

204 Bittler, D., et al. (1982) *Angew. Chem. (ANCEAD)* **94** (9), 718.

205 Schering AG, EP 75 189 (appl. 8.9.1982; D-prior. 21.9.1981).

206 Kuhnz, W., et al. (1995) *Contraception (CCPTAY)* **51** (2), 131.

207 Cygnus, WO 9 640 355 (19.12.1996; USA-prior. 7.6.1995).

208 Gastaud, J.M., DOS 2 522 533 (21.5.1975; GB-prior. 21.5.1974).

209 Théramex, EP 157 842 (appl. 4.10.1984; F-prior. 4.10.1983).

210 Brückner, K., et al. (1961) *Chem. Ber. (CHBEAM)* **94**, 1225.

211 E. Merck AG, DE 1 075 114 (appl. 29.4.1958).

212 Hübner, M., et al. (1978) *Pharmazie (PHARAT)* **33**, 792.

213 VEB Jenapharm, DE 2 718 872 (22.12.1977; appl. 28.4.1977; DDR-prior. 14.6.1976).

214 Nominé, G., et al. (1965) *C. R. Hebd. Seances Acad. Sci. (COREAF)* **260**, 4545.

215 Roussel-Uclaf, US 3 257 278 (21.6.1966; F-prior. 5.7.1963, 4.10.1963).

216 Nickisch, K., et al., US 20 130 123 523 (16.5.2013; appl. 10.11.2011).

217 Pohl, O., et al. (2013) *J. Clin. Pharm. Ther.* **38** (4), 314–320.

218 Heyl, W.F. and Herr, M.E. (1953) *J. Am. Chem. Soc. (JACSAT)* **75**, 1918.

219 Upjohn (1957), US 2 793 218 (prior. 1955).

220 Ruzicka, L. (1935) *Helv. Chim. Acta (HCACAV)* **18**, 1487.

221 Ciba, (1939) US 2 143 453 (CH-prior. 1935).

222 Ciba, (1943) US 2 308 833 (CH-prior. 1935).

223 Schering AG, DE 1 152 100 (appl. 23.12.1960).

224 Schering, US 2 927 921 (8.3.1960; prior. 19.5.1954, 24.1.1952).

225 Inhoffen, H.H., et al. (1938) *Ber. Dtsch. Chem. Ges. (BDCGAS)* **71**, 1024.

226 Ciba, DRP 702 063 (appl. 1938; CH-prior. 1937).

227 Inhoffen, H.H. and Zühlsdorff, G. (1941) *Ber. Dtsch. Chem. Ges. (BDCGAS)* **74**, 1911.

228 Schering Corp., (1937) US 2 096 744 (D-prior. 1932).

229 Gallagher, T.F. (1954) *J. Am. Chem. Soc. (JACSAT)* **76**, 2943.

230 Merrell, (1947) US 2 430 891 (prior. 1941).

231 Sih, C., et al. (1965) *J. Am. Chem. Soc. (JACSAT)* **87**, 2765.

232 Hoechst, EP 37 973 (appl. 2.4.1981; D-prior. 12.4.1980).

233 Sogeras, DE 2 215 499 (appl. 29.3.1972; GB-prior. 21.4.1971).

234 Colton, F.B., et al. (1957) *J. Am. Chem. Soc. (JACSAT)* **79**, 1123.

235 Searle, (1954) US 2 666 769 (appl. 1952).

236 Boots Pure Drug, GB 566 881 (appl. 1943).

237 Boots (1949) US 2 464 203 (GB-prior. 1943).

238 Dodds, E.C. (1938) *Nature (London) (NATUAS)* **141**, 247.

239 Lilly, (1946) US 2 392 852 (prior. 1941).

240 Horeau, A., et al. (1948) *Bull. Soc. Chim. Fr. (BSCFAS)* 711; 1955, 955.

241 Horeau, A. (1951) US 2 547 123 (F-prior. 1947).

242 Organon, US 3 340 279 (5.9.1967; NL-prior. 16.6.1964).

243 Popper, A., et al. (1969) *Arzneim.-Forsch. (ARZNAD)* **19**, 352.

244 Schering AG, DE 1 074 582 (appl. 24.9.1958).

245 Schering AG, (1956) US 2 753 360 (D-prior. 1953).

246 Oppenauer, R. (1937) *Recl. Trav. Chim. Pays-Bas Belg. (RTCPB4)* **56**, 137.

247 Schering Corp., (1945) US 2 379 832 (D-prior. 1936).

248 Westerhof, P. and Reerink, E.H. (1960) *Recl. Trav. Chim. Pays-Bas (RTCPA3)* **79**, 771.

249 North American Philips, US 3 198 792 (3.8.1965; prior. 8.4.1959, 12.6.1962).

250 Deghenghi, R. and Gaudry, R. (1961) *J. Am. Chem. Soc. (JACSAT)* **83**, 4668.

251 American Home Products, US 3 133 913 (19.5.1964; appl. 11.9.1961).

252 Théramex, EP 157 842 (appl. 4.10.1984; F-prior. 4.10.1983).

253 Joly, R., et al. (1973) *Bull. Soc. Chim. Fr. (BSCFAS)* 2694.

254 Roussel-Uclaf, US 3 453 267 (1.7.1969; F-prior. 31.12.1964, 25.2.1965).

255 Roussel-Uclaf, US 3 679 714 (25.7.1972; F-prior. 20.2.1970).

256 Organon, GB 841 411 (appl. 2.4.1958; NL-prior. 10.4.1957).

257 Ciba, (1942), US 2 272 131 (CH-prior. 1937).

258 Syntex, (1956) US 2 744 122 (Mex-prior. 1951).

259 Schering AG, BE 721 825 (appl. 4.10.1968; D-prior. 4.10.1967).

260 Cole, L.A. (2009) *Reprod. Biol. Endocrinol.* **7**, 8.

261 Lunenfeld, B. (2009) *Human Reproduct. Update* **10** (6), 453–467.

262 Murphy, B.D. (1991) *Endocr. Rev.* **12** (4), 27–44.

263 Van Wely, M., et al. (2005) *Treatments Endocrinol.* **4** (3), 155–165.

264 Goa, K.L., et al. (1998) *BioDrugs* **9** (3), 235–260.

265 Koechling, W., et al. (2017) *Endocr. Connect.* **6** (5), 297–305.

266 Dhillon, S., et al. (2008) *Drugs* **68** (11), 1529–1540.

267 Genzyme, US 767 251 (16.61998; USA-prior. 2.11.1983).

268 Loudradis, D., et al. (2009) *Curr. Opin. Invest. Drugs* **10** (4), 372–380.

269 Bosch, E., et al. (2019) *Reprod. BioMed. Online* **38** (2), 195–205.

270 Olsson, K.G. et al. US 3 287 397 (22.11.1966; GB-prior. 22.11.1960).

271 Merrell, US 2 914 563 (24.11.1959; prior. 6.8.1957).

272 Organon, NL 95 257 (appl. 1958).

273 Schering AG, US 3 234 093 (8.2.1966; appl. 24.4.1962; D-prior. 29.4.1961).

274 Clinton, R.O., et al. (1961) *J. Am. Chem. Soc. (JACSAT)* **83**, 1478.

275 Sterling Drug, GB 905 844 (valid from 1959; USA-prior. 1958).

276 Roussel-Uclaf, DE 1 593 307 (appl. 1966; F-prior. 1965).

277 Roussel-Uclaf, EP 57 115 (appl. 8.1.1982; F-prior. 9.1.1981).

278 Roussel-Uclaf, US 4 386 085 (31.5.1983; appl. 10.6.1982; F-prior. 9.1.1981).

279 Vicenzi, J.T., et al. (1999) *Org. Proc. Res. Dev. (OPRDFK)* **3**, 56–59.

280 Lilly & Co., EP 62 504 (appl. 1.4.1982; USA-prior. 3.4.1981).

281 Gruber, C., et al. (2008) *J. Bone Miner. Res.* **23** (4), 525–535.

282 Gennari, L., et al. (2006) *Exp. Opin. Investig. Drugs* **15** (9), 1091–1103.

283 Singh, M.M. (2001) *Med. Res. Rev.* **21** (4), 302–347.

284 Fermion, WO 2014 060 640 (24.04.2014; US-prior. 19.10.2012).

285 Rosenkranz, G., et al. (1956) *J. Org. Chem. (JOCEAH)* **21**, 520.

286 Parke Davis, (1943) US 2 335 616 (prior. 1941).

287 Procter and Gamble, US 3 963 573 (15.6.1976; appl. 3.3.1975).

288 KabiVitrum, US 4 536 495 (20.8.1985; appl. 4.3.1983; SE-prior. 12.3.1982).

289 Recordati, US 2 921 070 (12.1.1960; CH-prior. 5.11.1957).

290 Yakhontov, L.N., et al. (1981) *Pharm. Chem. J.* **15**, 546–561.

291 Mead Johnson, GB 940 540 (appl. 25.7.1961; USA-prior. 26.7.1960, 20.6.1961).

292 Aktiebolaget Recip., DE 1 170 417 (appl. 8.11.1961; GB-prior. 8.11.1960).

293 Klosa, J. and Delmar, G. (1962) *J. Prakt. Chem. (JPCEAO)* **16**, 71–82.

294 Starke, C. et al., DD 106 643 (appl. 12.7.1973; DD-prior. 12.7.1973).

295 Pharmacia, US 5 382 600 (17.1.1995; S-prior. 22.1.1988).

296 Yamanouchi Pharmaceutical, EP 801 067 (15.10.1997; J-prior. 28.12.1994).

297 Chem. Fabrik Dr. R. Pfleger, US 3 480 626 (25.11.1969; appl. 18.5.1967; prior. 2.3.1964).

298 Novartis, WO 2 003 080 599 (2.10.2003; GB-prior. 26.3.2002).

299 Schwarz Pharma, US 20 090 192 224 (30.7.2009; appl. 6.6.2007; EP-prior. 12.6.2006).

300 Yamanouchi Pharm., US 6 346 532 (12.02.2002; JP-prior. 17.10.1997).

301 Ratiopharm, US 20 150 152 044 (4.6.2015; appl. 9.12.2014; US-prior. 1.5.2013).

302 *Ullmanns Encykl. Tech. Chem.*, 3 Aufl., Vol. **3**, 201 f.

303 *Ullmanns Encykl. Tech. Chem.*, 4 Aufl., Vol. **7**, 151.

304 Tilford, C.H., et al. (1947) *J. Am. Chem. Soc. (JACSAT)* **69**, 2903.

305 Merrell Comp. (1949) US 2 474 796 (prior. 1946).

306 Smith Kline & French, (1952) US 2 599 000 (prior. 1950).

307 Wolf, E. (1934) US 1 962 224 (D-prior. 1930).

308 Vivus Inc., US 5 773 020 (30.6.1998; appl. 28.10.1997; USA-prior.25.4.1990).

309 C.E.R.M., US 3 823 234 (9.7.1974; F-prior. 16.5.1971).

310 Dale, D.J., et al. (2000) *Org. Proc. Res. Dev. (OPRDFK)* **4**, 17–22.

311 Pfizer, US 5 250 534 (5.10.1993; GB-prior. 20.6.1990).

312 Saint-Ruf, G. et al., US 3 940 387 (24.2.1976; appl. 5.5.1972; FR-prior. 6.5.1971)

313 Costa, P., et al. (1992) *J. Clin. Pharmacol.* **81** (12), 1223–1226.

314 Mallinckrodt Inc., US 20 100 228 032 (9.9.2010; appl. 19.9.2008; USA-prior. 31.3.2006).

315 Lab. Glaxo S.A., Fr., US 6 140 329 (GB-prior. 14.7.1995).

316 Icos Corp., US 6 143 746 (11.7.2001; GB-prior. 21.1.1994).

317 Haning, H., et al. (2002) *Bioorg. Med. Chem. Lett. (BMCLE8)* **12**, 865–868.

318 Bayer AG, DE 19 750 085 (20.5.1999; D-prior. 12.11.1997).

319 Tanabe Seiyaku, US 6 797 709 (28.9.2004; appl. 1.5.2003; JP-prior. 16.9.1999).

320 Dong-A Pharm., US 20 110 306 762 (15.12.2011; appl. 17.2.2010; KR-prior. 18.2.2009).

321 Fishbein, W., et al. (1983) *J. Biol. Chem.* **240** (3), 321–322.

322 Shreve, R.N., et al. (1943) *J. Am. Chem. Soc. (JACSAT)* **65**, 2241.

323 Boehringer, (1927) DRP 515 781.

324 Ozaki, H., et al. (2015) *Food Chem. Toxicol.* **86**, 116–123.

325 David, N.A. (1972) *Annu. Rev. Pharmacol.* **12**, 353–374.

326 Torre, O., et al. (2006) *Tetrahedron: Asymmetry (TASYE3)* **17**, 860–866.

327 Eli Lilly, WO 0 117 521 (15.3.2001; appl. 22.8.2000; USA-prior. 3.9.1999).

328 Santen, FR 1 491 204 (appl. 10.8.1962; J-prior. 2.11.1961).

329 Santen, GB 1 023 003 (appl. 14.9.1962).

330 Verges, J., et al. (2004) *Proc. West. Pharmacol. Soc.* **47**, 50–53.

331 Eli Lilly, EP 273 658 (18.12.1987; USA-prior. 22.12.1986).

332 Eli Lilly, EP 457 559 (15.5.1991; USA-prior. 17.5.1990).

333 ICI, GB 705 838 (appl. 1951; valid from 1952).

334 Otsuka Pharmaceutical Co., WO 2 007 026 971 (appl. 1.9.2006; J-prior. 2.9.2005).

335 Manoury, P.M., et al. (1986) *J. Med. Chem. (JMCMAR)* **29**, 19.

336 Synthelabo, US 4 315 007 (9.2.1982; F-prior. 6.2.1978, 29.12.1978).

337 Yamanouchi, EP 257 787 (2.3.1988; appl. 21.7.1987; J-prior. 21.7.1986).

338 Yamanouchi, US 4 731 478 (15.3.1988; appl. 27.11.1985; J-prior. 8.2.1980).

339 Abbott, US 4 026 894 (31.5.1977; prior. 14.10.1975).

340 Abbott, US 4 112 097 (5.9.1978; prior. 21.1.1977).

341 Kissei Pharmac., EP 600 675 (8.7.1998; J-prior. 2.12.1992).

342 Kissei Pharmac., US 5 387 603 (7.2.1995; J-prior. 2.12.1992).

343 Pfizer, US 4 188 390 (12.2.1980; GB-prior. 5.11.1977).

344 Bhattacharya, A., et al. (1988) *J. Am. Chem. Soc. (JACSAT)* **110**, 3318.

345 Merck & Co., US 4 760 071 (26.7.1988; appl. 21.11.1985; USA-prior. 27.2.1984).

346 Rasmussen, G.H., et al. (1986) *J. Med. Chem. (JMCMAR)* **29**, 2298.

347 Bakshi, R.K., et al. (1995) *J. Med. Chem. (JMCMAR)* **38**, 3189.

348 Glaxo, WO 2 002 046 207 (13.6.2002; appl. 2.11.2001; GB-prior. 3.11.2000).

349 Shore, N. (2010) *Exp. Opin. Invest. Drugs* **19** (2), 305–310.

350 Medipolar Oy, US 4 153 622 (8.5.1979; prior. 18.5.1978).

351 Bungardt, E. and Mutschler, E. (2020) Spasmolytics, *Ullmann's Encyclopedia of Industrial Chemistry*, Wiley-VCH Verlag, Weinheim, Electronic Release, June.

Hormone Therapeutics – Systemic Hormonal Preparations, Excluding Reproductive Hormones and Insulins (H)

HEINZ WEINBERGER, Neulußheim, Germany

ULLMANN'S
ENCYCLOPEDIA
OF INDUSTRIAL
CHEMISTRY

1. Introduction

The following chapters review the different drugs affecting hormones except for sexual hormones and insulin. They are arranged according to the Anatomical Therapeutical Chemical (ATC) Classification System – a system of alphanumeric codes developed by the WHO for the classification of drugs and other medical products. Here the active substances are divided into different groups according to the organ or system on which they act, and their therapeutic, pharmacological, and chemical properties.

ATC group H comprises all hormonal preparations for systemic use, except:

- Insulins and Analogs, see A10A
- Anabolic Steroids, see A14A
- Catecholamines, see C01C and R03C
- Sex Hormones, see G03
- Sex Hormones used in Treatment of Neoplastic Diseases, see L02 → Antineoplastic and Immunomodulating Agents (L)

Ullmann's Pharmaceuticals. Edited by Axel Kleemann and Bernhard Kutscher.
© 2022 Wiley-VCH GmbH
Set ISBN: 978-3-527-34252-5/ DOI: 10.1002/14356007.v13_v01

- Metreleptin used for Treatment of Complications of Leptin Deficiency in Patients with Generalised Lipodystrophy, classified in A16AA.

In the tables of the particular chapters, the individual drug substances are characterized by their INN, synonyms, brand or trade names, ATC number, structures, CAS registry number, formula, molecular weight, mode of action, medical use, formulation, DDD (Defined Daily Dose), originator, country and year of approval, and references. The DDDs are generally based on the treatment or diagnosis of endocrine disorders. The references for each drug include the basic patents and publications, in which the relevant synthetic or biosynthetic processes for production are described. Additionally, the reader is referred to the standard reference source [1].

Pituitary and hypothalamic hormones and analogs encompass anterior pituitary lobe hormones, extracts, purified natural hormones, and synthetic analogs. The hormones produced by the pituitary gland and hypothalamus are key regulators of metabolism, growth, and reproduction. Preparations of these hormones are used in the treatment of a variety of endocrine disorders, such as acromegaly and gigantism, dwarfism, prolactin deficiency, hypergonadotropism, hypothyroidism, or Cushing disease [2–4].

Gonadotropins are glycoprotein polypeptide hormones secreted by gonadotrope cells of the anterior pituitary of vertebrates. They were first introduced in therapy in the early 1960s and have been used in ovarian stimulation cycles to induce multiple follicular development, particularly between 1970 and 2000, in women undergoing in vitro fertilization (IVF) treatment. Starting in the early 1980s, the use of gonadotropin-releasing hormone (GnRH) agonists in ovarian stimulation greatly improved the success rate of IVF. Later, GnRH antagonists with high potency and fewer side effects, such as cetrorelix and ganirelix were introduced into IVF and have emerged as an alternative in preventing premature luteinizing hormone (LH) surges. Unlike GnRH agonists, these potent GnRH antagonists cause immediate, rapid gonadotropin suppression by competitively blocking GnRH receptors in the anterior pituitary gland, thereby preventing endogenous GnRH from inducing LH and follicle-stimulating hormone (FSH) release from the pituitary cells. Furthermore, GnRH antagonist suppression of gonadotropin secretion can be quickly reversed [5–7].

Growth hormone (GH) is a protein hormone synthesized and secreted by cells called somatotrophs in the anterior pituitary. GH regulates somatic growth, substrate metabolism, and body composition. Its actions are elaborated through the GH receptor (GHR), both directly by tyrosine kinase activation and indirectly by induction of insulin-like growth factor 1. Excessive activation of the GHR by circulating GH results in gigantism and acromegaly, whereas cell transformation and cancer can occur in response to autocrine activation of the receptor. Owing to the fact that GH has a short half-life, several approaches have been taken to create long-acting GHR agonists. This includes the pegylation, sustained release formulations, and ligand-receptor fusion proteins. Pegylation of a GH analog (pegvisomant), which binds but does not activate signal transduction, forms the basis of a new successful approach to the treatment of acromegaly, a disorder resulting from excess GH secretion after epihyseal fusion [8–11].

Corticosteroids represent a class of drugs based on hormones formed in the adrenal gland or derived synthetically. They are created from cholesterol and classified as either glucocorticoids (anti-inflammatory) or mineralocorticoids (salt retaining). Common glucocorticoids are cortisol, prednisolone, dexamethasone, and corticosterone. A crucial mineralocorticoid in humans is aldosterone. However, the term "corticosteroids" is generally used to refer to glucocorticoids. Named for their effect in carbohydrate metabolism, glucocorticoids regulate diverse cellular functions including development, homeostasis, metabolism, cognition, and inflammation. The secretion of these hormones increases during stress related to anxiety and severe injury. Due to their profound immune modulatory actions, glucocorticoids belong to the most widely prescribed drugs in the world. Glucocorticoids have become a clinical mainstay for the treatment of numerous inflammatory and autoimmune diseases, such as asthma, allergy, septic shock, rheumatoid arthritis, inflammatory bowel disease, and multiple sclerosis. They are also used to help prevent organ rejection in transplant recipients. Unfortunately, their therapeutic benefits are limited by the adverse side effects that are associated with high dose (used in the treatment of systemic vasculitis and systemic lupus erythematosus (SLE)) and long-term use. Chronic overproduction of these substances is associated with various disorders, such as Cushing syndrome [12–14].

Primary *hypothyroidism* or thyroid hormone deficiency due to abnormality in the thyroid gland is the most common endocrine disease. Clinical manifestations of hypothyroidism range from life threatening to no signs or symptoms. The most common symptoms in adults are fatigue, lethargy, cold intolerance, weight gain, constipation, change in voice, and dry skin, but clinical presentation can differ with age and sex, among other factors. Thyroid hormone replacement has been used for more than 100 years in the treatment of hypothyroidism, and there is no doubt about its overall efficacy. The standard treatment is thyroid hormone replacement therapy with levothyroxine. It has a long (7 day) half-life, allowing once-daily administration. Consequently, all of the common guidelines of major endocrine societies recommend levothyroxine monotherapy for first line use in hypothyroidism [15–19].

Hyperthyroidism is characterized by high levels of serum thyroxine and triiodothyronine, and low levels of thyroid-stimulating hormone. The main causes of hyperthyroidism are Graves disease, toxic multinodular goitre, and toxic adenoma. Besides radioactive iodine ablation of the thyroid gland or surgical thyroidectomy, there is consensus that antithyroid drugs, such as carbimazole, propylthiouracil, and thiamazole are effective in treating hyperthyroidism [20–22].

Hypoglycemia is a common yet serious event that may be associated with significant neurocognitive and physical impairment. Glucagon therapy is the treatment of choice in the management of severe hypoglycemia in insulin-treated patients with either type 1 or type 2 diabetes, if intravenous glucose is not immediately available. Glucagon is a 29-amino-acid peptide synthesized and secreted from α-cells of the islets of Langerhans, which are located in the endocrine portion of the pancreas. From there the hormone is secreted into the hepatic portal vein from which it acts on G protein-coupled receptors in the liver to stimulate glucose production. Glucagon is a potent and effective agent that can be administered intravenously, intramuscularly, or subcutaneously, but the intramuscular route is generally recommended for the treatment of severe hypoglycaemia. Recombinant glucagon was approved in 1998 and has an excellent safety profile [23, 24].

Hypoparathyroidism (HypoPT) is rare disease characterized by hypocalcemia with inappropriate low serum levels of parathyroid hormone (PTH). Most cases occur as a complication to thyroid or parathyroid surgery, during which the parathyroid glands are accidentally damaged. There are no formal guidelines for the management of hypoPT. In the acute setting, intravenous calcium may be necessary. Traditionally, hypoPT is treated with calcium supplements and activated vitamin D analogs at varying doses, based on clinical judgment. However, maintaining serum calcium levels can be a challenge. In addition, concerns exist regarding hypercalciuria and ectopic calcifications that can be associated with such treatment. HypoPT remains the only classic endocrine deficiency disease for which the missing hormone, PTH, is not yet an approved treatment. Two formulations of PTH were investigated in hypoPT, namely the full-length molecule, PTH(1–84), and the fully active but truncated amino-terminal fragment, PTH(1–34) (teriparatide). As the first PTH replacement therapy for hypoparathyroid patients with hypocalcemia, rhPTH (1–84) is an effective regimen, has generally acceptable tolerability and represents an important advance for the management of hypoparathyroidism. The parathyroid hormone analog teriparatide, a potent stimulator of bone remodeling, increases hip and spine bone mineral density and reduces the risk of vertebral and non-vertebral fractures in postmenopausal osteoporotic women [25–28].

Secondary hyperparathyroidism (SHPT) is a common, serious, and progressive complication in chronic kidney disease. It is characterized by elevated serum PTH, parathyroid gland hyperplasia, and mineral metabolism abnormalities. Calcium sensing receptor is an important target for the treatment of SHPT. Currently, various treatment options are available. They include vitamin D analogs, such as calcitriol, doxercalciferol and paricalcitol; calcimimetics, such as cinacalcet and etelcalcetide hydrochloride; and parathyroidectomy.

These treatment options have contributed to the successful control of SHPT, and recent clinical studies have provided evidence suggesting that effective treatment of SHPT leads to improved survival [29–33].

2. H01 Pituitary and Hypothalamic Hormones and Analogs

2.1. H01A Anterior Pituitary Lobe Hormones and Analogs

2.1.1. H01AA Adrenocorticotropic Hormone (Table 1)

Table 1. H01AA Adrenocorticotropic hormone

INN (Synonyms) [Brand names] ATC	Structure (Remarks)	[CAS-No.] Formula MW (g/mol)	Target (if known) Medical use Formulation DDD	Originator(s) Approval (country/year) References
Corticotropin (ACTH, Adrenocorticotrop(h)in, Corticotrophin) [Acethropan, Acortan, Acthar, Acton, Cortiphyson, Cortrophin, Isactid] H01AA01	SYSMEHFRWGKP VGKKRRPVKV YPNGAEDESAEAFPLEF	*[9002-60-2]* $C_{207}H_{308}N_{56}O_{58}S$ 4541.14	anterior pituitary hormone stimulating the adrenal cortex and its production of corticosteroids treatment of many different conditions such as multiple sclerosis, psoriatic or rheumatoid arthritis, ankylosing spondylitis, lupus, severe allergic reactions, breathing disorders, and inflammatory conditions of the eyes, infantile spasms in children younger than 2 years old. gel 40 units/mL; powder 40 units/vial; vial 80 units/mL. 25 units, P.	Sanofi Aventis USA/1950 [34, 35]
Tetracosactide (ACTH (1-24), Cosyntropin) [Cortrosyn, Synacthen] H01AA02	SYSMEHFRWGKPVG KKRRPVKVYP	*[16960-16-0]* $C_{136}H_{210}N_{40}O_{31}S$ 2933.49 Hexaacetate: *[22633-88-1]* $C_{148}H_{234}N_{40}O_{43}S$ 3293.75	adrenocorticotropic hormone receptor agonist for use as a diagnostic agent in the screening of patients presumed to have adrenocortical insufficiency. vial w/powder for solution for inj. 0.25 mg; amp. 0.25 mg/mL, 1 mg/mL (depot). 0.25 mg, P.	Amphastar Pharmaceuticals USA/1970 [36]

2.1.2. H01AB Thyrotropin (Table 2)

Table 2. H01AB Thyrotropin

INN (Synonyms) [Brand names] ATC	Structure (Remarks)	[CAS-No.] Formula MW (g/mol)	Target (if known) Medical use Formulation DDD	Originator(s) Approval (country/year) References
Thyrotropin alfa (Thyroid stimulating hormone, TSH, Thyrotrophin-alfa) [Thyrogen] H01AB01, V04CJ01	α-chain: APDVQDCPECTLQENPFF SQPGAPILQCMG CCFSRAYPTPLRSKK TMLVQKNVTSESTCC VAKSYNRVTVMGGFK VENHTACHCSTCYYHKS β-chain: FCIPTEYTMHIERRECA YCLTINTTICAGYCM TRDINGKLFLPKYAL SQDVCTYRDFIYRTV EIPGCPLHVAPYFSYP VALSCKCGKCNTDYS DCIHEAIKTNYCTKPQKSY	[194100-83-9] $C_{975}H_{1513}N_{267}O_{304}S_{26}$ 22672.90	thyrotropin receptor agonist for detection of residual or recurrent thyroid cancer (stimulating radioactive iodine uptake for better radiodiagnostic imaging). vial w/powder for solution for inj. 1.1 mg. 0.9 mg, P.	Sanofi Aventis; Genzyme USA/1953 [37, 38]

2.1.3. H01AC Somatropin and Somatropin Agonists (Table 3)

Table 3. H01AC Somatropin and somatropin agonists

INN (Synonyms) [Brand names] ATC	Structure (Remarks)	[CAS-No.] Formula MW (g/mol)	Target (if known) Medical use Formulation DDD	Originator(s) Approval (country/year) References
Somatropin (Somatotropin, Growth hormone, GH) [Genotropin, Nutropin, Humatrope] H01AC01	FPTIPLSRLFDNAMLRAHRL HQLAFDTYQE FEEAYIPKEQKYSFLQN PQTSLCFSESIPTP SNREETQQKSNLELLRIS LLLIQSWLEPVQ FLRSVFANSLVYGASD SNVYDLLKDLEE GIQTLMGRLEDGSPRTG QIFKQTYSKFDT NSHNDDALLKNYGLL YCFRKDMDKVETFL RIVQCRSVEGSCGF (disulfide bridges: 53–165, 182–189)	[12629-01-5] $C_{990}H_{1529}N_{263}O_{299}S_7$ 22124.08	GHR agonist for treatment of dwarfism, acromegaly and prevention of HIV-induced weight loss. vial w/powder for solution for SC inj. 0.2 mg, 0.4 mg, 0.6 mg, 0.8 mg, 1 mg, 1.2 mg, 1.4 mg, 1.6 mg, 1.8 mg, 2 mg, 5 mg, 12 mg. 2 U, P.	Serono USA/1976 [39–41]
Mecasermin (Insulin-like growth factor 1, IGF-1, Somatomedin-C) [Increlex, Iplex] H01AC03	GPETLCGAELVDALQFVCGD RGFYFNKPT GYGSSSRRAPQTGIVDEC CFRSCDLRRLE MYCAPLKPAKSA (disulfide bridge: 6–48; 18–61; 47–52)	[68562-41-4] $C_{331}H_{512}N_{94}O_{101}S_7$ 7648.71	IGF-1 receptor agonist for the long-term treatment of growth failure in pediatric patients with primary IGFD or with GH gene deletion who have developed neutralizing antibodies to GH. vial w/powder for solution for inj. 10 mg/mL; 40 mg/4 mL. 2 mg, P.	Ipsen Biopharmaceuticals USA/2005 [42, 43]
Sermorelin (Somatoliberin) [Geref, Groliberin] H01AC04, V04CD03	YADAIFTNSYRKVLGQL SARKLLQDIMSRQ	[86168-78-7] $C_{149}H_{246}N_{44}O_{42}S$ 3357.93 Acetate: [114466-38-5] $C_{149}H_{246}N_{44}O_{42}S$ $\cdot x\,C_2H_4O_2 \cdot y\,H_2O$	growth hormone-releasing hormone receptor agonist for the treatment of dwarfism, prevention of HIV-induced weight loss. vial 0.5 mg, 1.0 mg, P.; amp. 0.05 mg.	EMD Serono (now Merck Serono) USA/1990 [44, 45]

(continued)

Table 3. (*Continued*)

INN (Synonyms) [Brand names] ATC	Structure (Remarks)	[CAS-No.] Formula MW (g/mol)	Target (if known) Medical use Formulation DDD	Originator(s) Approval (country/year) References
Mecasermin rinfabate [Iplex] H01AC05	complex of recombinant human insulin-like growth factor 1 and insulin-like growth factor-binding protein-3	[478166-15-3] $C_{1562}H_{2485}N_{465}O_{485}S_{27}$ 36404.02	IGF-1 receptor agonist treatment of growth failure in children with severe primary IGF-1 deficiency or with GH gene deletion who have developed neutralizing antibodies to GH. vial 36 mg 0.6 mL, P.	Insmed USA/2005 [46–48]
Tesamorelin (GHRH(1-44), TH9507) [Egrifta] H01AC06	YADAIFTNSYRKVLGQ LSARKLLQDIMSR QQGESNQERGARARL	[218949-48-5] $C_{221}H_{366}N_{72}O_{67}S$ 5135.86 Acetate: [901758-09-6] $C_{221}H_{366}N_{72}O_{67}S \cdot x$ $C_2H_4O_2$ ($x \approx 7.4$)	growth hormone-releasing hormone receptor agonist for the reduction of excess abdominal fat in HIV-infected patients with lipodystrophy. vial 2 mg, P. 2 mg, P.	Theratechnologies USA/2010 [49, 50]

2.1.4. H01AX Other Anterior Pituitary Lobe Hormones and Analogs (Table 4)

Table 4. H01AX Other anterior pituitary lobe hormones and analogs

INN (Synonyms) [Brand names] ATC	Structure (Remarks)	[CAS-No.] Formula MW (g/mol)	Target (if known) Medical use Formulation DDD	Originator(s) Approval (country/year) References
Pegvisomant (GH, GH-N, pituitary growth hormone) [Somavert] H01AX01	FPTIPLSRLFDNAMLR AHRLHQLAFDTY QEFEEAYIPKEQKYSFL QNPQTSLCFSE SIPTPSNREETQQKSNL ELLRISLLLIQSW LEPVQFLRSVFANSLVYG ASDSNVYDLL KDLEEGIQTLMGRLEDGSP RTGQIFKQT YSKFDTNSHNDDALLK NYGLLYCFRK DMDKVETFLRIVQC RSVEGSCGF	[218620-50-9] $C_{990}H_{1532}N_{262}O_{300}S_7$ 22129.0	GHR antagonist for the treatment of acromegaly in patients who have had an inadequate response to surgery or radiation therapy, or for whom these therapies are not appropriate. vial w/powder for solution for inj. 10 mg, 15 mg, 20 mg, 25 mg, 30 mg. 10 mg, P.	Pharmacia and Upjohn EU/2002, USA/2003 [51]

2.2. H01B: Posterior Pituitary Lobe Hormones

2.2.1. H01BA Vasopressin and Analogs (Table 5)

Table 5. H01BA Vasopressin and analogs

INN (Synonyms) [Brand names] ATC	Structure (Remarks)	[CAS-No.] Formula MW (g/mol)	Target (if known) Medical use Formulation DDD	Originator(s) Approval (country/year) References
Vasopressin (8-L-arginine vasopressin, argipressin, ADH, AVP) [Pitressin, Pressyn] H01BA01		[11000-17-2] $C_{46}H_{65}N_{15}O_{12}S_2$ 1084.23	arginine vasopressin receptor agonist for prevention and treatment of postoperative abdominal distention, in abdominal roentgenography to dispel interfering gas shadows, and in diabetes insipidus. vials 10 USP units/0.5 mL, 20 USP units/mL, P. 4 U, P.	Parke Davis Canada/1971 [52–56]
Desmopressin (DDAVP, 1-Deamino-8-D-arginine vasopressin) [Adiuretin, DesmoMelt] H01BA02		[16679-58-6] $C_{46}H_{64}N_{14}O_{12}S_2$ 1069.22 Acetate: [62288-83-9] $C_{48}H_{68}N_{14}O_{14}S_2$ 1129.27	arginine vasopressin receptor agonist used to prevent and control excessive thirst, urination, and dehydration caused by injury, surgery, specific types of diabetes insipidus and conditions after head injury or pituitary surgery. vial 4 µg/mL, P.; tabl. 0.1 mg, 0.2 mg; nasal spray 0.1 µg/mL. 25 µg, N.; 0.4 mg, O.; 4 µg, P.; 0.24 mg, SL.	Ferring Pharmaceuticals USA/1978 [57–59]
Lypressin (Lysipressin, Lysine vasopressin, Syntopressin) [Diapid] H01BA03		[50-57-7] $C_{46}H_{65}N_{13}O_{12}S_2$ 1056.22	vasopressin receptor agonist treatment of partial central diabetes insipidus or improvement of vasomotor tone and blood pressure. spray 0.185 mg/mL. 20 U, N.; 20 U, P.	Novartis USA/1982 [60–64]
Terlipressin (Triglycyl-8-lysine-vasopressin) [Glypressin, Glycylpressin, Teripress, Remestyp] H01BA04		[14636-12-5] $C_{52}H_{74}N_{16}O_{15}S_2$ 1227.38 Acetate (diacetate pentahydrate): $C_{56}H_{82}N_{16}O_{19}S_2$ · 5 H_2O 1437.56	vasopressin receptor (V1a, V1b, V2) agonist vasoactive drug in the management of hypotension, also treatment of uterine and esophageal bleeding. vial w/powder for solution for inj. 1 mg (as acetate salt); vials 0.12 mg/mL, 0.2 mg/mL (as acetate salt). 12 mg, P.	Ferring Pharmaceuticals EU/2010 [65, 66]

(continued)

Table 5. (*Continued*)

INN (Synonyms) [Brand names] ATC	Structure (Remarks)	[CAS-No.] Formula MW (g/mol)	Target (if known) Medical use Formulation DDD	Originator(s) Approval (country/year) References
Ornipressin (ornithine 8 vasopressin, orpressin) [Por-8] H01BA05		[3397-23-7] $C_{45}H_{63}N_{13}O_{12}S_2$ 1042.19 Acetate: $C_{47}H_{67}N_{13}O_{14}S_2$ 1102.25	vasopressin V1 receptor agonist used as a local vasoconstrictor and hemostatic. amp. 2.5 U/0.5 mL, 2.5 U/5 mL. 5 U, P.	Ferring Pharmaceuticals [67, 68]

2.2.2. H01BB Oxytocin and Analogs (Table 6)

Table 6. H01BB Oxytocin and analogs

INN (Synonyms) [Brand names] ATC	Structure (Remarks)	[CAS-No.] Formula MW (g/mol)	Target (if known) Medical use Formulation DDD	Originator(s) Approval (country/year) References
Demoxytocin (ODA-914, Desaminooxytocin) [Sandopart, Odeax, Sandopral] H01BB01		[113-78-0] $C_{43}H_{65}N_{11}O_{12}S_2$ 992.18	oxytocin receptor agonist used to induce labor, promote lactation, and to prevent and treat postpartum breast inflammation. tabl. 50 U. 100 U, O.	Sandoz [69–71]
Oxytocin (Orasthin, Oxystin, Partocon, Perlacton) [Pitocin, Syntocinon, Vagitocin] H01BB02		[50-56-6] $C_{43}H_{66}N_{12}O_{12}S_2$ 1007.19	oxytocin receptor agonist to assist in labor, elective labor induction, uterine contraction induction. vials 3 U/mL, 5 U/mL, 10 U/mL. 15 U, N.; 200 U, O.; 15 U, P.	Novartis USA/1980 [72, 73]
Carbetocin (Deamino-2-O-methyltyrosine-1-carbaoxytocin) [Decomoton, Depotocin, Duratocin, Pabal] H01BB03		[37025-55-1] $C_{45}H_{69}N_{11}O_{12}S$ 988.17	oxytocin receptor agonist used to control postpartum hemorrhage and bleeding after giving birth. vial 0.1 mg, P. 0.1 mg, P.	Ferring Pharmaceuticals UK/2007 [74, 75]

2.3. H01C Hypothalamic Hormones

2.3.1. H01CA Gonadotropin-Releasing Hormones (Table 7)

Table 7. H01CA Gonadotropin-releasing hormones

INN (Synonyms) [Brand names] ATC	Structure (Remarks)	[CAS-No.] Formula MW (g/mol)	Target (if known) Medical use Formulation DDD	Originator(s) Approval (country/year) References
Gonadorelin (GnRH, LHRH, Ru 19847) [Factrel, Lutrelef, Lutrepulse, Fertagyl] H01CA01 V04CM01		[33515-09-2] $C_{55}H_{75}N_{17}O_{13}$ 1182.29 Acetate: [52699-48-6] $C_{57}H_{81}N_{17}O_{16}$ 1260.38 Hydrochloride: [51952-41-1] $C_{55}H_{76}ClN_{17}O_{13}$ 1218.77	GnRH receptor agonist stimulates the synthesis and secretion of both pituitary gonadotropins, luteinizing hormone and follicle stimulating hormone. vial w/powder for solution for inj. 0.1 mg, 0.5 mg, 0.8 mg, 3.2 mg.	A.V. Schally USA/1982 [76–79]
Nafarelin (RS-94991-298, Nafarelin acetate) [Synarel, Nasanyl] H01CA02		[76932-56-4] $C_{66}H_{83}N_{17}O_{13}$ 1322.47 Acetate hydrate: [86220-42-0] $C_{68}H_{89}N_{17}O_{16}$ 1400.56	GnRH receptor agonist for treatment of central precocious puberty (true precocious puberty, GnRH-dependent precocious precocity, complete isosexual precocity) in children of both sexes, treatment of endometriosis. nasal spray 2 mg/mL. 0.4 mg, N.	G.D. Searle LLC USA/1990 [80–82]

2.3.2. H01CB Somatostatin and Analogs (Table 8)

Table 8. H01CB Somatostatin and analogs

INN (Synonyms) [Brand names] ATC	Structure (Remarks)	[CAS-No.] Formula MW (g/mol)	Target (if known) Medical use Formulation DDD	Originator(s) Approval (country/year) References
Somatostatin (GH-RIF, Growth hormone-release inhibiting factor, SRIF-14) [Aminopan, Modustatine, Stilamin] H01CB01	AGCKNFFWKTFTSC (disulfide bridge: 3–14)	[38916-34-6] $C_{76}H_{104}N_{18}O_{19}S_2$ 1637.90 Acetate: [54472-66-1] $C_{78}H_{108}N_{18}O_{21}S_2$ 1697.95	somatostatin receptor agonist for treatment of severe, acute hemorrhage of gastro-duodenal ulcers, treatment of erosive or hemorrhagic gastritis amp. w/powder for solution for inj. 3 mg. 6 mg, P.	unknown [83–88]
Octreotide (SMS 201-995) [Longastatin, Sandostatin] H01CB02		[83150-76-9] $C_{49}H_{66}N_{10}O_{10}S_2$ 1019.25 Acetate: [79517-01-4] $C_{51}H_{70}N_{10}O_{12}S_2$ 1079.30	somatostatin receptor agonist for treatment of acromegaly and reduction of side effects from cancer chemotherapy. amp. for solution for inj. 50, 100, 200, 500, 1000 µg/mL. 0.7 mg, P.	Novartis USA/1988 [89–91]
Lanreotide (BIM-23014C; DC-13-116) [Somatuline] H01CB03		[108736-35-2] $C_{54}H_{69}N_{11}O_{10}S_2$ 1096.33 Acetate: [127984-74-1] $C_{56}H_{73}N_{11}O_{12}S_2$ 1156.39	somatostatin receptor agonist for treatment of neuroendocrine tumors and acromegaly. inj. (depot) 60 mg, 90 mg, 120 mg, P. 3 mg, P.	Ipsen Biopharmaceuticals USA/2007 [92, 93]
Vapreotide (RC-160, BMY-41606) [Octastatin, Sanvar] H01CB04		[103222-11-3] $C_{57}H_{70}N_{12}O_9S_2$ 1131.38 Acetate: [849479-74-9] $C_{59}H_{74}N_{12}O_{11}S_2$ 1191.43	somatostatin receptor agonist for treatment of esophageal variceal bleeding in patients with cirrhotic liver disease; efficacy in treatment of patients with AIDS-related diarrhea. acromegaly, Crohn's disease, neuroendocrine tumors.	Debiovision Inc. [94]

Table 8. (*Continued*)

INN (Synonyms) [Brand names] ATC	Structure (Remarks)	[CAS-No.] Formula MW (g/mol)	Target (if known) Medical use Formulation DDD	Originator(s) Approval (country/year) References
Pasireotide (SOM-230) [Signifor] H01CB05		[396091-73-9] $C_{58}H_{66}N_{10}O_9$ 1047.21 Diaspartate: [820232-50-6] $C_{66}H_{80}N_{12}O_{17}$ 1313.41 Pamoate: [396091-79-5] $C_{81}H_{82}N_{10}O_{15}$ 1435.60 Acetate: [396091-76-2] $C_{60}H_{70}N_{10}O_{11}$ 1107.28	somatostatin receptor agonist for treatment of Cushing disease. vial 0.3 mg, 0.6 mg, 0.9 mg, P.; vial w/powder for suspension 40 mg, 60 mg, P. 1.2 mg, P.	Novartis USA/2012 [95]

2.3.3. H01CC Anti-Gonadotropin-Releasing Hormones (Table 9)

Table 9. H01CC Anti-gonadotropin-releasing hormones

INN (Synonyms) [Brand names] ATC	Structure (Remarks)	[CAS-No.] Formula MW (g/mol)	Target (if known) Medical use Formulation DDD	Originator(s) Approval (country/year) References
Ganirelix (Org-37462, RS 26306) [Orgaluran, Antagon] H01CC01		*[124904-93-4]* $C_{80}H_{113}ClN_{18}O_{13}$ 1570.35 Acetate: *[129311-55-3]* $C_{84}H_{121}ClN_{18}O_{17}$ 1690.45	GnRH receptor antagonist inhibition of premature luteinizing hormone surges in women undergoing controlled ovarian hyperstimulation. vial for inj. 0.25 mg/0.5 mL. 0.25 mg, P.	Organon USA Inc. USA/1999 [82, 96–99]
Cetrorelix (SB-75, D-20761) [Cetrotide] H01CC02		*[120287-85-6]* $C_{70}H_{92}ClN_{17}O_{14}$ 1431.06 Acetate: *[145672-81-7]* $C_{70}H_{92}ClN_{17}O_{14} \cdot x$ $C_2H_4O_2$ unspecified	GnRH receptor antagonist inhibition of premature LH surges in women undergoing controlled ovarian stimulation. vial w/powder 0.25 mg, 3 mg (acetate) for inj. sol. 0.25 mg, P.	Serono USA/2000 [94, 100–102]

3. H02 Corticosteroids Systemic

3.1. H02A Corticosteroids for Systemic Use, Plain

3.1.1. H02AA Mineralocorticoids (Table 10)

Table 10. H02AA Mineralocorticoids

INN (Synonyms) [Brand names] ATC	Structure (Remarks)	[CAS-No.] Formula MW (g/mol)	Target (if known) Medical use Formulation DDD	Originator(s) Approval (country/year) References
Fludrocortisone (Fluorocortisone) [Astonin H, Florinef] H02AA02, S01CA06, S02CA07, S03CA05 **WHO essential medicine**		*[127-31-1]* $C_{21}H_{29}FO_5$ 380.46 Acetate: *[514-36-3]* $C_{23}H_{31}FO_6$ 422.49	mineralocorticoid receptor and GR agonist for partial replacement therapy for primary and secondary adrenocortical insufficiency in Addison's disease and for the treatment of salt-losing adrenogenital syndrome. ear drops 8 mg/8 mL in comb. with polymyxin B; ointment 0.001%; tabl. 0.1 mg (as acetate). 0.1 mg, O.	unknown USA/1955 [103–108]
Desoxycortone (11-decorticosterone, deoxycortone acetate) [Cortexone, Syncortyl] H02AA03		*[64-85-7]* $C_{21}H_{30}O_3$ 330.47 Acetate: *[56-47-3]* $C_{23}H_{32}O_4$ 372.50 Pivalate: *[808-48-0]* $C_{26}H_{38}O_4$ 414.59	mineralocorticoid receptor agonist treatment of adrenocortical insufficiency especially in multiple sclerosis, congenital cerebral palsy, polyarteritis nodosa, and rheumatoid arthritis, treatment of Addison's disease. amp. 10 mg/mL. 5 mg, O.; 5 mg, P.	Organon USA/1939 (acetate) [109–112]

3.1.2. H02AB Glucocorticoids (Table 11)

Table 11. H02AB Glucocorticoids

INN (Synonyms) [Brand names] ATC	Structure (Remarks)	[CAS-No.] Formula MW (g/mol)	Target (if known) Medical use Formulation DDD	Originator(s) Approval (country/year) References
Betamethasone (NSC-39470, Sch-4831) [Celestone, Rinderon, Diprosone] A07EA04, C05AA05, D07AC01, D07XC01, H02AB01, R01AD06, R03BA04, S01BA06, S01CB04, S02BA07, S03BA03		[378-44-9] $C_{22}H_{29}FO_5$ 392.47 Acetate: [987-24-6] $C_{24}H_{31}FO_6$ 434.50 Benzoate: [22298-29-9] $C_{29}H_{33}FO_6$ 496.58 Dipropionate: [5593-20-4] $C_{28}H_{37}FO_7$ 504.60 Sodium phosphate: [151-73-5] $C_{22}H_{28}FNa_2O_8P$ 516.41 Valerate: [2152-44-5] $C_{27}H_{37}FO_6$ 476.59	corticosteroid hormone receptor agonist for the treatment of edocrine, rheumatic and hematologic disorders; collagen, dermatological, respiratory, gastrointestinal, neoplastic and ophthalmic diseases; allergic and edematous states; tuberculous meningitis and trichinosis. powder 0.1%; suppos. 0.5 mg, 1 mg; syrup 0.01%, 0.6 mg/5 mL; tabl. 0.5 mg, 0.6 mg, 1 mg. 1.5 mg, O.; 1.5 mg P.; 0.4 mg, P. (depot).	Merck Sharp Dohme USA/1961 [113–117]
Dexamethasone (FT-4145, ENV-1105) [Decadron, DexaSite, LenaDex, Maxidex] A01AC02, C05AA09, D07AB19, D07XB05, D10AA03, H02AB02, R01AD03, S01BA01, S01CB01, S02BA06, S03BA01		[50-02-2] $C_{22}H_{29}FO_5$ 392.47 Acetate: [1177-87-3] $C_{24}H_{31}FO_6$ 434.50 Phosphate: [312-93-6] $C_{22}H_{30}FO_8P$ 472.44 Propionate: [15423-89-9] $C_{25}H_{33}FO_6$ 448.53	corticosteroid hormone receptor agonist used as anti-inflammatory or immunosuppressive agent, able to penetrate the CNS, used to manage cerebral edema. amp. 0.4%/5 mL, 0.45%/5 mL (as acetate); 2.5 mg (as palmitate); cream 0.1% (as propionate); f. c. tabl. 0.5 mg, 0.75 mg, 1.5 mg. 1.5 mg, O.; 1.5 mg, P.	Merck Sharp Dohme USA/1958 [118–122]
Fluocortolone (SH-742) [Ultralan] C05AA08, D07AC05, D07XC05, H02AB03, S01CA04		[152-97-6] $C_{22}H_{29}FO_4$ 376.47 Acetate: [1176-82-5] $C_{24}H_{31}FO_5$ 418.51 Caproate: [303-40-2] $C_{28}H_{39}FO_5$ 474.61 Pivalate: [29205-06-9] $C_{27}H_{37}FO_5$	GR agonist anti-inflammatory activity used topically for various skin disorders. cream 2.5 mg/g; lotion 2.5 mg/g; ointment 2.5 mg/g; tabl. 5 mg, 20 mg, 50 mg. 10 mg, O.	unknown [123–127]

Table 11. (*Continued*)

INN (Synonyms) [Brand names] ATC	Structure (Remarks)	[CAS-No.] Formula MW (g/mol)	Target (if known) Medical use Formulation DDD	Originator(s) Approval (country/year) References
Methylprednisolone (NSC-19987) [Depo-Medrol, Medrone, Urbason] D07AA01, D10AA02, H02AB04 **WHO essential medicine**		[83-43-2] $C_{22}H_{30}O_5$ 374.48 Acetate: [53-36-1] $C_{24}H_{32}O_6$ 416.51 Succinate: [2921-57-5] $C_{26}H_{34}O_8$ 474.55 Sodium succinate: [2375-03-3] $C_{26}H_{33}NaO_8$ 496.53	corticosteroid hormone receptor agonist used for its anti-inflammatory effects. amp. 20 mg, 30 mg, 40 mg, 60 mg, 80 mg; cream 0.1%; ointment 0.1%; tabl. 2 mg, 4 mg, 6 mg, 24 mg, 60 mg. 7.5 mg, O.; 20 mg, P.	Pharmacia & Upjohn USA/1957 [128–131]
Paramethasone (CS-1483) [Cortidene, Haldrone, Monocortin, Stemex] H02AB05		[53-33-8] $C_{22}H_{29}FO_5$ 392.47 Acetate: [1597-82-6] $C_{24}H_{31}FO_6$ 434.50 Disodium phosphate: [2145-14-4] $C_{22}H_{28}FNa_2O_8P$ 516.41	GR agonist for the treatment of all conditions in which corticosteroid therapy is indicated except adrenal-deficiency states. amp. 20 mg/mL (as acetate); tabl. 2 mg, 6 mg (as acetate) 4 mg, O.; 4 mg, P.	Lilly USA [132]
Prednisolone (NCS-9120) [Predonine, Prelone, Delta-Cortef, Blephamide] A07EA01, C05AA04, D07AA03, D07XA02, H02AB06, R01AD02, S01BA04, S01CB02, S02BA03, S03BA02		[50-24-8] $C_{21}H_{28}O_5$ 360.45 Acetate: [52-21-1] $C_{23}H_{30}O_6$ 402.49 Sodium succinate: [1715-33-9] $C_{25}H_{31}NaO_8$ 482.51	GR agonist for the treatment of primary or secondary adrenocortical insufficiency, such as congenital adrenal hyperplasia, thyroiditis. Also used to treat psoriatic arthritis, rheumatoid arthritis, ankylosing spondylitis, bursitis, acute gouty arthritis, epicondylitis, systemic lupus erythematosus, pemphigus and acute rheumatic carditis. amp. 10 mg/mL, 25 mg/mL, 50 mg/mL (as acetate); cream 0.5%; eye drops 1.2 mg/mL, 10 mg/mL (as acetate); eye ointment 0.25%; ointment 0.5%, 100 mg/g; 1%, powder 97-102%; suppos. 100 mg (as acetate); syrup 15 mg/5 mL; tabl. 1 mg, 5 mg, 20 mg. 10 mg, O.; 10 mg, P.	Schering USA/1955 (as acetate) [133, 134]
Prednisone (NSC-10023, Dehydrocortisone) [Decortin, Cortancyl, Deltacortene, Meticorten] A07EA03, H02AB07		[53-03-2] $C_{21}H_{26}O_5$ 358.43 Acetate: [125-10-0] $C_{23}H_{28}O_6$ 400.47	GR agonist used to treat allograft rejection, asthma, systemic lupus erythematosus, and many other inflammatory states. It has some mineralocorticoid activity and thus may affect ion exchange in the kidney. cream 5 mg/g; eye drops 2 mg/g in comb. with chloramphenicole; suppos. 5 mg, 10 mg, 30 mg, 100 mg; syrup 5 mg/5 mL, 25 mg/mL, 50 mg/mL; tabl. 1 mg, 5 mg, 20 mg, 50 mg. 10 mg, O.	unknown USA/1955 [133]

(*continued*)

Table 11. (*Continued*)

INN (Synonyms) [Brand names] ATC	Structure (Remarks)	[CAS-No.] Formula MW (g/mol)	Target (if known) Medical use Formulation DDD	Originator(s) Approval (country/year) References
Triamcinolone (CL-19823, Fluoxyprednisolone) [Aristocort, Aristospan, Kenacort, Ledercort, Volon] A01AC01, C05AA12, D07AB09, D07XB02, H02AB08, R01AD11, R03BA06, S01BA05		[124-94-7] $C_{21}H_{27}FO_6$ 394.44 Acetonide: [76-25-5] $C_{24}H_{31}FO_6$ 434.50 Diacetate: [67-78-7] $C_{25}H_{31}FO_8$ 478.51 Hexacetonide: [5611-51-8] $C_{30}H_{41}FO_7$ 532.65	corticosteroid hormone receptor agonist for the treatment of perennial and seasonal allergic rhinitis. Acetonide: syrup 0.4 mg, 2 mg; tabl. 2 mg, 4 mg; cream 0.25 mg, 1 mg, 5 mg; liquid 25 mg, 40 mg; ointment 0.1%, 0.25 mg, 1 mg; inj. suspension 5 mg/mL, 10 mg/mL, 20 mg/mL, 40 mg/mL; spray 55 µg. 7.5 mg, O.; 7.5 mg, P.; 0.22 mg, N.	USA/1957 Apothecon USA/1960 (as acetonide) Sandoz USA/1969 (as hexacetonide) [135–138]
Hydrocortisone (Cortisol) [Hydrocortone, Cortef, Solu-Cortef, Cortifoam] A01AC03, A07EA02, C05AA01, D07AA02, D07XA01, H02AB09, S01BA02, S01CB03, S02BA01		[50-23-7] $C_{21}H_{30}O_5$ 362.47 Acetate: [50-03-3] $C_{23}H_{32}O_6$ 404.50 17-butyrate: [13609-67-1] $C_{25}H_{36}O_6$ 432.56	GR agonist used either as an injection or topically, in the treatment of inflammation, allergy, collagen diseases, asthma, adrenocortical deficiency, shock, and some neoplastic conditions. cream 0.1% (17-butyrate), 0.5%, 1% (acetate); eye drops 0.5% (acetate); ointment 0.1% (17-butyrate), 0.5% (acetate), 1%, 2.5%; lotion 0.5%, 1 mg/g (17-butyrate); suppos. 3.3 mg (acetate); tabl. 10 mg. 30 mg, O.; 30 mg, P.	Merck USA/1951 (as acetate) Precision Dermatology USA/1982 (as 17-butyrate) [139–142]
Cortisone (NSC-49420, 17-hydroxy-11-dehydrocorticosterone) [Cortone, Cortogen, Cortisyl, Cortistab, Ricortex] H02AB10, S01BA03		[53-06-5] $C_{21}H_{28}O_5$ 360.45 Acetate: [50-04-4] $C_{23}H_{30}O_6$ 402.49	corticosteroid hormone receptor agonist used in replacement therapy for adrenal insufficiency and as an anti-inflammatory agent. ointment 0.5%, 1%; tabl. 5 mg, 25 mg, 50 mg; vial 25 mg (2.5 mg/mL), 500 mg (50 mg/mL) (as acetate). 37.5 mg, O.; 37.5 mg, P.	unknown USA/1950 [143–146]
Rimexolone (Org-6216, trimexolone) [Rimexel, Vexol] H02AB12, S01BA13		[49697-38-3] $C_{24}H_{34}O_3$ 370.53	GR agonist treatment of inflammation in the eye. suspension 1%; eye drops 1%. 20 mg, P.	Alcon USA/1994 [147, 148]
Deflazacort (Azacort, Oxazacort, DL-458-IT) [Calcort, Deflan, Dezacor, Emfalza, Flantadin] H02AB13		[14484-47-0] $C_{25}H_{31}NO_6$ 441.52	GR agonist anti-inflammatory and immunosuppressant agent, treatment of Duchenne muscular dystrophy (DMD) in patients 5 years of age and older. tabl. 6 mg, 18 mg, 30 mg, 36 mg; suspension 22.75 mg/mL. 15 mg, O.	Marathon Pharm. USA/2017 (for DMD) [149–153]
Cloprednol (RS-4691) [Cloradryn, Novacort, Syntestan] H02AB14		[5251-34-3] $C_{21}H_{25}ClO_5$ 392.88	GR agonist treatment of asthma and rheumatoid arthritis. tabl. 2.5 mg, 5 mg, 10 mg.	unknown [154, 155]
Cortivazol (H-3625, MK-650) [Altim, Diaster, Dilaster] H02AB17		[1110-40-3] $C_{32}H_{38}N_2O_5$ 530.67	GR agonist/treatment of musculoskeletal and joint disorders. syringe 3.75 mg.	unknown [156, 157]

4. H03 Thyroid Therapy

4.1. H03A Thyroid Preparations

4.1.1. H03AA Thyroid Hormones (Table 12)

Table 12. H03AA Thyroid hormones

INN (Synonyms) [Brand names] ATC	Structure (Remarks)	[CAS-No.] Formula MW (g/mol)	Target (if known) Medical use Formulation DDD	Originator(s) Approval (country/year) References
Levothyroxine sodium (L-T4) [Berlthyrox, Euthyrox, Levothroid, Synthroid, Thyradin] H03AA01		*[55-03-8]* $C_{15}H_{10}I_4NNaO_4$ 798.85	thyroid hormone receptor agonist treatment of hypothyroidism, thyroid cancer, and thyrotoxicosis. tabl. 0.025 mg, 0.05 mg, 0.075 mg, 0.1 mg, 0.125 mg, 0.15 mg, 0.175 mg, 0.2 mg, 0.3 mg. 0.15 mg, P., 0.15 mg, O.	Jerome Stevens Pharmaceuticals USA/2000 [158–160]
Liothyronine sodium (BCT-303, sodium L-triiodothyronine) [Cytomel, Cytobin, Tertroxin, Thyromax, Triostat] H03AA02		*[55-06-1]* $C_{15}H_{11}I_3NNaO_4$ 672.95 Hydrochloride: *[6138-47-2]* $C_{15}H_{13}I_3ClNO_4$ 687.43	thyroid hormone receptor agonist treatment of hypothyroidism. tabl. 5 μg, 0.02 mg, 0.025 mg. 0.05 mg, 0.1 mg; vial 0.01 mg/mL, 0.1 mg/mL. 60 μg, O.; 60 μg, P.	King Pharmaceuticals R&D USA/1956 [161]

4.2. H03B Antithyroid Preparations

4.2.1. H03BA Thiouracils (Table 13)

Table 13. H03BA Thiouracils

INN (Synonyms) [Brand names] ATC	Structure (Remarks)	[CAS-No.] Formula MW (g/mol)	Target (if known) Medical use Formulation DDD	Originator(s) Approval (country/year) References
Methylthiouracil (6-methyl-2-thiouracil; 4-methyl-2-thiouracil; MTU) [Alkiron, Methiacil, Muracil] H03BA01		*[56-04-2]* $C_5H_6N_2OS$ 142.18	thyroid hormone synthesis inhibitor treatment of hyperthyroidism. tabl. 25 mg, 100 mg. 100 mg, O.	unknown [162, 163]
Propylthiouracil (6-propyl-2-thiouracil) [Propacil, Propycil, Propyl-Thyracil, Thyreostat II] H03BA02		*[51-52-5]* $C_7H_{10}N_2OS$ 170.23	thyroid hormone synthesis inhibitor treatment of hyperthyroidism. tabl. 25 mg, 50 mg. 100 mg, O.	DAVA Pharmaceuticals USA/1947 [163]

4.2.2. H03BB Sulfur-Containing Imidazole Derivatives (Table 14)

Table 14. H03BB Sulfur-containing imidazole derivatives

INN (Synonyms) [Brand names] ATC	Structure (Remarks)	[CAS-No.] Formula MW (g/mol)	Target (if known) Medical use Formulation DDD	Originator(s) Approval (country/year) References
Carbimazole (Athyromazole) [Basolest, Neomercazole, Neo-Thyreostat] H03BB01	H₃C–N N O CH₃ (S, O)	[22232-54-8] C₇H₁₀N₂O₂S 186.23	thyroid peroxidase inhibitor treatment of hyperthyroidism and thyrotoxicosis; also used to prepare patients for thyroidectomy. tabl. 5 mg, 10 mg. 15 mg, O.	unknown USA/1959 [164–166]
Thiamazole (NSC-38608, Methimazole) [Favistan, Northyx, Tapazole, Thyrozol] H03BB02	H₃C–N NH (S)	[60-56-0] C₄H₆N₂S 114.17	thyroid peroxidase inhibitor treatment of hyperthyroidism, goiter, Graves disease and psoriasis. amp. 10 mg/mL, 40 mg/mL; tabl. 5 mg, 10 mg, 20 mg. 10 mg, O.	Pfizer USA/1950 [166, 167]

4.2.3. H03BX Other Antithyroid Preparations (Table 15)

Table 15. H03BX Other antithyroid preparations

INN (Synonyms) [Brand names] ATC	Structure (Remarks)	[CAS-No.] Formula MW (g/mol)	Target (if known) Medical use Formulation DDD	Originator(s) Approval (country/year) References
Diiodotyrosine (DIT, iodogorgoic acid) [Agontan] H03BX01	COOH NH₂ HO (I, I)	(L)-form: [300-39-0] (DL)-form: [66-02-4] C₉H₉I₂NO₃ 432.98	thyroid inhibitor treatment of Graves disease and other disorders.	unknown [168, 169]

5. H04 Pancreatic Hormones

5.1. H04AA Glycogenolytic Hormones (Table 16)

Table 16. H04AA Glycogenolytic hormones

INN (Synonyms) [Brand names] ATC	Structure (Remarks)	[CAS-No.] Formula MW (g/mol)	Target (if known) Medical use Formulation DDD	Originator(s) Approval (country/year) References
Glucagon (hyperglycemic-glycogenolytic factor, HGF) [Glucagon] H04AA01	HSQGTFTSDYSKYLDSRRA QDFVQWLMNT	*[9007-92-5]* $C_{153}H_{225}N_{43}O_{49}S$ 3482.75	glucagon receptor agonist treatment of severe hypoglycemia. vial w/powder for solution for inj. 1 mg. 1 mg, P.	Eli Lilly USA/1998 purification: [170, 171] amino acid sequence: [172]

6. H05 Calcium Homeostasis

6.1. H05AA Parathyroid Hormones and Analogs (Table 17)

Table 17. H05AA Parathyroid hormones and analogues

INN (Synonyms) [Brand names] ATC	Structure (Remarks)	[CAS-No.] Formula MW (g/mol)	Target (if known) Medical use Formulation DDD	Originator(s) Approval (country/year) References
Teriparatide (hPTH 1-34, MN-10T) [Parathar, Forteo] H05AA02	SVSEXQLMHNLGKHLNSMERVEWL RKKLQDVHNF	*[52232-67-4]* $C_{181}H_{291}N_{55}O_{51}S_2$ 4117.77 Acetate: *[99294-94-7]* $C_{181}H_{291}N_{55}O_{51}S_2 \cdot y$ $C_2H_4O_2 \cdot xH_2O$	PTH/PTH-related peptide receptor agonist treatment of osteoporosis in patients with a high risk for bone fracture. vial 250 μg/mL. 20 μg, P.	Eli Lilly USA/2002 [173–175]
Parathyroid hormone (ALX-111, rhPTH-1-84, NPSP-558) [Natpara, Preotact, Natpar] H05AA03	SVSEIQLMHNLGKHLNSMERVEWLRKKLQD VHNFVALGAPLAPRDAGSQRPRKKEDNVLV ESHEKSLGEADKADVNVLTKAKSQ	*[68893-82-3]* $C_{408}H_{674}N_{126}O_{126}S_2$ 9424.62	PTH/PTH-related peptide receptor agonist treatment of osteoporosis, control of hypocalcemia in patients with hypoPT. vial w/powder for solution for inj. 25 μg, 50 μg, 75 μg, 0.1 mg; prefilled pen 0.1 mg. 0.1 mg, P.	NPS Pharmaceuticals EU/2006, USA/2015 [176]

6.2. H05B: Anti-Parathyroid Agents

6.2.1. H05BA Calcitonin Preparations (Table 18)

Table 18. H05BA Calcitonin preparations

INN (Synonyms) [Brand names] ATC	Structure (Remarks)	[CAS-No.] Formula MW (g/mol)	Target (if known) Medical use Formulation DDD	Originator(s) Approval (country/year) References
Calcitonin (salmon synthetic) (Salmotonin) [Calcimar, Miacalcin] H05BA01	CSNLSTCVLGKLSQELHKLQTY PRTNTGSGTP (disulfide bridge: 1–7)	*[47931-85-1]* $C_{145}H_{240}N_{44}O_{48}S_2$ 3431.90	calcitonin receptor agonist treatment of post-menopausal osteoporosis, hypercalcemia, and Paget disease. vial 200 U/mL; nasal spray 200 U/ actuation, amp. 50 U/mL, 100 U/mL. 100 U, P., 200 U, N.	Mylan Ireland Ltd. USA/1991 [177]
Calcitonin (human synthetic) (BA-47175) [Cibacalcin] H05BA03	CGNLSTCMLGTYTQDFNKFHTF PQTAIGVGAP (disulfide bridge: 1–7)	*[21215-62-3]* $C_{151}H_{226}N_{40}O_{45}S_3$ 3417.85	calcitonin receptor agonist treatment of post-menopausal osteoporosis, hypercalcemia, and Paget disease. vial 200 U/mL; nasal spray 200 U/ actuation, amp. 50 U/mL, 100 U/mL. 100 U, P.	Novartis USA [178, 179]

6.2.2. H05BX Other Anti-Parathyroid Agents (Table 19)

Table 19. H05BX Other anti-parathyroid agents

INN (Synonyms) [Brand names] ATC	Structure (Remarks)	[CAS-No.] Formula MW (g/mol)	Target (if known) Medical use Formulation DDD	Originator(s) Approval (country/year) References
Cinacalcet (AMG-073, KRN-1493) [Mimpara, Sensipar] H05BX01		[226256-56-0] C$_{22}$H$_{22}$F$_3$N 357.41 Hydrochloride: [364782-34-3] C$_{22}$H$_{23}$ClF$_3$N 393.88	calcium-sensing receptor agonist for treatment of SHPT in patients with CKD who are on hemodialysis or peritoneal dialysis, for treatment of hypercalcemia in patients with parathyroid carcinoma. tabl. 30 mg, 60 mg, 90 mg. 60 mg, O.	Amgen USA/2004 [180]
Paricalcitol (Paracalcin, ABT-358) [Zemplar] H05BX02		[131918-61-1] C$_{27}$H$_{44}$O$_3$ 416.64	vitamin D receptor agonist for prevention and treatment of SHPT associated with chronic renal failure. amp. 5 μg/mL; 1 mL, 2 mL, 5 mL. 2 μg, O.: 2 μg, P.	Abbott USA/1998 [181, 182]
Doxercalciferol (TSA-840) [Hectorol, Redispar] H05BX03		[54573-75-0] C$_{28}$H$_{44}$O$_2$ 412.65	vitamin D receptor agonist for treatment of SHPT in patients with CKD on dialysis, with stage 3 or stage 4 CKD. cps. 0.5 μg, 1 μg, 2.5 μg; vial for solution for inj. 2 μg/mL, 4 μg/2 mL.	Genzyme USA/1999 [183, 184]

(continued)

Table 19. (*Continued*)

INN (Synonyms) [Brand names] ATC	Structure (Remarks)	[CAS-No.] Formula MW (g/mol)	Target (if known) Medical use Formulation DDD	Originator(s) Approval (country/year) References
Etelcalcetide (AMG 416, KAI-4169) [Parsabiv] H05BX04		[1262780-97-1] $C_{38}H_{73}N_{21}O_{10}S_2$ 1048.25 Hydrochloride: [1334237-71-6] $C_{38}H_{74}ClN_{21}O_{10}S_2$ 1084.72	calcium-sensing receptor agonist for treatment of SHPT in adult patients with CKD on hemodialysis. vial for solution for inj. 2.5 mg/0.5 mL, 5 mg/mL, 10 mg/2 mL. 5 mg. P.	Amgen EU/2016 [185]

Abbreviations

Amp.	ampoule(s)
ATC	Anatomical Therapeutical Chemical Classification
CKD	chronic kidney disease
cps.	capsules
DDD	defined daily dose
FSH	follicle-stimulating hormone
GH	growth hormone
GHR	growth hormone receptor
GnRH	gonadotropin-releasing hormone
GR	glucocorticoid receptor
hypoPT	hypoparathyroidism
IGF-1	insulin-like growth factor 1
IGFD	insulin growth factor deficiency
Inj.	injection
INN	international nonproprietary name
IVF	in vitro fertilization
LH	luteinizing hormone
N	nasal
NCE	new chemical entity
O	oral
P	parenteral
PTH	parathyroid hormone
SC	subcutaneous
SHPT	secondary hyperparathyroidism
SL	sublingual/buccal
suppos.	suppositories
tabl.	tablets
U	unit

References

1 Kleemann, A. *et al.* (2009) *Pharmaceutical Substances*, 5th edn, Thieme Verlag, Stuttgart, and/or its online version.

2 Trevor, A.J. *et al.* (2012) *Katzung & Trevor's Pharmacology Examination & Board Review*, 10th edn, McGraw-Hill Education, New York.

3 Antoniazzi, F. *et al.* (2015) *Minerva Endocrinologica*, **40**, 129.

4 Harvey, S. (2013) *General and Comparative Endocrinology*, **190**, 3.

5 Copperman, A.B. and Benadiva, C. (2013) *Reproductive Biology and Endocrinology*, **11**, 20.

6 Huirne, J.A. and Lambalk, C.B. (2001) *Lancet*, **358**, 1793.

7 Pierce, J.G. and Parsons, T.F. (1981) *Annual Review of Biochemistry*, **50**, 465.

8 Birzniece, V. *et al.* (2009) *Reviews in Endocrine & Metabolic Disorders*, **10**, 145.

9 Brooks, A.J. and Waters, M.J. (2010) *Nature Reviews Endocrinology*, **6**, 515.

10 Higham, C.E. and Trainer, P.J. (2008) *Experimental Physiology*, **93**, 1157.

11 Tritos, N.A. and Biller, B.M. (2017) *Pituitary*, **20**, 129.

12 Gomez-Sanchez, E. and Gomez-Sanchez, C.E. (2014) *Comprehensive Physiology*, **4**, 965.

13 Nussey, S. and Whitehead, S. (2001) *Endocrinology: An Integrated Approach*, BIOS Scientific Publishers, Oxford.

14 Ramamoorthy, S. and Cidlowski, J.A. (2016) *Rheumatic Disease Clinics of North America*, **42**, 15.

15 Biondi, B. and Wartofsky, L. (2014) *Endocrine Reviews*, **35**, 433.

16 Chakera, A.J. *et al.* (2012) *Drug, Design, Development and Therapy*, **6**, 1.

17 Chaker, L. *et al.* (2017) *Lancet*, **390**, 1550.

18 Hennessey, J.V. (2017) *Endocrine*, **55**, 6.

19 Wiersinga, W.M. (2001) *Hormone Research*, **56** (Suppl 1), 74.

20 Nygaard, B. (2010) *BMJ Clinical Evidence*, **2010**.

21 De Leo, S. *et al.* (2016) *Lancet*, **388**, 906.

22 Kravets, I. (2016) *American Family Physician*, **93**, 363.

23 Cryer, P.E. (2012) *Endocrinology*, **153**, 1039.

24 Chung, S.T. and Haymond, M.W. (2015) *Journal of Diabetes Science and Technology*, **9**, 44.

25 Leder, B.Z. (2017) *Current Osteoporosis Reports*, **15**, 110.

26 Rejnmark, L. *et al.* (2015) *Endocrinology and Metabolism (Seoul)*, **30**, 436.

27 Kim, E.S. and Keating, G.M. (2015) *Drugs*, **75**, 1293.

28 Cusano, N.E. *et al.* (2013) *Journal of Endocrinological Investigation*, **36**, 1121.

29 Hamano, N. *et al.* (2017) *Expert Opinion on Pharmacotherapy*, **18**, 529.

30 Komaba, H. *et al.* (2017) *Clinical and Experimental Nephrology*, **21**, 37.

31 Komaba, H. *et al.* (2010) *Expert Opinion on Biological Therapy*, **10**, 1729.

32 Cozzolino, M. *et al.* (2015) *Expert Opinion on Emerging Drugs*, **20**, 197.

33 Cunningham, J. *et al.* (2011) *Clinical Journal of the American Society of Nephrology*, **6**, 913.

34 Pickering, B.T. *et al.* (1963) *Biochimica et Biophysica Acta*, **74**, 763.

35 Upjohn (1964) US 3 124 509, US-prior. 1952.

36 Kappeler, H. and Schwyzer, R. (1961) *Helvetica Chimica Acta*, **44**, 1136.

37 Cole, E.S. *et al.* (1993) *Biotechnology (N Y)*, **11**, 1014.

38 Genzyme (31.08.1993) US 5 240 832, US-prior. 1989.

39 Olson, K.C. *et al.* (1981) *Nature*, **293**, 408.

40 The Regents of the University of Minnesota (1978) US 4 124 448, US-prior. 1976.

41 Genentech (1987) US 4 634 677, US-prior. 1979.

42 Celtrix Pharma (06.04.1993) US 5 200 509, US-prior. 1987.

43 Rosenbloom, A.L. (2009) *Advances in Therapy*, **26**, 40.

44 Ling, N. *et al.* (1984) *Biochemical and Biophysical Research Communications*, **123**, 854.

45 The Salk Institute for Biological Studies (1982) US 4 703 035, US-prior. 1987.

46 Adis Editorial (2005) *Drugs in R&D*, **6**, 120.

47 Kemp, S.F. (2007) *Drugs of Today (Barcelona, Spain: 1998)*, **43**, 149.

48 Williams, R.M. *et al.* (2008) *Expert Opinion on Drug Metabolism and Toxicology*, **4**, 311.

49 Theratechnologies (19.01.1999) US 5 861 379, US-prior. 26.05.1995.

50 Theratechnologies (08.01.2008) US 7 316 997, US-prior. 29.05.2003.

51 Genentech (15.12.1998) US 5 849 535, US-prior. 21.09.1995.

52 American Home Products Corp. (1978) US 4 093 610, US-prior. 1977.

53 du Vigneaud, V. *et al.* (1958) *Journal of the American Chemical Society*, **80**, 3355.

54 Bodanszky, M. *et al.* (1964) *Journal of the American Chemical Society*, **86**, 4452.

55 Meienhofer, J. *et al.* (1970) *Journal of the American Chemical Society*, **92**, 7199.

56 Jones, D.A. Jr. *et al.* (1973) *The Journal of Organic Chemistry*, **38**, 2865.

57 Ceskoslovenska Akad. Ved (1970) US 3 497 491, CZ-prior. 1966.

58 Huguenin, R.L. and Boissonnas, R.A. (1966) *Helvetica Chimica Acta*, **49**, 695.

59 Zaoral, M. *et al.* (1967) *Collection of Czechoslovak Chemical Communication*, **32**, 1250.

60 du Vigneaud, V. *et al.* (1953) *Journal of the American Chemical Society*, **75**, 4880.

61 Bartlett, C. *et al.* (1956) *Journal of the American Chemical Society*, **78**, 2905.

62 Bodanszky, M. *et al.* (1960) *Journal of the American Chemical Society*, **82**, 3195.

63 Boissonnas, R.A. and Huguenin, R.L. (1960) *Helvetica Chimica Acta*, **43**, 182.

64 Meienhofer, J. and Sano, Y. (1968) *Journal of the American Chemical Society*, **90**, 2996.

65 Kasafirek, E. *et al.* (1966) *Collection of Czechoslovak Chemical Communication*, **31**, 4581.

66 Procházka, Z. *et al.* (1978) *Collection of Czechoslovak Chemical Communication*, **43**, 1285.

67 Huguenin, R.L. and Boissonnas, R.A. (1963) *Helvetica Chimica Acta*, **46**, 1669.

68 Sandoz (1967) US 3 299 036, CH-prior. 1963.

69 du Vigneaud, V. *et al.* (1960) *The Journal of Biological Chemistry*, **235**, PC64.

70 Hope, D.B. *et al.* (1962) *The Journal of Biological Chemistry*, **237**, 1563.

71 Takashima, H. *et al.* (1968) *Journal of the American Chemical Society*, **90**, 1323.

72 Roussel-Uclaf (1960) US 2 938 891, F-prior. 1956.

73 Roussel-Uclaf (1963) US 3 076 797, F-prior. 1957.

74 Fric, I. *et al.* (1974) *Collection of Czechoslovak Chemical Communication*, **39**, 1290.

75 Ceskoslovenska Akad. Ved (1978) DE 2 732 175, CZ-prior. 1976.

76 Schally, A.V. *et al.* (1971) *Biochemical and Biophysical Research Communications*, **43**, 393.

77 Matsuo, H. *et al.* (1971) *Biochemical and Biophysical Research Communications*, **43**, 1334.

78 Hoechst (1985) EP 156 280, DE-prior. 1984.

79 Hoechst (1973) DE 2 213 737, DE-prior. 1972.

80 Syntex (1980) US 4 234 571, US-prior. 1979.

81 Nestor, J.J. Jr. *et al.* (1982) *Journal of Medicinal Chemistry*, **25**, 795.

82 Arzeno, H.B. *et al.* (1993) *International Journal of Peptide and Protein Research*, **41**, 342.

83 Rivier, J. *et al.* (1973) *Comptes Rendus des Seances de l'Academie des Sciences. Serie D, Sciences Naturelles*, **276**, 2737.

84 Sarantakis, D. and McKinley, W.A. (1973) *Biochemical and Biophysical Research Communications*, **54**, 234.

85 Felix, A.M. *et al.* (1980) *International Journal of Peptide and Protein Research*, **15**, 342.

86 Coy, D.H. *et al.* (1973) *Biochemical and Biophysical Research Communications*, **54**, 1267.

87 Genentech (1982) US 4 356 270, US-prior. 1977.

88 Genentech (1982) US 4 366 246, US-prior. 1977.

89 Sandoz (1981) EP 29 579, CH-prior. 1980.

90 Sandoz (1983) US 4 395 403, CH-prior. 1979.

91 Bauer, W. *et al.* (1982) *Life Sciences*, **31**, 1133.

92 Tulane Educational Fund (1987) EP 215 171, US-prior. 1985.

93 Tulane Educational Fund (1989) US 4 853 371, US-prior. 1985.

94 Schally A.V. Cai R.Z. (1987) US 4 650 787, US-prior. 1985.

95 Novartis (01.05.2001) US 4 356 270, US-prior. 29.06.1995.

96 Syntex (1988) EP 312 052, US-prior. 1987.

97 Syntex (1988) EP 301 850, US-prior. 1987.

98 Syntex (1989) US 4 801 577, US-prior. 1987.

99 Syntex (18.05.1993) US 5 212 288, US-prior. 1990.

100 Tulane Univeristy (30.03.1993) US 5 198 533, US-prior. 1987.

101 ASTA Medica (1988) EP 299 402, US-prior. 1987.

102 Bajusz, S. *et al.* (1988) *International Journal of Peptide and Protein Research*, **32**, 425.

103 Fried, J. and Sabo, E.F. (1954) *Journal of the American Chemical Society*, **76**, 1455.

104 Olin Mathieson (1954) GB 792 224, US-prior. 1954.

105 Merck & Co. (1956) DE 1 035 133, US-prior. 1955.

106 Merck & Co. (1959) US 2 894 007,

107 Upjohn (1956) US 2 771 475, US-prior. 1953.

108 Schering AG (1958) DE 1 028 572, DE-prior. 1957.

109 Schindler, W. *et al.* (1941) *Helvetica Chimica Acta*, **24**, 371.

110 Schering AG (1953) DE 875 353, DE-prior. 1938.

111 Hoechst AG (1953) DE 871 153, DE-prior. 1937.

112 Ciba Pharm. Prod (1957) US 2 778 776, US-prior. 1955.

113 Schering Corp. (1965) US 3 164 618, US-prior. 1957.

114 Oliveto, E.P. *et al.* (1958) *Journal of the American Chemical Society*, **80**, 4428.

115 Oliveto, E.P. *et al.* (1958) *Journal of the American Chemical Society*, **80**, 6687.

116 Roussel-Uclaf (1963) US 3 104 246, F-prior. 1961.

117 Julian, P.L. *et al.* (1955) *Journal of the American Chemical Society*, **77**, 4601.

118 Arth, G.E. *et al.* (1958) *Journal of the American Chemical Society*, **80**, 3160.

119 Merck & Co. (1958) DE 1 113 690, US-prior. 1957.

120 Oliveto, E.P. *et al.* (1958) *Journal of the American Chemical Society*, **80**, 4431.

121 Olin Mathieson (1958) US 2 852 511, US-prior. 1953.

122 Lab. Franç. de Chimiothérapie (1961) US 3 007 923, F-prior. 1959.

123 Schering AG (1962) DE 1 135 899, DE-prior. 1960.

124 Schering AG (1966) US 3 232 839, DE-prior. 1961.

125 Doménico, A. *et al.* (1965) *Arzneim-Forsch*, **15**, 46.

126 Kieslich, K. *et al.* (1969) *Justus Liebigs Annalen der Chemie*, **726**, 168.

127 Schering AG (1962) BE 614 196, DE-prior. 1961.

128 Upjohn (1959) US 2 897 218, US-prior. 1956.

129 Speero, G.B. *et al.* (1956) *Journal of the American Chemical Society*, **78**, 6213.

130 Speero, G.B. *et al.* (1957) *Journal of the American Chemical Society*, **79**, 1515.

131 Schering Corp. (1962) DE 3 053 832, US-prior. 1957.

132 Edwards, J.A. *et al.* (1960) *Journal of the American Chemical Society*, **40**, 2318.

133 Schering Corp. (1959) US 2 897 216, US-prior. 1952.

134 Upjohn (1977) US 4 041 055, US-prior. 1975.

135 American Cyanamid (1957) US 2 789 118, US-prior. 1956.

136 Bernstein, S. *et al.* (1956) *Journal of the American Chemical Society*, **78**, 5693.

137 Bernstein, S. *et al.* (1959) *Journal of the American Chemical Society*, **81**, 1689.

138 Squibb & Sons Inc. (1970) US 3 536 586, US-prior. 1968.

139 Wendler, N.L. *et al.* (1950) *Journal of the American Chemical Society*, **72**, 5793.

140 Upjohn (1953) US 2 649 401, US-prior. 1950.

141 Pfizer (1953) US 2 658 023, US-prior. 1952.
142 Upjohn (1957) US 2 794 816, US-prior. 1954.
143 Sarett, L.H. (1946) *The Journal of Biological Chemistry*, **162**, 601.
144 Sarett, L.H. (1948) *Journal of the American Chemical Society*, **70**, 1454.
145 Upjohn (1952) US 2 602 769, US-prior. 1950.
146 Upjohn (1956) US 769 823, US-prior. 1954.
147 Akzona Inc. (1976) US 3 947 478, GB-prior. 1972.
148 Cairns, J. *et al.* (1981) *Journal of the Chemical Society-Perkin Transactions*, **1**, 2306.
149 Lepetit (1965) GB 1 077 393.
150 Lepetit (1969) US 3 436 389, GB-prior. 1965.
151 Nathanson, G. *et al.* (1967) *Journal of Medicinal Chemistry*, **10**, 799.
152 Griggs, R.C. *et al.* (2016) *Neurology*, **87**, 2123.
153 Traynor, K. (2017) *American Journal of Health-System Pharmacy*, **74**, 368.
154 Syntex (1966) US 3 232 965, US-prior. 1958.
155 Syntex (1958) GB 890 835, Mex-prior. 1957.
156 Fried, J.H. *et al.* (1963) *Journal of the American Chemical Society*, **85**, 236.
157 Merck & Co (1962) US 3 067 194, US-prior. 1961.
158 Nahm, H. and Siedel, W. (1963) *Chemische Berichte*, **96**, 1.
159 Hoechst (1955) DE 1 067 826.
160 Hoechst (1958) DE 1 077 673.
161 Roche, J. *et al.* (1953) *Biochimica et Biophysica Acta*, **11**, 215.
162 List, R. *et al.* (1886) *Justus Liebigs Annalen der Chemie*, **236**, 1.
163 Anderson, G.W. *et al.* (1945) *Journal of the American Chemical Society*, **67**, 2197.
164 Nat. Res. Dev. Corp. (1954) US 2 671 088, GB-prior. 1951.
165 Nat. Res. Dev. Corp. (1957) US 2 815 349, GB-prior. 1955.
166 Baker, J.A. (1958) *Journal of the Chemical Society*, 2387.
167 Wohl, A. *et al.* (1889) *Berichte der Deutschen Chemischen Gesellschaft*, **22**, 1354.
168 Basic Inc. (1958) US 2 835 700, US-prior. 1954.
169 Jurd, L. (1955) *Journal of the American Chemical Society*, **77**, 5747.
170 Staub, A. *et al.* (1953) *Science*, **117**, 628.
171 Staub, A. *et al.* (1955) *The Journal of Biological Chemistry*, **214**, 619.
172 Thomsen, J. *et al.* (1972) *FEBS Lett*, **21**, 315.
173 Andreatta, R.H. *et al.* (1973) *Helvetica Chimica Acta*, **56**, 470.
174 Armour Pharma (1977) DE 2 649 727, DE-prior. 1975.
175 Armour Pharma (1978) US 4 086 196, US-prior. 1975.
176 Armour Pharmaceutical Co. (1978) US 4 105 602, US-prior. 1975.
177 Lipotec (18.06.1996) US 5 527 881, ES-prior. 10.12.1992.
178 Sieber, P. *et al.* (1968) *Helvetica Chimica Acta*, **51**, 2057.
179 Hirt, J. *et al.* (1979) *Recueil des Travaux Chimiques*, **98**, 143.
180 Nps Pharmaceuticals (03.04.2001) US 6 211 244, US-prior. 21.10.1994.
181 Wisconsin Alumni Research Foundation (19.01.1994) EP 387 077, DE-prior. 1989.
182 Wisconsin Alumni Research Foundation (24.12.1996) US 5 587 497, DE-prior. 1989.
183 Lam, H.Y. *et al.* (1974) *Science*, **186**, 1038.
184 Paaren, H.E. *et al.* (1980) *The Journal of Organic Chemistry*, **45**, 3253.
185 Amgen (29.09.2016) WO 2016/ 154580, US-prior. 26.03.2015.

Antiinfectives for Systemic Use, 1. Antibacterials (J01)

Axel Kleemann, Hanau, Germany

1. Introduction

In this and two further articles on antiinfectives (→ Antiinfectives for Systemic Use, 2. Antimycotics (Antifungals) and Antimycobacterials), the different antiinfective drug classes are reviewed and arranged according to their Anatomical Therapeutical Chemical (ATC) classification of the WHO, in which they are arranged according to the organs or organ systems on which they act, and their chemical, pharmacological, and therapeutic properties. For each drug, the ATC classification number is given, except in the very rare cases where such an ATC numbering is still pending.

Biopharmaceuticals (e.g., antibodies, polypeptides) are not included in the tables, only so-called "small molecules". The references for each drug include the basic patents and publications, in which the relevant synthetic or biosynthetic processes for production are described. In addition, the reader is referred to the standard reference source by Kleemann et al. [1] of which the online version is updated 1–2 times per year and so contains the most recent drug substances with their syntheses.

For syntheses of older compounds, the reader is referred to earlier versions of the antiinfectives chapters (antibiotics, antimycotics, HIV and AIDS therapeutics, chemotherapeutics). An overview on the synthesis of antibiotic drugs.

For detailed information about pharmacology, pharmacokinetics, metabolism, toxicology, interactions with other drugs, etc., the reader is referred to the available standard references [2, 3].

Drugs that have been withdrawn from the market or drugs that have become economically insignificant are not included, even if they are still contained in the ATC list.

In the tables of the particular chapters, the individual drug substances are characterized by their INN, synonyms, brand or trade names, ATC number, structures, CAS-no., formula, molecular mass, mode of action, medical use, year of approval, formulation, DDD (Defined Daily Dose), originator, and references.

Antiinfectives are also classified in the following other ATC groups in case of nonsystemic use or in combinations:

- **A01AB** (local oral treatment)
- **A02BD** (combination for eradication of *Helicobacter pylori*)
- **A07A** (intestinal antiinfectives)
- **D01** (antifungals for dermatological use)

Ullmann's Pharmaceuticals. Edited by Axel Kleemann and Bernhard Kutscher.
© 2022 Wiley-VCH GmbH
Set ISBN: 978-3-527-34252-5/ DOI: 10.1002/14356007.a02_467.pub4

- **D06** (antibiotics and chemotherapeutics for dermatological use)
- **G01** (gynecological antiinfectives and antiseptics)
- **P** (antiparasitic products, insecticides and repellents)
- **R02AB** (throat preparations, antibiotics)
- **R05X** (other cold preparations)
- **S01, S02, S03** (eye and ear preparations containing antiinfectives)

Antibiotics (synonymous to antimicrobials and also antibacterials) are used in the treatment and prevention of bacterial infections, some are also effective against fungi and protozoans, but not against viruses. Antibiotics are an important subgroup of antiinfectives.

By far the most antibiotic active substances are either natural products or semisynthetic derivatives thereof, only a minority is produced by total synthesis (e.g., sulfonamides, quinolones, and oxazolidinones) [4–9]. In many cases the natural source (e.g., fermentation of respective microorganisms) is mentioned under remarks. Whenever possible, the basic patent(s) (mainly the US patents) with preparation process and relevant publications are given in the references. For a more detailed description of the production methods and processes for antibiotics see the corresponding references or special monographs.

In contrast to antibacterials, practically all antivirals (with the exception of vaccines) are synthetic products.

2. Antibiotic Resistance

Within the last two decades more and more bacterial pathogens became resistant to the commonly used antibiotics [10, 11]. There are predominantly three classes of antibiotic-resistant bacterials, which are emerging as major threats to public health:

1. Methicillin-resistant *Staphylococcus aureus* (MRSA), and more and more also vancomycin-resistant *S. aureus* (VRSA).
2. Multidrug-resistant (MDR) and pandrug-resistant (PDR) gram-negative bacteria, namely the strains of *Acinetobacter*

baumanii, Escherichia coli, Klebsiella pneumoniae, and *Pseudomonas aeruginosa.*
3. MDR and extensively drug-resistant (XDR) strains of *Mycobacterium tuberculosis* (MDR-TB and XDR-TB), which are partly very difficult to treat and sometimes requires 2-year long treatment.

According to a recent WHO report [10], infection diseases are today's second cause of death worldwide. The fight against antimicrobial resistance has to be considered as a global crisis project.

According to estimates, 25 000 people in the EU die from infections with antibiotic-resistant bacteria, this figure being comparable with the situation in the USA. The governments of USA, UK, and Germany are currently undertaking efforts to beef up the pipeline of new and effective antibiotics and diagnostics and to reduce the overprescribing and unregulated use of antibiotics and their use in raising and feeding animals for food production (livestock), which contributes to the rising resistance.

Many antibiotics of the classes of penicillins, cephalosporins and others are no longer effective because many pathogens have become resistant or are connected with serious unwanted side effects. This is the reason why a considerable part of antibiotics, e.g., more than 50 β-lactam drugs, are no longer in use, respectively on the market. And in the following lists they have, of course, been omitted.

Since the 1980s the pharmaceutical industry almost stopped the development of new antibiotics, leading to a lack of new antimicrobial drugs in development. They have set their priorities on remedies for chronic diseases with lifelong medications and much better prospects for revenues and earnings. Most infections require only 1−2 weeks of treatment, so it is obvious that the expected revenues will not pay off for the development costs.

The situation has slightly improved recently, when the following new antibiotic drugs were approved: Linezolid (approval in 2000), daptomycin (2003), telavancin (2008), ceftobiprole (2013), ceftolozane (2014, combination with tazobactam), dalbavancin (2014), oritavancin (2014), and tedizolid (2014). Part of these new types of antibacterials cannot offer a solution on their own, they are too expensive and unaffordable in developing countries.

New antibiotics with new modes of action are urgently needed to combat resistant strains. The CDC estimates for the USA that more than 2 million people are sickened every year with antibiotic-resistant bacterial infections, with at least 23 000 dying as a result (thereof 11 285 alone from MRSA) [12]. The pharmaceutical industry has to be incentivized to develop new antibiotics, and maybe the regulatory hurdles by FDA and EMA have to be modified for new antibiotics, at least such new antibiotic drugs should be considered for orphan drug status [13–25]. All references cited in the headlines of the following chapters are general references, giving an in-depth introduction into the respective class of antiinfectives.

3. Individual Antibiotics

3.1. J01A Tetracyclines [26] (Table 1)

Table 1. J01AA Tetracyclines

INN (Synonyms) [Brand names] ATC	Structure	[CAS-No.] Formula MW (g/mol)	Target (if known) Medical use Formulation DDD	Approval (country/year) Originator(s)	Production method References
Demeclocycline (demethylchlortetracycline; RP-10192) [Ledermycin] J01AA01; D06AA01.		[127-33-3] $C_{21}H_{21}ClN_2O_8$ 464.86 Hydrochloride: [64-73-3] $C_{21}H_{21}ClN_2O_8 \cdot HCl$ 501.32	impairment of protein synthesis in bacteria by binding to the 30S and 50S ribosomal subunits/bacteriostatic activity against gram-positive and gram-negative bacteria; resistance is very common. tbl. 150, 300 mg; cps. 150 mg; ointment 0.5%. 600 mg, O.	USA/1958 American Cyanamid	prepared by fermentation of mutants of *Streptomyces aureofaciens* [27–29]
Doxycycline (GS-3065) [Vibramycin, Bassado, Clinofug, Spanor, Unacil, Vibravenös] J01AA02; A01AB22. **WHO essential medicine**		[564-25-0] $C_{22}H_{24}N_2O_8$ 444.44 Monohydrate: [17086-28-1] $C_{22}H_{24}N_2O_8 \cdot H_2O$ 462.46 Hyclate: [24390-14-5] $C_{22}H_{25}ClN_2O_8 \cdot$ $\frac{1}{2}C_2H_6O \cdot \frac{1}{2}H_2O$	same mechanism as all tetracyclines/broad application as antiinfective; see special literature. tbl. 50, 100, 200 mg; amp. 100 mg/5 mL. 100 mg, O, P.	USA/1966 American Cyanamid; Pfizer	semisynthetic from oxytetracycline [30–33]
Chlortetracycline [Aureomycin] J01AA03		[57-62-5] $C_{22}H_{23}ClN_2O_8$ 478.89 Hydrochloride: [64-72-2] $C_{22}H_{23}ClN_2O_8 \cdot HCl$ 515.35	same mechanism as all tetracyclines/today only very limited use in humans (ophthalmic and dermal use; and as veterinary drug). ointments 1% and 3%.	USA/1948 American Cyanamid	prepared by fermentation of *Streptomyces aureofaciens*; was the first tetracycline isolated in 1945 by B.M. Duggar at Lederle Labs. [34–36]
Lymecycline [Tetralysal] J01AA04		[992-21-2] $C_{29}H_{38}N_4O_{10}$ 602.64	same mechanism as all tetracyclines/moderate to severe acne. cps. 150, 300 mg. 600 mg, O, P.	USA/1963 Pfizer; Galderma	semisynthetic from tetracycline [37]

Name	Structure	Formula / CAS	Description	Origin	Notes
Oxytetracycline [Terramycin] J01AA06		[79-57-2] $C_{22}H_{24}N_2O_9$ 460.44 Hydrochloride: [2058-46-0] $C_{22}H_{24}N_2O_9 \cdot HCl$ 496.90	same mechanism as all tetracyclines/treatment of *Chlamydia* and *Mycoplasma* infections and of acne. s. 250, 500 mg; ophthalmic ointment 1%. 1000 mg, O. P.	USA/1950 Pfizer	prepared by fermentation of *Streptomyces rimosus* [38, 39]
Tetracycline [Achromycin, Hostacyclin, Sumycin, Tefilin] J01AA07		[60-54-8] $C_{22}H_{24}N_2O_8$ 444.44 Hydrochloride: [64-75-5] $C_{22}H_{24}N_2O_8 \cdot HCl$ 480.90	same mechanism as all tetracyclines/broad spectrum of antibiotic action. cps. 250, 500 mg, 1000 mg O. P.	USA/1955 Pfizer; American Cyanamid	prepared by fermentation of *Streptomyces aureofaciens* or *viridifaciens* or by catalytic hydrogenation of chlortetracycline [40–44]
Minocycline [Dynacin, Minocin, Solodyn, Minakne. Udima, Klinomycin] J01AA08		[10118-90-8] $C_{23}H_{27}N_3O_7$ 457.48 Hydrochloride: [13614-98-7] $C_{23}H_{27}N_3O_7 \cdot HCl$ 493.94	same mechanism as all tetracyclines/broad spectrum antibiotic with long half-life for treatment of skin infections and acne vulgaris. f.c.tbl. 50 mg; cps. 50, 100 mg. 200 mg, O.	USA/1972 American Cyanamid (Lederle Labs.)	semisynthetic [45–48]
Tigecycline (TBG-MINO; GAR-936) [Tygacil] J01AA12		[220620-09-7] $C_{29}H_{39}N_5O_8$ 585.66	same mechanism as all tetracyclines/ glycylcycline antibiotic, active against MRSA-resistant *S.aureus* and *Acinetobacter baumannii*; derivative of minocycline. vials 50 mg for i. v. infusion 100 mg. P.	USA/2005 American Cyanamid/Wyeth	semisynthetic; derivative of minocycline [49, 50]

3.2. J01B Amphenicols (Table 2)

Table 2. J01BA Amphenicols

INN (Synonyms) [Brand names] ATC	Structure	[CAS-No.] Formula MW (g/mol)	Target (if known) Medical use Formulation DDD	Approval (country/year) Originator(s)	Production method References
Chloramphenicol (chlornitromycin) [Chloromycetin, Posifenicol, Kemicetine, Minims] J01BA01; D06AX02; D10AF03, G01AA05; S01AA01; S02AA01; S03AA08; **WHO essential medicine**		[56-75-7] $C_{11}H_{12}Cl_2N_2O_5$ 323.13 Succinate sodium salt: [982-57-0] $C_{15}H_{15}Cl_2N_2NaO_8$ 445.18	stops bacterial growth by inhibition of protein synthesis (inhibits peptidyl transferase activity of the bacterial ribosome)/broad activity spectrum, but can exert serious adverse effects like aplastic anemia, bone marrow suppression, hypersensitivity and neurotoxic reactions. cps. 250 mg; ear drops 5 g/100 mL; eye drops 5 mg, 10 mg; ointment 1 and 2%; tbl. 50, 250 mg. 3 g, O, P.	USA/1949 Parke Davis	originally isolated from *Streptomyces venezuelae*; synthesis is not very complicated and much cheaper [51–55]

3.3. J01C β-Lactam Antibacterials, Penicillins [56–58] (Tables 3–7)

Table 3. J01CA Penicillins with extended spectrum

INN (Synonyms) [Brand names] ATC	Structure	[CAS-No.] Formula MW (g/mol)	Target (if known) Medical use Formulation DDD	Approval (country/ year) Originator(s)	Production method References
Ampicillin (AY-6108; BRL-1341; P-50) [Amcill, Albipen, Amblosin, Amfipen, Amplital, Binotal, Omnipen, Penbritin, Penicline, Totapen, Unacim, Unasyn] J01CA01; S01AA19 **WHO essential medicine**		[69-53-4] $C_{16}H_{19}N_3O_4S$ 349.41 Trihydrate: [7177-48-2] $C_{16}H_{19}N_3O_4S \cdot 3H_2O$ 403.46 Sodium salt: [69-52-3] $C_{16}H_{18}N_3NaO_4S$ 371.39	inhibitor of bacterial cell wall synthesis by irreversible inhibition of the enzyme D-*alanintranspeptidase*/ broad activity spectrum, but more and more bacteria have become resistant. All species of *Pseudomonas* and most of *Klebsiella* and *Aerobacter* show resistance to ampicillin. f.c. tbl. 1000 mg vials with powder for inj. 500, 1000, 2000 mg. suppos. 125, 250 mg. 2000 mg O, P, R	USA/1961 UK/1961 Beecham	semisynthetic [59, 60]
Amoxicillin (amoxycillin; BRL-2333) [Agram, Amocilline, Amoxil, Amoxypen, Betamox, Bristamox, Clamoxyl, Pasetocin, Penimox, Sawacillin, Supramox, Widecillin, Zamocilline, Zimox] J01CA04; see also J01CR02 (comb. with clavulanic acid) **WHO essential medicine**		[26787-78-0] $C_{16}H_{19}N_3O_5S$ 365.40 Trihydrate: [61336-70-7] $C_{16}H_{19}N_3O_5S \cdot 3H_2O$ Sodium salt: [34642-77-8] $C_{16}H_{18}N_3NaO_5S$ 387.39	inhibitor of bacterial cell wall synthesis; often combined with the β-lactamase inhibitor clavulanic acid in order to overcome bacterial resistance/used in treatment of many bacterial infections. tbl. and f.c. tbl. 500, 750, 1000 mg; effervescent tbl. 1000 mg; granulate for suspension 250, 500, 1000 mg. 1000 mg, O.P.	USA/1972 UK/1972 Beecham	semisynthetic [61, 62]

(continued)

Table 3. (*continued*)

INN (Synonyms) [Brand names] ATC	Structure	[CAS-No.] Formula MW (g/mol)	Target (if known) Medical use Formulation DDD	Approval (country/ year) Originator(s)	Production method References
Bacampicillin [Ambacamp, Penglobe, Spectrobid] J01CA06		[50972-17-3] $C_{21}H_{27}N_3O_7S$ 465.52 Hydrochloride: [37661-08-8] $C_{21}H_{27}N_3O_7S \cdot HCl$ 501.98	inhibitor of bacterial cell wall synthesis; prodrug of ampicillin with improved oral bioavailability/broad spectrum antibiotic. f.c. tbl. 400, 600, 800 mg, O.	GB/1980 DE/1981 USA/1975 Astra	semisynthetic-made from azidocillin [63, 64]
Mezlocillin [Baypen, Mezlin, Mezlocillin Carino] J01CA10		[51481-65-3] $C_{21}H_{25}N_5O_8S_2$ 539.58 Sodium salt: [42057-22-7] $C_{21}H_{24}N_5NaO_8S_2$ 561.56	inhibitor of bacterial cell wall synthesis/ broad spectrum antibiotic, active against many gram-negative and some gram-positive bacteria. vials with 1, 2, 4 g mezlocillin 6 g, P.	DE/1977 Bayer	synthesized from ampicillin [65]
Mecillinam (amdinocillin; FL-1060; Ro-10-9070) [Coactin; Selexid] J01CA11		[32887-01-7] $C_{15}H_{23}N_3O_3S$ 325.43	inhibitor of bacterial cell wall synthesis/ active against gram-negative bacteria. 1200 mg, P.	Leo Pharma	[66]
Piperacillin (CL-227193; T-1220) [Isipen, Pentcillin, Pipracil, Pipril] J01CA12; see also combin. with tazobactam: J01CG02, J01CR05.		[61477-96-1] $C_{23}H_{27}N_5O_7S$ 517.56 Sodium salt: [59703-84-3] $C_{23}H_{26}N_5NaO_7S$ 539.54	inhibitor of bacterial cell wall synthesis/broad spectrum antibiotic, derivative of ampicillin. vials with powder 1, 2, 4 g for inj.sol. 14 g, P.	USA/1982 Toyama	synthesized from ampicillin [67, 68]
Ticarcillin (BRL-2288) [Monapen, Ticar, Ticarpen, Ticillin] J01CA13; see also combin. with potassium clavulanate: J01CR03.		[34787-01-4] $C_{15}H_{16}N_2O_6S_2$ 384.42 Disodium salt: [4697-14-7] $C_{15}H_{14}N_2Na_2O_6S_2$ 428.38	inhibitor of bacterial cell wall synthesis/ mainly used against gram-negative bacteria, especially *Pseudomonas aeroginosa*; main use in combination with potassium clavulanate. vials 1, 3, 6 g, P. 15 g, P.	USA/1975 Beecham	semisynthetic carboxypeni-cillin [69]

Table 4. J01CE β-Lactamase sensitive penicillins

INN (Synonyms) [Brand names] ATC	Structure	[CAS-No.] Formula MW (g/mol)	Target (if known) Medical use Formulation DDD	Approval (country/year) Originator(s)	Production method References
Benzylpenicillin (penicillin G) [Crystapen, Penilevel, Falapen, Megacillin, Pentids, Pfizerpen] J01CE01 **WHO essential medicine**		[61-33-6] $C_{16}H_{18}N_2O_4S$ 334.39 Sodium salt: [69-57-8] $C_{16}H_{17}N_2NaO_4S$ 356.37 Potassium salt: $C_{16}H_{17}KN_2O_4S$ 372.48	inhibitor of bacterial cell wall synthesis/ narrow spectrum penicillin for i.v. or i.m. administration; mainly active against gram-positive bacteria. vials with powder 1, 5, 10×10^6 I.E. for inj. or infusion solution. 3.6 g. P.	USA and UK 1942-1944 Alexander Fleming (discovery in 1928); consortium of US and UK companies 1942-1945	production by fermentation of *Penicillium chrysogenum* under addition of phenylacetate: today one gets yields of more than 100 g/L! [70]
Phenoxymethylpenicillin (penicillin V) [Antibiocin, Arcasin, Calcipen, Cliacil, Fenospen, Infectocillin, Isocillin, Ispenoral. Megacillin, Oracilline, Ospen, Pen-Vee, Primcillin, Veetids, V-Cillin, V-Pen, V-Tablopen] J01CE02 benzathine: J01CE10 **WHO essential medicine**		[87-08-1] $C_{16}H_{18}N_2O_5S$ 350.39 Potassium salt: [132-98-9] $C_{16}H_{17}KN_2O_5S$ 388.48 Calcium salt: [147-48-8] $C_{32}H_{34}CaN_4O_{10}S_2$ 738.84 Salt with benzathine 2:1: [5928-84-7] $(C_{16}H_{18}N_2O_5S)_2 \cdot C_{16}H_{20}N_2$ 941.13	inhibitor of bacterial cell wall peptidoglycan synthesis/activity is similar to that of benzylpenicillin against gram-positive bacteria. but less active against gram-negative bacteria. It can be administered orally (first oral penicillin). f.c. tbl. 653.6, 980.4 mg (K salt), i.e. 1×10^6, 1.5×10^6 I.E.; syrup and solution 0.3×10^6 I.E.	AT/1952 USA/1952 Biochemie Kundl GmbH (1952)	production by fermentation of *Penicillium notatum* or *Penicillium chrysogenum Thom* under addition of phenoxyacetate or phenoxyethanol [71, 72]
Benzathine (as salt with benzylpenicillin or phenoxymethylpenicillin) [Benzethacil, DBED-penicillin. Beacillin, Bicillin L-A, Extencilline. Lentopenil, Megacillin susp.. Penidural. Permapen. Tardocillin] J01CE08 **WHO essential medicine**		Salt with benzylpenicillin benzathine 2:1 [1538-09-6] $C_{48}H_{56}N_6O_8S_2$ 909.13	repository form of benzylpenicillin. hydrolyzed to benzylpenicillin in vivo after i.m. injection with prolonged antibiotic action over more than 2 weeks/medical use in cases of rheumatic fever and early or latent syphilis. vials with 1.2×10^6 I.E. powder or susp. for inj. 3.6 g. P.	USA/1953 Wyeth	[73, 74]

Table 5. J01CF β-Lactamase resistant penicillins

INN (Synonyms) [Brand names] ATC	Structure	[CAS-No.] Formula MW (g/mol)	Target (if known) Medical use Formulation DDD	Approval (country/ year) Originator(s)	Production method References
Dicloxacillin (BRL-1702) [Dynapen, InfectoStaph, Maclicine] J01CF01		[3116-76-5] $C_{19}H_{17}Cl_2N_3O_5S$ 470.32 Sodium salt monohydrate: [13412-64-1] $C_{19}H_{16}Cl_2N_3NaO_5S \cdot H_2O$ 510.32	inhibitor of bacterial cell wall synthesis/narrow-spectrum semisynthetic penicillin for treatment of infections with gram-positive β-lactamase producing bacteria (e.g. S.aureus). cps. 250, 500 mg 2 g, O, P.	USA/1968 Beecham	[75]
Cloxacillin (BRL-1621) [Cloxapen, Cloxacap, Tegopen, Orbenin] J01CF02 **WHO essential medicine**		[61-72-3] $C_{19}H_{18}ClN_3O_5S$ 435.88 Sodium salt monohydrate: [7081-44-9] $C_{19}H_{17}ClN_3NaO_5S \cdot H_2O$ 475.88	inhibitor of bacterial cell wall synthesis/used against β-lactamase producing bacteria (e.g. S. aureus). cps. 250, 500 mg 2 g, O, P.	USA/1974 UK/1962 Beecham	[76, 77]
Meticillin (methicillin; BRL-1241; X-1497) [no longer in clinical use] J01CF03		[61-32-5] $C_{17}H_{20}N_2O_6S$ 380.42 Sodium salt: [132-92-3] $C_{17}H_{19}N_2NaO_6S$ 402.40	inhibitor of bacterial cell wall synthesis/narrow-spectrum β-lactam, previously used in treatment of gram-positive bacteria (e.g. S. aureus), but no longer used in therapy, only as test substance. Meticillin-resistant *Staphylococcus aureus* (**"MRSA"**) means strains which are resistant to all penicillins.	UK/1959 USA/1960 Beecham	[78, 79]
Oxacillin (BRL-1400; penicillin P-12) [Bactocill, Bristopen, Cryptocillin, Penstapho, Stapenor] J01CF04		[66-79-5] $C_{19}H_{19}N_3O_5S$ 401.44 Sodium salt monohydrate: [7240-38-2] $C_{19}H_{18}N_3NaO_5S \cdot H_2O$ 441.43	inhibitor of bacterial cell wall synthesis/narrow-spectrum β-lactamase resistant penicillin for treatment of S. aureus, but many strains have become resistant (MRSA). vials 1, 2, 10 g powder for inj. i.v. or i.m. 2 g, O, P.	USA/1971 Beecham	[80, 81]
Flucloxacillin (Floxacillin) [Abboflox, Floxapen, Flopen, Stafoxil, Staphylex] J01CF05		[5250-39-5] $C_{19}H_{17}ClFN_3O_5S$ 453.87 Sodium salt monohydrate: [34214-51-2] $C_{19}H_{16}ClFN_3NaO_5S \cdot H_2O$ 493.87	inhibitor of bacterial cell wall synthesis/narrow-spectrum β-lactamase resistant penicillin for treatment of infections with gram-positive bacteria as, e.g., S. aureus, but not effective against MRSA. cps. 250, 500 mg; oral suspensions 125 mg/5 mL and 250 mg/5 mL; vials with 250, 500, 1000 mg powder for reconstitution. 3 g, O, P.	Beecham	[82, 83]

Table 6. J01CG β-Lactamase inhibitors

INN (Synonyms) [Brand names] ATC	Structure	[CAS-No.] Formula MW (g/mol)	Target (if known) Medical use Formulation DDD	Approval (country/year) Originator(s)	Production method References
Sulbactam (CP-45899) [Betamaze] J01CG01		[68373-14-8] $C_8H_{11}NO_5S$ 233.24 Sodium salt: [69388-84-7] $C_8H_{10}NNaO_5S$ 255.22	β-lactamase inhibitor for combination with β-lactam antibiotics, e.g., ampicillin or cefoperazone.	USA/1986 DE/1987 Pfizer	[84, 85]
Ampicillin/ sulbactam mixture of sodium salts [Loricin, Unacid, Unacim, Unasyn] J01CR01			vials with injectable solution (1.5, 3 g; 1 g ampicillin + 0.5 g sulbactam, 2 g of ampicillin + 1 g of sulbactam). 2 g, P (as ampicillin).		
Cefoperazone/ sulbactam mixture of sodium salts [Sulperazon] J01DD62	For the formula of cefoperazone see Table 10.		vials with dry powder for injection 250, 500, 1000 mg (of each cefoperazone and sulbactam).		
Tazobactam (YTR-830; CL-29874; CL-307579) J01CG02		[89786-04-9] $C_{10}H_{12}N_4O_5S$ 300.29 Sodium salt: [89785-84-2] $C_{10}H_{11}N_4NaO_5S$ 322.27	β-lactamase inhibitor for combination with β-lactam antibiotics, e.g., piperacillin.	USA/1992 Taiho Pharmac.	[86, 87]
Piperacillin/ tazobactam combination of sodium salts [Tazobac, Tazocin, Zosyn] J01CR05		[123683-33-0]	vials 2 g of piperacillin/0.25 g of tazobactam and 4 g of piperacillin and 0.5 g of tazobactam.		
Avibactam (NXL-104; AVE-1330A) [Avycaz] J01CG		[1192500-31-4] $C_7H_{11}N_3O_6S$ 265.24 Sodium salt: [1192491-61-4] $C_7H_{10}N_3NaO_6S$ 287.22	β-lactamase inhibitor for combination with β-lactam antibiotics, e.g., with ceftazidime.	USA/2015 Aventis, Novexel, AstraZeneca, Cerexa, Forest Labs.	[88, 89]
Ceftazidime (see J01DD02)/ avibactam sodium J01DD52			fix combination; indicated in combination with metronidazole for complicated intraabdominal infections and complicated urinary tract infections. Single dose vials with powder for solution for i.v. infusion containing 2.0 g of ceftazidime and 0.5 g of avibactam sodium.	USA/2015	[90–94]

Table 7. J01CR Combinations of penicillins, incl. β-lactamase inhibitors

INN (Synonyms) [Brand names] ATC	Structure	[CAS-No.] Formula MW (g/mol)	Target (if known) Medical use Formulation DDD	Approval (country/year) Originator(s)	Production method References
Clavulanic acid (MM14151) J01CR		[58001-44-8] $C_8H_9NO_5$ 199.16 Potassium salt: [61177-45-5] $C_8H_9KNO_5$ 237.25	suicide inhibitor of β-lactamase/combination of the potassium salt with amoxicillin and ticarcillin to overcome resistance.	DE/1982; USA/1984; GB/1984 Beecham	made by fermentation of *Streptomyces clavuligerus* [95–97]
Potassium clavulanate/ amoxicillin comb. [Augmentan, Augmentin, Infecto-Supramox] J01CR02 **WHO essential medicine**			f.c. tbl. 574 or 1004.5 mg of amoxicillin trihydrate w/148.9 mg of K-clavulanate; powder for suspension 400 mg of amoxicillin trihydrate, 57 mg of K-clavulanate.		[98]
Potassium clavulanate/ Ticarcillin comb. [Timentin] J01CR03			vials for i.v. inj. w/ticarcillin disodium and K-clavulanate (equivalent to 3 g ticarcillin and 100 mg clavulanic acid).	USA/1986	

3.4. J01D Other β-Lactam Antibacterials (Tables 8–14)

Table 8. J01DB First-generation cephalosporins

INN (Synonyms) [Brand names] ATC	Structure	[CAS-No.] Formula MW (g/mol)	Target (if known) Medical use Formulation DDD	Approval (country/ year) Originator(s)	Production method References
Cefalexin (cephalexin; LY-061188) [Ceporex, Keflex, Keforal, Larixin] J01DB01 **WHO essential medicine**		[*15686-71-2*] $C_{16}H_{17}N_3O_4S$ 347.39 Monohydrate: [*23325-78-2*] $C_{16}H_{17}N_3O_4S \cdot H_2O$ 365.41	β-lactam antibiotic, inhibits synthesis of bacterial cell wall/high activity against gram-positive and moderate activity against some gram-negative bacteria. granulate for suspension 263 mg of cefalexin monohydrate; f.c. tbl. 526 and 1052 mg of monohydrate. 2 g, O; children: 1 g, O.	USA/1970 Eli Lilly	[99–104]
Cefazolin (Cephazolin; SKF-41558) [Ancef, Basocef, Cefamezin, Gramaxin, Kefzol, Totacef, Zolicef] J01DB04 **WHO essential medicine**		[*25953-19-9*] $C_{14}H_{14}N_8O_4S_3$ 454.50 Sodium salt: [*27164-46-1*] $C_{14}H_{13}N_8NaO_4S_3$ 476.48	inhibitor of bacterial cell wall synthesis/clinically effective against gram-positive *Staphylococci pneumoniae* and *Streptococci* found on human skin. also against *E.coli*. vials with powder for infusion solution 1048 and 2096 mg of Na-salt. 3 g, P.	USA/1973 Fujisawa	[105–107]
Cefadroxil (BL-S578; MJF-11567-3) [Baxan, Duricef, Grüncef] J01DB05		[*50370-12-2*] $C_{16}H_{17}N_3O_5S$ 363.39 Monohydrate: [*66592-87-8*] $C_{16}H_{17}N_3O_5S \cdot H_2O$ 381.40	inhibitor of bacterial cell wall synthesis/broad-spectrum cefalosporin against gram-positive and -negative bacteria. vials with powder for suspension 524.8 mg of monohydrate; tbl. 1049.6 mg of monohydrate. 2 g, O; children 1 g, O.	USA/1978 Bristol-Myers	[108, 109]

Table 9. J01DC Second-generation cephalosporins

INN (Synonyms) [Brand names] ATC	Structure	[CAS-No.] Formula MW (g/mol)	Target (if known) Medical use Formulation DDD	Approval (country/year) Originator(s)	Production method References
Cefuroxime (cefuroxime axetil; CCI-15641) [Elobact, Curoxim, Zinacef, Zinnat] J01DC02		[55268-75-2] $C_{16}H_{16}N_4O_8S$ 424.38 Sodium salt: [56238-63-2] $C_{16}H_{15}N_4NaO_8S$ 446.37 Cefuroxime axetil (1-acetoxyethyl ester): [64544-07-6] $C_{20}H_{22}N_4O_{10}S$ 510.47	inhibitor of bacterial cell wall synthesis/effective against *Haemophilus influenzae, Neisseria gonorrhoeae* and *Lyme disease*; Cefuroxime axetil is an orally active prodrug. vials with powder for inj.solution 263 and 789 mg (sodium salt); f.c. tbl. 300.72 and 601.44 mg of axetil. 0.5 g, O; 3 g, P.	USA/1983 Glaxo	[110, 111]
Cefaclor [Alfatil, Ceclor, Distaclor, Keflor, Panoral] J01DC04		[53994-73-3] $C_{15}H_{14}ClN_3O_4S$ 367.80 Monohydrate: [70356-03-5] $C_{15}H_{14}ClN_3O_4S\cdot H_2O$ 385.82	inhibitor of bacterial cell wall synthesis/treatment of bacterial infections like pneumonia and of ear, lung, skin, throat, and urinary tract. f.c. tbl. 524.48 mg (monohydrate); granulate for oral suspension 131 and 262 mg; effervescent tbl. 262.24, 524.48, 1048.96 mg; cps. 524.48 mg. 1 g, O; children: 0.75 g, O.	USA/1979 Eli Lilly	[112, 113]

Table 10. J01DD Third-generation cephalosporins

INN (Synonyms) [Brand names] ATC	Structure	[CAS-No.] Formula MW (g/mol)	Target (if known) Medical use Formulation DDD	Approval (country/year) Originator(s)	Production method References
Cefotaxime (HR-756; RU-24756) [Cefotax, Claforan] J01DD01 **WHO essential drug**		[63527-52-6] $C_{16}H_{17}N_5O_7S_2$ 455.46 Sodium salt: [64485-93-4] $C_{16}H_{16}N_5NaO_7S_2$ 477.44	inhibitor of bacterial cell wall synthesis/broad spectrum for use against gram-positive and -negative bacteria; crosses blood-brain barrier (use in meningitis). vials with powder for inj. 500, 1000, 2000 mg (as sodium salt) 4 g, P.	DE/1980 Roussel-Uclaf	[114]
Ceftazidime (GR-20263) [Fortaz, Fortum, Glazidim, Modacin, Spectrum] J01DD02 **WHO essential drug**		[72558-82-8] $C_{22}H_{22}N_6O_7S_2$ 546.52 Pentahydrate: [78439-06-2] $C_{22}H_{22}N_6O_7S_2 \cdot 5H_2O$ 636.65	inhibitor of bacterial cell wall synthesis/active against Pseudomonas aeruginosa and other gram-negative and some gram-positive aerobes, but not against MRSA. vials with powder for inj. 500, 1000, 2000 mg (as pentahydrate) 4 g, P.	USA/1983 DE/1984 Glaxo	[115, 116]
Ceftriaxone (Ro-13-9904/001) [Cefotrix, Rocephin] J01DD04 **WHO essential drug**		[73384-59-5] $C_{18}H_{18}N_8O_7S_3$ 554.57 Disodium salt: [74578-69-1] $C_{18}H_{16}N_8Na_2O_7S_3$ 598.55 Disodium salt hemiheptahydrate: [104376-79-6] $C_{18}H_{16}N_8Na_2O_7S_3 \cdot 3.5H_2O$ 661.59	inhibitor of bacterial cell wall synthesis/broad spectrum against gram-positive and -negative bacteria for use in pneumonia and bacterial meningitis; haemophilus influenzae and others. vials with powder for inj. 500, 1000, 2000 mg (as disodium hemiheptahydrate) 2 g, P.	DE/1982 Roche	[117, 118]
Cefixime (FK-027; FR-17027; CL-284635) [Cephoral, Cefixoral, Cefspan, Suprax, Uro-Cephoral] J01DD08 **WHO essential drug**		[79350-37-1] $C_{16}H_{15}N_5O_7S_2$ 453.44 Trihydrate: [125110-14-7] $C_{16}H_{15}N_5O_7S_2 \cdot 3H_2O$ 507.49	inhibitor of bacterial cell wall synthesis/broad spectrum against E.coli, Haemophilus influenzae, and Proteus mirabilis for use against infections of the ear, urinary tract, and upper respiratory tract. f.c.tbl. 200, 400 mg; granulate 100 mg/5 mL. 400 mg, O; 200 mg, O (children).	USA/1987 DE/1991 Fujisawa	[119–122]

(continued)

Table 10. (continued)

INN (Synonyms) [Brand names] ATC	Structure	[CAS-No.] Formula MW (g/mol)	Target (if known) Medical use Formulation DDD	Approval (country/year) Originator(s)	Production method References
Cefoperazone (CP-52640; T-1551) [Cefazone, Cefobid, Cefina-SB] J01DD12		[62893-19-0] $C_{25}H_{27}N_9O_8S_2$ 645.67 Sodium salt: [62893-20-3] $C_{25}H_{26}N_9NaO_8S_2$ 667.65	inhibitor of bacterial cell wall synthesis/broad spectrum against *Haemophilus influenzae*, *Staphylococcus aureus*, *Streptococcus pneumoniae* for use against infections of respiratory and urinary tract, skin and female genital tract; combination with sulbactam. vials with 0.25, 0.5, 1 and 2 g (as sodium salt) 4 g, P.	USA/1986 EU/1998 Toyama; Pfizer (combination with Sulbactam)	[123–125]
Cefpodoxime proxetil (CS-807, R-3763, U-76252, U-76253) [Cefodox, Orelox, Podomexef, Vantin] J01DD13		[87239-81-4] $C_{21}H_{27}N_5O_9S_2$ 557.59 Cefpodoxime: [80210-62-4] $C_{15}H_{17}N_5O_6S_2$ 427.45	inhibitor of bacterial cell wall synthesis/active against most gram-positive and gram-negative bacteria and used in community acquired pneumonia, uncomplicated skin and urinary tract infections; prodrug of cefpodoxime. f.c.tbl. 100, 200 mg; 40 mg/5 mL suspension. 0.4 g, O; 0.2 g, O (children).	USA/1989 Fujisawa (Cefpodoxime) Sankyo (Cefpodoxime proxetil)	[126–128]
Ceftibuten (7432-S; Sch-39720) [Cedax, Keimax] J01DD14		[97519-39-6] $C_{15}H_{14}N_4O_6S_2$ 410.42 Dihydrate: [118081-34-8] $C_{15}H_{14}N_4O_6S_2 \cdot 2H_2O$ 446.45	inhibitor of bacterial cell wall synthesis/used to treat acute bacterial exacerbations of chronic bronchitis, otitis media, pharyngitis, and tonsilitis. cps. 200, 400 mg; oral suspension 70, 140 mg; (as dihydrate). 0.4 g, O.	USA/1992 Shionogi; Schering-Plough	[129–131]

Table 11. J01DE Fourth-generation cephalosporins

INN (Synonyms) [Brand names] ATC	Structure	[CAS-No.] Formula MW (g/mol)	Target (if known) Medical use Formulation DDD	Approval (country/ year) Originator(s)	Production method References
Cefepime (BMY-28142) [Maxipime, Neopime, Maxcef] J01DE01		[88040-23-7] C$_{19}$H$_{24}$N$_6$O$_5$S$_2$ 480.56 Hydrochloride monohydrate: [123171-59-5] C$_{19}$H$_{24}$N$_6$O$_5$S$_2$ · HCl·H$_2$O 571.49	inhibitor of bacterial cell wall synthesis with extended spectrum of activity/ treatment of moderate to severe nosocomial pneumonia, infections caused by multiple drug-resistant microorganisms, e.g. *P. aeroginosa*. vials for inj. 1.0 g, 2.0 g (as hydrochloride monohydrate) 2 g, P.	USA/1993 BMS	[132]

Table 12. J01DF Monobactams

INN (Synonyms) [Brand names] ATC	Structure	[CAS-No.] Formula MW (g/mol)	Target (if known) Medical use Formulation DDD	Approval (country/ year) Originator(s)	Production method References
Aztreonam (azthreonam; SQ-26776) [Azactam (inj.), Cayston (inhal.)] J01DF01		[78110-38-0] C$_{13}$H$_{17}$N$_5$O$_8$S$_2$ 435.43 L-Lysine salt: [827611-49-4] C$_{13}$H$_{17}$N$_5$O$_8$S$_2$ · C$_6$H$_{14}$N$_2$O$_2$ 581.62	inhibitor of bacterial cell wall synthesis; bactericidal; active against gram-negative bacteria, e.g. *P.aeruginosa, Citrobacter, Enterobacter, E.coli, Haemophilus, Klebsiella, Proteus* and *Serratia*. vials 0.5, 1, 2 g lyoph. f. inj. vials 75 mg + solvent amp. f. inhal. (as L-lysine salt). 0.225 g, Inhal.; 4 g, P.	USA/1986 DE/1985 Squibb	[133–137]

Table 13. J01DH Carbapenems

INN (Synonyms) [Brand names] ATC	Structure	[CAS-No.] Formula MW (g/mol)	Target (if known) Medical use Formulation DDD	Approval (country/year) Originator(s)	Production method References
Meropenem (ICI-194660, SM-7338) [Meronem, Merrem] J01DH02		[96036-03-2] $C_{17}H_{25}N_3O_5S$ 383.46 Trihydrate: [119478-56-7] $C_{17}H_{25}N_3O_5S \cdot 3H_2O$ 437.51	inhibitor of bacterial cell wall synthesis/ highly resistant to β-lactamases and stable against dehydropeptidase-1/ultra-broad spectrum antibiotic. vials for inj. 500, 1000 mg. 2 g. P.	USA/1996 Sumitomo Pharmaceuticals	[138, 139]
Ertapenem (MK-826; ZD-443) [Invanz] J01DH03		[153832-46-3] $C_{22}H_{25}N_3O_7S$ 475.52 Sodium salt: [153773-82-1] $C_{22}H_{24}N_3NaO_7S$ 497.50	inhibitor of bacterial cell wall synthesis, similar properties like meropenem, but longer half-life/ vials for inj. (i.v. or i.m.) 1 g. 1 g. P.	USA/2002 Zeneca; Merck & Co.	[140, 141]
Doripenem (S-4661) [Doribax, Finibax] J01DH04		[148016-81-3] $C_{15}H_{24}N_4O_6S_2$ 420.50 Hydrate: [364622-82-2] $C_{15}H_{24}N_4O_6S_2 \cdot H_2O$ 438.51	inhibitor of bacterial cell wall synthesis, similar properties like meropenem and ertapenem, but more active especially against *Pseudomonas aeruginosa*. vials for inj. 500 mg. 1.5 g. P.	USA/2007; EU/2008; JP/2005 Shionogi	[142–144]
Imipenem (MK-787) comb.w/cilastatin [Imipem, Primaxin, Tenacid, Tienam, Zienam] J01DH51		[64221-86-9] $C_{12}H_{17}N_3O_4S$ 299.35 Monohydrate: [74431-23-5] $C_{12}H_{17}N_3O_4S \cdot H_2O$ 317.36 [82009-34-5] $C_{16}H_{26}N_2O_5S$ 358.45 Sodium salt: [81129-83-1] $C_{16}H_{25}N_2NaO_5S$ 380.44	inhibitor of bacterial cell wall synthesis, was the first member of the class of carbapenem antibiotics/highly resistant to β-lactamase, but is rapidly degraded by dehydropeptidase-1 and has therefore to be coadminstered with the dehydropeptidase inhibitor cilastatin. Imipenem is a stable derivative of the natural thienamycin, which is readily unstable in aqueous solution. vials for inj. 500 mg. 2 g. P (imipenem).	DE/1985 Merck & Co.	[145–154]

Table 14. J01DI Other cephalosporins and penems

INN (Synonyms) [Brand names] ATC	Structure	[CAS-No.] Formula MW (g/mol)	Target (if known) Medical use Formulation DDD	Approval (country/year) Originator(s)	Production method References
Ceftobiprole medocaril (BAL-9141; Ro-63-9141; sodium salt: BAL-5788; Ro-65-5788) [Zeftera] J01DI01		[376653-43-9] $C_{26}H_{26}N_8O_{11}S_2$ 690.66 Sodium salt: [252188-71-9] $C_{26}H_{25}N_8NaO_{11}S_2$ 712.64	inhibitor of bacterial cell wall synthesis/active against MRSA, Penicillin-resistant *Streptococcus pneumoniae*, *Pseudomonas aeruginosa* and *Enterococci* (resistant to staphylococcal β-lactamase). vials for i.v. infusion 666.6 mg. 1.5 g, P.	EU/2013 Roche; Basilea	[155–157]
Ceftaroline fosamil (PPI-0903; TAK-599) [Teflaro, Zinfloro] J01DI02		[229016-73-3] $C_{22}H_{21}N_8O_8PS_4$ 684.67 Monoacetate: [400827-46-5] $C_{22}H_{21}N_8O_8PS_4 \cdot C_2H_4O_2$ 744.72	inhibitor of bacterial cell wall synthesis/broad-spectrum cephalosporin for treatment of acute bacterial skin and skin structure infections and community-acquired bacterial pneumonia; active against MRSA. vials for i.v. infusion 600 mg	USA/2010 EU/2012 Takeda; Forest Labs	[158–160]
Ceftolozane (CXA-101; FR-264205) [Zerbaxa] J01DI54 (comb. w/tazobactam)		[689293-68-3] $C_{23}H_{30}N_{12}O_8S_2$ 666.69 Sulfate: [936111-69-2] $C_{23}H_{30}N_{12}O_8S_2 \cdot H_2SO_4$ 764.76	inhibitor of bacterial cell wall synthesis/in combination with the β-lactamase inhibitor tazobactam (J01CG02) it is used for treatment of complicated abdominal infections and complicated urinary tract infections. vials for infusion with 1 g of ceftolozane and 0.5 g of tazobactam	USA/2014 Astellas Pharma; Cubist Pharmaceuticals	[161]

3.5. J01E Sulfonamides and Trimethoprim (Tables 15–18)

Table 15. J01EA Trimethoprim and derivatives [162]

INN (Synonyms) [Brand names] ATC	Structure	[CAS-No.] Formula MW (g/mol)	Target (if known) Medical use Formulation DDD	Approval (country/year) Originator(s)	Production method References
Trimethoprim [Cotrim, Eusaprim, Kepinol, Infectotrimet, Bactrim, Polytrim, Motrim, Triprim-mainly combinations with Sulfonamides] J01EA01 **Cotrimoxazole** (= comb. w/sulfamethoxazole see J01EE01) **WHO essential drug**		[738-70-5] $C_{14}H_{18}N_4O_3$ 290.32	inhibitor of bacterial dihydrofolate reductase/active against a broad spectrum of gram-negative and gram-positive bacteria; most often it is used as cotrimoxazol, a fixed combination with sulfamethoxazole. tbl. 100, 150, 200 mg (trimethoprim only): oral suspension 50, 100 mg (trimethoprim). 0.4 g, O, P; children 0.15 g, O.	UK/1969 Burroughs Wellcome	[163, 164]

Table 16. J01EC Intermediate-acting sulfonamides (biological half-life approximately 11–12 hours)

INN (Synonyms) [Brand names] ATC	Structure	[CAS-No.] Formula MW (g/mol)	Target (if known) Medical use Formulation DDD	Approval (country/year) Originator(s)	Production method References
Sulfamethoxazole (SMZ; SMX) [as **Cotrimoxazole**: Bactrim, Drylin, Eusaprim, Kepinol] J01Ec1		[723-46-6] $C_{10}H_{11}N_3O_3S$ 253.28	competitive inhibitor of dihydropteroate synthetase (DHPS), an enzyme involved in folate synthesis; antimetabolite of para-aminobenzoic acid (PABA)/bacteriostatic antibiotic; in most countries only used in fixed combination w/trimethoprim. 2 g, O.	Shionogi	[165]
Sulfadiazine (sulphadiazine) [Sulfadiazin-Heyl, Adiazine; Silver salt: Flammazine, InfectoFlam, Physiotulle, Silvadene] J01EC02		[68-35-9] $C_{10}H_{10}N_4O_2S$ 250.28 Silver salt: [22199-08-2] $C_{10}H_9AgN_4O_2S$ 357.14	competitive inhibitor of dihydropteroate synthetase (DHPS), an enzyme involved in folate synthesis; antimetabolite of para-aminobenzoic acid (PABA)/bacteriostatic antibiotic. tbl. 500 mg (sulfadiazine). cream (10 mg of Ag-salt/1 g); dermal admin. 0.6 g, O.	Sharp & Dohme; American Cyanamid	[166, 167]

Table 17. J01ED Long-acting sulfonamides (biological half-life approximately 35 hours or more)

INN (Synonyms) [Brand names] ATC	Structure	[CAS-No.] Formula MW (g/mol)	Target (if known) Medical use Formulation DDD	Approval (country/year) Originator(s)	Production method References
Sulfadimethoxine [Albon, Di-Methox] J01ED01 QJ01EQ09 (Veterin.)		[122-11-2] $C_{12}H_{14}N_4O_4S$ 310.33 Sodium salt: [1037-50-9] $C_{12}H_{13}N_4NaO_4S$ 332.31	competitive inhibitor of dihydropteroate synthetase (DHPS), an enzyme involved in folate synthesis; antimetabolite of para-aminobenzoic acid (PABA)/bacteriostatic antibiotic; rarely in human use, mostly as veterinary drug. 0.5 g, O.	Chemie Linz	[168]

Table 18. J01EE Combinations of sulfonamides and trimethoprim, incl. derivatives

INN (Synonyms) [Brand names] ATC	Structure	[CAS-No.] Formula MW (g/mol)	Target (if known) Medical use Formulation DDD	Approval (country/year) Originator(s)	Production method References
Cotrimoxazole [Bactrim, Drylin, Eusaprim, Kepinol] J01EE01 **WHO essential drug**	Trimethoprim: Sulfamethoxazole 1 : 5	[8064-90-2]	mechanism of action see the single components. tbl. 80/400 mg and 160/800 mg; oral suspension 40/200 and 80/400 mg. 0.320/1.6 g, O; children 0.16/0.8 g.		

3.6. J01F Macrolides, Lincosamides, and Streptogramins [169–171] (Tables 19–21)

Table 19. J01FA Macrolides

INN (Synonyms) [Brand names] ATC	Structure	[CAS-No.] Formula MW (g/mol)	Target (if known) Medical use Formulation DDD	Approval (country/year) Originator(s)	Production method References
Erythromycin [Erycin, Erycinum, Erythrocin, Infectomycin, E.E.S., Ilosone] J01FA01 (D10AF02; D10AF52; S01AA17) **WHO essential drug**	Erythromycin A (main component besides B and C)	[114-07-8] $C_{37}H_{67}NO_{13}$ 733.94 (2')-Ethylsuccinate: [41342-53-4] $C_{41}H_{75}NO_{16}$ 862.06 (2')-Estolate: [3521-62-8] $C_{52}H_{97}NO_{18}S$ 1056.40 (2')-Stearate: [643-22-1] $C_{55}H_{103}NO_{15}$ 1018.42 Lactobionate salt (1:1): [3847-29-8] $C_{49}H_{89}NO_{25}$ 1092.23	inhibitor of microsomal protein synthesis by inhibition of the translocation process in bacteria (specific binding to the 50S subunit or 70S ribosome; no binding to the stable 80S mammalian ribosome)/active against many gram-positive and some gram-negative bacteria, mycoplasmas and chlamydia. vials f.infusion 500, 1000 mg (744, 1488 mg as lactobionate); f.c.tbl.500 mg (588 mg of ethylsuccinate or 693.8 mg of stearate); 2 g, O, P; children 1 g, O.	USA/1952 Eli Lilly	made by fermentation of *Streptomyces erythreus*. [172–177]
Spiramycin (RP-5337) [Rovamycine, Selectomycin] J01FA02	Spiramycin I: R = H Spiramycin II: R = COCH₃ Spiramycin III: R = COCH₂CH₃	[8025-81-8] Spiramycin I: [24916-50-5] $C_{43}H_{74}N_2O_{14}$ 843.07 Spiramycin II: [24916-51-6] $C_{45}H_{76}N_2O_{15}$ 885.10 Spiramycin III: [24916-52-7] $C_{46}H_{78}N_2O_{15}$ 899.13	mechanism of action same as erythromycin (inhibition of protein synthesis/active against gram-positive *cocci* and *rods*, gram-negative *cocci, legionellae, mycoplasmas, chlamydiae*. f.c.tbl. 166.67 mg, 375 mg. 3 g, O.	FR/1955 Rhone-Poulenc	made by fermentation of *Streptomyces ambofaciens* [178–180]

(continued)

Roxithromycin
(RU-28965; RU-965)
[Rulid, Forilin, Claramid]
J01FA06

[80214-83-1]
$C_{41}H_{76}N_2O_{15}$
837.06

mechanism of action same as erythromycin (inhibition of protein synthesis during translocation)/antibiotic spectrum similar to erythromycin.
f.c.tbl. 150, 300 mg.
0.3 g, O; children 0.15 g, O.

DE/1987
Roussel Uclaf

semisynthetic from erythromycin [181]

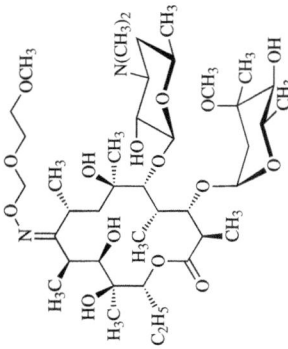

Clarithromycin
(A-56268; TE-031)
[Klacid, Biaxin, Binoclar, Clarith]
J01FA09

[81103-11-9]
$C_{38}H_{69}NO_{13}$
747.96

mechanism of action same as erythromycin (inhibition of protein synthesis during translocation)/broad spectrum, treatment a.o. pneumonia, *Helicobacter pylori*, strep throat, Lyme disease, toxoplasmosis.
f.c.tbl. 250, 500 mg; vials w/powder for infusion 500 mg; granulate for oral susp.
0.5 g, O; 1 g, P: children 0.375 g, O.

JP/1991, USA/1991
Taisho

6-*O*-methylerythromycin, semisynthetic from erythromycin [182]

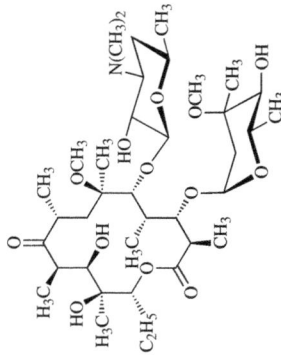

Azithromycin
(CP-62993; XZ-450)
[Zithromax, Ultreon, Sumamed]
J01FA10
WHO essential drug

[83905-01-5]
$C_{38}H_{72}N_2O_{12}$
749.00
Dihydrate:
[117772-70-0]

mechanism of action same as erythromxcin (inhibition of protein synthesis during translocation)/acute bacterial exacerbations of COPD, acute bacterial sinusitis, community-acquired pneumonia, acute otitis media, pharyngitis or tonsillitis, skin and skin structure infections, urethritis, and cervicitis.
f.c.tbl. 250, 500 mg: cps. 250, 500 mg; vials for infusion powder 500 mg.

USA/1991 Central & Eastern Europe/1988
Pliva

semisynthetic from erythromycin [183–185]

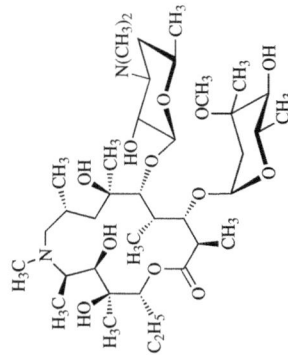

Table 19. (*continued*)

INN (Synonyms) [Brand names] ATC	Structure	[CAS-No.] Formula MW (g/mol)	Target (if known) Medical use Formulation DDD	Approval (country/year) Originator(s)	Production method References
Telithromycin (HMR-3647; RU-66647) [Ketek] J01FA15	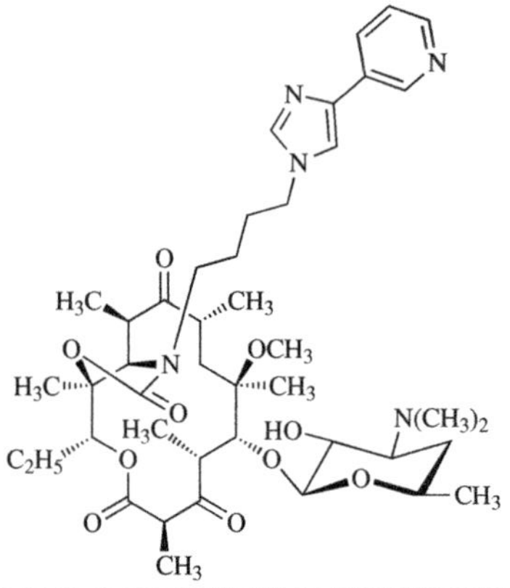	[*191114-48-4*] $C_{43}H_{65}N_5O_{10}$ 812.02	binds to the subunit 50S of the bacterial ribosome and by this blocks the progression of the growing polypeptide chain/treatment of community-acquired pneumonia. f.c.tbl. 400 mg 0.8 g, O.	EU/2001 USA/2004 Roussel Uclaf The only drug of the class of ketolides that is approved	semisynthetic from erythromycin via clarithromycin [186, 187]

Table 20. J01FF Lincosamides (Acylaminopyranosides)

INN (Synonyms) [Brand names] ATC	Structure	[CAS-No.] Formula MW (g/mol)	Target (if known) Medical use Formulation DDD	Approval (country/ year) Originator(s)	Production method References
Clindamycin (U-21251; U-28508) [Cleocin, Basocin, Dalacin, Sobelin, Zindaclin] J01FF01 (D10AF01; G01AA10) **WHO essential drug**		[18323-44-9] $C_{18}H_{33}ClN_2O_5S$ 424.98 Hydrochloride: [21462-39-5] $C_{18}H_{33}ClN_2O_5S \cdot HCl$ 461.44 2-Dihydrogen phosphate: [24729-96-2] $C_{18}H_{34}ClN_2O_8PS$ 504.96	inhibitor of bacterial protein synthesis, binds to the 50S *r*RNA of the ribosome subunit and inhibits ribosomal translocation/treatment of anaerobic infections incl. dental infections, respiratory tract infections, skin and soft tissue, peritonitis, and bone and joint infections. cps. 75, 150, 300 mg (as hydrochloride); amp. 300, 600, 900 mg of clindamycin (as 2-dihydrogen phosphate); f.c.tbl. 150, 300, 600 mg (as hydrochloride). 1.2 g, O; 1.8 g, P; children: 0.45 g, O.	USA/1970 Upjohn	semisynthetic from lincomycin [188, 189]
Lincomycin (U-10149; NSC-70731) [Albiotic, Lincocin] J01FF02		[154-21-2] $C_{18}H_{34}N_2O_6S$ 406.54 Hydrochloride monohydrate: [7179-49-9] $C_{18}H_{34}N_2O_6S \cdot HCl \cdot H_2O$ 461.01	similar mechanism as clindamycin/narrow spectrum antibiotic with activities against some gram-positive bacteria. cps. 125 mg; vials 1.8 g, O, P.	CH/1978 Upjohn	made by fermentation from *Streptomyces lincolnensis* [190–192]

Table 21. J01FG Streptogramins

INN (Synonyms) [Brand names] ATC	Structure	[CAS-No.] Formula MW (g/mol)	Target (if known) Medical use Formulation DDD	Approval (country/year) Originator(s)	Production method References
Pristinamycin (pristinamycin IA + IIA 30:70; RP-7293) [Pyostacine] J01FG01	Pristinamycin IIA Pristinamycin IA	[270076-60-3] **Pristinamycin IA** (Streptogramin B; Mikamycin IA): [3131-03-1] $C_{45}H_{54}N_8O_{10}$ 866.97 **Pristinamycin IIA** (Virginiamycin M_1; Mikamycin A; Staphylomycin M_1; Streptogramin A; Vernamycin A) [21411-53-0] $C_{28}H_{35}N_3O_7$ 525.60	synergistic combination against infections with *Staphylococci* and *Streptococci*; each component binds to the bacterial 50S ribosomal subunit and inhibits the elongation process of protein synthesis/also active against MRSA. tbl. 250, 500 mg. 2 g. O.	Rhone-Poulenc	made by fermentation of *Streptomyces pristinaespiralis* NRRL 2958 [193–195]
Quinupristin/ Dalfopristin (comb. 30:70; RP-59500) [Synercid] J01FG02		[126602-89-9]	inhibition of bacterial protein synthesis; synergistic mixture 30:70 (w/w)/treatment of complicated skin and skin structure infections caused by *Staphylococcus aureus* and *Streptococcus pyogenes*.	USA/1999 Rhone-Poulenc	
Quinupristin (RP-57669)		[120138-50-3] $C_{53}H_{67}N_9O_{10}S$ 1022.23	vials with lyophilisate 500 mg (150 mg of quinupristin and 350 mg of dalfopristine) for infusion. 1.5 g. P.		semisynthetic from pristinamycin IA [196]

Dalfopristin
(RP-54476)

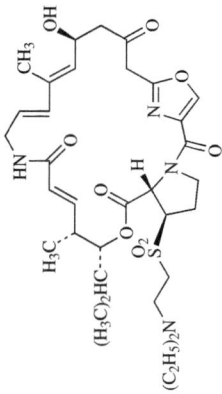

[*112362-50-2*]
$C_{34}H_{50}N_4O_9S$
690.85

semisynthetic from
pristinamycin IIA
[197]

3.7. J01G Aminoglycoside Antibacterials [198–200] (Tables 22, 23)

Table 22. J01GA Streptomycins

INN (Synonyms) [Brand names] ATC	Structure	[CAS-No.] Formula MW (g/mol)	Target (if known) Medical use Formulation DDD	Approval (country/year) Originator(s)	Production method References
Streptomycin [Streptomycin generics] J01GA01 **WHO essential drug**		[*57-92-1*] $C_{21}H_{39}N_7O_{12}$ 581.58 Sesquisulfate: [*3810-74-0*] $(C_{21}H_{39}N_7O_{12})_2 \cdot 3H_2SO_4$ 1457.39	S. acts by binding to the 30S ribosomal subunit of susceptible organisms and disrupts the initiation and elongation steps in protein synthesis; bactericidal/use in treatment of infections with *Mycobacterium tuberculosis* or with bacteria that have been proven to be susceptible to streptomycin and which are not amenable to the use of less potentially toxic agents; risk of severe neurotoxicity/ototoxicity/nephrotoxicity. vials with 1 g (sesquisulfate) for i.m. inj. 1 g, P.	S.A. Waksman, A. Schatz (discovery; 1943) Merck & Co.	made by fermentation of *Streptomyces griseus* [201–203]

Table 23. J01GB Other aminoglycosides

INN (Synonyms) [Brand names] ATC	Structure	[CAS-No.] Formula MW (g/mol)	Approval (country/year) Originator(s)	Target (if known) Medical use Formulation DDD	Production method References
Tobramycin (nebramycin factor 6, NF 6) [Bramitop, Gernebcin, Nebcin, TOBI, Tobramaxin, Tobrazid] J01GB01		[32986-56-4] $C_{18}H_{37}N_5O_9$ 467.52 Sulfate: $(C_{18}H_{37}N_5O_9)_2 \cdot 5H_2SO_4$ 1425.45	USA/1967 Eli Lilly	inhibition of protein synthesis by binding to the 30S and 50S ribosomal subunits/bactericidal; potential for ototoxicity, nephrotoxicity, and neuromuscular blockade - as all aminoglycosides/effective against infections with gram-negative bacteria. e.g. *Pseudomonas*. amp. for i.m. or i.v. inj./inf. 40 mg/1 mL, 80 or 160 mg/2 and 3 mL, 300 mg/5 mL; cps. for Podhaler/inhalation. 240 mg. P.	made by fermentation of *Streptomyces tenebrarius* ATCC 17920 or 17921' and separation resp. purification of the nebramycin complex or semisynthetically by hydrolysis of nebramycin factor V' or from bekanamycin [204–206]
Gentamicin complex (gentamycin; mixture of C_1, C_2 and C_{1a}) [Alcomicin, Cidomycin, Garamycin, Gentacin, Gentocin, Gentogram, Gent-Ophtal, Ophtagram, Refobacin, Septopal, Sulmycin] J01GB03 and D06AX; S01AA11; S02AA14; S03AA06. **WHO essential drug**		complex: [1403-66-3] Sulfate: [1405-41-0] Gentamicin C_1: [25876-10-2] $C_{21}H_{43}N_5O_7$ 477.60 Gentamicin C_2: [25876-11-3] $C_{20}H_{41}N_5O_7$ 463.58 Gentamicin C_{1a}: [26098-04-4] $C_{19}H_{39}N_5O_7$	USA/1971 Schering Corp.	inhibition of protein synthesis by binding to the 30S ribosomal subunit like other aminoglycosides/active against gram-negative bacteria like *Pseudomonas*, *Proteus*, *Serratia* and gram-positive *Staphylococcus*. amp. 40 or 80 mg/2 mL gentamicin as sulfate for i.m. or i.v. inj.; vials 80 mL with 80 or 240 mg and 120 mL with 360 mg of gentamicin (as sulfate); PMMA-chain implant with 2.8 or 7.5 mg of sulfate. 240 mg. P.	made by fermentation of *Micromonospora purpurea* or *M. echinospora* [207–209]

Gentamicin C_1 $R_1 = R_2 = CH_3$
Gentamicin C_2 $R_1 = CH_3$, $R_2 = H$
Gentamicin C_{1a} $R_1 = R_2 = H$

Kanamycin A
[Kanamycin-POS,
Kanamytrex, Cantrex,
Resistomycin, Kantrex]
J01GB04 and A07AA08,
S01AA24.

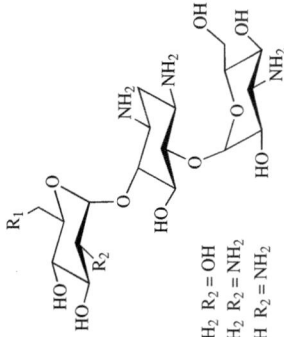

Kanamycin A: R₁ = NH₂ R₂ = OH
Kanamycin B: R₁ = NH₂ R₂ = NH₂
Kanamycin C: R₁ = OH R₂ = NH₂

Kanamycin A:
[59-01-8]
$C_{18}H_{36}N_4O_{11}$
484.50
Sulfate:
[25389-94-0]
$C_{18}H_{36}N_4O_{11} \cdot H_2SO_4$
582.58

inhibition of protein synthesis by
binding to the 30S ribosomal
subunit like other aminoglycosides/
active against gram-negative
bacteria like *E.coli, Proteus spp.,
Serratia marcescens* and *Klebsiella
pneumoniae*.
amp. 1 g/4 mL (sulfate); eye
ointment 5 mg/1 g, eye drops
5 mg/1 mL (both as Sulfate).
1 g. P.

H. Umezawa/Inst. of
Microbial Chemistry,
Tokyo, JP

produced by fermentation
of *Streptomyces
kanamyceticus* as a
complex of the 3
components; major
component with min.
75% is Kanamycin A
[210–213]

Neomycin B
(framycetin; EF 185)
[Cysto-Myacine,
Myacine, Uro-Nebacetin,
Vagicillin, Fraquinol,
Neosporin] J01GB05 and
A01AB08, A07AA01,
B05CA09, B06AC04,
G01AA14, R02AB01,
S01AA03, S02AA07,
S03AA01.

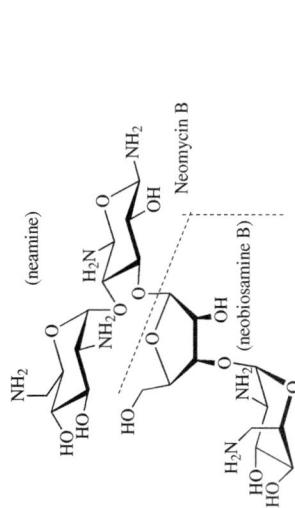

(neamine)
Neomycin B
(neobiosamine B)

[119-04-0]
$C_{23}H_{46}N_6O_{13}$
614.65
Sulfate:
[1405-10-3]
$C_{23}H_{46}N_6O_{13} \cdot 3H_2SO_4$
908.88

like all aminoglycosides, it binds
to the 30S ribosomal subunit and
blocks protein synthesis/mainly for
topical uses and preferably against
gram-negative bacteria. it is not
absorbed from the gastrointestinal
tract and finds use in many topical
combination preparations. Can
cause allergies. cream 0.5%; drops
1.25%; eye drops 0.5%; ointment
0.35%.
1 g. O.

S.A.
Waksman/Rutgers
Res. Found.; Upjohn;
Merck & Co.

produced by fermentation
of *Streptomyces fradiae*
as complex of neomycin
A, B, and C [214–219]

Amikacin [Biklin,
Amikin, Lukadin,
Mikavir, BB-K8]
J01GB06 and D06AX12,
S01AA21.
WHO essential drug

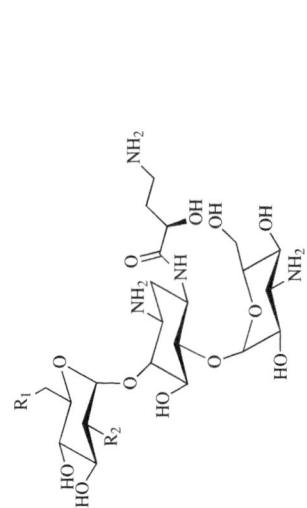

[37517-28-5]
$C_{22}H_{43}N_5O_{13}$
585.61
Sulfate:
[39831-55-5]
$C_{22}H_{43}N_5O_{13} \cdot 2H_2SO_4$
781.75

inhibition of protein synthesis like
the other aminoglycosides/broad
spectrum and unique resistance to
aminoglycosideinactivating
enzymes; preferred agent for initial
treatment of serious nosocomial
gram-negative bacterial infections
in hospitals; can cause oto- and
nephrotoxicity.
vials 100 mL with 500 mg of
amikacin (as 667 mg sulfate) for
infusion.
1 g. P.

USA/1976
DE/1976
Bristol-Myers

semisynthetic from
Kanamycin A [220–222]

(continued)

Table 23. (*continued*)

INN (Synonyms) [Brand names] ATC	Structure	[CAS-No.] Formula MW (g/mol)	Target (if known) Medical use Formulation DDD	Approval (country/year) Originator(s)	Production method References
Netilmicin (Sch-20569) [Certomycin, Netillin, Netromycine, Nettacin] J01GB07 and S01AA23		[56391-56-1] $C_{21}H_{41}N_5O_7$ 475.59 Sulfate: [5639-57-2] $(C_{21}H_{41}N_5O_7)_2 \cdot 5H_2SO_4$ 1441.53	like all aminoglycosides netilmicin inhibits protein synthesis/it finds limited use in treatment of serious infections which are resistant to gentamicin. amp. 10, 25, 50, 100 mg of netilmicin/mL (as sulfate) for inj. or infusion. 350 mg, O, P.	DE/1980 Schering Corp.	semisynthetic from sisomicin [223, 224]
Sisomicin (antibiotic 6640; Sch-13475) [Baymicin, Extramycin, Mensiso, Siseptin, Sisobiotic, Sisolline, Sisomin] J01GB08		[32385-11-8] $C_{19}H_{37}N_5O_7$ 447.53 Sulfate: [53179-09-2] $(C_{19}H_{37}N_5O_7)_2 \cdot 5H_2SO_4$ 1385.43	inhibitor of protein synthesis like all aminoglycosides/similar activity as gentamicin; limited medical use. solution of sulfate for i.v. infusion 240 mg, P.	Schering Corp.	made by fermentation of *Micromonospora inyoesis NRRL 3292* [225, 226]

Paromomycin
(1600 antibiotic; Fl-5853;
R-400; monomycin;
aminosidine)
[Farmiglucin,
Gabbromicina, Humatin,
Pargonyl, Paramicina]
A07AA06
WHO essential drug

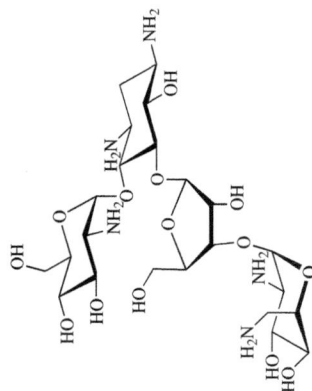

[7542-37-2]
$C_{23}H_{45}N_5O_{14}$
615.63
Sulfate:
[1263-89-4]
$C_{23}H_{45}N_5O_{14} \cdot H_2SO_4$
713.71

inhibitor of protein synthesis by
binding to 16S ribosomal
RNA/treatment of intestinal infections
(cryptosporidiosus, amoebiasis,
leishmaniasis)
cps. 250 mg; syrup 125 mg/5 mL.
3 g. O.

USA/1960
Parke Davis

made by fermentation of
Streptomyces rimosus
forma paromomycinus or
Streptomyces
krestomyceticus N.C.L.B.
8995 [227]

3.8. J01M Quinolone Antibacterials [228] (Table 24)

Table 24. J01MA Fluoroquinolones

INN (Synonyms) [Brand names] ATC	Structure	[CAS-No.] Formula MW (g/mol)	Target (if known) Medical use Formulation DDD	Approval (country/year) Originator(s)	Production method References
Ofloxacin (DL-8280; HOE-280) [Exocin, Flobacin, Floxil, Floxin, Monoflocet, Ocuflox, Oflocet, Oflocin, Tarivid] J01MA01		[82419-36-1] $C_{18}H_{20}FN_3O_4$ 361.37 Hydrochloride: [118120-51-7] $C_{18}H_{20}FN_3O_4 \cdot HCl$ 397.83	inhibitor of DNA gyrase, types II and IV topoisomerase, and so inhibits bacterial cell division/broad spectrum antibiotic against gram-positive and -negative bacteria/treatment of acute bacterial exacerbations of chronic bronchitis, commun.-acquired pneumonia, skin and skin-structure infections, infections of urethra and cervix, uncomplicated cystitis, complicated urinary tract infections, et al. f.c.tbl. 100, 200, 400 mg; ear drops 0.3%; eye drops 0.3%; vials 100 mg/50 mL, 200 mg/100 mL, 400 mg/200 mL (as hydrochloride). 400 mg, O. P.	USA/1990 Daiichi Seiyaku	[229, 230]
Ciprofloxacin (Bay q 3939; Bay o 9867) [Ciprobay, Ciflox, Ciloxan, Ciproxin, Uniflox] J01MA02 and S01AE03, S02AA15, S03AA07.		[85721-33-1] $C_{17}H_{18}FN_3O_3$ 331.35 Hydrochloride: [93107-08-5] $C_{17}H_{18}FN_3O_3 \cdot HCl$ 367.81	gyrase inhibitor (types II and IV topoisomerase)/broad spectrum against gram-positive (S.pneumoniae, S.epidermitis, Enterococcus faecalis, S.pyogenes) and -negative bacteria (E.coli, H.influenzae, Klebsiella pneumoniae, Legionella pneumophila, Moraxella catarrhalis, Proteus mirabilis, P.aeruginosa) vials 100 mg/50 mL, 200 mg/100 mL, 300 mg/150 mL for infus.; f.c.tbl. 250, 500, 750 mg. 1 g. O: 0.5 g. P.	USA/1987 DE/1987 Bayer	[231, 232]
Pefloxacin (EU-5306; 1589RB; AM-725) [Peflacine, Peflox] J01MA03		[70458-92-3] $C_{17}H_{20}FN_3O_3$ 333.36 Methanesulfonate dihydrate: [149676-40-4] $C_{17}H_{20}FN_3O_3 \cdot CH_3SO_3H \cdot 2H_2O$ 465.49	inhibitor of gyrase/broad spectrum against gram-positive and -negative bacteria in uncomplicated gonococcal urethris, infections in the gastrointestinal system, genitourinary tract infections. tbl. 400 mg; vials for infusion 400 mg/100 mL. 0.8 g, O: 0.8 g. P.	FR/1985 Roger Bellon/Dainippon	[233]

Enoxacin (AT-2266; CI-919; PD-107779) [Enoxor, Flumark] J01MA04		[74011-58-8] $C_{15}H_{17}FN_4O_3$ 320.32	gyrase inhibitor/broad spectrum, treatment of urinary tract infections and gonorrhea, infections of respiratory tract, ear and nose, skin and skin structure. f.c. tbl. 200, 300, 400 mg. 0.8 g, O.	Roger Bellon/Dainippon	[234, 235]
Norfloxacin (AM-715; MK-366) [Baccidal, Barazan, Chibroxin, Floxacin, Noroxin, Sebercim, Utinor] J01MA06 and S01AE02		[70458-96-7] $C_{16}H_{18}FN_3O_3$ 319.34	inhibitor of DNA gyrase and topoisomerases II and IV/broad spectrum with activity against gram-positive and -negative bacteria/treatment of urinary tract infections including cystitis and prostatitis. f.c. tbl. 400 mg. 0.8 g, O.	USA/1986 Kyorin	[236, 237]
Levofloxacin ((S)-ofloxacin; DR-3355; HR-355; RWJ-25213) [Cravit, Levaquin, Tavanic, Quixin]	 (S)-enantiomer of ofloxacin	[100986-85-4] $C_{18}H_{20}FN_3O_4$ 361.37 Hemihydrate: [138199-71-0] $(C_{18}H_{20}FN_3O_4) \cdot H_2O$ 740.76	inhibitor of DNA gyrase and topoisomerase IV like all quinolone antibiotics/broad spectrum against gram-positive and -negative bacteria/treatment of pneumonia, respiratory and urinary tract infections and abdominal infections. f.c. tbl. 250, 500 mg; vials for infusion with 250 mg/50 mL and 500 mg/100 mL. 0.5 g; O: P.	USA/1996 DE/1998 JP/1992 Daiichi Seiyaku	[238–240]
Moxifloxacin (Bay-12-8039) [Avalox, Avelon, Izilox, Octegra, Vigamox] J01MA14 and S01AX22		[151096-09-2] $C_{21}H_{24}FN_3O_4$ 401.44 Hydrochloride: [186826-86-8] $C_{21}H_{24}FN_3O_4 \cdot HCl$ 437.90	inhibitor of bacterial DNA gyrase (bacterial topoisomerases II and IV), so-called "fourth generation fluoroquinolone antibiotic" with broad spectrum of activities/treatment of acute exacerbations of chronic bronchitis and bacterial sinusitis, pneumonia, skin and skin-structure infections, intra-abdominal infections. f.c. tbl. 400 mg; vials 400 mg/250 mL for i.v. infusion. 400 mg, O: P.	USA/1999 DE/1999 Bayer	[241, 242]

3.9. J01X Other Antibacterials (Tables 25–30)

Table 25. J01XA Glycopeptide antibacterials [243–245]

INN (Synonyms) [Brand names] ATC	Structure	[CAS-No.] Formula MW (g/mol)	Target (if known) Medical use Formulation DDD	Approval (country/year) Originator(s)	Production method References
Vancomycin [Vancocin, Vancocina] J01XA01 and A07AA09 **WHO essential drug**		[1404-90-6] $C_{66}H_{75}Cl_2N_9O_{24}$ 1449.27 Hydrochloride: [1404-93-9] $C_{66}H_{75}Cl_2N_9O_{24} \cdot HCl$ 1485.72	vancomycin is active against gram-positive bacteria by inhibition of the cell wall synthesis; it binds strongly to the D-alanyl-D-alanine terminus of cell wall precursor units./It is used as **last resort medication** in septicemia and lower respiratory tract and bone infections by *Staphylococcus aureus* including MRSA, *Staphylococcus epidermitis*, *Staphylococcus pyogenes*, *Staphylococcus pneumoniae*. vials with powder 500 or 1000 mg for solution for i.v. infusion; cps. 250 mg. 2 g. P.	USA/1958 Eli Lilly	made by fermentation of *Amycolatopsis orientalis*, formerly designated as *Nocardia orientalis* NRRL 2450, 2451, 2452 [246–249]
Teicoplanin (teichomycin; MDL–507) [Targocid; Targosid] J01XA02		[61036-64-4] mixture of 5 components	inhibition of bacterial cell wall synthesis/treatment of serious infections by gram-positive bacteria including *Staphylococcus aureus* and *Enterococcus faecalis*;similar activity as vancomycin. vials 100, 200, and 400 mg (inj.) 400 mg. P.	DE/1989 USA/1988 Lepetit	glycopeptide antibiotic complex produced by fermentation of *Actinoplanes teichomyceticus* nov.sp.; mixture of teicoplanin A_2-(1–5) [250–253]

(continued)

Teicoplanin A$_2$-1
[91032-34-7]
C$_{88}$H$_{95}$Cl$_2$N$_9$O$_{33}$
1877.66

Teicoplanin A$_2$-2
[91032-26-7]
C$_{88}$H$_{97}$Cl$_2$N$_9$O$_{33}$
1879.67

Teicoplanin A$_2$-3
[91032-36-9]
C$_{88}$H$_{97}$Cl$_2$N$_9$O$_{33}$
1879.67

Teicoplanin A$_2$-4
[91032-37-0]
C$_{89}$H$_{99}$Cl$_2$N$_9$O$_{33}$
1893.70

Teicoplanin A$_2$-5
[91032-38-1]
C$_{89}$H$_{99}$Cl$_2$N$_9$O$_{33}$
1893.70

R side chain

(Z)-4-decenoic acid

8-Methylnonanoic acid

n-Decanoic acid

R side chain

8-Methyldecanoic acid

9-Methyldecanoic acid

Table 25. (*continued*)

INN (Synonyms) [Brand names] ATC	Structure	[CAS-No.] Formula MW (g/mol)	Target (if known) Medical use Formulation DDD	Approval (country/year) Originator(s)	Production method References
Telavancin (TD-6424) [Vibativ] J01XA03		[372151-71-8] $C_{80}H_{106}Cl_2N_{11}O_{27}P$ 1755.65 Monohydrochloride: [560130-42-9] Hydrochloride unspecified (1–3 x HCl): [380636-75-9]	similar to vancomycin it binds to D-Ala-D-Ala of the peptidoglycan of the growing cell wall and so inhibits bacterial cell wall synthesis, and further disrupts bacterial membranes by depolarization/treatment of complicated skin and skin structure infections and hospital-acquired and ventilator-associated bacterial pneumonia (*S.aureus*). single-use vials 250 or 750 mg	USA/2008 EU/2011 Theravance	semisynthetic derivative of vancomycin [254–261]
Dalbavancin (MDL-63397; BI-397; VER-001) [Dalvance, Xydalba] J01XA04		[171500-79-1] $C_{88}H_{100}Cl_2N_{10}O_{28}$ 1816.71	mechanism of action is the same as with telavancin or vancomycin by inhibition of bacterial cell wall synthesis/active against gram-positive bacteria including MRSA and methicillin-resistant *S.epidermis* (MRSE); treatment of acute skin and skin structure infections. single-use vials 500 mg for infusion (1000 mg on day 1 and 500 mg on day 8).	USA/2014 EU/2015 Lepetit/Vicuron Pharmaceut./ Durata	semisynthetic from a teicoplanin-like antibiotic A-40,926, made by fermentation of *Nonomuria sp.* [262–268]

Oritavancin
(LY-333328)
[Orbactiv]
J01XA05

[171099-57-3]
$C_{86}H_{97}Cl_3N_{10}O_{26}$
1793.12
Diphosphate:
[192564-14-0]
$C_{86}H_{97}Cl_3N_{10}O_{26} \cdot 2H_3PO_4$
1989.10

inhibits bacterial cell wall formation by blocking the transglycosylation step in peptidoglycan biosynthesis like other glycopeptides; it is active against glycopeptide-susceptible and -resistant gram-positive infections/treatment of MRSA and Vancomycinresistant *Enterococcus spp.* (VRE) as "antibiotic of last resort"; vials with powder 400 mg for reconstitution to solution with 10 mg/1 mL for infusion.

USA/2014
EU/2015
Eli Lilly/
InterMune/
Targanta
Therap./The
Medicines
Company

semisynthetic *N-p-*chlorophenylbenzyl derivative of chloroeremomycin [LY-264826; A-82846B;- chloroorienticin A), an vancomycin analogue made by fermentation of *Nocardia orientalis* strains NRRL 18098 or NRRL 18099 and NRRL 18100 [269–274]

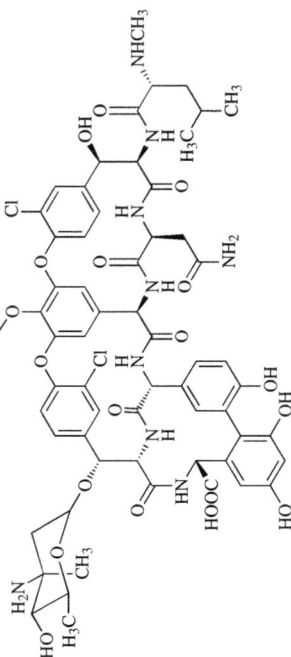

Table 26. J01XB Polymyxins [275]

INN (Synonyms) [Brand names] ATC	Structure	[CAS-No.] Formula MW (g/mol)	Target (if known) Medical use Formulation DDD	Approval (country/year) Originator(s)	Production method References
Colistin (polymyxin E; colistimethate sodium) [Diaront, ColFin, Colobreathe, Colomycin, Coly-mycin, Promixin] J01XB01 (A07AA10)	γ-NH$_2$, L-DAB → D-Leu → Leu, R → L-DAB → Thr → L-DAB → L-DAB, γ-NH$_2$, Thr → L-DAB → L-DAB, γ-NH$_2$, γ-NH$_2$ **Colistin A R** = (+)-6-methyloctanoyl **Colistin B R** = 6-methylheptanoyl DAB = α,γ-diaminobutyric acid	[1066-17-7] complex mixture with mainly colistins A and B colistin A: [7722-44-3] C$_{53}$H$_{100}$N$_{16}$O$_{13}$ 1169.48 Colistin B: [7239-48-7] C$_{52}$H$_{98}$N$_{16}$O$_{13}$ 1155.46 Colistin sodium methanesulfonate (made from complex by treating with formaldehyde and sodium bisulfite): [8068-28-8] Sulfate: [1264-72-8]	the polycationic regions of colistin interact with the bacterial outer membranes of gram-negative bacteria and displace counter ions in the lipopolysaccharide, acting like a detergent/used against infections with *Pseudomonas aeruginosa, Escherichia, and Klebsiella*. It is also used as a last resort antibiotic in infections with carbapenem-resistant *Acinetobacter baumannii.* tbl. 95 mg (Sulfate); vials with solution 5.95 mg of sulfate/1 mL for oral appl.: vials with 80 or 160 mg of colistimethate sodium solution for nebulizer; vials with 80 mg of colistimethate sodium powder for inj. or infusion. 3×10^6 I.U. ($\cong$240 mg) for inhalation.	USA and other countries/1959 Kayaku	cyclic polypeptide from *Bacillus polymyxa subsp. colistinus* [276–278]
Polymyxin B (mainly B$_1$+B$_2$) [Aerosporin, Mastimyxin] J01XB02 (A07AA05, S01AA18, S02AA11, S03AA03)	γ-NH$_2$, L-DAB → D-Phe → Leu, R → L-DAB → Thr → L-DAB → L-DAB, γ-NH$_2$, Thr → L-DAB → L-DAB, γ-NH$_2$, γ-NH$_2$ **Polymyxin B$_1$ R** = (+)-6-methyloctanoyl **Polymyxin B$_2$ R** = 6-methylheptanoyl DAB = α,γ-diaminobutyric acid	Polymyxin B (B$_1$+B$_2$): [1404-26-8] Sulfate: [1405-20-5] Polymyxin B$_1$: [4135-11-9] C$_{56}$H$_{98}$N$_{16}$O$_{13}$ 1203.50 Polymyxin B$_2$: [34503-87-2] C$_{55}$H$_{96}$N$_{16}$O$_{13}$ 1189.47	same mechanism of action like colistin (detergent-like)/treatment of multidrug-resistant gram-negative pathogens: *Pseudomonas aeruginosa, Acinetobacter baumannii,* and *Klebsiella pneumonia.* vials with 0.5×10^6 units (ca. 50 mg) of lyophilized powder (as sulfate) for reconstitution. 150 mg. P.	Burroughs Wellcome	cyclic polypeptide from *Bacillus polymyxa* strains [279–282]

Table 27. J01XC Steroid antibacterials

INN (Synonyms) [Brand names] ATC	Structure	[CAS-No.] Formula MW(g/mol)	Target (if known) Medical use Formulation DDD	Approval (country/year) Originator(s)	Production method References
Fusidic acid (ZN-6; ramycin) [Fucidin, Fucidine, Fucithalmic, Fusicutan] J01Xc1 (D06AX01; D09AA02; S01AA13)		[6990-06-3] $C_{31}H_{48}O_6$ 516.72 Sodium salt: [751-94-0] $C_{31}H_{47}NaO_6$ 538.70	inhibitor of protein synthesis by preventing the turnover of elongation factor G from the ribosome/bacteriostatic against gram-positive bacteria like MRSA and *Streptococci*; also used for topical treatment of acne vulgaris. tbl. 250 mg; creams and ointments for topical use. 1.5 g O. P.	Leo (DK)	made by fermentation of *Fusidium coccineum* [283–285]

Table 28. J01XD Imidazole derivatives

INN (Synonyms) [Brand names] ATC	Structure	[CAS-No.] Formula MW (g/mol)	Target (if known) Medical use Formulation DDD	Approval (country/year) Originator(s)	Production method References
Metronidazole (Bayer-5630; RP-8823; SC-326421) [Anabact, Arilin, Clont, Flagyl, MetroGel, Trichozide] J01XD01 (A01AB17; D06BX01. G01AF01. P01AB01) **WHO essential Drug**		[443-48-1] $C_6H_9N_3O_3$ 171.16 Hydrochloride: [69198-10-3] $C_6H_9N_3O_3 \cdot HCl$ 207.61	inhibitor of DNA of microbial cells in anaerobic bacteria and protozoa/used for treatment of bacterial vaginosis, pelvic inflammatory, pseudomembranous colitis, aspiration pneumonia, rosacea, fungating wounds, intra-abdominal infections, periodontitis, amoebiasis, trichomoniasis. f.c.tbl. 250. 400. 500 mg; solution for infusion 5 mg/1 mL; cremes and emulsions for topical application. 1.5 g, P.	USA/1963 Rhone-Poulenc	[286]
Tinidazole (CP-12574) [Fasigyn, Simplotan, Sporinex, Tindamax] J01XD02 (P01AB02)		[19387-91-8] $C_8H_{13}N_3O_4S$ 247.27	mechanism of action as well as medical use are quite similar to metronidazole. tbl. 250. 500mg.	USA/1972 Pfizer	[287, 288]

Table 29. J01XE Nitrofuran derivatives

INN (Synonyms) [Brand names] ATC	Structure	[CAS-No.] Formula MW (g/mol)	Target (if known) Medical use Formulation DDD	Approval (country/year) Originator(s)	Production method References
Nitrofurantoin [Furadantine, Furalan, Furadoine, Ituran, Macrodantin, Uro-Tablinen, Nifuretten] J01XE01		[67-20-9] $C_8H_6N_4O_5$ 238.16	in the bacterial cell, the drug is reduced by flavoproteins (*nitrofuran reductase*) to reactive intermediates, which attack ribosomal proteins and DNA. Mainly used for treatment of urinary tract infections. tbl. and f.c. tbl. 25, 50, 100, 150 mg; cps. 100 mg; suspension 25 mg/5 mL.	USA/1953 Eaton Labs.	[289]

Table 30. J01XX Other antibacterials

INN (Synonyms) [Brand names] ATC	Structure	[CAS-No.] Formula MW (g/mol)	Target (if known) Medical use Formulation DDD	Approval (country/year) Originator(s)	Production method References
Fosfomycin (MK-955; phosphomycin; phosphonomycin) [Fosfuro, Infectofos, Monuril, Monurol, Fosfocine, Fosmicin] J01XX01		[23155-02-4] $C_3H_7O_4P$ 138.06 Calcium salt monohydrate: [26016-98-8] $C_3H_5CaO_4P \cdot H_2O$ 194.14 Tromethamine salt: [78964-85-9] $C_7H_{18}NO_7P$ 259.19	inhibitor of bacterial cell wall synthesis by inactivating the enzyme UDP-N-acetylglucosamine-3-enolpyruvyltransferase (MurA), which catalyzes the bridging of glycan- and peptide residues to form the peptidoglycan/broad antibacterial activities against gram-positive and -negative pathogens. Main use for treatment of urinary tract infections and cystitis. sachets 3 g (5.6 g as tromethamine salt); vials 2, 3, 5, 8 g for infusions. 3 g, O; 8 g, P.	USA/1972 Merck & Co; Cepa, Spain	made by fermentation of *Streptomyces fradiae* ATCC 2196-21099 [290–298]
Nitroxoline [Enterocol, Nibiol, Nilox, Noxibiol, Uritrol, Urocoli] J01XX07		[4008-48-4] $C_9H_6N_2O_3$ 190.16	decrease of biofilm density of *P. aeruginosa* infections by chelating Fe^{2+}- and Zn^{2+}-ions from the biofilm matrix/use in treatment of acute and chronic urinary tract infections. e.g. cystitis. cps. 80, 150, 250 mg. 1 g. O.	FR/1963	[299]
Linezolid (PNU-100766; U-100766) [Zyvox, Zyvoxid] J01XX08		[165800-03-3] $C_{16}H_{20}FN_3O_4$ 337.35	inhibitor of protein synthesis by disrupting translation of messenger RNA (*mRNA*) into proteins in the ribosome./active against most gram-positive bacterial pathogens, including *Streptococci*, vancomycin-resistant *Enterococci* (VRE), and MRSA. Used for treatment of serious infections of skin and soft tissue and pneumonia. f.c.tbl. 600 mg; suspension 100 mg/5 mL; solution for infusion 600 mg/300 mL. 1.2 g, O, P.	USA/2000 EU/2001 Pharmacia & Upjohn	[300, 301]

(continued)

Table 30. (continued)

INN (Synonyms) [Brand names] ATC	Structure	[CAS-No.] Formula MW (g/mol)	Target (if known) Medical use Formulation DDD	Approval (country/year) Originator(s)	Production method References
Daptomycin (LY-146032) [Cubicin] J01XX09	N-decanoyl-L-tryptophyl-D-asparaginyl-L-aspartyl-L-threonyl-glycyl-L-ornithyl-L-aspartyl-D-alanyl-L-aspartyl-glycyl-D-seryl-threo-3-methyl-L-glutamyl-3-anthraniloyl-L-alanin-ε- 1-lacton (cyclic lipopeptide)	[103060-53-3] $C_{72}H_{101}N_{17}O_{26}$ 1620.69	daptomycin mode of action is different to other approved antibiotics. It disrupts several cell membrane functions and rapidly kills gram-positive pathogens of the strains *Staphylococcus aureus* (also MRSA), *Streptococcus pyogenes, agalactiae, dysgalactiae subsp., equisimilis,* and *Enterococcus faecalis*/used for treatment of skin and skin structure infections and *S. aureus* bacteraemia and endocarditis. vials 350 and 500 mg powder for solution for inj. or infusion 280 mg, P.	USA/2003 EU/2006 Eli Lilly; Cubist Pharmaceuticals	made by fermentation of *Streptomyces roseosporus NRRL 11379* under feeding with decanoic acid [302–309]
Tedizolid (torezolid; TR-700; DA-7157; prodrug: TR-701 FA) [Sivextro] J01XX11		[858866-72-3] $C_{17}H_{15}FN_6O_3$ 370.34 Phosphate: [856867-55-5] $C_{17}H_{16}FN_6O_6P$ 450.32 Phosphate disodium: [856867-39-5] $C_{17}H_{14}FN_6Na_2O_6P$ 494.29	inhibition of protein synthesis by binding to the 50S ribosomal subunit; bacteriostatic against gram-positive bacterial pathogens/treatment of complicated skin and skin structure infections (cSSSIs), caused by e.g. MRSA. Similar to linezolid. also same structural class. f.c.tbl. 200 mg; vials with powder 200 mg (as phosphate disodium salt) for i.v. inj.	USA/2014 EU/2015 Dong-A Pharmaceuticals/ Trius/Cubist Pharmaceut.	[310–312]

Abbreviations

AIDS	acquired immunodeficiency syndrome
ATC	anatomical therapeutical chemical (classification of the WHO)
CAS no.	Chemical Abstracts Service registry number
CDC	Center for Disease Control (USA)
COPD	chronic obstructive pulmonary disease
cps	capsules
DDD	defined daily dose
DHPS	dihydropteroate synthase
EMA	European Medicines Agency
f.c.tbl	film-coated tablet
FDA	US Food and Drug Administration
HIV	human immune-deficiency virus
Hyclate	USAN contraction for monohydrochloride hemiethanolate hemihydrate
INN	international nonproprietary name
MDR	multidrug resistance
MDR-TB	multidrug-resistant TB
MRSA	methicillin-resistant Staphylococcus aureus
MRSE	methicillin-resistant Staphylococcus epidermisNCE
O	oral application
P	parenteral application
PDR	pandrug-resistant
TB	tuberculosis
tbl	tablet
VRE	vancomycin resistant enterococci
VRSA	vancomycin-resistant S. aureus
WHO	World Health Organization
XDR	extensively drug-resistant

References

1 Kleemann-Engel-Kutscher-Reichert (2009) *Pharmaceutical Substances*, 5th edn, Thieme Verlag, Stuttgart; and online version.

2 Goodman & Gilman's (2011) *The Pharmacological Basis of Therapeutics*, 12th edn, McGraw Hill Medical, N.Y. *et al.*; Section VII, Chemotherapy of Microbial Diseases, chapters 48–59, p. 1363–1663.

3 Mutschler (2012) *Arzneimittelwirkungen*, 10th edn, Wissenschaftliche Verlagsgesellschaft, Stuttgart, chapter 21, Therapie von Infektionskrankheiten.

4 Nussbaum, F.v. *et al.* (2006) Antibacterial natural products in medicinal chemistry – exodus or revival? *Angew. Chem. Int. Ed.*, **45**, 5072–5129.

5 Nicolaou, K.C. *et al.* (2009) Recent advances in the chemistry and biology of naturally occuring antibiotics. *Angew. Chem. Int. Ed.*, **48**, 660–719.

6 Szychowski, J. *et al.* (2014) Natural products in medicine: transformational outcome of synthetic chemistry. *J. Med. Chem. (Perspective)*, **57**, 9292–9308.

7 Wright, P.M. *et al.* (2014) The evolving role of chemical synthesis in antibacterial drug discovery. *Angew. Chem. Int. Ed.*, **53**, 8840–8869.

8 Sneader, W. (2005) *Drug Discovery – A History*, John Wiley & Sons, Chichester, UK.

9 Brown, E.D. and Wright, G.D. (2005) New targets and screening approaches in antimicrobial drug discovery. *Chem. Rev.*, **105**, 759–774.

10 WHO Report, "Antimicrobial Resistance 2014".

11 Reardon, S. (2014) Antibiotic Resistance sweeping developing world. *Nature*, **509**, 141.

12 "Antibiotic Resistance Threats in the United States 2013".

13 Fernandes, P. (2006) Antibacterial discovery and development – the failure of success? *Nature Biotechnol.*, **24**, 1497–1503.

14 Christoffersen, R.E. (2006) Antibiotics – an investment worth making? *Nature Biotechnol.*, **24**, 1512–1514.

15 Fox, J.I. (2006) The business of developing antibacterials. *Nature Biotechnol.*, **24**, 1521–1528.

16 Chu, D.T.W. *et al.* (1996) New directions in antibacterial research. *J. Med. Chem. (Perspectives)*, **39**, 3853–3874.

17 Coates, A. *et al.* (2002) The future challenges facing the development of new antimicrobial drugs. *Nature Rev./Drug Discov.*, **1**, 895–910.

18 Walsh, Chr.T. and Wright, G. Editorial (2005) Introduction: antibiotic resistance. *Chem. Rev.*, **105**, 391–394.

19 Fisher, J.F. *et al.* (2005) Bacterial resistance to ß-lactam antibiotics: compelling opportunism, compelling opportunity. *Chem. Rev.*, **105**, 395–424.

20 Silver, L.L. (2007) *Nature Rev./Drug Discov.*, **6**, 41–55.

21 Fischbach, M.A. and Walsh, Chr.T. (2009) Antibiotics for emerging pathogens. *Science*, **325**, 1089–1093.

22 Meer, J.W.M.v.d. *et al.* (2014) Antimicrobial innovation: combining commitment, creativity and coherence. *Nature Rev./Drug Discov.*, **13**, 709–710.

23 Lewis, K. (2013) Platforms for antibiotic discovery. *Nature Rev./Drug Discov.*, **12**, 371–387.

24 O'Connell, K.M.G. *et al.* (2013) Combating multidrug-resistant bacteria: current strategies for the discovery of novel antibacterials. *Angew. Chem. Int. Ed.*, **52**, 10706–10733.

25 PhRMA (2013) Medicines in Development Report: Infectious Diseases (www.phrma.org).

26 Chopra and Roberts, M. (2001) Tetracycline antibiotics: mode of action, applications, molecular biology, and epidemiology of bacterial resistance. *Microbiol. and Molecular Biology Reviews*, **65**, 232–260.

27 McCormick, J.R.D. *et al.* (1957) *J. Am. Chem: Soc.*, **79**, 4561.

28 American Cyanamid Co. (1959) US 2 878 289, USA-prior. 1956.

29 American Cyanamid Co. (1961) US 3 012 946, USA-prior. 1960.

30 Stephens, C.R. *et al.* (1958) *J. Am. Chem. Soc.*, **80**, 5324.

31 Blackwood, R.K. *et al.* (1966) *J. Org. Chem.*, **31**, 613.

32 American Cyanamid Co. (1962) US 3 019 260, USA-prior. 1959.

33 Pfizer (1965) US 3 200 149, USA-prior. 1960.

34 Duggar, B.M. *et al.* (1948) *Ann. N. Y. Acad. Sci.*, **51**, 175.

35 American Cyanamid Co. (1949) US 2 482 055 USA-prior. 1948.

36 American Cyanamid Co. (1962) US 3 050 446 USA-prior. 1960.
37 Pfizer (1962) US 3 042 716, USA-prior. 1961.
38 Finlay, A.C. et al. (1950) Science, **111**, 85.
39 Pfizer (1950) US 2 516 080 USA-prior. 1949.
40 Boothe, J.H. et al. (1953) J. Am. Chem. Soc., **75**, 4621.
41 Conover, L.H. et al. (1953) J. Am. Chem. Soc., **75**, 4622.
42 American Cyanamid (1956) US 2 734 018, USA-prior. 1953.
43 Pfizer (1955) US 2 699 054, USA-prior. 1953.
44 American Cyanamid (1961) US 3 005 023, USA-prior. 1957.
45 Martell, M.J. Jr. and Boothe, J.H. (1967) J. Med. Chem., **10**, 44.
46 Church, R.F.R. et al. (1971) J. Org. Chem., **36**, 723.
47 American Cyanamid (1964) US 3 148 212, USA-prior. 1961.
48 American Cyanamid (1965) US 3 226 436, USA-prior. 1961.
49 Sum, P.E. and Petersen, P. (1999) Bioorg. Med. Chem. Lett., **9**, 1459.
50 American Cyanamid, (1996) US 5 494 903, USA-prior. 1991, 1992.
51 Rebstock, M.C. et al. (1949) J. Am. Chem. Soc., **71**, 2458.
52 Long, L.M. and Troutman, H.D. (1949) J. Am. Chem. Soc., **71**, 2469, 2473.
53 Parke Davis (1949) US 2 483 871, US-prior. 1948.
54 Parke Davis (1949) US 2 483 884, US-prior. 1948.
55 Parke Davis (1949) US 2 483 892, US-prior. 1948.
56 Sammes, P.G. (1976) Chem. Rev., **76**, 113–155.
57 Dürckheimer, W. et al. (1985) Recent developments in the field of ß-lactam antibiotics. Angew. Chem. Int. Ed., **24**, 180–202.
58 Kirrstetter, R. and Dürckheimer, W. (1989) Development of new ß-lactam antibiotics derived from natural and synthetic sources. Die Pharmazie, **44**, 177–185.
59 Doyle, F.P. et al. (1962) J. Chem Soc., 1440.
60 Beecham (1961) US 2 985 648, GB-prior. 1958.
61 Long, A.A.W. et al. (1971) J. Chem. Soc., 1920.
62 Beecham (1965) US 3 192 198, GB-prior. 1962.
63 Astra (1971) US 3 873 521, S-prior. 1970.
64 Astra (1976) US 3 939 270, S-prior. 1970.
65 Bayer (1976) US 3 974 142, DE-prior. 1971.
66 Lövens Kemiske Fabrik (1976) US 3 957 764, GB-prior. 1969, 1970.
67 Toyama (1978) US 4 087 424, JP-prior. 1974, 1975.
68 Toyama (1978) US 4 112 090, JP-prior. 1974.
69 Beecham, (1966) US 3 282 926, GB-prior. 1963.
70 Fleming, A. (1929) Br. J. Exp. Pathol., **10**, 226.
71 Brandl, E. et al. (1953) Wien. Med. Wochenschr., 602.
72 American Home Products (1958) US 2 820 789, AT-prior. 1954.
73 Szabo, J.L. et al. (1951) Antibiot. Chemother., **1**, 499.
74 Wyeth (1953) US 2 627 491, USA-prior. 1950.
75 Beecham (1966) US 3 239 507, GB-prior. 1962.
76 Doyle, F.P. et al. (1963) J. Chem. Soc., 5838.
77 Beecham (1961) US 2 996 501, GB-prior. 1960.
78 Doyle, F.P. et al. (1962) J. Chem. Soc., 1457.
79 Beecham (1960) US 2 951 839, GB-prior. 1959.
80 Doyle, F.P. et al. (1961) Nature, **192**, 1183.
81 Beecham (1961) US 2 996 501, GB-prior. 1960.
82 Beecham (1961) US 2 996 501, GB-prior. 1960.
83 Beecham (1966) US 3 239 507, GB-prior. 1962.
84 Volkmann, R.A. et al. (1982) J. Org. Chem., **47**, 3344.
85 Pfizer (1980) US 4 234 579, USA-prior. 1977.
86 Micetich, R.G. et al. (1987) J. Med. Chem., **30**, 1469.
87 Taiho (1985) US 4 562 073, JP-prior. 1982.
88 Ehmann, D.E. et al. (2012) (AstraZeneca). PNAS, **109** (29), 11663–11668.
89 Aventis Pharma, (26.09.2006) US 7 112 592, FR-prior. 01.8.2000.
90 Novexel, (03.11.2009) US 7 612 087, FR-prior. 28.01.2002.
91 AstraZeneca, (03.04.2012) US 8 148 540, FR-prior. 08.06.2001.
92 Lampilas, M. et al., (15.05.2012) US 8 178 554, FR-prior. 01.08.2000.
93 AstraZeneca (03.03.2015) US 8 969 566, USA-prior. 17.06.2011, 15.06.2012, 14.02.2014.
94 Bonnefoy, A. et al. (2004) (Aventis Pharma). J. Antimicrob. Chemother., **54**, 410–417.
95 Beecham, (1985) US 4 525 352, GB-prior. 1974.
96 Beecham, (1985) US 4 529 720, GB-prior. 1974, 1975.
97 Glaxo, (1083) US 4 367 175, GB-prior. 1975.
98 Beecham, (1981) US 4 301 149, GB-prior. 1977.
99 Ryan, C.W. et al. (1969) J. Med. Chem., **12**, 310.
100 Eli Lilly (1966) US 3 275 626, USA-prior. 1962.
101 Eli Lilly (1970) US 3 507 861, USA-prior. 1962, 1966.
102 Eli Lilly (1970) US 3 531 481, USA-prior. 1969.
103 Eli Lilly (1972) US 3 655 656, USA-prior. 1970.
104 Bristol-Myers (1975) US 3 862 186, USA-prior. 1972.
105 Kariyone, K. et al. (1970) J. Antibiot., **23**, 131.
106 Fujisawa (1970) US 3 516 997, JP-prior. 1967.
107 Eli Lilly (1978) US 4 104 470, USA-prior. 1977.
108 Bristol-Myers (1970) US 3 489 752, USA-prior. 1967.
109 Takeda (1974) US 3 816 253, JP-prior. 1971.
110 Glaxo (1976) US 3 974 153, GB-prior. 1971. 1973.
111 Glaxo (1981) US 4 267 320, GB-prior. 1976.
112 Chauvette, R.R. et al. (1975) J. Med. Chem., **18**, 403.
113 Lilly (1975) US 3 925 372, USA-prior. 1973, 1974.
114 Roussel-Uclaf (1977) US 4 152 432, FR-prior, 1976.
115 Glaxo (1981) US 4 258 041, GB-prior. 1978.
116 Glaxo (1982) US 4 329 453, GB-prior. 1979.
117 Reiner, R. et al. (1980) J. Antibiot., **33**, 783.
118 Roche (1982) US 4 327 210, CH-prior. 1978.
119 Yamanaka, H. et al. (1985) J. Antibiot., **38**, 1738.
120 Kawabata, K. et al. (1986) Chem. Pharm. Bull., **34**, 3458.
121 Fujisawa (1983) US 4 409 214, GB-prior. 1979, 1980.
122 Fujisawa (1980) EP 30 630, GB-prior. 1979, 1980.
123 Toyama (1983) US 4 410 522, JP-prior. 1974.
124 Toyama (1978) US 4 110 327, JP-prior. 1974.
125 Pfizer (1981) US 4 276 285, USA-prior. 1977, 1978.
126 Fujimoto, K. et al. (1987) J. Antibiot., **40**, 370.
127 Fujisawa (1983) US 4 409 215, UK-prior. 1979, 1980.
128 Sankyo (1984) US 4 486 425, JP-prior. 1980, 1981.
129 Shionogi (1987) US 4 634 697, JP-prior. 1983, 1984.
130 Schering Corp., (1997) US 5 599 557, USA-prior. 1993.
131 Bernasconi, E. et al. (2002) Org. Proc. Res. & Dev., **6**, 152, 158, 169, 178.
132 Bristol-Myers (1983) US 4 406 899, 3USA-prior. 1982.
133 Squibb (1983) US 4 386 034, USA-prior. 1982.
134 Squibb (1985) US 4 529 698, USA-prior. 1981.
135 Squibb (1986) US 4 625 022, USA-prior. 1981.
136 Squibb (1988) US 4 775 670, USA-prior. 1980.
137 Corus Pharma (2007) US 7 262 293, USA-prior. 2003, 2004.
138 Sunagawa, M. et al. (1990) J. Antibiot., **43**, 519.
139 Sumitomo (1990) US 4 943 569, JP-prior. 1983.
140 Zeneca (1995) US 5 478 820, GB-prior. 1992.
141 Williams, J.M. et al. (2005) J. Org. Chem., **70**, 7479.
142 Shionogi (1994) US 5 317 016, JP-prior. 1991.
143 Iso, Y. et al. (1996) J. Antibiot., **49**, 199; 478.
144 Nishino, Y. et al. (2003) Org. Process Res. Dev., **7**, 649. dito **7**, 846 (2003); dito **8**, 408 (2004).
145 Leanza, W.J. et al. (1979) J. Med. Chem., **22**, 1435.
146 Merck & Co (1980) US 4 194 047, USA-prior. 1975.
147 Johnston, D.B.R. et al. (1978) J. Am. Chem. Soc., **100**, 313.
148 Melillo, D.G. et al. (1980) Tetrahedron Lett., **21**, 2783.

149 Melillo, D.G. *et al.* (1986) *J. Org. Chem.*, **51**, 1498.
150 Karady, S. *et al.* (1981) *J. Am. Chem. Soc.*, **103**, 6765.
151 Graham, D.W. *et al.* (1987) *J. Med. Chem.*, **30**, 1074.
152 Merck & Co EP 10 573, USA-prior. 1978.
153 Merck & Co EP 48 301, USA-prior. 1980.
154 Merck & Co (1985) US 4 539 208, USA-prior. 1978.
155 Roche (09.11.1999) US 5 981 519, EP-prior. 19.12.1996, 07.11.1997.
156 Roche (15.05.2001) US 6 232 306, EP-prior. 15.06.1998, 10.09.1998.
157 Basilea (12.05.2009) US 7 531 650, EP-prior. 27.03.2003.
158 Ishikawa, T. *et al.* (2003) *Bioorg. Med. Chem.*, **11**, 2427.
159 Takeda (09.07.2002) US 6 417 175, JP-prior. 19.12.1997.
160 Laudano, J.B. (2011) *J. Antimicrob. Chemother.*, **66** (Suppl 3), iii111–18.
161 Astellas (31.10.2006) US 7 129 232, AU-prior. 30.10.2002, 04.09.2003.
162 Kompis, I.M. *et al.* (2005) DNA and RNA synthesis: antifolates. *Chem. Rev.*, **105**, 593–620.
163 B. Wellcome (1959) US 2 909 522, GB-prior. 1957.
164 B. Wellcome (1962) US 3 049 544, GB-prior. 1959.
165 Shionogi (1959) US 2 888 455, JP-prior. 1956.
166 Sharp & Dohme (1946) US 2 407 966, USA-prior. 1940.
167 Am. Cyanamid (1946) US 2 410 793, USA-prior. 1940.
168 Chemie Linz (1955) US 2 703 800, AU-prior. 1951
169 Katz, L. and Ashley, G.W. (2005) Translation and protein synthesis: macrolides. *Chem. Rev.*, **105**, 499–527.
170 Mukhtar, T.A. and Wright, G.D. (2005) Streptogramins, oxazolidinones, and other inhibitors of bacterial protein synthesis. *Chem. Rev.*, **105**, 529–542.
171 McDaniel, R. *et al.* (2005) Genetic approaches to polyketide antibiotics. *Chem. Rev.*, **105**, 543–558.
172 McGuire, J.M. *et al.* (1952) *Antibiot. Chemother.*, **2**, 281.
173 Eli Lilly (1953) US 2 653 899, USA-prior. 1952 (ethylsuccinate).
174 Abbott (1961) US 2 967 129, USA-prior. 1956 (estolate).
175 Eli Lilly (1961) US 3 000 874, USA-prior. 1959 (stearate).
176 Abbott (1959) US 2 881 163, USA-prior. 1953 (lactobionate).
177 Abbott (1956) US 2761859, USA-prior. 1953.
178 Rhone-Poulenc (1960) US 2 943 023, FR-prior. 1956.
179 Rhone-Poulenc (1961) US 3 000 785, FR-prior. 1953.
180 Rhone-Poulenc (1961) US 3 011 947, FR-prior. 1955.
181 Roussel Uclaf (1982) US 4 349 545, FR-prior. 1980.
182 Taisho (1982) US 4 331 803, JP-prior. 1980.
183 Pliva (1985) US 4 517 359, YU-prior. 1981.
184 Pfizer (1984) US 4 474 768, USA-prior. 1982.
185 Pfizer (2001) US 6 268 489, WO-prior. 1987.
186 Roussel Uclaf (1997) US 5 635 485, FR-prior. 1994.
187 Denis, A. *et al.* (1999) *Bioorg. Med. Chem. Lett.*, **9**, 3075.
188 Upjohn (1969) US 3 475 407, USA-prior. 1967.
189 Upjohn (1970) US 3 496 163, USA-prior. 1965.
190 Mason, D.J. *et al.* (1962) *Antimicrob. Agents Chemother.*, 555.
191 Herr, R.R. and Bergy, A.M. (1962) *Antimicrob. Agents Chemother.*, 560.
192 Upjohn (1963) US 3 086 912, USA-prior. 1961.
193 Rhone Poulenc (1964) US 3 154 475, FR-prior. 1961.
194 Celmer, W.D. and Sobin, B.A. (1955–1956) *Antibiot. Annual*, 437.
195 Preud'homme, J. *et al.* (1968) *Bull. Soc. Chim. Fr.*, 585.
196 Rhone-Poulenc (1989) US 4 798 827, FR-prior. 1986.
197 Rhone-Poulenc (1987) US 4 668 669, FR-prior. 1985.
198 Becker, B. and Cooper, M.A. (2013) Aminoglycoside antibiotics in the 21st century. *ACS Chemical Biology*, **8**, 105–115.
199 Magnet, S. and Blanchard, J.S. (2005) Molecular insights into aminoglycoside action and resistance. *Chem. Rev.*, **105**, 477–498.
200 Busscher, G.F. *et al.* (2005) 2-Deoxystreptamine: central scaffold of aminoglycoside antibiotics. *Chem. Rev.*, **105**, 775–791.
201 Rutgers Research and Endowment Foundation (1948) USP 2 449 866, USA-prior. 1945.
202 Olin Mathieson (1959) USP 2 868 779, USA-prior. 1956.
203 Olin Mathieson (1956) USP 2 765 302, USA-prior. 1953.
204 Eli Lilly (1972) USP 3 691 279, USA-prior. 1965, 1969.
205 Thompson, R.Q. *et al.* (1967) *Antimicrob. Agents Chemother.*, 314–348.
206 Okutani, T. *et al.* (1977) *J. Am. Chem. Soc.*, **99**, 1278.
207 Weinstein, M.J. *et al.* (1963) *Antimicrob. Agents Chemother.*, 1.
208 Schering Corp (1963) USP 3 091 572, USA-prior. 1962.
209 Schering Corp (1964) USP 3 136 704, USA-prior. 1962.
210 Umezawa, H. *et al.* (1957) *J. Antibiot.*, **10A**, 181–188.
211 Umezawa, H., Maeda, K., and Meda, M. (1960) US 2 931 798, JP-prior. 1956.
212 Bristol-Myers (1960) US 2 936 307, USA-prior. 1957.
213 Bristol-Myers (1961) US 2 967 177, USA-prior. 1958.
214 Waksman, S.A. *et al.* (1949) *Science*, **109**, 305.
215 Rutgers Research and Endowment Foundation (1957) US 2 799 620, USA-prior. 1956.
216 Upjohn (1958) US 2 848 365, USA-prior. 1950.
217 Upjohn (1963) US 3 108 996, USA-prior. 1962.
218 Merck & Co (1961) US 3 005 815, USA-prior. 1955, 1957.
219 Penick, S.B. (1962) US 3 022 228, USA-prior. 1960.
220 Kawaguchi, H. *et al.* (1972) *J. Antibiot.*, **25**, 695.
221 Bristol-Myers (1973) US 3 781 268, USA-prior. 1971.
222 Bristol-Myers (1976) US 3 974 137, USA-prior. 1973, 1974.
223 Wright, J.J. *et al.* (1976) *Chem. Commun.*, 206.
224 Schering Corp (1977) US 4 002 742, USA-prior. 1974.
225 Weinstein, M.J. *et al.* (1970) *J. Antibiot.*, **23**, 551.
226 Schering Corp (1974) US 3 832 286, USA-prior. 1968.
227 Parke Davis (1959) US 2 916 485, USA-prior. 1959.
228 Mitscher, L.A. (2005) Bacterial topoisomerase inhibitors: quinolone and pyridone antibacterial agents. *Chem. Rev.*, **105**, 559–592.
229 Egawa, H. *et al.* (1986) *Chem. Pharm. Bull.*, **34**, 4098.
230 Daiichi Seiyaku (1983) US 4 382 892, JP-prior. 1980.
231 Krohe, K. and Heitzer, H. (1987) *Liebigs Ann.*, 29.
232 Bayer (1987) US 4 670 444, DE-prior. 1980.
233 Roger Bellon/Dainippon (1981) US 4 292 317, GB-prior.1977, 1978.
234 Matsumoto, J. *et al.* (1984) *J. Med. Chem.*, **27**, 292.
235 Dainippon/Roger Bellon (1982) US 4 352 803, JP-prior. 1978.
236 Koga, H. *et al.* (1980) *J. Med. Chem.*, **23**, 1358.
237 Kyorin (1979) US 4 146 719, JP-prior. 1977.
238 Atarashi, S. *et al.* (1987) *Chem. Pharm. Bull.*, **35**, 1896.
239 Daiichi (1983) US 4 382 892, JP-prior. 1980.
240 Daiichi (1991) US 5 053 407, JP-prior. 1985, 1986.
241 Bayer (1991) US 4 990 517, DE-prior. 1988, 1989.
242 Bayer (1997) US 5 607 942, DE-prior. 1988, 1989.
243 Nicolaou, K.C. *et al.* (1999) Chemistry, biology, and medicine of the glycopeptide antibiotics. *Angew. Chem. Int. Ed.*, **38**, 2096–2152.
244 Kahne, D. *et al.* (2005) Glycopeptide and lipoglycopeptide antibiotics. *Chem. Rev.*, **105**, 425–448.
245 Hubbard, B.K. and Walsh, Chr.T. (2003) Vancomycin assembly: nature's way. *Angew. Chem. Int. Ed.*, **42**, 730–765.
246 McCormick, M.H. *et al.* (1955 –56). *Antibiot. Ann.*, 606.
247 Eli Lilly (1962) US 3 067 099, USA-prior. 1955.
248 Harris, C.M. *et al.* (1983) *J. Am. Chem. Soc.*, **105**, 6915.

249 Nicolaou, K.C. *et al.* (1999) *Angew. Chem. Int. Ed.*, **38**, 240 (total synthesis).

250 Lepetit (1980) US 4 239 751, GB-prior. 1975.

251 Parenti, F. *et al.* (1978) *J. Antibiot.*, **31**, 276.

252 Lepetit (1985) US 4 542 018, GB-prior. 1982.

253 Borghi, A. *et al.* (1984) *J. Antibiot.*, **37**, 615.

254 Leadbetter, M. *et al.* (2004) *J. Antibiot.*, **57**, 326.

255 Theravance (21.10.2003) US 6 635 618, USA-prior. 22.06.2000, 01.05.2001.

256 Theravance (29.03.2005) US 6 872 701, USA-prior. 22.06.2000, 01.05.2001, 12.05.2003.

257 Theravance (07.03.2006) US 7 008 923, USA-prior. 22.06.2000, 01.05.2001, 24.09.2003, 18.01.2005.

258 Theravance (24.04.2007) US 7 208 471, USA-prior. 22.06.2000, 01.05.2001, 24.09.2003, 18.01.2005, 02.11.2005

259 Theravance (01.04.2008) US 7 351 691, USA-prior. 22.06.2000, 01.05.2001, 24.09.2003, 18.01.2005, 02.11.2005, 23.10.2006.

260 Theravance (21.03.2006) US 7 015 307, USA-prior. 24.08.2001, 23.08.2002

261 Theravance (23.08.2011) US 8 003 755, USA-prior. 22.10.2003, 21.10.2004

262 Malabarba, A. *et al.* (1995) *J. Antibiot.*, **48**, 869.

263 Malabarba, A. *et al.* (2005) *J. Antimicrob. Chemotherapy*, **55** (Suppl S2), ii15–ii20.

264 Gruppo Lepetit (12.05.1998) US 5 750 509, EP-prior. 29.07.1991, 12.06.1992.

265 Vicuron (31.05.2005) US 6 900 175, USA-prior. 18.11.2002, 08.07.2003, 13.08.2003, 19.08.2003, 14.11.2003

266 Vicuron (03.10.2006) US 7 115 564, USA-prior. 18.11.2002, 08.07.2003, 13.08.2003, 19.08.2003,14.11.2003, 16.04.2004

267 Vicuron (10.10.2006) US 7 119 061, USA-prior. 18.11.2002, 08.07.2003, 13.08.2003, 19.08.2003,14.11.2003, 27.04.2004

268 Vicuron (27.03.2012) US 8 143 212, USA-prior. 14.11.2003, 27.04.2004, 26.04.2005, 02.06.2009

269 Allen, N.E. and Nicas, T.I. (2003) *FEMS Microbiology Reviews*, **26**, 511–532.

270 Das, B. *et al.* (2013) *Pak. J, Pharm. Sci.*, **26**, 1045.

271 Eli Lilly (24.11.1998) US 5 840 684, USA-prior. 28.01.1994, 15.12.1994, 24.03.1995

272 Eli Lilly (07.12.1999) US 5 998 581, USA-prior. 21.11.1996, 12.11.1997, 12.04.1999

273 Eli Lilly (17.05.1994) US 5 312 738, USA-prior. 19.09.1986, 02.12.1987, 09.05.1990.

274 The Medicines Company (16.04.2013) US 8 420 592, USA-prior. 02.09.2008, 30.08.2008, 29.08.2009.

275 Velkov, T. *et al.* (2010) Structure-activity relationships of polymyxin antibiotics. *J. Med. Chem. (Perspective)*, **53**, 1898–1916.

276 Koyama, Y. *et al.* (1950) *J. Antibiot.*, **3**, 457.

277 Suzuki, T. *et al.* (1963) *J. Biochem. (Tokyo)*, **54**, 25.

278 Barnett, M. *et al.* (1964) *Br. J. Pharmacol. Chemother.*, **23**, 552 (colistimethate sodium).

279 Burroughs Wellcome (1951) US 2 565 057, GB-prior. 1946.

280 American Cyanamid (1952) US 2 595 605, USA-prior. 1948.

281 Biotika (2011) US 7 951 913, SK-prior. 2006.

282 Zavascki, A.P. *et al.* (2007) *J. Antimicrob. Chemother.*, **60**, 1206–1215.

283 Lovens Kemiske (1966) US 3 230 240, GB-prior. 1962.

284 Godtfredsen, W.O. *et al.* (1962) *Nature*, **193**, 987.

285 Godtfredsen, W.O. and Vangedal, S. (1962) *Tetrahedron*, **18**, 1029.

286 Rhone-Poulenc (1960) US 2 944 061, FR-prior. 1957.

287 Pfizer (1968) US 3 376 311, USA-prior. 1964.

288 Miller, M.W. *et al.* (1970) *J. Med. Chem.*, **13**, 849.

289 Eaton Labs (1950) US 2 610 181, USA-prior. 1950.

290 Hendlin, D. *et al.* (1969) *Science*, **166**, 122.

291 Christensen, B.G. *et al.* (1969) *Science*, **166**, 123.

292 Merck & Co (1972) US 3 639 590, USA-prior. 1967.

293 Merck & Co (1975) US 3 914 231, USA-prior. 1967.

294 Merck & Co (1972) US 3 641 063, USA-prior. 1968.

295 Zambon (1989) US 4 863 908, IT-prior. 1979.

296 Zambon (1992) US 5 162 309, IT-prior. 1989.

297 Zambon (1993) US 5 191 094, IT-prior. 1989.

298 Glamkowski, E.J. *et al.* (1970) *J. Org. Chem.*, **35**, 3510 (synthesis).

299 Desvignes, A. and Leguen, P. (1963) *Ann. Pharm. Fr.*, **21**, 803.

300 Brickner, S.J. *et al.* (1996) *J. Med. Chem.*, **39**, 673.

301 Pharmacia & Upjohn (1997) US 5 688 792, PCT-prior. 1994.

302 Raja, A. *et al.* (2003) *Nature Revs. Drug Discov.*, **2**, 943.

303 Steenbergen, J.N. *et al.* (2005) *J. Antimicrob. Chemother.*, **55**, 283.

304 Baltz, R.H. *et al.* (2005) *Nat. Prod. Rep.*, **22**, 717.

305 Eli Lilly (1985) US 4 537 717, USA-prior. 1982, 1983, 1984.

306 Cubist Pharmaceut (24.02.2004) US 6 696 412, USA-prior. 20.01.2000, 28.11.2000.

307 Cubist Pharmaceut (06.03.2012) US 8 129 342, USA-prior. 20.01.2000, 28.11.2000, 29.12.2003, 24.04.2007, 22.09.2010.

308 Cubist Pharmaceut (30.09.2014) US 8 846 610, USA-prior. 18.12.2000, 09.03.2001, 13.12.2001, 17.12.2001, 18.04.2005, 26.08.2008, 31.07.2013.

309 Lohani, C.R. *et al.* (2015) *Org. Lett.*, **17**, 748 (synthesis).

310 Bin Im, W. *et al.* (2011) *Eur. J. Med. Chem.*, **40**, 1027.

311 Dong-A Pharm (19.10.2010) US 7 816 379, KR-prior. 18.12.2003, 27.07.2004.

312 Dong-A Pharm (16.04.2013) US 8 420 676, KR-prior. 18.12.2003, 27.07.2004.

Antiinfectives for Systemic Use, 2. Antimycotics (Antifungals) (J02-04) and Antimycobacterials

AXEL KLEEMANN, Hanau, Germany

1. Introduction

The general topic "Antiinfectives" is reviewed in this and two further articles (Antiinfectives for Systemic Use, 1. Antibacterials, and → Antiinfectives for Systemic Use, 3. Antivirals). For an introduction on the general topic Antiinfectives for Systemic Use see → Antiinfectives for Systemic Use, 1. Antibacterials. All references cited in the headlines of the following chapters are general references, giving an in-depth introduction into the respective class of antiinfectives.

2. J02 Antimycotics [1–8]

In the international literature this drug class is most often classified as Antifungals. Besides for systemic use (oral and parenteral application) many antifungals are in use for topical (A01AB, A07A), dermatological (D01), and gynecological (G01) administration. Only three drug classes were in use for systemic application in 2016, namely polyene antibiotics (amphotericin B), triazole antifungals (fluconazole, itraconazole, voriconazole, and posaconazole), and echinocandins (caspofungin, micafungin, and anidulafungin). The world market for these drugs amounts to about US $ 6 billion, voriconazole (Vfend/Pfizer) being the market leader with US $ 0.7 billion annual sales (ex works in 2015).

With the increasing number of immune compromised patients (because of, e.g., cancer, organ transplants, and use of highly potent drugs) opportunistic fungal infections especially in the respiratory tract are on the rise. Like with antibiotics, drug resistance is a steadily increasing issue, especially with candida and aspergillus species, and this needs new and effective drugs, optionally new drug classes with new mechanisms of action (Tables 1–4).

Ullmann's Pharmaceuticals. Edited by Axel Kleemann and Bernhard Kutscher.
© 2022 Wiley-VCH GmbH
Set ISBN: 978-3-527-34252-5/ DOI: 10.1002/14356007.a03_077.pub2

Table 1. J02AA Antibiotics

INN (Synonyms) [Brand names] ATC	Structure	[CAS-No.] Formula MW (g/mol)	Target (if known) Medical use Formulation DDD	Approval (country/year) Originator(s)	Production method References
Amphotericin B [Fungizone; Fungilin; Ampho-Moronal; liposomal complex[1]: AmBisome; lipid complex[2]: Abelcet] J02AA01 and A01AB04; A07AA07; D01AA10; G01AA03. **WHO essential drugs**		[1397-89-3] $C_{47}H_{73}NO_{17}$ 924.09	by binding to ergosterol of the fungus cell membrane, a transmembrane channel is formed which leads to leakage of K^+ and Na^+ so causing fungal cell death/treatment of systemic fungal infections incl. cryptococcal meningitis and also visceral leishmaniasis. vials with 50 mg powder for infusion solutions; suspension for local appln. in mouth; tbl. 100 mg. 35 mg. P.	USA/1992; AmBisome: USA/1997; Abelcet: USA/2010. Squibb; Olin Mathieson	Polyen antibiotic, made by fermentation of *Streptomyces nodosus* M4575 [9]

[1] unilamellar liposomes containing phosphatidylcholine, cholesterol, distearoylphosphatidylglycerol and amphotericin B in a molar ratio of 2:1:0.8:0.4.
[2] complex with L-α-dimyristoyl phosphatidylcholine and L-α-dimyristoyl phosphatidylglycerol in a 7:3 molar ratio.

Table 2. J02AB Imidazole derivatives (not for systemic use!)

INN (Synonyms) [Brand names] ATC	Structure	[CAS-No.] Formula MW (g/mol)	Target (if known) Medical use Formulation DDD	Approval (country/year) Originator(s)	Production method References
Miconazole (R-14889) [Daktar, Daktarin, Florid, Fungoid, Loramyc, Albistat, Epi-Monistat, Miconal, Micotef, Monistat, Prilagin] J02AB01 (no longer for systemic application) A01AB09; A07AC01; D01AC02; G01AF04; S02AA13. **WHO essential drug**		[22916-47-8] $C_{18}H_{14}Cl_4N_2O$ 416.12 Nitrate: [22832-87-7] $C_{18}H_{14}Cl_4N_2O \cdot HNO_3$ 479.14	inhibitor of ergosterol biosynthesis/antifungal, mainly used externally. cream 1%; lotion 1%; ointment 1%; oral gel 2%; powder 2%; vaginal cream 2%.	Janssen	[10–12]
Ketoconazole (R-41400) [Fungoral, Ketoderm, Nizoral, Terzolin, Triatop] J02AB02 (in many countries no longer for oral application) and D01AC08; G01AF11.		[65277-42-1] $C_{26}H_{28}Cl_2N_4O_4$ 531.43	inhibitor of ergosterol biosynthesis/broad-spectrum antimycotic, mainly topical administration against fungal infections of the skin and mucous membranes. cream 2%; shampoo 2%; solution 2%; suspension 100 mg; tbl. 200 mg. 200 mg. O.	USA/1981 Janssen	[13–16]

Table 3. J02AC Triazole derivatives

INN (Synonyms) [Brand names] ATC	Structure	[CAS-No.] Formula MW (g/mol)	Target (if known) Medical use Formulation DDD	Approval (country/year) Originator(s)	Production method References
Fluconazole (UK-49858) [Diflucan, Flucoderm, Elazor, Fungata, Triflucan] J02AC01 and D01AC15 **WHO essential drug**		[86386-73-4] $C_{13}H_{12}F_2N_6O$ 306.28	inhibitor of ergosterol biosynthesis from lanosterol; inhibits the fungal cytochrome P450 enzyme 14α-demethylase/antifungal for systemic use against infections with candida and *cryptococcal meningoencephalitis*. cps. 50, 100, 200 mg; vials 100 mg/50 mL, 200 mg/100 mL, 400 mg/200 mL for i.v. 200 mg, O. P.	USA/1990 Pfizer	[17, 18]
Itraconazole (R-51211) [Itraderm, Itrizole, Sempera, Siros, Sporanox, Triasporin] J02AC02		[84625-61-6] $C_{35}H_{38}Cl_2N_8O_4$ 705.64	inhibitor of ergosterol biosynthesis; P450 inhibitor/treatment of systemic infections like apergillosis, candidiasis, and cryptococcosis. cps. 50, 100 mg; amp. 250 mg concentrate solution for infus. 200 mg O. P.	USA/1988 Janssen	[19, 20]
Voriconazole (UK-109496) [Vfend] J02AC03		[137234-62-9] $C_{16}H_{14}F_3N_5O$ 349.32	inhibitor of ergosterol biosynthesis/used in treatment of immunocompromised patients with invasive aspergillosis, invasive candidiasis, and emerging fungal infections. vials with suspension for oral application; f.c. tbl. 50, 200 mg; vials with powder 200 mg for reconstitution sol. for i.v. application. 400 mg O. P.	USA/2002 Pfizer	[21–23]

Posaconazole
(Sch-56592)
[Noxafil]
J02AC04

[171228-49-2]
$C_{37}H_{42}F_2N_8O_4$
700.79

inhibition of the enzyme lanosterol
14α-demethylase and thus ergosterol
biosynthesis is inhibited/treatment of
invasive aspergillosis and candidiasis.
f.c. tbl. 100 mg; vials with concentrated
solution for preparation of infusion
solution; suspension 40 mg/ mL for oral
administration.
800 mg, O.

USA/2005   [24, 25]
EU/2005
Schering-Plough

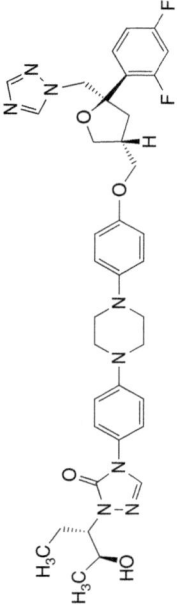

Isavuconazole
(BAL-4815;
RO-0094815)
[Cresemba]
J02AC05

[241479-67-4]
$C_{22}H_{17}F_2N_5OS$
437.47

inhibition of ergosterol
biosynthesis/treatment of invasive
aspergillosis and invasive mucomycosis.
(as isovuconazonium sulfate vials 372 mg
lyophilized powder (corresponding to 200
mg isovuconazole)

USA/2015   [26–30]
Basilea; Astellas
Pharma

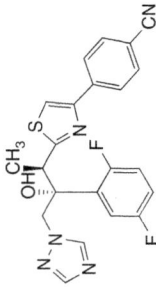

**Isavuconazonium
sulfate (prodrug)**
(BAL-8557-002)

[946075-13-4]
$C_{35}H_{35}F_2N_8O_5S \cdot HSO_4$
814.84

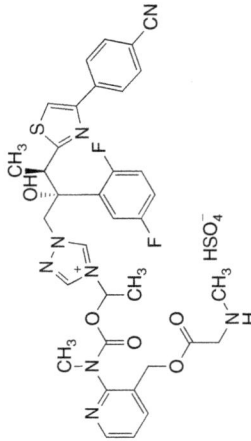

Table 4. J02AX Other Antimycotics for systemic use [31]

INN (Synonyms) [Brand names] ATC	Structure	[CAS-No.] Formula MW (g/mol)	Target (if known) Medical use Formulation DDD	Approval (country/year) Originator(s)	Production method References
Flucytosine (Ro-2-9915; 5-FC) [Ancobon, Ancotil] J02AX01 and D01AE21 **WHO essential drug**	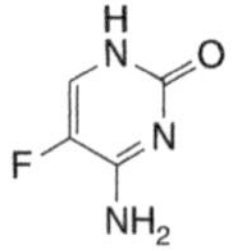	[2022-85-7] $C_4H_4FN_3O$ 129.09	after metabolic conversion in fungi into fluorouracil and further into 5-fluorouridinetriphosphate it interacts with RNA biosynthesis and in addition after conversion into 5-fluorodeoxyuridinemonophosphate, it inhibits fungal DNA synthesis/treatment of serious infections with *Candida* or *Cryptococcus neoformans*, often in combination with amphotericin B. vials with infusion solution 2.5 g/250 mL; cps. 250, 500 mg. 10 g, O; P.	Roche	[32–34]
Caspofungin (MK-991; MK0991) [Cancidas] J02AX04	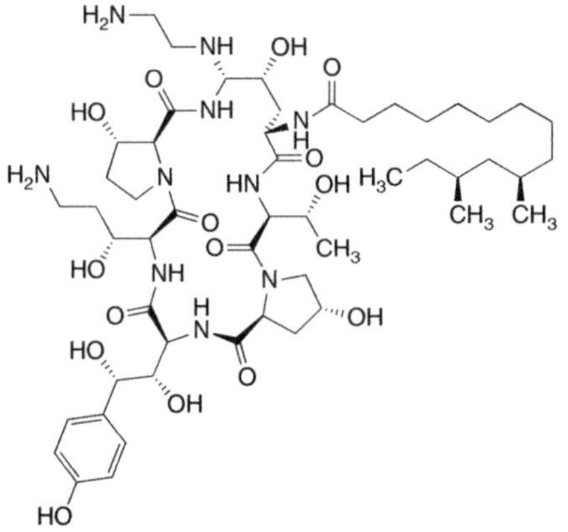	[162808-62-0] $C_{52}H_{88}N_{10}O_{15}$ 1093.33 Acetate: [179463-17-3] $C_{52}H_{88}N_{10}O_{15} \cdot 2C_2H_4O_2$ 1213.44	inhibitor of 1,3-β-glucan synthase (β-1,3-D-glucan is necessary for formation of the fungal cell wall)/treatment of invasive aspergillosis and candidiasis. vials with powder 50 and 70 mg for preparation of infusion solutions. 50 mg, P.	USA/2001 DE/2002 Merck & Co.	semisynthetic lipopeptide made from echinocandin B_0 [35–38]

Micafungin
(Pneumocandin AO; FK-463)
[Funguard, Mycamine]
J02AX05

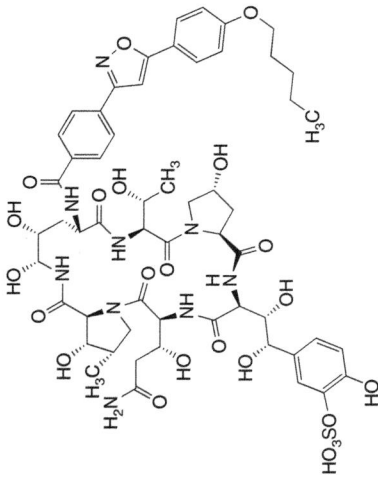

[235114-32-6]
$C_{56}H_{71}N_9O_{23}S$
1270.28
Sodium salt:
[208538-73-2]
$C_{56}H_{70}N_9NaO_{23}S$
1292.27

inhibitor of 1,3-β-D-glucan synthase; leads to fungal cell lysis./treatment of candidemia, Candida peritonitis, vials with powder 50 and 100 mg (Na salt) for inj. solution. 100 mg. P.

USA/2005
EU/2008
Fujisawa (Astellas Pharma)

semisynthetic echinocandin lipopeptide [39–41]

Anidulafungin
(V-echinocandin; LY-303366; VER-002)
[Ecalta, Eraxis]
J02AX06

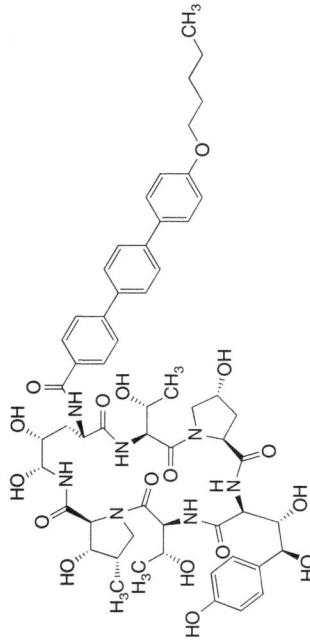

[166663-25-8]
$C_{58}H_{73}N_7O_{17}$
1140.25

inhibitor of 1,3-β-D-glucan synthase; leads to fungal cell lysis./treatment of invasive candidiasis. vials with powder 100 mg for inj. solution. 100 mg. P.

USA/2006
EU/2008
Eli Lilly; Vicuron; Pfizer

semisynthetic echinocandin lipopeptide [42–46]

3. J04 Antimycobacterials

3.1. J04A Drugs for Treatment of Tuberculosis [47–54]

The best sources for information about the important aspects of tuberculosis are the WHO Global Tuberculosis Report [47], Goodman & Gilman's "The Pharmacological Basis of Therapeutics", chapter 56 [48], and also Wikipedia.

According to WHO, "TB is a treatable and curable disease, but remains a major global health problem."

Tuberculosis (TB) and HIV (human immune-deficiency virus) are still leading causes of death worldwide, in spite of considerable advances in diagnosis, treatment, and prevention in the years between 1990 and 2015: in 2014, 1.5 million people were killed by TB, thereof 1.1 million HIV-negative and 0.4 million HIV-positive among 890 000 men, 480 000 women and 140,000 children; an estimated 190 000 people died of multidrug-resistant TB (MDR-TB). The HIV death figure amounted to an estimated 1.2 million, including the 0.4 million TB deaths among HIV-positive people.

The number of new TB cases in 2014 were 9.6 million people with 5.4 million men and 3.2 million women; 12% of these new TB cases were HIV-positive; an estimated 480 000 people contracted MDR-TB. The main geographic areas of the 9.6 million new TB cases were Southeast Asia and Western Pacific regions with 58%, and the African region with 28% contribution – but highest rate per capita.

TB is mainly caused by *Mycobacterium tuberculosis* (MTB), an aerobic bacillus with more than 60% content of lipids in its cell wall, which prevents many drug molecules from getting to the cell wall membrane and penetrate into the cytosol. These pathogens are slow grower and divide by a very slow rate of every 12 to 18 hours. In comparison, other bacteria divide in less than 1 hour. The MTB complex knows 4 TB-causing strains, namely *M. bovis* (often in nonpasteurized milk), *M. africanum* (cause of TB in Africa), *M. canetti* (rare, a few cases in Africa), and *M. microti* (rare, found in immunodeficient patients). In addition there are three other strains of mycobacteria, which have to be observed, namely *M. leprae*, and the not causing TB, but pulmonary diseases, *M. avium*, and *M. kansasii*. Normally, only 5-10% of people with MTB infection develop active TB in their lifetime, whereas about 30% of HIV-positive patients develop active TB. The latter fact illustrates the important risk factor HIV.

As in practically all infections, also MDR-TB is an increasing problem and therefore more therapeutic agents optionally with new mechanisms of action are necessary [49–56].

For treatment of active TB, combinations of 2 or better 3 antibiotics are employed in order to avoid or reduce the risk of development of

Table 5. J04AA Aminosalicylic acid and derivatives

INN (Synonyms) [Brand names] ATC	Structure	[CAS-No.] Formula MW (g/mol)	Target (if known) Medical use Formulation DDD	Approval (country/year) Originator(s)	Production method References
4-Aminosalicylic acid (*p*-aminosalicylic acid; PAS) [PAS, PASER] J04AA01 **WHO essential drug**		[65-49-6] $C_7H_7NO_3$ 153.14 Sodium salt: [133-10-8] $C_7H_6NNaO_3$ 175.12 Sodium salt dihydrate: [6018-19-5] $C_7H_6NNaO_3 \cdot 2\,H_2O$ 211.15	inhibits dihydropteroate synthase and by that folate biosynthesis/used as second-line agent in treatment of drug-resistant TB in conjunction with other antituberculosis drugs; also in use for treatment of ulcerative colitis and Crohn's disease. Granule 100 mg; tbl. 250 mg; vials for infusion 13.49 (sodium salt). 12 g, O (acid); 14 g, O; P (sodium salt).	USA/1944 J. Lehmann (SE)	made by carboxylation of 3-aminophenol [58–65]

Table 6. J04AB Antibiotics

INN (Synonyms) [Brand names] ATC	Structure	[CAS-No.] Formula MW (g/mol)	Target (if known) Medical use Formulation DDD	Approval (country/year) Originator(s)	Production method References
Cycloserine (D-cycloserine; orientomycin; PA-94) [Seromycin] J04AB01 **WHO essential drug**		[68-41-7] $C_3H_6N_2O_2$ 102.09	inhibitor of D-alanine racemase and ligase, resulting in inhibition of peptidoglycan synthesis (cell-wall biosynthesis)/second-line drug for treatment of multiple drug-resistant TB. tbl. and cps. 250 mg. 750 mg. O.	USA/1957 (?) Pfizer; Merck & Co.; Eli Lilly	technically made by synthesis from D-serine [66–72]
Rifamycins				firstly isolated by P. Sensi et al. in the company Gruppo Lepetit.	a group of antibiotics that are made by fermentation of *Streptomyces mediterranei* (later classified as *Amycolatopsis, editerranei*) and belong to the family of ansamycins. The most important and stable component of the mixture (A, B, C …) is rifamycin B, which has a rather low activity, but serves as starting material for the semisynthetic derivates rifampicin, rifabutin, and rifapentin. [73–77]
Rifampicin (Rifampin) [Eremfat, Rifa, Rifadin, Rimactan] J04AB02 **WHO essential drug**		[13292-46-1] $C_{43}H_{58}N_4O_{12}$ 822.95	inhibitor of DNA-dependent RNA-polymerase by binding to the β subunit, whereby the initiation of RNA biosynthesis is suppressed/active against *M. tuberculosis* and many gram-positive and gram-negative bacteria. Oral administration either alone or as fixed-dose combinations with isoniazid and/or pyrazinamide (see J04AM) or parenteral alone. f.c. tbl. 159, 300, 450, and 600 mg; cps. 150 and 300 mg; vials with lyophilized powder for inj. 600 mg. 0.6 g O,P: 0.3 g O for children.	USA/1971 Lepetit	semisynthetic from rifamycin B via rifamycin SV and 3-formylrifamycin SV [78–80]

Rifamycin B

(continued)

Table 6. (*Continued*)

INN (Synonyms) [Brand names] ATC	Structure	[CAS-No.] Formula MW (g/mol)	Target (if known) Medical use Formulation DDD	Approval (country/year) Originator(s)	Production method References
Rifabutin (LM-427) [Ansatipine, Mycobutin] J04AB04 **WHO essential drug**		[72559-06-9] $C_{46}H_{62}N_4O_{11}$ 847.02	same mechanism of action as rifampicin: inhibits DNA-dependent RNA polymerase/treatment of *Mycobacterium avium* complex disease, e.g., in patients with HIV-related tuberculosis. cps. 150 mg.	USA/1992 Archifar (Carlo Erba)	semisynthetic via 3-aminorifamycin S [81–83]
Rifapentin (MDL-473; R-773) [Priftin] J04AB05 **WHO essential drug**		[61379-65-5] $C_{47}H_{64}N_4O_{12}$ 877.05	same mechanism of action as rifampicin, and also similar medical use. tbl. 150 mg.	USA/1998 Lepetit; Sanofi	analog to synthesis of rifampicin [84–86]
Capreomycin (NRRL-2773) [Capastat] J04AB30 **WHO essential drug**	Capreomycin IA: R = OH Capreomycin IB: R = H	mixture of IA and IB [11003-38-6] Disulfate: [1405-36-3] Capreomycin IA: [37280-35-6] $C_{25}H_{44}N_{14}O_8$ 668.72 Capreomycin IB: [33490-33-4] 652.72	this cyclic polypeptide inhibits protein synthesis (interbridge B2a between 30S and 50S ribosomal subunits/second-line drug, given with other antibiotics. vials with powder 1 g for inj. (as disulfate). 1 g, P.	USA/1971 Eli Lilly	Made by fermentation of *Streptomyces capreolus* [87, 88]

Table 7. J04AC Hydrazides

INN (Synonyms) [Brand names] ATC	Structure	[CAS-No.] Formula MW (g/mol)	Target (if known) Medical use Formulation DDD	Approval (country/year) Originator(s)	Production method References
Isoniazid (INH: RP-5015; FSR-3) [Isonex, Isozid, Neoteben, Nicozid, Nydrazid, Rifater, Rimifon, Tebesium] J04AC01 **WHO essential Drug**		[54-85-3] $C_6H_7N_3O$ 137.14	inhibitor of mycolic acid synthesis by binding to the enoyl-acyl carrier reductase (InhA); bactericidal against rapidly dividing *Mycobacteria*, and bacteriostatic to slow-growing ones./first-line agent for prevention and treatment of latent and active tuberculosis; component of several combinations (regimens) tbl. 50, 100, 200, 300 mg; vials 500 mg with powder for inj. 300 mg O, P.	USA/1952 (Roche: Rimifon) DE/1952 (Bayer: Neoteben) Roche	[89, 90]

Table 8. J04AD Thiocarbamide derivatives

INN (Synonyms) [Brand names] ATC	Structure	[CAS-No.] Formula MW (g/mol)	Target (if known) Medical use Formulation DDD	Approval (country/year) Originator(s)	Production method References
Protionamide **(Prothionamide)** [Ektebin, Peteha] J04AD01		[14222-60-7] $C_9H_{12}N_2S$ 180.27	inhibitor of mycolic acid synthesis/second-line antitubercular drug for drug-resistant cases after first-line drugs were not effective. tbl. 125, 250 mg; f.c.tbl. 250 mg. 750 mg, O.	FR/1956 (?) Theraplix	[91–93]
Ethionamide **(Etionamide;** TH-1314; Bay-5312) [Trecator, Trescatyl, Tubermin] J04AD03 **WHO essential drug**		[536-33-4] $C_8H_{10}N_2S$ 166.24	inhibitor of mycolic acid synthesis/second-line drug used like protionamide. tbl. 100, 250 mg. 750 mg, O.	Theraplix	[94–96]

resistance. The standard regimen consists of isoniazid, rifampicin and pyrazinamide for 2 months, followed by intermittent rifampicin plus isoniazid (2–3 x weekly) for 4 months. If there is sign of resistance to isoniazid, ethambutol, or streptomycin can be used. The mentioned five drugs are first-line agents. Second-line agents are ethionamid, *p*-aminosalicylic acid, cycloserine, amikacin, kanamycin and capreomycin, and the recently approved and WHO-accepted new drugs bedaquiline (2013) and delamanid (2014) [57] (Tables 5–9).

Note that plain streptomycin has been classified as J01GA01. Thalidomide, which is also used for treatment of lepra, is classified as L04AX02.

Table 9. J04AK Other drugs for treatment of tuberculosis

INN (Synonyms) [Brand names] ATC	Structure	[CAS-No.] Formula MW (g/mol)	Target (if known) Medical use Formulation DDD	Approval (country/year) Originator(s)	Production method References
Pyrazinamide (D-50) [Pyrafat, Rifater, Tebesium TRIO, Tebrazid, Unipyranamide, Zinamide] J04AK01 WHO essential drug		[98-96-4] $C_5H_5N_3O$ 123.12	target is S1 component of 30S ribosomal subunit: Inhibits translation and trans-translation and acidifies cytoplasm/first-line drug used in combination with isoniazid and rifampicin. tbl. 150, 400, 500 mg 1.5 g. O.	USA/1954 American Cyanamid (Lederle Labs.)	[97–102]
Ethambutol (EMB) [Dexambutol, Ebutol, Etibi, Myambutol, Servambutol] J04AK02 WHO essential drug		[74-55-5] $C_{10}H_{24}N_2O_2$ 204.31 Dihydrochloride: [1070-11-7] $C_{10}H_{24}N_2O_2 \cdot 2HCl$ 277.23	inhibits arabinogalactan biosynthesis (target: arabinosyl transferase)/first-line drug to treat tuberculosis in combination with isoniazid, rifampicin and pyrazinamide. tbl. 125, 250 mg; f.c.tbl. 100, 250, 400, 500 mg; amp. 400 mg/4 mL, 1000 mg/10 mL. 1.2 g O, P.	USA/1961 (discovery date) American Cyanamid (Lederle Labs.)	[103–106]
Terizidone (prodrug of cycloserin/J04AB01) [Terizidon] J04AK03		[25683-71-9] $C_{14}H_{14}N_4O_4$ 302.23	target and medical use: See cycloserin/second-line drug. cps. 250 mg		made by condensation of cycloserin with terephthaldialdehyde [107]

Bedaquiline
(TMC-207; R-207910)
[Sirturo]
J04AK05
WHO essential drug

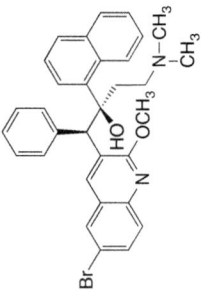

[843663-66-1]
$C_{32}H_{31}BrN_2O_2$
555.52
Fumarate:
[845533-86-0]
$C_{32}H_{31}BrN_2O_2 \cdot C_4H_4O_4$
671.59

inhibitor of mycobacterial F_1F_0-adenosine triphosphate (ATP) synthase/treatment of multi-drugresistant tuberculosis, second-line drug.
tbl. 100 mg.

USA/2012
EU/2014
Johnson & Johnson/Janssen/Tibotec

[108–112]

Delamanid
(OPC-67683)
[Deltyba]
J04AK06
WHO essential drug

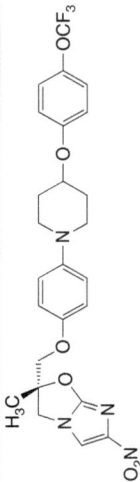

[681492-22-8]
$C_{25}H_{25}F_3N_4O_6$
534.48

inhibitor of mycolic acid biosynthesis/second-line drug to be used in combination with other TB drugs.
f.c.tbl. 50 mg.

EU/2014 (Orphan)
Otsuka

[113–115]

Table 10. J04BA Drugs for treatment of lepra (leprosy)

INN (Synonyms) [Brand names] ATC	Structure	[CAS-No.] Formula MW (g/mol)	Target (if known) Medical use Formulation DDD	Approval (country/year) Originator(s)	Production method References
Clofazimine (G30320; B-663) [Lamprene] J04BA01 **WHO essential medicine**		[2030-63-9] $C_{27}H_{22}Cl_2N_4$ 473.40	mechanism of action not quite clear; targets appear to be the outer membrane and the bacterial respiratory chain and ion transporters/treatment of lepromatous leprosy, including dapsone-resistant leprosy complicated by erythema nodosum leprosum in combination with dapsone and/or rifampicin; orphan status for use in multidrug-resistant strains of *Mycobacterium tuberculosis*. cps. 50, 100 mg. 100 mg O.	USA/1969 Geigy	[116–118]
Dapsone (DDS; Diaphenylsulfone; DADPS; 1358F) [Avlosulfon, Diphenasone, Disulone,] J04BA02 **WHO essential medicine**		[80-08-0] $C_{12}H_{12}N_2O_2S$ 248.30	target is dihydropteroate synthase and dihydrofolic acid synthesis is inhibited/treatment of leprosy in combination with rifampicin and clofazimine. tbl. 25, 50, 100 mg: gel 5%. 50 mg, O.	USA/1945 I.G. Farben; Wellcome Res.; Parke Davis	[119–121]

3.2. J04B Drugs for Treatment of Lepra (Leprosy)

The prevalence of leprosy has declined considerably by about 90% since 1985, thanks to the activities of the WHO. The treatment of this disease relies on multi-drug standard regimens with the use of rifampicin (J04AB02), clofazimine, and dapsone (Table 10).

Abbreviations

cps	capsules
DNA	deoxyribonucleic acid
f.c. tbl.	
HIV	human immunodeficiency virus
MDR	multidrug resistance
MDR-TB	multidrug-resistant TB
MTB	mycobacterium tuberculosis
O	oral application
P	parenteral application
TB	tuberculosis
tbl	tablet form
WHO	World Health Organization

References

1 Suerbaum, S. *et al.* (eds.) (2012) *Medizinische Mikrobiologie und Infektiologie*, 7th edn Springer, [G. Haase: Mykologie, p. 591-634 and Antimykotika, p. 753–762].

2 Kayser, F.H. *et al.* (eds.) (2010) *Taschenlehrbuch Medizinische Mikrobiologie*, 12 edn, Thieme, [F.H. Kayser and E.C. Böttger, Mykologie, p. 360–387].

3 Lartey, P.A. and Moehle, Ch.M. (1997) Recent advances in antifungal agents. *Annual Reports in Medicinal Chemistry*, **32**, 151–160.

4 Santo, R.Di. (2006) Recent developments in antifungal drug discovery. *Annual Reports in Medicinal Chemistry*, **41**, 299–315.

5 Ostrosky-Zeichner, L. *et al.* (2010) An insight into the antifungal pipeline: selected new molecules and beyond. *Nature Reviews/Drug Discovery*, **9**, 719–727.

6 Denning, D.W. and Bromley, M.J. (2015) How to bolster the antifungal pipeline. *Science*, **347**, 1414–1416.

7 Monk, B.C. and Goffeau, A. (2008) Outwitting multidrug resistance to antifungals. *Science*, **321**, 367–369.

8 Bell, A.S. (2007) Triazole Antifungals: Itraconazole, Fluconazole, Voriconazole, and Fosfluconazole in *The Art of Drug Synthesis* (ed. D.S. Johnson and J.J. Li), Wiley, p.71–82.

9 Olin Mathieson (1958) US 2 908 611, USA-prior. 1954.

10 Godefroi, E.F. *et al.* (1969) *J. Med. Chem.*, **12**, 784.

11 Janssen (1973) US 3 717 655, USA-prior. 1968.

12 Janssen (1974) US 3 839 574, USA-prior. 1969.

13 Heeres, J. *et al.* (1979) *J. Med. Chem.*, **22**, 1003.

14 Janssen (1979) US 4 144 346, USA-prior. 1977.

15 Janssen (1980) US 4 223 036, USA-prior. 1977.

16 Janssen (1982) US 4 335 125, USA-prior. 1977.

17 Pfizer (1983) US 4 404 216, GB-prior. 1981.

18 Pfizer (1983) US 4 416 682, GB-prior. 1980.

19 Meeres, J. *et al.* (1984) *J. Med. Chem.*, **27**, 894.

20 Janssen (1981) US 4 267 179, USA-prior. 1978.

21 Dickinson, R.P. *et al.* (1996) *Bioorg. Med. Chem. Lett.*, **6**, 2031.

22 Pfizer (1994) US 5 278 175, GB-prior. 1990.

23 Pfizer (1998) US 5 773 443, GB-prior. 1990.

24 Saksena, A.K. *et al.* (1996) *Tetrahedron Lett.*, **37**, 5657.

25 Schering-Plough (1997) US 5 661 151, USA-prior. 1994.

26 Basilea (2001) US 6 300 353, EP-prior. 1998, 1999.

27 DPLA, inventor MSoukup (2012) US 8 207 352, USA-prior. 08.10.2009.

28 Basilea (2004) US 6 812 238, USA-prior. 2000.

29 Basilea (2006) US 7 151 182, EP-prior. 1999.

30 Basilea (2008) US 7 459 561, EP-prior. 1999.

31 Denning, D.W. (2003) Echinocandin antifungal drugs. *The Lancet*, **362**, 1142–1151.

32 Duschinsky, R. *et al.* (1957) *J. Am. Chem. Soc.*, **79**, 4559.

33 Roche (1960) US 2 945 038, USA-prior. 1956.

34 Roche (1968) US 3 368 938, USA-prior. 1962, 1966.

35 Leonard, W.R. *et al.* (2007) *J. Org. Chem.*, **72**, 2335.

36 Merck & Co. (1995) US 5 378 804, USA-prior. 1993.

37 Merck & Co. (1999) US 5 936 062, USA-prior. 1997.

38 Merck & Co. (2007) US 9 214 768, USA-prior. 2002.

39 Tomishima, M. *et al.* (1999) *J. Antibiot.*, **52**, 674.

40 Fujisawa (2000) US 6 107 458, UK-prior. 1994, 1995.

41 Xellia Pharmaceut. (2011) US 9 132 163, USA-prior. 20.04.2011.

42 Debono, M. *et al.* (1995) *J. Med. Chem.*, **38**, 3271.

43 Eli Lilly (1999) US 5 932 543, USA-prior. 10.3.1992.

44 Eli Lilly (2005) US 6 916 784, L USA-prior. 19.03.1992.

45 Eli Lilly (2002) EP 561 639, USA-prior. 19.03.1992.

46 Vicuron (2007) US 7 198 796, USA-prior. 19.03.2002.

47 WHO "Global Tuberculosis Report 2015", 20th Ed, WHO Library Cataloguing-in-Publication Data; ISBN 978 92 4 156505 9.

48 Gumbo, T. (2011) Chemotherapy of Tuberculosis, *Mycobacterium Avium Complex Disease, and Leprosy*, in *Goodman & Gilman's The Pharmacological Basis of Therapeutics*, 12th edn (eds L.L. Brunton, B.A. Chabner, and B.C. Knollmann), McGraw-Hill, p. 1549–1570.

49 Russell, D.G. *et al.* (2010) Tuberculosis: What we don't know, can, and does, hurt us. *Science*, **328**, 852–856.

50 Dye, Chr. and Williams, B.G. (2010) The population dynamics and control of tuberculosis. *Science*, **328**, 856–861.

51 Nature Outlook (2013) "Tuberculosis", *Nature* **502** (7470) S1–S17.

52 Gutierrez-Lugo, M.-T. and Bewley, C. (2008) Natural products, small molecules, and genetics in tuberculosis drug development. *J. Med. Chem.*, **51**, 2606–2612.

53 Dover, L.G. and Coxon, G.D. (2011) Current status and research strategies in tuberculosis drug development. *J. Med. Chem.*, **54**, 6157–6165.

54 Jones, C.L. *et al.* (2015) Drug discovery for the developing world: progress at the Novartis Institute for Tropical Diseases. *Nature Reviews/Drug Discovery*, **14**, 442–444.

55 Katsuno, K. *et al.* (2015) Hit and lead criteria in drug discovery for infectious diseases of the developing world. *Nature Reviews/Drug Discovery*, **14**, 751–758.

56 Arinaminpathy, N. and Dowdy, D. (2015) Understanding the incremental value of novel diagnostic tests for tuberculosis" in *Nature* Special (Modelling the impact of improved diagnostics), Supplement to Nature **528** (7580), S60–S67.

57 Zumla, A. *et al.* (2013) Advances in the development of new tuberculosis drugs and treatment regimens. *Nature Reviews/Drug Discovery*, **12**, 388–404.

58 Lehmann, J. (1946) *Lancet*, **1**, 6384.

59 Sheehan, J.C. (1948) *J. Amer. Chem. Soc.*, **70**, 1665.

60 v. Heyden Nachf. (1890) US 427 564, DE-prior. 1889.

61 Parke Davis (1951) US 2 540 104, USA-prior. 1949.

62 Parke Davis (1953) US 2 640 854, USA-prior. 1950.

63 American Cyanamid (1953) US 2 644 011, USA-prior. 1950.

64 Monsanto (1953) US 2 658 073, USA-prior. 1950.

65 Miles Labs. (1958) US 2 844 625, USA-prior. 1954.

66 Kuehl, F.A. *et al.* (1955) *J. Am. Chem. Soc.*, **77**, 2344; 2345; 2346.

67 Plattner, P.A. *et al.* (1957) *Helv. Chim. Acta*, **40**, 1531.

68 Merck & Co. (1956) US 2 772 280, USA-prior. 1954.

69 Merck & Co. (1958) US 2 840 565, USA-prior. 1954.

70 Pfizer (1956) US 2 773 878, USA-prior. 1952.

71 Commercial Solvents (1957) US 2 789 983, USA-prior. 1954.

72 Merck & Co. (1958) US 2 845 433, USA-prior. 1955.

73 Sensi, P. *et al.* (1959) *Ed. Sci.*, **14**, 146.

74 Floss, H.G. and Yu, T.-W. (2005) Rifamycin – mode of action, resistance, and biosynthesis. *Chem. Rev*, **105**, 621–632.

75 Lepetit (1964) US 3 150 046, GB-prior. 1958.

76 Lepetit (1975) US 3 871 965, IT-prior. 1973.

77 Lepetit (1981) US 4 267 274, IT-prior. 1973.

78 Maggi, N. and Sensi, P. (1966) *Chemotherapia*, **11**, 285.

79 Lepetit (1967) US 3 342 810, GB-prior. 1964.

80 Holco Inv. (1979) US 4 174 320, GB-prior. 1977.

81 Archifar (1980) US 4 219 478, IT-prior. 1975.

82 Archifar (1977) US 4 017 481, IT-prior. 1975.

83 Archifar (1980) US 4 217 277, GB-prior. 1978.

84 Traxler, P. *et al.* (1990) *J. Med. Chem.*, **33**, 552.

85 Lepetit (1977) US 4 002 752, GB-prior. 1975.

86 Lepetit (1994) US 5 306 715, GB-prior. 1988.

87 Eli Lilly (1964) US 3 143 468, USA-prior. 1962.

88 Bristol-Myers (1977) US 4 026 766, USA-prior. 1976.

89 Roche (1952) US 2 596 069, USA-prior. 1952.

90 Distillers (1958) US 2 830 994, GB-prior. 1955.

91 Libermann, D. *et al.* (1956) *Hebd. Séances Acad. Sci.*, **242**, 2187, 2409.

92 Chimie et Atomistique Paris (1959) US 2 901 488, FR-prior. 1956.

93 Libermann, D. *et al.* (1958) *Bull. Soc. Chim. France*, 687, 694.

94 Libermann, D. *et al.* (1956) *Hebd. Séances Acad. Sci.*, **242**, 2187, 2409.

95 Chimie et Atomistique Paris (1959) US 2 901 488, FR-prior. 1956.

96 Libermann, D. *et al.* (1958) *Bull. Soc. Chim. France*, 687, 694.

97 Kushner, S. *et al.* (1949) *J. Lab. Clin. Med*, **33**, 1249.

98 Muschenheim, C. *et al.* (1954) *Amer. Rev. Tubercul.*, **70**, 743. **72**, 851 (1955).

99 American Cyanamid (1954) US 2 677 641, USA-prior 1952.

100 American Cyanamid (1957) US 2 780 624, USA-prior.1953.

101 Kushner, S. *et al.* (1952) *J. Am. Chem. Soc.*, **74**, 3617.

102 Hall, S.A. *et al.* (1940) *J. Am. Chem. Soc.*, **62**, 664.

103 Wilkinson, R.G. *et al.* (1961) *J. Am. Chem. Soc.*, **83**, 2212.

104 Wilkinson, R.G. *et al.* (1962) *J. Med. Chem.*, **5**, 835.

105 American Cyanamid (1965) US 3 176 040, USA-prior. 1960.

106 American Cyanamid (1971) US 3 553 257, USA-prior. 1966.

107 Bonati, F. *et al.* (1965) *Farmaco (Pavia) Ediz. prat.*, **20**, 381.

108 Andries, K. *et al.* (2005) *Science*, **307**, 223.

109 Janssen (2009) US 7 498 343, USA-prior. 25.07.2002.

110 Janssen (2010) EP 1 527 050, USA-prior. 25.07.2002.

111 Janssen (2011) US 8 039 628, EP-prior. 25.05.2005.

112 Janssen (2013) US 8 546 428, EP-prior. 05.12.2006.

113 Sasaki, H. *et al.* (2006) *J. Med. Chem.*, **49**, 7854.

114 Otsuka Pharm. (2013) EP 1 555 267 B1, JP-prior. 05.12.2002.

115 Cler, M.T. *et al.* (2012) *N. Engl. J. Med.*, **366**, 2151.

116 Barry, V.C. *et al.* (1957) *Nature*, **179**, 1013.

117 Geigy (1960) US 2 948 726, CH-prior. 1956.

118 Cholo, M.C. *et al.* (2011) *J. Antimicrob. Chemother.* doi: 10.1093/jac/dkr444.

119 I.G. Farben (1939) GB 506 227, appln. 1938.

120 American Cyanamid (1940) US 2 227 400, USA-prior. 1939.

121 Lundbeck (2009) US 7 531 694, PCT-prior. 07.07.2004.

Antiinfectives for Systemic Use, 3. Antivirals (J05)

Axel Kleemann, Hanau, Germany

1. Introduction

The general topic "Antiinfectives" is reviewed in this and two further articles (→ Antiinfectives for Systemic Use, 1. Antibacterials and → Antiinfectives for Systemic Use, 2. Antimycotics (Antifungals) and Antimycobacterials). For an introduction on the general topic Antiinfectives for Systemic Use, see → Antiinfectives for Systemic Use, 1. Antibacterials. All references cited in the headlines of the following chapters are general references, giving an in-depth introduction into the respective class of antivirals.

Since 2016, several new drugs entered the market, and some Anatomical Therapeutical Chemical classifications (ATC) changed. The current chapter also describes relatively new drugs for treatment of hepatitis C virus (HCV) infections under ATC code **J05AP**, which have formerly been part of the ATC code J05AX (Other Antivirals).

2. J05A Direct Acting Antivirals [1–4]

Since the early 1980s, when the dimension of the human immunodeficiency virus (HIV) and the acquired immunodeficiency syndrome (AIDS) became apparent, the search for effective antivirals became a main area in drug discovery. The discovery of aciclovir as an effective remedy against herpes simplex virus infections (HSV-1 and -2) by Gertrude Elion 1977 in the labs of Burroughs Wellcome as well as the development of follow-up drugs (nucleoside analogues of ATC group J05AB) with activity also against the cytomegalovirus (CMV) and the varicella zoster virus (VZV) encouraged some labs to look for antiretroviral drugs in order to combat HIV. Within short, the first clinical candidate out of Wellcome's collection of about 100 already available nucleoside analogues was identified with the nucleoside reverse transcriptase inhibitor (NRTI)

Ullmann's Pharmaceuticals. Edited by Axel Kleemann and Bernhard Kutscher.
© 2022 Wiley-VCH GmbH
Set ISBN: 978-3-527-34252-5/ DOI: 10.1002/14356007.c13_c01.pub4

zidovudine (azidothymidine; AZT), which was approved for treatment of HIV in 1987. Within the 1990s, a handful of follow-up drugs of this class of NRTIs were approved, and in parallel, the HIV protease inhibitors and somewhat later the nonnucleoside reverse transcriptase inhibitors (NNRTIs) were developed and approved. This made the development of combinations between the different classes possible in order to overcome resistance. Until today, HIV drugs of two other classes, namely viral entry or fusion inhibitors and integrase inhibitors, were added. With the available modern drug arsenal — 30 drugs in 5 different classes and effective combinations — HIV is no longer a death sentence as it had been in the first years after its identification; it cannot (yet) be healed, but it can be managed to be a chronic disease. And more new chemical entities (NCEs) and combinations are in advanced clinical studies and will reach the market in the second and third decades of the twenty-first century [5–12].

According to WHO, in 2014 a total of 36.9 million people were living with HIV, thereof 34.3 million adults, 17.4 million women, and 2.6 million children (<15 years). The number of newly infected people in 2014 was in total 2.0 million, thereof 1.8 million adults and 0.2 million children. And the AIDS deaths in 2014 amounted to 1.2 million in total, thereof 1.0 million adults and 0.15 million children. The new figures of WHO for the end of 2020 show that the situation did not change significantly: An estimated 37.7 million people were living with HIV, two-thirds of whom are living in the WHO African Region, and 680 000 people died from HIV-related causes, and an estimated 1.5 million people acquired HIV. Anyway, these facts show that HIV/AIDS is still a big issue.

In the following tables, 75 antiviral drug substances are listed, in addition >10 combination regimens. Quite a number of these drugs are on the WHO list of essential medicines, currently 27 mono substances and 13 combinations. Further antivirals are found in ATC classes D06BB (dermatological use) (→ Dermatologicals (D), 3. Antipsoriatics (D05), Antibiotics and Chemotherapeutics (D06), and Corticosteroids (D07) for Dermatological Use) and S01A (ophthalmological use).

Table 8 also contains the most modern effective hepatitis C (HCV) drugs which are on the market since 2014, after sofosbuvir paved the way for HCV cure. These drugs are able to eradicate this infection with healing rates of 90–100% within 12 weeks — this being a real breakthrough [13–17]. According to WHO, worldwide about 160 million people are chronically infected. In 2015, worldwide sales were dominated by just three new drugs, namely Harvoni ($13.86 10^9/Gilead), Sovaldi ($5.27 10^9/Gilead), and Viekira Pak ($1.639 10^9/AbbVie). These three drugs, which are all included on the WHO essential medicines list, stand for about 50% of the 2015 worldwide sales of antiviral agents. The other part with somewhat <50% is assigned to HIV/AIDS drugs, whereas the residual sales are contributed to aciclovir and follow-up drugs of group J05AB (with expired patents). The total sales of antiviral drugs in 2015 amounted to almost 45×10^9.

2.1. J05AB Nucleosides and Nucleotides Excluding Reverse Transcriptase Inhibitors (RTIs, Table 1) [46, 47]

Table 1. J05AB Nucleosides and nucleotides excl. reverse transcriptase inhibitors [46, 47]

INN (Synonyms) [Brand names] ATC	Structure	[CAS-No.] Formula MW, g/mol	Target (if known) Medical use Formulation DDD	Approval (country/year) Originator(s) Production method References
Aciclovir (acyclovir; acycloguanosine; BW-248U; Wellcome 248U) [Activir, Cycloviran, Zoliparin, Zovirax] J05AB01 and D06BB03; S01AD03. (→ Dermatologicals (D), 3. Antipsoriatics (D05), Antibiotics and Chemotherapeutics (D06), and Corticosteroids (D07) for Dermatological Use) **WHO essential medicine**		[59277-89-3] $C_8H_{11}N_5O_3$ 225.21 sodium salt: [69657-51-8] $C_8H_{10}N_5NaO_3$ 248.20	aciclovir is phosphorylated to its monophosphate by viral thymidine kinase and subsequently by host cell kinases to its triphosphate, which inhibits viral DNA synthesis. Treatment of herpes simplex virus (HSV) infections, chickenpox, and shingles. tbl. 200, 400, 800 mg; amp. 125, 250 mg; cream 5%; eye ointment 3%. 4 g, O, P.	USA/1983 [18, 19] DE/1983 Burroughs Wellcome
Idoxuridine (IdU; IdUR; IdUrd; 2'-deoxy-5-iodouridine) [Herplex; Herpid; Virunguent] J05AB02 and D06BB01; S01AD01 (→ Dermatologicals (D), 3. Antipsoriatics (D05), Antibiotics and Chemotherapeutics (D06), and Corticosteroids (D07) for Dermatological Use) (→ Sensory Organs – Ophthalmologicals (S01))		[54-42-2] $C_9H_{11}IN_2O_5$ 354.10	antimetabolite nucleoside, blocks base pairing in viral DNA replication. Only used in topical form as ophthalmic to treat herpes simplex keratitis; withdrawn in many countries. eye drops 0.1%; ointment 0.25, 0.5%.	USA/1962 [20, 21] Roussel-Uclaf

(continued overleaf)

Table 1. (*Continued*)

INN (Synonyms) [Brand names] ATC	Structure	[CAS-No.] Formula MW, g/mol	Target (if known) Medical use Formulation DDD	Approval (country/year) Originator(s) Production method References
Vidarabine (ara-A; CI-673; adenine arabinoside) [Arasena-A; Vira-A] J05AB03 and S01AD06 (→ Sensory Organs – Ophthalmologicals (S01))		[*5536-17-4*] $C_{10}H_{13}N_5O_4$ 267.25	antimetabolite nucleoside, inhibits viral DNA synthesis. Originally developed as anticancer, the antiviral activity was found in 1964 and it was first drug for treatment of HSV infections; by 2016 it is discontinued in most countries and only limited use in ophthalmics. cream and ointment 3%.	USA/1977 Parke Davis; Pfizer	besides by total synthesis it can also be made by fermentation of *Streptomyces antibioticus* NRRL 3238 [22–24]
Ribavirin (guanosine analogue; RTCA; tribavirin; ICN-1229) [Copegus, Rebetol, Viramid, Virazide, Virazole] J05AB04 **WHO essential medicine**		[*36791-04-5*] $C_8H_{12}N_4O_5$ 244.21	guanosine antagonist, blocks DNA and RNA synthesis. Treatment of hepatitis C in combination with peginterferon-α-*2a* or -*2b* or interferon-α-*2a* or -*2b* and of respiratory syncytial virus (RSV) infection. f.c. tbl. 200, 400 mg; cps. 200 mg; powder for inhalation solution 6 g. 1 g O.	USA/1973 (?) ICN	[25–27]
Ganciclovir (analogue of 2′-deoxyguanosine; DHPG; 2′NDG; BIOLF-62; BW-B759U; BW-759; RS-21592) [Cymevan, Cymevene, Cytovene, Denosine, Virgan, Vitrasert, Zirgan] J05AB06 and S01AD09 (→ Sensory Organs – Ophthalmologicals (S01))		[*82410-32-0*] $C_9H_{13}N_5O_4$ 255.23 sodium salt: [*107910-75-8*] $C_9H_{12}N_5NaO_4$ 277.22	competitive inhibitor of deoxyguanosine, blocks viral DNA polymerases. Treatment of cytomegalovirus infections and herpetic keratitis. cps. 250, 500 mg; vials 500 mg for i.v. inj.; ophthalmic gel 0.15%; vitrasert implants 4.5 mg. 3 g O, 0.5 g P.	USA/1988 Syntex	[28–31]

Table 1. (*Continued*)

INN (Synonyms) [Brand names] ATC	Structure	[CAS-No.] Formula MW, g/mol	Target (if known) Medical use Formulation DDD	Approval (country/year) Originator(s) Production method References
Famciclovir (BRL-42810; FCV) [Famvir, Oravir] J05AB09 and S01AD07 (→ Sensory Organs – Oph- thalmologicals (S01))		[*104227-87-4*] $C_{14}H_{19}N_5O_4$ 321.34	diacetylester prodrug of 6-deoxy penciclovir. Active against herpes simplex virus (HSV) and varicellazoster virus (VZV; shingles). f.c. tbl. 125, 250, 500 mg. 1.5 g O.	USA/1994 DE/1996 Beecham
Valaciclovir (BW-256U; Vala-ACV) [Valtrex, Zelitrex] J05AB11		[*124832-26-4*] $C_{13}H_{20}N_6O_4$ 324.34 hydrochloride: [*124832-27-5*] $C_{13}H_{20}N_6O_4$ ·HCl 360.80	prodrug of aciclovir with improved bioavailability. Treatment of herpes simplex and herpes zoster infections. f.c. tbl. 250, 500 and 1000 mg. 3 g O.	USA/1995 DE/1996 Burroughs Wellcome
Cidofovir (HPMPC; GS-504) [Vistide] J05AB12		[*113852-37-2*] $C_8H_{14}N_3O_6P$ 279.19 dihydrate: [*149394-66-1*] $C_8H_{14}N_3$-O_6P ·2 H$_2$O 315.22	selective inhibitor of viral DNA polymerase. Treatment of cytomegalovirus (CMV) retinitis in AIDS patients. vials 5 mL (75 mg/mL) for i,v, inj. (5 mg/kg).	USA/1996 EMA/1997 Ceskoslovenska Akad. Ved.; Gilead
Penciclovir (BRL-39123; PCV) [Denavir, Vectavir] J05AB13 and D06BB06 (→ Dermatologicals (D), 3. Antipsoriatics (D05), Antibiotics and Chemotherapeu- tics (D06), and Corticosteroids (D07) for Dermatological Use)		[*39809-25-1*] $C_{10}H_{15}N_5O_3$ 253.26	guanosine analogue, inhibits viral polymerase. Topical treatment of herpes simplex (HSV) infections. cream 1%.	USA/1996 Beecham

(continued overleaf)

Table 1. (*Continued*)

INN (Synonyms) [Brand names] ATC	Structure	[CAS-No.] Formula MW, g/mol	Target (if known) Medical use Formulation DDD	Approval (country/year) Originator(s) Production method References
Valganciclovir (Ro-107-9070/194; RS-79070-194) [Cymeval, Darilin, Rovalcyte, Valcyte] J05AB14		[175865-60-8] $C_{14}H_{22}N_6O_5$ 354.37 hydrochloride: [175865-59-5] $C_{14}H_{22}N_6O_5$ ·HCl 390.83	prodrug of ganciclovir. Treatment of CMV infections. f.c. tbl. 450 g; vials with 5.5 g of hydrochloride for oral solution. 0.9 g O.	USA/2001 [40, 41] Roche; Syntex
Brivudine (BVDU) [Brivex, Brivirac, Nervinex, Zecovir, Zostex] J05AB15		[69304-47-8] $C_{11}H_{13}BrN_2O_5$ 333.14	analogue of thymidine, blocks viral DNA polymerase. Treatment of herpes zoster (VZV). tbl. 125 mg. 125 mg O.	Univ. [42] of Birmingham; Berlin-Chemie
Remdesivir (GS-5734; prodrug of GS-441524) [Veklury] J05AB16		[1809249-37-3] $C_{27}H_{35}N_6O_8P$ 602.59	broad-spectrum antiviral for treatment of ebola and since 2020 emergency use for postinfection treatment of COVID-19. vials 100 mg powder for infusion sol. 0.1 g P.	Gilead Sciences USA/2016; EU/2020 synthesis [43–45]

2.2. J05AC Cyclic Amines (Table 2)

Table 2. J05AC Cyclic amines

INN (Synonyms) [Brand names] ATC	Structure	[CAS-No.] Formula MW, g/mol	Target (if known) Medical use Formulation DDD	Approval (country/year) Originator(s) Production method References
Rimantadine (EXP-126) [Flumadine, Roflual] J05AC02		[13392-28-4] $C_{12}H_{21}N$ 179.31 hydrochloride: [1501-84-4] $C_{12}H_{21}N$·HCl 215.77	mechanism of action not fully clear, possibly the uncoating of the virus is inhibited. Treatment of infections by influenza A virus. f.c. tbl. 100 mg.	USA/1994 [48–50] DuPont

Table 2. (*Continued*)

INN (Synonyms) [Brand names] ATC	Structure	[CAS-No.] Formula MW, g/mol	Target (if known) Medical use Formulation DDD	Approval (country/year) Originator(s) Production method References
Tromantadine (D-41) [Viru-Merz, Viruserol] J05AC03 and D06BB02 (→ Dermatologicals (D), 3. Antipsoriatics (D05), Antibiotics and Chemotherapeutics (D06), and Corticosteroids (D07) for Dermatological Use)		[*53783-83-8*] $C_{16}H_{28}N_2O_2$ 280.41 hydrochloride: [*41544-24-5*] $C_{16}H_{28}N_2O_2 \cdot HCl$ 316.87	inhibits early and late steps in herpes simplex virus replication cycle and inhibits penetration of the virus. Topical treatment of HSV infections. ointment 1%.	Merz [51, 52]
Amantadine (EXP-105-1; NSC-83653) [Amixx, Mantadan, Mantadix, PK-Merz, Symmetrel] J05AC04 (old); new: N04BB01		[*768-94-5*] $C_{10}H_{17}N$ 151.25 hydrochloride: [*665-66-7*] $C_{10}H_{17}N \cdot HCl$ 187.71 sulfate: [*31377-23-8*] $C_{10}H_{17}N \cdot$ 1/2 H_2SO_4 400.58	antiviral activity by interference with the viral protein M2; antiparkinson activity by dopaminergic, noradrenergic and serotonergic activity and blocking NMDA receptors. Originally in use as prophylactic agent against Asian influenza, since 1969 (in USA) only for treatment of Parkinson's disease. f.c. tbl. 100, 150, 200 mg; vials 200 mg/500 mL for infusion. 200 mg O.	USA/1966 [53, 54] DuPont/Endo; Merz & Co.

2.3. J05AD Phosphonic Acid Derivatives (Table 3)

Table 3.   J05AD Phosphonic acid derivatives

INN (Synonyms) [Brand names] ATC	Structure	[CAS-No.] Formula MW, g/mol	Target (if known) Medical use Formulation DDD	Approval (country/year) Originator(s) Production method References
Foscarnet sodium (A-29622) [Foscavir, Triapten] J05AD01		[*63585-09-1*] CNa_3O_5P 191.95 hexahydrate: [*34156-56-4*] $CNa_3O_5P \cdot 6 H_2O$ 300.04	the phosphonoformic acid foscarnet acts at the pyrophosphate binding site of the viral DNA polymerase and prevents chain elongation. Second-line treatment for resistant herpes virus infections; treatment also of CMV retinitis in AIDS patients. vials with 6 g for i.v. infusion; cream 2%. 6.5 g P.	USA/1989 [55–57] Astra AB

2.4. J05AE Protease Inhibitors [58–61] (Table 4)

Table 4.   J05AE Protease inhibitors [58, 60, 61]

INN (Synonyms) [Brand names] ATC	Structure	[CAS-No.] Formula MW, g/mol	Target (if known) Medical use Formulation DDD	Approval (country/year) Originator(s) Production method References
Saquinavir (Ro-31-8959, Ro-31-8959-003) [Invirase, Fortovase] J05AE01 **WHO essential medicine**		[127779-20-8] $C_{38}H_{50}N_6O_5$ 670.86 methane-sulfonate: [149845-06-7] $C_{38}H_{50}N_6O_5 \cdot CH_4O_3S$ 766.96	inhibitor of HIV-1- and -2-proteases. First protease inhibitor for HIV on market; antiretroviral treatment of HIV-1 infection in combination with ritonavir or other antiretroviral drugs. f.c. tbl. 500 mg.	USA/1995 EU/1995 Roche [62, 63]
Indinavir (MK-639) [Crixivan] J05AE02 **WHO essential medicine**		[150378-17-9] $C_{36}H_{47}N_5O_4$ 613.80 sulfate: [157810-81-6] $C_{36}H_{47}N_5O_4 \cdot H_2SO_4$ 711.88	HIV-protease inhibitor. Treatment of HIV-1 infection in combination with other antiretroviral drugs (e.g., NRTI and NNRTI). cps 200, 400 mg (sulfate) 2.4 g O.	USA/1996 EU/1996 Merck & Co. [64, 65]
Ritonavir (A-84538; ABT-538) [Norvir, Kaletra] J05AE03 **WHO essential medicine**		[155213-67-5] $C_{37}H_{48}N_6O_5S_2$ 720.96	the drug inhibits not only HIV protease, but also cytochrome P450-3A4 (CYP3A4), which metabolizes other protease inhibitors and so boosting other protease inhibitors. This is reason for using ritonavir together with other drugs like lopinavir (in kaletra) in relatively low doses. Antiretroviral treatment of HIV-1 infections. f.c. tbl. 100 mg: solution 600 mg/7.5 mL. 1.2 g O.	USA/1996 Abbott [66-69]

Nelfinavir
(AG-1346;
AG-1343)
[Viracept]
J05AE04

[159989-64-7]
$C_{32}H_{45}N_3O_4S$
567.79
methansulfonate:
[159989-65-8]
$C_{32}H_{45}N_3O_4S \cdot CH_4O_3S$
663.90

HIV protease inhibitor.
Treatment of HIV-1 infections
in combination with other
retroviral drugs—discontinued
by Roche worldwide since 2013.
tbl. 250 mg.
2.25 g O.

USA/1997
EU/1998
Agouron

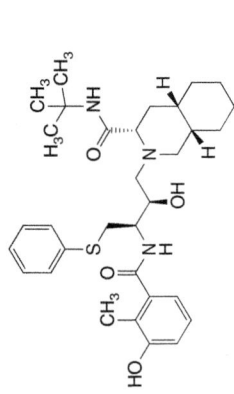

[70, 71]

Amprenavir
(141W94;
KVX-478;
VX-478)
[Agenerase,
Prozei]
J05AE05

[161814-49-9]
$C_{25}H_{35}N_3O_6S$
505.63

HIV-1 protease inhibitor. was
used for treatment of HIV, but
discontinued since end of 2004
in favor of its prodrug
fosamprenavir (see J
05AE07). cps.
50, 150 mg;
solution for oral use 15 mg/mL.

USA/1999
EU/2000
Vertex

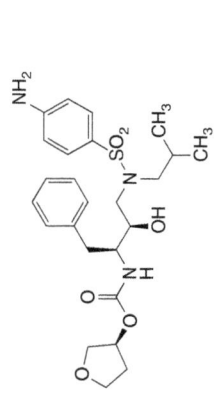

[72]

Lopinavir
(ABT-378;
A-157378)
[kaletra(comb.
with ritonavir)]
J05AE06: combo:
J05AR10
**WHO essential
medicine**
(combination
lopinavir+
ritonavir)

[192725-17-0]
$C_{37}H_{48}N_4O_5$
628.81

HIV-1 protease inhibitor.
Treatment of HIV infections
in combination with ritonavir or
other antiretroviral drugs.
f.c. tbl. lopinavir
100 mg/ritonavir 25 mg
and 200/50 mg: solution (oral
use) 5 mL with 400 mg
lopinavir+100 mg ritonavir.
0.8 g O.

USA/2000
EU/2001
Abbott

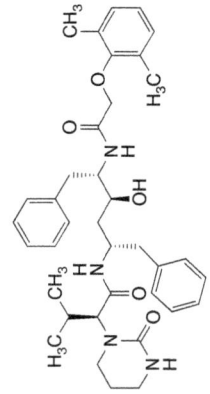

[73–75]

Fosamprenavir
(VX-175;
GW-433908)
[Lexiva, Telzir]
J05AE07

[226700-79-4]
$C_{25}H_{36}N_3-O_9PS$
585.61
calcium salt:
[226700-81-8]
$C_{25}H_{34}CaN_3-O_9PS$
623.67

HIV protease inhibitor: prodrug
of amprenavir. Treatment
of HIV-1 infections, in general
in combination with other
retroviral drugs. e.g., ritonavir.
f.c. tbl. 700 mg (Ca salt); oral
suspension 50 mg/mL.
1.4 g O.

USA/2003
EU/2004
Vertex;
Glaxo

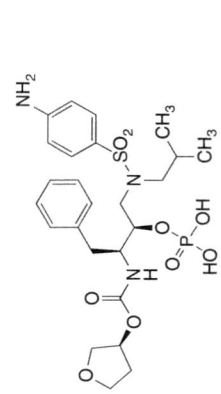

[76, 77]

(continued overleaf)

Table 4. (*Continued*)

INN (Synonyms) [Brand names] ATC	Structure	[CAS-No.] Formula MW, g/mol	Target (if known) Medical use Formulation DDD	Approval (country/year) Originator(s) Production method References
Atazanavir (BMS-232632; CGP-73547) [Reyataz] J05AE08 **WHO essential medicine**		[198904-31-3] $C_{38}H_{52}N_6O_7$ 704.87 sulfate: [229975-97-7] $C_{25}H_{34}N_6O_7 \cdot H_2SO_4$ 802.94	HIV-protease inhibitor. Treatment of HIV-1 infections, also in combination with other retroviral drugs; it can be applied once-daily. cps. 150, 200, 300 mg. 300 mg O.	USA/2003 EU/2004 Novartis Bristol-Myers Squibb [78–80]
Tipranavir (PNU-140690) [Aptivus] J05AE09		[174484-41-4] $C_{31}H_{33}F_3N_2O_5S$ 602.67	HIV protease inhibitor. Treatment of HIV-1 infections with coadministration of ritonavir; active against viruses that are resistant to other protease inhibitors. cps. 250 mg; oral solution 100 mg/1 mL. 1 g O; children 640 mg O.	USA/2005 EU/2005 Pharmacia & Upjohn; Boehringer Ing. [81–84]
Darunavir (TMC-114; UIC-94017) [Prezista] J05AE10 **WHO essential medicine**		[206361-99-1] $C_{27}H_{37}N_3O_7S$ 547.67 ethanolate: [635728-49-3] $C_{27}H_{37}N_3O_7S \cdot C_2H_5OH$ 593.74	second-generation HIV protease inhibitor. Treatment of HIV-1 infections in coadministration with ritonavir, even in patients with multiple resistance mutations. f.c. tbl. 75, 150, 400, 600, 800 mg; oral suspension 100 mg/mL; all forms contain ethanolate. 1.2 g O.	USA/2006 EU/2007 Tibotec [85–87]

Drug	Structure	CAS / Formula	Description	Approval	Ref.
Telaprevir (VX-950; VRT-111950; LY-570310) [Incivek, Incivo] J05AE11	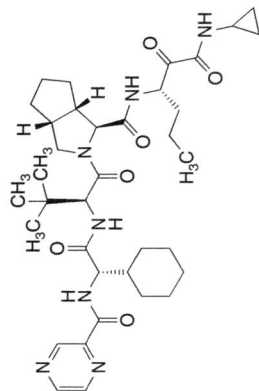	[402957-28-2] $C_{36}H_{53}N_7O_6$ 679.86	inhibitor of hepatitis C viral enzyme NS3.4A serine protease. Treatment of hepatitis C genotype 1 viral infections in combination with pegylated interferon and ribavirin; since August 2014, incivek was discontinued on the market because of heavy competition from more effective drugs.	USA/2011 EU/2011 Vertex; Johnson & Johnson	[88]
Boceprevir (Sch-503034) [Victrelis] J05AE12	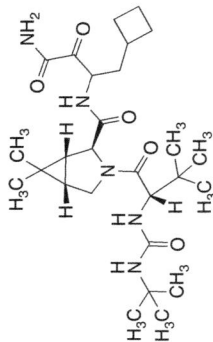	[394730-60-0] $C_{27}H_{45}N_5O_5$ 519.69	inhibitor of hepatitis C viral enzyme NS3.4A serine protease. Treatment of hepatitis C genotype 1 viral infections in combination with pegylated interferon and ribavirin; since December 2015 victrelis is discontinued on the market because of heavier competition with more effective drugs (e.g., sofosbuvir).	USA/2011 EU/2011 Schering–Plough; Merck & Co.	[89–91]
Simeprevir (TMC-435; TMC-435350) [Olysio, SOVRIAD] J05AE14 **WHO essential medicine**	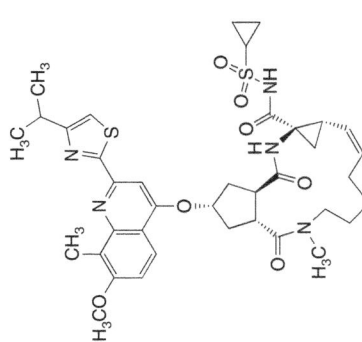	[923604-59-5] $C_{38}H_{47}N_5O_7S_2$ 749.9	inhibitor of hepatitis C virus enzyme NS3.4A serine protease. Treatment of hepatitis C genotype 1 viral infections in combination with peginterferon-alfa and ribavirin. cps. 150 mg. 150 mg O.	USA/2013 EU/2014 JP/2013 Medivir; Janssen Pharmaceutica	[92, 93]

2.5. J05AF Nucleoside and Nucleotide Reverse Transcriptase Inhibitors (NRTIs, Table 5)

Table 5.   J05AF Nucleoside and nucleotide reverse transcriptase inhibitors (NRTIs)

INN (Synonyms) [Brand names] ATC	Structure	[CAS-No.] Formula MW, g/mol	Target (if known) Medical use Formulation DDD	Approval (country/year) Originator(s) Production method References
Zidovudine (azidothymidine; AZT; BW-A509U; 3′-azido-3′-deoxythymidine) [Retrovir] J05AF01 **WHO essential medicine**		[30516-87-1] $C_{10}H_{13}N_5O_4$ 267.25	nucleoside analogue reverse transcriptase inhibitor (NRTI), preventing viral DNA synthesis. First AIDS therapeutic at all; HIV are susceptible to become resistant to AZT, therefore it is usually used in combination with other anti-HIV drugs (e.g., combivir, trizivir). cps. 100, 250 mg; oral solution 10%; vial 200 mg/20 mL for infusion. 600 mg O, P.	USA/1987 made from EU/1987 thymidine Burroughs [94–97] Wellcome
Didanosine (dideoxyinosine; ddI; DDI; 2′, 3′-dideoxyinosine; BMY-40900; NSC-612049) [Videx EC] J05AF02 **WHO essential medicine**		[69655-05-6] $C_{10}H_{12}N_4O_3$ 236.23	NRTI; nucleoside analogue of adenosine; slower development of resistance in comparison to zidovudine. Treatment of HIV/AIDS in combination with other drugs as part of HAART. enteric-coated cps.125, 200, 250, 400 mg; vials with powder 2 g for oral solution. 400 mg O.	USA/1991 made from EU/1992 dideoxya-Bristol–Myers denosine Squibb [98–101]
Zalcitabine (Dideoxycytidine; ddC; Ro-24-2027) [Hivid] J05AF03		[7481-89-2] $C_9H_{13}N_3O_3$ 211.22	NRTI for treatment of HIV/AIDS in combination with zidovudine. discontinued in 2006. f.c. tbl. 0.375, 0.750 mg. 2.25 mg O.	USA/1992 [102, 103] NCI; Roche
Stavudine (d4T; D4T; BMY-27857) [Zerit] J05AF04 **WHO essential medicine**		[3056-17-5] $C_{10}H_{12}N_2O_4$ 224.22	NRTI for treatment of HIV/AIDS in combination with other antiretroviral drugs. cps. 20, 30, 40 mg; vials with powder for oral solution 200 mg. 80 mg O.	USA/1994 [104, 105] Bristol–Myers Squibb

Table 5. (*Continued*)

INN (Synonyms) [Brand names] ATC	Structure	[CAS-No.] Formula MW, g/mol	Target (if known) Medical use Formulation DDD	Approval (country/year) Originator(s) Production method References
Lamivudine (3TC; BCH-189; GR-109714X); [Epivir, Zefix; comb.: Combivir, Kivexa, Triumeq, Trizivir, Epzicom] J05AF05 **WHO essential medicine**		[*134678-17-4*] $C_8H_{11}N_3O_3S$ 229.25	NRTI for treatment of hepatitis B and HIV/AIDS preferably in combination with other drugs. f.c. tbl. 100 mg; oral solution 5 mg/mL. 300 mg O.	USA/1995 [106, 107] IAF BioChem International; Glaxo SmithKline
Abacavir (1592U89) [Ziagen; comb.: Kivexa, Trizivir, Triumeq] J05AF06 **WHO essential medicine**		[*136470-78-5*] $C_{14}H_{18}N_6O$ 286.34 sulfate: [*188062-50-2*] $(C_{14}H_{18}N_6O)_2$ $\cdot H_2SO_4$ 670.75	NRTI for treatment of HIV/AIDS in combination with other antiretroviral agents. f.c. tbl. 300 mg; oral solution 20 mg/mL. 600 mg O.	USA/1998 [108–110] EU/1999 Burroughs Wellcome; Glaxo
Tenofovir disoproxil (Tenofovir DF; (*R*)-bis(POC)PMPA; GS-4331-05) [Viread; combinations: Atripla, Eviplera, Stribild, Truvada, Complera] J05AF07 **WHO essential medicine**		[*201341-05-1*] $C_{19}H_{30}N_5O_{10}P$ 519.45 fumarate: [*202138-50-9*] $C_{19}H_{30}N_5O_{10}$-P $\cdot C_4H_4O_4$ 635.52 Tenofovir: [*147127-20-6*] $C_9H_{14}N_5O_4P$ 287.22	NRTI for treatment of chronic hepatitis B and HIV/AIDS in combination with other antiretroviral drugs; tenofovir disoproxil is prodrug of the active molecule tenofovir. f.c. tbl. 123, 163, 204, 245 mg (fumarate). 245 mg O.	USA/2001 [111, 112] Gilead
Adefovir dipivoxil (bis-POM PMEA; GS-840) [Hepsera] J05AF08		[*142340-99-6*] $C_{20}H_{32}N_5O_8P$ 501.48 adefovir: $C_8H_{12}N_5O_4P$ 273.19	NRTI for treatment of hepatitis B; adefovir dipivoxil is prodrug of the active substance adefovir. tbl. 10 mg. 10 mg O.	USA/2002 [113–116] EU/2003 Ceskoslo- venska akad. vet; Gilead
Emtricitabine (coviracil; (−)-FTC; BW-524W91) [Emtriva; combin.: Atripla, Complera, Eviplera, Stribild, Truvada] J05AF09 **WHO essential medicine**		[*143491-57-0*] $C_8H_{10}FN_3O_3S$ 247.24	NRTI for treatment of HIV/AIDS infection in combination with other antiretroviral drugs, e.g., in the fixed-dose combinations atripla, stribild, truvada. cps. 200 mg; oral solution 10 mg/mL. 200 mg O.	USA/2003 [117–122] EU/2003 Emory University; Triangle Pharmaceu- ticals; Gilead Sciences

(*continued overleaf*)

Table 5. (*Continued*)

INN (Synonyms) [Brand names] ATC	Structure	[CAS-No.] Formula MW, g/mol	Target (if known) Medical use Formulation DDD	Approval (country/year) Originator(s) Production method References
Entecavir (BMS-200475; SQ-34676) [Baraclude] J05AF10 **WHO essential medicine**		[142217-69-4] $C_{12}H_{15}N_5O_3$ 277.28 monohydrate: [209216-23-9] $C_{12}H_{15}N_5O_3$ ·H_2O	NRTI for treatment of chronic hepatitis B (HBV) infection; deoxyguanosine analogue. Not for use in HIV. f.c. tbl. 0.5, 1.0 mg; oral solution 0.05 mg/mL. 0.5 mg O.	USA/2005 [123–126] EU/2006 Bristol–Myers Squibb (BMS)
Telbivudine (L-dT; L-deoxythymidine) [Sebivo, Tyzeka] J05AF11		[3424-98-4] $C_{10}H_{14}N_2O_5$ 242.23	NRTI for treatment of chronic HBV infection; the active form is the triphosphate, which is formed by kinases of the host cells and acts as antagonist of thymidine-5′-triphosphate. tbl. 600 mg. 0.6 g O.	EU/2007 [127–129] Idenix; Novartis
Tenofovir alafenamide (GS-7340; TAF; prodrug of tenofovir) [Descovy, Genvoya, Odefsey, Vemlidy] J05AF13		[379270-37-8] $C_{21}H_{29}N_6O_5P$ 476.5 fumarate: [1392275-56-7] $C_{21}H_{29}N_6O_5P$ ·$C_4H_4O_4$ 534.5	reverse transcriptase inhibitor; prodrug of tenofovir. treatment of chronic hepatitis B (HBV); component of several fixed dose combinations for treatment of HIV infections (see J05AR17, J05AR18, J05AR19, J05AR20, and J05AR22). f.c. tabl. 10 mg, 25 mg. 25 mg O.	Gilead [130–132] Sciences USA/2015; EU/2015 synthesis

2.6. J05AG Nonnucleoside Reverse Transcriptase Inhibitors (NNRTIs, Table 6)

Table 6.   J05AG Nonnucleoside reverse transcriptase inhibitors (NNRTIs)

INN (Synonyms) [Brand names] ATC	Structure	[CAS-No.] Formula MW, g/mol	Target (if known) Medical use Formulation DDD	Approval (country/year) Originator(s) Production method References
Nevirapine (BI-RG-587; NVP) [Viramune] J05AG01 **WHO essential medicine**		[129618-40-2] $C_{15}H_{14}N_4O$ 266.30	NNRTI for treatment of HIV-1 infection, optionally in combination with other antiretroviral agents. s.r. tbl. 100, 400 mg; tbl. 200 mg; oral suspension 50 mg/5 mL. 0.4 g O; children 0.3 g O.	USA/1996 EU/1997 Boehringer Ing. [133, 134]
Delavirdine (U-90152S) [Rescriptor] J05AG02		[136817-59-9] $C_{22}H_{28}N_6O_3S$ 456.57 methane-sulfonate: [147221-93-0] $C_{22}H_{28}N_6O_3S$ ·CH_4O_3S 552.67	NNRTI for treatment of HIV-1 infection as second-line therapy and in combination with other antiretroviral agents. tbl. 100, 200 mg (as mesylate).	USA/1997 Upjohn [135, 136]
Efavirenz (DMP-266; L-743726) [Stocrin, Sustiva] J05AG03 **WHO essential medicine**		[154598-52-4] $C_{14}H_9ClF_3NO_2$ 315.68	NNRTI for treatment of HIV-1 infection in combination with other antiretroviral agents as HAART (e.g., atripla, a combination of efavirenz with emtricitabine and tenofovirdiso-proxil). cps. 50, 100, 200 mg; f.c.tbl. 600 mg; oral solution 30 mg/mL. 0.6 g O.	USA/1998 EU/1999 Merck & Co.; Bristol-Myers Squibb [137–141]
Etravirine (R-165335; TMC-125) [Intelence] J05AG04		[269055-15-4] $C_{20}H_{15}BrN_6O$ 435.29	NNRTI for treatment of HIV-1 infection in combination with other antiretroviral agents. tbl. 25, 100, 200 mg. 0.4 g O.	USA/2008 EU/2008 Janssen (JnJ) [142, 143]

(continued overleaf)

Table 6. (*Continued*)

INN (Synonyms) [Brand names] ATC	Structure	[CAS-No.] Formula MW, g/mol	Target (if known) Medical use Formulation DDD	Approval (country/year) Originator(s) Production method References
Rilpivirine (TMC-278; R-278474) [Edurant; combin.: complera, eviplera] J05AG05		[500287-72-9] $C_{22}H_{18}N_6$ 366.43 hydrochloride: [700361-47-3] $C_{22}H_{18}N_6 \cdot HCl$ 402.89	NNRTI for treatment of HIV-1 infection in combination with other antiretroviral agents (e.g., emtricitabine and tenofovir as in complera and eviplera). f.c.tbl. 25 mg. 25 mg O.	USA/2011 [144–146] EU/2011 Tibotec; Janssen (JnJ)
Doravirine (MK-1439) [Pifeltro, Delstrigo) J05AG06		[1338225-97-0] $C_{17}H_{11}ClF_3N_5O_3$ 425.7	nonnucleoside reverse transcriptase inhibitor for treatment of HIV infection alone and in comb. with lamivudine (300 mg) and tenofovir disoproxil fumarate (300 mg). tabl. 100 mg. 0.1 g O.	Merck & Co. [147–150] USA/2018; EU/2018 synthesis

2.7. J05AH Neuraminidase Inhibitors [151, 152] (Table 7)

Table 7. J05AH Neuraminidase inhibitors [151, 152]

INN (Synonyms) [Brand names] ATC	Structure	[CAS-No.] Formula MW, g/mol	Target (if known) Medical use Formulation DDD	Approval (country/year) Originator(s) Production method References
Zanamivir (GG-167; GR-121167X) [Relenza] J05AH01		[139110-80-8] $C_{12}H_{20}N_4O_7$ 332.31	neuraminidase inhibitor for treatment and prophylaxis of influenza from influenza A and B viruses. Blister with 5 mg for inhalation with diskhaler.	USA/1999 made from EU/1999 N-acetylneura- CSIRO/Biota minic acid (both [153, 154] Australia); Glaxo
Oseltamivir (GS-4071; GS-4104) [Tamiflu] J05AH02 **WHO essential medicine**		[196618-13-0] $C_{16}H_{28}N_2O_4$ 312.41 phosphate: [204255-11-8] $C_{16}H_{28}N_2O_4$ $\cdot H_3PO_4$ 410.40	neuraminidase inhibitor for treatment and prophylaxis of influenza (A and B viruses). cps. 30, 45, 75 mg; oral suspension 6 mg/mL. 150 mg O.	USA/1999 semisynthetic EU/2002 from Gilead; (—)-shikimic Roche acid [155]
Peramivir (BCX-1812; RWJ-270201) [Rapivab, Rapiacta] J05AH03		[229614-55-5] $C_{15}H_{28}N_4O_4$ 328.4	neuraminidase inhibitor for treatment of influenza infection. vials 200 mg/20 mL for i.v. inj. 0.6 g P.	BioCryst synthesis Pharm. [156–158] USA/2014

2.8. J05AP Antivirals for Treatment of HCV Infections (Table 8)

Table 8. J05AP Antivirals for treatment of HCV infections

INN (Synonyms) [Brand names] ATC	Structure (Remarks)	[CAS No.] Formula MW, g/mol	Target (if known) Medical use Formulation DDD	Originator Approval (country/year) Production method [References]
Ribavirin (guanosine analogue; ICN-1229) [Copecus, Repetol, Viramid, Virazide, Virazole] J05AP01 **WHO essential medicine**		[36791-04-5] $C_8H_{12}N_4O_5$ 244.21	guanosine antagonist, blocks DNA and RNA synthesis. treatment of hepatitis C in combination with peginterferon-α-2a or -2b or interferon-α-2a or -2b and of respiratory syncytial virus (RSV) infection. f.c. tabl. 200 mg, 400 mg; cps. 200 mg; powder for inhalation solution. 1 g O.	ICN USA/1973 [25–27]
Telaprevir (VX-950; VRT-111950; LY-570310) [Incivek, Incivo] J05AP02		[402957-28-2] $C_{36}H_{53}N_7O_6$ 679.86	inhibitor of hepatitis C viral enzyme NS3.4A serine protease. treatment of hepatitis C genotype 1 viral infections in combination with pegylated interferon and ribavirin; since August 2014, Incivek was discontinued on the market because of heavy competition from more effective drugs.	Vertex/Johnson & Johnson USA/2011; EU/2011 [88]
Boceprevir (Sch-503034) [Victrelis] J05AP03		[394730-60-0] $C_{27}H_{45}N_5O_5$ 519.69	inhibitor of hepatitis C viral enzyme NS3.4A serine protease. treatment of hepatitis C genotype 1 viral infections in combination with pegylated interferon and ribavirin; since December 2015, Victrelis is discontinued on the market because of heavier competition with more effective drugs (e.g., sofosbuvir).	Schering-Plough; Merck & Co USA/2011; EU/2011 [89–91]

(continued overleaf)

Table 8. (*Continued*)

INN (Synonyms) [Brand names] ATC	Structure (Remarks)	[CAS No.] Formula MW, g/mol	Target (if known) Medical use Formulation DDD	Originator Approval (country/year) Production method [References]
Simeprevir (TMC-435; TMC-435350) [Olysio, Sovriad] J05AP05 **WHO essential medicine**		[923604-59-5] $C_{38}H_{47}N_5O_7S_2$ 749.9	inhibitor of hepatitis C virus enzyme NS3.4A serine protease. Treatment of hepatitis C genotype 1 viral infections in combination with peginterferon-α and ribavirin. cps. 150 mg. 150 mg O.	Medivir/Janssen USA/2013; EU/2014; JP/2013 [92, 93]
Daclatasvir (BMS-790052) [Daklinza] J05AP07 **WHO essential medicine**		[1009119-64-5] $C_{40}H_{50}N_8O_6$ 738.9 dihydrochloride: [1009119-65-6] $C_{40}H_{50}N_8O_6 \cdot 2\,HCl$ 811.3	inhibitor of the HCV nonstructural protein NS5A. treatment of hepatitis C genotype 3 infection. optional in combination with peginterferon and ribavirin or with sofosbuvir. f.c. tabl. 30 mg, 60 mg. 60 mg O.	BMS USA/2015; EU/2014 [159–161]
Sofosbuvir (PSI-7977; GS-7977) [Sovaldi; comb: Harvoni] J05AP08 **WHO essential medicine**		[1190307-88-0] $C_{22}H_{29}FN_3O_9P$ 529.45	nucleotide analogue, binds to NS5B protein and so acts as polymerase inhibitor. treatment of HCV genotype 2 infection alone or in combination with other drugs (e.g., with ledipasvir in Harvoni). f.c. tabl. 400 mg. 0.4 g O.	Pharmasset/ Gilead USA/2013; EU/2014 [162–165]

Dasabuvir
(ABT-333)
[Exviera; comb.: Viekira Pak]
J05AP09
WHO essential medicine

[1132935-63-7]
$C_{26}H_{27}N_3O_5S$
493.58

NSS5B inhibitor, acts as RNA polymerase inhibitor in HCV. treatment of HCV genotype 1 infection in comb. with ombitasvir, paritaprevir. and ritonavir.
f.c. tabl. 250 mg.
0.5 g O.

AbbVie
USA/2014; EU/2015
[166, 167]

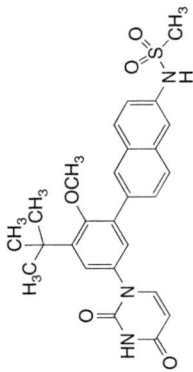

Elbasvir
(MK-8742)
[Zepatier]
J05AP10

[1370468-36-2]
$C_{49}H_{55}N_9O_7$
882.0

hepatitis C virus nonstructural protein SA inhibitor used in combination with grazoprevir (J05AP11) for treatment of chronic HCV genotypes 1 or 4 infection.
f.c. tabl. 50 mg + 100 mg grazoprevir.
50 mg O.

Merck & Co.
USA/2016; EU/2016
synthesis
[168–171]

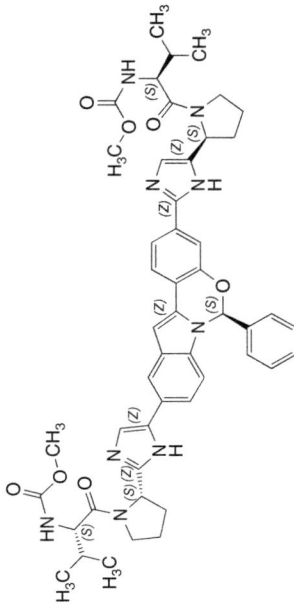

(methyl N-1-[(2S)-1-[(2S)-2-[5-[(6S)-3-[2-[(2S)-1-[(2S)-2-(methoxycarbonylamino)-3-methylbutanoyl]pyrrolidin-2-yl]-1H-imidazol-5-yl]-6-phenyl-6H-indolo[1,2-c][1,3]benzoxazin-10-yl]-1H-imidazol-2-yl]pyrrolidin-1-yl]-3-methyl-1-oxobutan-2-yl]carbamate)

Grazoprevir
(MK-5172)
[Zepatier]
J05AP11

[1350514-68-9]
$C_{38}H_{50}N_6O_9S$
766.9
sodium salt:
[1425038-27-2]
$C_{38}H_{49}N_6NaO_9S$
789.9

hepatitis C virus NS3/4A protease inhibitor; used in comb. with elbasvir (J05AP10).
f.c. tabl. 100 mg + 50 mg elbasvir.
0.1 g O.

Merck & Co,
USA/2016; EU/2016
[170–174]

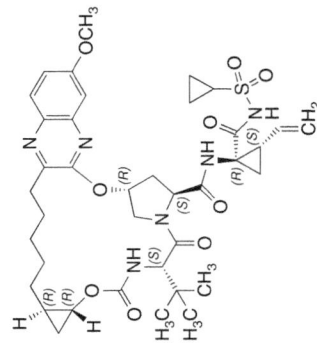

[(1R,18R,20R,24S,27S)-24-tert-butyl-N-[(1R,2S)-1-(cyclopropylsulfonyl)carbamoyl)-2-ethenylcyclopropyl]-7-methoxy-22,25-dioxo-2,21-dioxa-4,11,23,26-tetrazapentacyclo[24.2.1.0^{3,12}.0^{5,10}.0^{18,20}]nonacosa-3,5(10),6,8,11-pentaene-27-carboxamide]

(continued overleaf)

Table 8. (*Continued*)

INN (Synonyms) [Brand names] ATC	Structure (Remarks)	[CAS No.] Formula MW, g/mol	Target (if known) Medical use Formulation DDD	Originator Approval (country/year) Production method [References]

The following drug substances ledipasvir, ombitasvir, paritaprevir, velpatasvir, voxilaprevir, glecaprevir, and pibrentasvir do not have an own classification, but they are components of the classified fixed combinations J05AP51–J05AP57.

J05AP51: sofosbuvir + ledipasvir (**WHO essential medicine**)
J05AP52: dasabuvir + ombitasvir + paritaprevir + ritonavir
J05AP53: ombitasvir + paritaprevir + ritonavir (**WHO essential medicine**)
J05AP54: elbasvir + grazoprevir
J05AP55: sofosbuvir + velpatasvir (**WHO essential medicine**)
J05AP56: sofosbuvir + velpatasvir + voxilaprevir
J05AP57: glecaprevir + pibrentasvir (**WHO essential medicine**)

INN (Synonyms) [Brand names] ATC	Structure (Remarks)	[CAS No.] Formula MW, g/mol	Target (if known) Medical use Formulation DDD	Originator Approval (country/year) Production method [References]
Ledipasvir (GS-5885) [Harvoni (comb)] J05AP51 **WHO essential medicine** (Harvoni)	 (methyl N-[(2S)-1-[(6S)-6-[5-[9,9-difluoro-7-[2-[(1R,3S,4S)-2-[(2S)-2-(methoxycarbonyl-amino)-3-methylbutanoyl]-2-azabicyclo[2.2.1]heptan-3-yl]-3H-benzimidazol-5-yl]fluoren-2-yl]-1H-imidazol-2-yl]-5-azaspiro[2.4]heptan-5-yl]-3-methyl-1-oxobutan-2-yl]carbamate)	[1256388-51-8] C$_{49}$H$_{54}$F$_2$N$_8$O$_6$ 889.0	inhibitor of the HCV NS5A protein. treatment of HCV genotype 1 patients in fixed comb. with sofosbuvir (J05AX15). f.c. tabl. 90 mg ledipasvir + 400 mg sofosbuvir.	Gilead USA/2014; EU/2014 [175–178]
Ombitasvir (ABT-267) [Viekira Pak, Technivie, Viekirax] J05AP52 (comb.); J05AP53 (comb).	 (methyl N-[(2S)-1-[(2S)-2-[[4-[(2S,5S)-1-(4-tert-butylphenyl)-5-[4-[[(2S)-1-[(2S)-2-(methoxycarbonylamino)-3-methylbutanoyl] pyrrolidine-2-carbonyl]amino]phenyl] pyrrolidin-2-yl]phenyl]carbamoyl]pyrrolidin-1-yl]-3-methyl-1-oxobutan-2-yl]carbamate)	[1258226-87-7] C$_{50}$H$_{67}$N$_7$O$_8$ 894.1	inhibitor of HCV NS5A protein. treatment of HCV genotype 1 patients in comb. with paritaprevir, ritonavir, and dasabuvir (Viekira Pak) and HCV genotype 4 in comb. with paritaprevir and ritonavir (Technivie); (Viekira Pak) and 2015 (Technivie) (Viekirax). tabl. 12.5 mg in comb. with 250 mg of paritaprevir, 75 mg of dasabuvir, 75 mg of paritaprevir, and 50 mg of ritonavir (copacket).	AbbVie USA/2014; EU/2015 [179–181]

Paritaprevir

(veruprevir; ABT-459)

[Viekira Pak, Technivie, Viekirax]

J05AP52 (comb.);
J05AP53 (comb.).

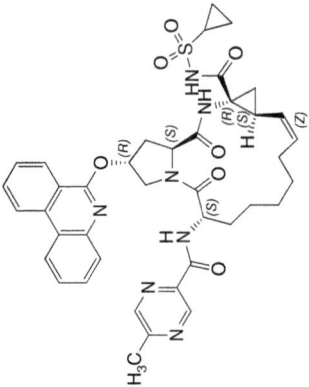

[1221573-85-8]
$C_{40}H_{43}N_7O_7S$
765.9

inhibitor of NS3-4A serine protease. treatment of HCV infection in comb. with ritonavir and ombitasvir or with ombitasvir, ritonavir, and dasabuvir.
tabl. 75 mg in comb. with 250 mg of dasabuvir + 12.5 mg of ombitasvir + 50 mg of ritonavir.

AbbVie
USA/2014; EU/2015
[182, 183]

[(1S,4R,6S,7Z,14S,18R)-N-cyclopropylsulfonyl-14-[(5-methylpyrazine-2-carbonyl)amino]-2,15-dioxo-18-phenanthridin-6-yloxy-3,16-diazatricyclo[14.3.0.0^{4,6}]nonadec-7-ene-4-carboxamide]

Velpatasvir

(GS-5816)

[Epclusa, Vosevi]

J05AP55 (comb.);
J05AP56 (comb.).

[1377049-84-7]
$C_{49}H_{54}N_8O_8$
883.0

inhibitor of HCV NS5A protein complex with potential activity against all HCV genotypes 1–6. recommended as first-line therapy in fixed comb. with sofosbuvir for all six genotypes of HCV.
tabl. 100 mg in comb. with 400 mg of sofosbuvir and tabl. 50 mg in comb. with 200 mg of sofosbuvir.

Gilead Sciences
USA/2016; EU/2016
[169, 184, 185]

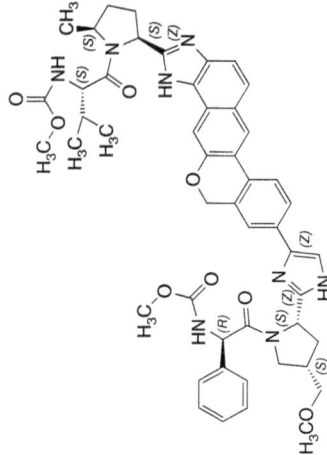

(methyl N-[(1R)-2-[(2S,4S)-2-[5-[6-[(2S,5S)-1-[(2S)-2-(methoxycarbonylamino)-3-methylbutanoyl]-5-methylpyrrolidin-2-yl]-21-oxa-5,7-diazapentacyclo[11.8.0.0^{3,11}.0^{4,8}.0^{14,19}]-henicosa-1(13),2,4(8),5,9,11,14(19),15,17-nonaen-17-yl]-1H-imidazol-2-yl]-4-(methoxymethyl)pyrrolidin-1-yl]-2-oxo-1-phenylethyl]carbamate)

(continued overleaf)

Table 8. (*Continued*)

INN (Synonyms) [Brand names] ATC	Structure (Remarks)	[CAS No.] Formula MW, g/mol	Target (if known) Medical use Formulation DDD	Originator Approval (country/year) Production method [References]
Voxilaprevir (GS-9857) [Vosevi] J05AP56 (comb.)	[[(1*R*,18*R*,20*R*,24*S*,27*S*,28*S*)-24-*tert*-butyl-*N*-[(1*R*,2*R*)-2-(difluoromethyl)-1-[[(1-methylcyclopropyl)sulfonylcarbamoyl]cyclopropyl]-28-ethyl-13,13-difluoro-7-methoxy-22,25-dioxo-2,21-dioxa-4,11,23,26-tetrazapentacyclo[24.2.1.0³,¹².0⁵,¹⁰.0¹⁸,²⁰]nonacosa-3,5(10),6,8,11-pentaene-27-carboxamide]	[1535212-07-7] $C_{40}H_{52}F_4N_6O_9S$ 868.9	inhibitor of HCV nonstructural protein 4A (NS3/NS4A) serine protease and prevents viral replication and function. treatment of HCV infections of all genotypes 1–6 in comb. with sofosbuvir and velpatasvir. f.c. tabl. 100 mg in comb. with 400 mg of sofosbuvir and 100 mg of velpatasvir.	Gilead Sciences USA/2017; EU/2017 [186, 187]
Glecaprevir (ABT-493) [Mavyret, Maviret] J05AP57 (comb.)	[[(1*R*,14*E*,18*R*,22*R*,26*S*,29*S*)-26-*tert*-butyl-*N*-[(1*R*,2*R*)-2-(difluoromethyl)-1-[[(1-methylcyclopropyl)sulfonylcarbamoyl]cyclopropyl]-13,13-difluoro-24,27-dioxo-2,17,23-trioxa-4,11,25,28-tetrazapentacyclo[26.2.1.0³,¹².0⁵,¹⁰.0¹⁸,²²]hentriaconta-3,5,7,9,11,14-hexaene-29-carboxamide]	[1365970-03-1] $C_{38}H_{46}F_4N_6O_9S$ 838.9	inhibitor of HCV nonstructural protein 4A (NS3/NS4A) serine protease. treatment of HCV infections of all genotypes 1–6 in fixed-dose comb. with NS5 inhibitor pibrentasvir. f.c. tabl. 100 mg with 40 mg of pibrentasvir.	AbbVie USA/2017; EU/2017 [188–190]

Pibrentasvir
(ABT-530)
[Mavyret, Maviret]
J05AP57 (comb.)

[1353900-92-1]
$C_{57}H_{65}F_5N_{10}O_8$
1113.2

inhibitor of HCV NS5A protein.
in fixed-dose comb. with glecaprevir
used for treatment of HCV infections
of all genotypes 1–6.
f.c. tabl. 40 mg with 100 mg
of glecaprevir.

AbbVie
USA/2017; EU/2017
[191, 192]

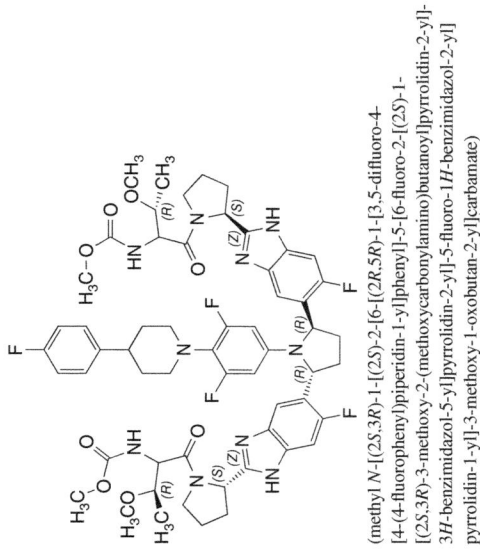

(methyl N-[(2S,3R)-1-[(2S)-2-[6-[(2R,5R)-1-[(2S)-2-[6-[(2S)-1-
[4-(4-fluorophenyl)piperidin-1-yl]phenyl]-5-[6-fluoro-2-[(2S)-1-
[(2S,3R)-3-methoxy-2-(methoxycarbonylamino)butanoyl]pyrrolidin-2-yl]-
3H-benzimidazol-5-yl]pyrrolidin-2-yl]-5-fluoro-1H-benzimidazol-2-yl]
pyrrolidin-1-yl]-3-methoxy-1-oxobutan-2-yl]carbamate)

2.9. J05AR Antivirals for Treatment of HIV Infections, Combinations (Table 9)

Table 9. J05AR Antivirals for treatment of HIV infections, combinations

ATC	Drug substances	Trade name/Company
J05AR01	zidovudine + lamivudine	Combivir/ViiV (**WHO essential medicine**)
J05AR02	lamivudine + abacavir	Kivexa; Epzicom/ViiV (**WHO essential medicine**)
J05AR03	tenofovir disoproxil + emtricitabine	Truvada/Gilead
J05AR04	zidovudine + lamivudine + abacavir	Trizivir/ViiV
J05AR05	zidovudine + lamivudine + nevirapine	Duovir/Cipla (**WHO essential medicine**)
J05AR06	emtricitabine + tenofovir disoproxil + efavirenz	Atripla/Gilead (**WHO essential medicine**)
J05AR07	stavudine + lamivudine + nevirapine	Emtri/Emcure
J05AR08	emtricitabine + tenofovir disoproxil + rilpivirine	Complera; Eviplera/Gilead
J05AR09	emtricitabine + tenofovir disoproxil + elvitegravir + cobicistat	Stribild/Gilead
J05AR10	lopinavir + ritonavir	Kaletra/Abb Vie (**WHO essential medicine**)
J05AR11	lamivudine + tenofovir disoproxil + efavirenz	−/−
J05AR12	lamivudine + tenofovir disoproxil	−/−
J05AR13	lamivudine + abacavir + dolutegravir	Triumeq/ViiV
J05AR14	darunavir + cobicistat	Prezcobix/Janssen

2.10. J05AX Other Antivirals [193, 194] (Table 10)

Table 10. J05AX Other antivirals [193, 194]

INN (Synonyms) [Brand names] ATC	Structure (Remarks)	[CAS No.] Formula MW, g/mol	Target (if known) Medical use Formulation DDD	Originator Approval (country/year) Production method [References]
Enfuvirtide (T-20; R-698; DP-178) [Fuzeon] J05AX07	N-acetyl-L-tyrosyl-L-threonyl-L-seryl-L-leucyl-L-isoleucyl-L-histidyl-L-seryl-L-leucyl-L-isoleucyl-L-threonyl-L-seryl-L-leucyl-L-isoleucyl-L-histidyl-L-seryl-L-leucyl-L-α-glutamyl-L-α-glutamyl-L-seryl-L-glutaminyl-L-asparaginyl-L-glutaminyl-L-glutaminyl-L-α-glutaminyl-L-lysyl-L-asparaginyl-L-α-glutamyl-L-glutaminyl-L-α-glutaminyl-L-leucyl-L-leucyl-L-α-glutamyl-L-leucyl-L-α-aspartyl-L-lysyl-L-tryptophyl-L-alanyl-L-seryl-L-leucyl-L-tryptophyl-L-asparaginyl-L-tryptophyl-L-phenylalaninamide Ac-Tyr-Thr-Ser-Leu-Ile-His-Ser-Leu-Ile-Glu-Glu-Ser-Gln-Asn-Gln-Gln-Glu-Lys-Asn-Glu-Gln-Glu-Leu-Leu-Glu-Leu-Asp-Lys-Trp-Ala-Ser-Leu-Trp-Asn-Trp-Phe-NH$_2$	[159519-65-0] C$_{204}$H$_{301}$N$_{51}$O$_{64}$ 4491.95	HIV-1 fusion inhibitor with CD4+ cells. treatment of HIV-1 infection in combination with other antiretroviral agents in case other treatments have failed. vials 108 mg powder for reconstitution to solution with 90 mg/mL for s.c. inj. 180 mg P.	Duke Univ. Trimeris; Roche USA/2003 [195–200]
Raltegravir (MK-0518) [Isentress] J05AX08 **WHO essential medicine**		[518048-05-0] C$_{20}$H$_{21}$FN$_6$O$_5$ 444.42 potassium salt: [871038-72-1] C$_{20}$H$_{20}$FKN$_6$O$_5$ 482.51	HIV-1 integrase inhibitor. treatment of HIV-1 infection in combination with other antiretroviral agents. f.c. tabl. 400 mg; chewing tabl. 25 mg, 100 mg; granulate for oral suspension 20 mg/mL. 0.8 g O.	Inst. Di Ricerche Di Biologia Molecolare (IRBM); Merck & Co. USA/2007; EU/2007 [201–204]
Maraviroc (UK-427857) [Celsentri, Selzentry] J05AX09		[376348-65-1] C$_{29}$H$_{41}$F$_2$N$_5$O 513.68	CCR5 receptor antagonist, entry inhibitor; blocking the HIV protein gp120 and thereby blocking the virus to enter human macrophages and T cells. treatment of HIV infection in combination with other antiretroviral agents. f.c. tabl. 150 mg, 300 mg. 0.6 g O.	Pfizer USA/2007; EU/2007 [205, 206]

(continued overleaf)

Table 10. (*Continued*)

INN (Synonyms) [Brand names] ATC	Structure (Remarks)	[CAS No.] Formula MW, g/mol	Target (if known) Medical use Formulation DDD	Originator Approval (country/year) Production method [References]
Elvitegravir (GS-9137; JTK-303; EVG) [Vitekta; Stribild (comb.)] J05AX11		[697761-98-1] $C_{23}H_{23}ClFNO_5$ 447.89	integrase inhibitor. treatment of HIV-1-infection, optionally in combination with other antiretroviral agents, e.g., with emtricitabine, tenofovir disoproxil, and cobicistat (see V03AX03) in Stribild. tabl. 85 mg, 150 mg.	Japan Tobacco/Gilead USA/2012 (Stribild); USA/2014 (Vitekta); EU/2013 (both prep.) [207–209]
Dolutegravir (GSK-1349572) [Tivicay, Triumeq] J05AX12 **WHO essential medicine**		[1051375-16-6] $C_{20}H_{19}F_2N_3O_5$ 419.38 sodium salt: [1051375-19-9] $C_{20}H_{18}F_2N_3NaO_5$ 441.37	HIV integrase inhibitor. treatment of HIV infection as monotherapy or in combination with other antiretroviral agents (e.g. abacavir and lamivudine in Triumeq). tabl. 50 mg. 50 mg O.	Shionogi/GSK USA/2013; EU/2014 [210–212]
Umifenovir (AR-1/9514; GTPL-11089) [Arbidol (marketed in RU and CN)] J05AX13		[131707-25-0] $C_{22}H_{25}BrN_2O_3S$ 477.41	membrane fusion inhibitor of influenza virus; prevents viral entry into the target cell. approved in China and Russia for prophylaxis and treatment of influenza (A- and B-type) and other respiratory viral infections. f.c. tabl. 50 mg, 100 mg. 0.8 g O.	Pharmstandard (RU) RU/1993 [213–215]

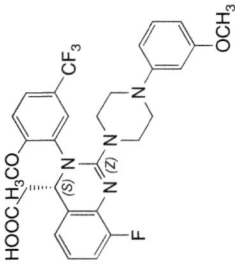

Name (codes) [Brand] ATC	Structure	CAS / Formula / MW	Description	Company / Country/Year / Refs
Letermovir (AIC-246; MK-8228; Bay-73-6327) [Prevymis] J05AX18		[917389-32-3] $C_{29}H_{28}F_4N_4O_4$ 572.6	cytomegalovirus (CMV) terminase inhibitor. prevention and treatment of CMV infections. tabl. 240 mg, 480 mg: vials for inj. 240 mg/12 mL; 480 mg/24 mL. 0.48 g O, P.	Bayer/AiCuris/ Merck & Co USA/2017; EU/2017 [216–220]
Tecovirimat (ST-246; SIGA-246) [Tpoxx] J05AX24		[869572-92-9] $C_{19}H_{15}F_3N_2O_3$ 376.34	indicated for the treatment of human smallpox disease caused by variola virus (e.g., in case of smallpox being used as bioweapon). First approved drug by FDA for smallpox. cps. 200 mg.	Siga Technologies Inc. USA/2018 [221–225]
Baloxavir marboxil (S-033188; S-033447; BXM) [Xofluza] J05AX25		[1985606-14-1] $C_{27}H_{23}F_2N_3O_7S$ 571.55	inhibitor of the influenza cap-dependent endonuclease enzyme. prevention and treatment of influenza A and B. tabl. 20 mg, 40 mg; oral suspension 2 mg/mL.	Shionogi/Roche USA/2018; JP/2018; AU/2020 [225–230]
Favipiravir (T-705) [Avigan] J05AX27		[259793-96-9] $C_5H_4FN_3O_2$ 157.10	selective inhibitor of RNA polymerase enzymes, which are necessary for the transcription and replication of viral genomes; the active form of the drug is its ribofuranosyl triphosphate derivative. approved in Japan for treatment of cases of influenza which were unresponsive to conventional treatment; has been under investigation against ebola and lassa virus as well as COVID-19. tabl. 200 mg. 1.6 g O.	Toyama Chemical/Fujifilm JP/2014 [231–234]

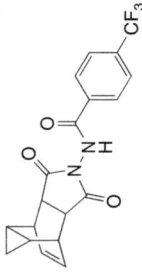

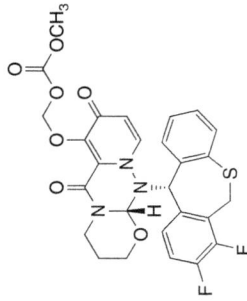

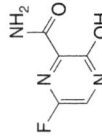

(continued overleaf)

Table 10. (*Continued*)

INN (Synonyms) [Brand names] ATC	Structure (Remarks)	[CAS No.] Formula MW, g/mol	Target (if known) Medical use Formulation DDD	Originator Approval (country/year) Production method [References]
Bulevirtide acetate (MyrB; myrcludex-B) [Hepcludex] J05AX28	N-tetradecanoylglycylglycyl-Thr-Asn-Leu-Ser-Val-Pro-Asn-Pro-Leu-Gly-Phe-Phe-Pro-Asp-His-Gln-Leu-Asp-Pro-Ala-Phe-Gly-Ala-Asn-Ser-Asn-Asn-Pro-Asp-Trp-Asp-Phe-Asn-Pro-Asn-Lys-Asp-His-Trp-Pro-Glu-Ala-Asn-Lys-Val-Gly-NH$_2$	[2012558-47-1] C$_{248}$H$_{355}$N$_{65}$O$_{72}$ 5398.95	entry blocker for hepatitis B (and D) to hepatocytes. treatment of hepatitis B (and D). vials with lyophilized powder 2 mg for s.c. inj. sol. 2 mg/mL. 2 mg P.	MYR GmbH EU/2020 synthesis of the linear 47-amino acid peptide by solid phase method (SPPS) on a 4-methyl-benzhydryl-amine (MBHA) resin and prep. chromatography and N-terminal acylation with myristic acid. [235–239]
Fostemsavir (BMS-663068; GSK-3684934; prodrug of temsavir) [Rukobia] J05AX29		[864953-29-7] C$_{25}$H$_{26}$N$_7$O$_8$P 583.5 tromethamine salt: [864953-39-9] C$_{29}$H$_{37}$N$_8$O$_{11}$P 704.6	HIV entry inhibitor, blocks the activity of gp120. treatment of patients with multidrug-resistant HIV-1 infection. e.r. tabl. EQ 600 mg base (as tromethamine salt).	BMS USA/2020; EU/2020 [240–245]

List of Abbreviations

AIDS	acquired immunodeficiency syndrome
ATC	Anatomical Therapeutical Chemical (classification)
AZT	azidothymidine = zidovudine
CCR5	chemokine CC-motif receptor type 5
CMV	cytomegalovirus
CD4	glycoprotein on surface of immune cells
comb.	combination
cps.	capsules
DDD	defined daily dose
DNA	deoxyribonucleic acid
e.r. tabl.	extended release
f.c. tabl.	film-coated tablets
gp120	glycoprotein on the surface of HIV
HAART	highly active antiretroviral therapy
HBV	hepatitis B virus
HCV	hepatitis C virus
HIV	human immunodeficiency virus
HSV	herpes simplex virus
INN	international nonproprietary name
NCE	new chemical entity
NMDA(R)	*N*-methyl-D-aspartate (receptor)
NNRTI	nonnucleoside reverse transcriptase inhibitor
NRTI	nucleoside reverse transcriptase inhibitor
NS5B	nonstructural protein 5 B (viral protein in HCV)
O	oral application
P	parenteral application
PEG-IFN	pegylated interferon
RNA	ribonucleic acid
RTI	reverse transcriptase inhibitor
sol.	solution
tabl.	tablet(s)
VZV	varicella zoster virus
wfm	withdrawn from market
WHO	World Health Organization

References

1 De Clercq, E. (2007) Three Decades of Antiviral Drugs, *Nat. Rev. Drug Discov.* **6**, 941.

2 De Clercq, E. (2007) The Design of Drugs for HIV and HCV, *Nat. Rev. Drug Discov.* **6**, 1001–1018.

3 Gable, J.E., et al. (2014) Current and Potential Treatments for Ubiquitous but Neglected Herpesvirus Infections, *Chem. Rev.* **114**, 11382–11412.

4 Jordheim, L.P., et al. (2013) Advances in the Development of Nucleoside and Nucleotide Analogues for Cancer and Viral Diseases, *Nat. Rev. Drug Discov.* **12**, 447–464.

5 Esposito, F., et al. (2012) HIV-1 Reverse Transcriptase Still Remains a New Drug Target: Structure, Function, Classical Inhibitors, and New Inhibitors with Innovative Mechanisms of Actions, *Mol. Biol. Internat.* 58640123. doi: 10.1155/2012/586401.

6 Flexner, C. (2007) HIV Drug Development: The Next 25 Years, *Nat. Rev. Drug Discov.* **6**, 959–966.

7 Fraser, C., et al. (2014) Virulence and pathogenesis of HIV-1 infection: An evolutionary perspective, *Science* **343**, 1243727-1-7. doi: 10.1126/science.1243727.

8 Mehellou, Y. and De Clercq, E. (2010) Twenty-Six Years of Anti-HIV Drug Discovery: Where Do We Stand and Where Do We Go? *J. Med. Chem.* **53**, 521–538.

9 Oversteegen, L., et al. (2007) HIV Combination Products, *Nat. Rev. Drug Discov.* **6**, 951–952.

10 Wong, A. (2014) The HIV Pipeline, *Nat. Rev. Drug Discov.* **13**, 649–650.

11 Zhan, P., et al. (2016) Anti-HIV Drug Discovery and Development: Current Innovations and Future Trends, *J. Med. Chem.* **59**, 2849–2878.

12 WHO (2015) *Consolidated Guidelines on the Use of Antiretroviral Drugs for Treating and Preventing HIV Infection; What's New*, WHO.

13 Meanwell, N.A. and Watkins, W.J. (2014) Introduction to Hepatitis C Virus (HCV) Therapies Special Thematic Issue, *J. Med. Chem.* **57**, 1625–1626. And following articles on new HCV therapeutics, e.g., Simeprevir, Asunaprevir, Danoprevir, Dasabuvir, Daclatasvir, Ledoipasvir and Ombitasvir.

14 Cannalire, R., et al. (2016) A Journey Around The Medicinal Chemistry of Hepatitis C Virus Inhibitors Targeting NS4B: From Target to Preclinical Drug Candidates, *J. Med. Chem.* **59**, 16–41.

15 Manns, M.P., et al. (2007) The Way Forward in HCV Treatment–Finding the Right Path, *Nat. Rev. Drug Discov.* **6**, 991–1000.

16 Manns, M.P. and von Hahn, T. (2013) Novel Therapies for Hepatitis C – One Pill Fits All? *Nat. Rev. Drug Discov.* **12**, 595–610.

17 Sofia, M.J., et al. (2012) Nucleoside, Nucleotide, and Non-Nucleoside Inhibitors of Hepatitis C Virus NS5B RNA-Dependent RNA-Polymerase, *J. Med. Chem.* **55**, 2481–2531.

18 Schaeffer, H.J., et al. (1978) *Nature* **272**, 583.

19 Wellcome, (1980) US 4 199 574, GB-prior. 1974.

20 Chang, P.K. and Welch, A.D. (1963) *J. Med. Chem.* **6**, 428.

21 Roussel-Uclaf, (1963) GB 1 024 156, FR-prior. 1962.

22 Lee, W.W., et al. (1960) *J. Am. Chem. Soc.* **82**, 2648.

23 Parke Davis, (1971) US 3 616 208, USA-prior. 1967.

24 Schabel, F.M. Jr. (1968) *Chemotherapy* **13**, 321.

25 Witkowski, J.T., et al. (1972) *J. Med. Chem.* **15**, 1150.

26 ICN, (1976) US 3 976 545, USA-prior. 1971.

27 ICN, (1979) US 4 138 547, USA-prior. 1977.

28 Martin, J.C., et al. (1983) *J. Med. Chem.* **26**, 759.

29 Syntex, (1983) US 4 355 032, USA-prior. 1981.

30 Syntex, (1983) US 4 423 050, USA-prior. 1981.

31 Syntex, (1985) US 4 507 305, USA-prior. 1981.

32 Harnden, M.R., et al. (1989) *J. Med. Chem.* **32**, 1738.

33 Beecham, (1993) US 5 246 937, GB-prior. 1985.

34 Co Pharma Corp, (1986) US 4 567 182, IT-prior. 1981.
35 Burroughs Wellcome (1990) US 4 957 924, USAprior. 1987.
36 Snoeck, R., et al. (1988) *Antimicrob, Agents Chemother.* **32**, 1839.
37 Ceskoslovenska Akad., (1992) US 5 142 051, CS-prior. 1986.
38 Harnden, M.R., et al. (1987) *J. Med. Chem.* **30**, 1636.
39 Beecham, (1991) US 5 075 445, GB-prior. 1983.
40 Roche, (1997) US 5 700 936; USA-prior. 1996.
41 Syntex, (2000) US 6 083 953; USA-prior. 1994.
42 Rega Inst.VZW, (1984) US 4 424 211, USA-prior. 1979.
43 Eastman, R.T., et al. (2020) Remdesivir: A Review of Its Discovery and Development, *ACS Central Sci.* **6**, 672–683.
44 Siegel, D., et al. (2017) *J. Med. Chem.* **60**, 1648–1661.
45 Gilead, (2017) US 9 724 360, USA-prior. 2014.
46 Coen, D.M. and Schaffer, P.A. (2003) Antiherpes Virus Drugs: A Promising Spectrum Of New Drugs And Drug Targets, *Nat. Rev. Drug Discov.* **2**, 278–288.
47 Elion, G.B. (1989) The Purine Path to Chemotherapy (Nobel Lecture), *Angew. Chem. Int. Ed.* **28**, 870–878. (History of Aciclovir Discovery).
48 Du Pont, (1967) US 3 352 912, USA-prior. 1963.
49 Du Pont, (1971) US 3 592 934, USA-prior. 1963.
50 Du Pont, (1985) US 4 551 552, USA-prior. 1984.
51 Merz& Co., (1971) DE 1 941 218, DE-prior. 1969.
52 Rosenthal, K.S., et al. (1982) *Antimicrob. Agents Chemother.* **22**, 1031.
53 Stetter, H., et al. (1960) *Chem. Ber.* **93**, 226.
54 DuPont, (1967) US 3 310 469, USA-prior. 1961.
55 Nylen, P. (1924) *Ber. Dtsch. Chem. Ges.* **57b**, 1023.
56 DuPont, (1977) US 4 018 854, USA-prior. 1973.
57 Astra, (1982) US 4 339 445, SE-prior. 1976.
58 Ghosh, A.K., et al. (2016) Recent Progress in the Development of HIV-1 Protease Inhibitors for the Treatment of HIV/AIDS, *J. Med. Chem. ASAP.* **59**, 5172–5208.
59 Ghosh, A.K., et al. (2012) Enhancing Protein Backbone Binding – A Fruitful Concept for Combating Drug-Resistant HIV, *Angew. Chem. Int. Ed.* **51**, 1778–1802.
60 Izawa, K. and Onishi, T. (2006) Industrial Syntheses of the Central Core Molecules of HIV Protease Inhibitors, *Chem. Rev.* **106**, 2811–2827.
61 Leung, D., et al. (2000) Protease Inhibitors: Current Status and Future Prospects, *J. Med. Chem.* **43**, 305–341.
62 Parkes, K.E.B., et al. (1994) *J. Org. Chem.* **59**, 3656.
63 Roche, (1993) US 5 196 438; GB-prior. 1989.
64 Dorsey, B.D., et al. (1994) *J. Med. Chem.* **37**, 3443.
65 Merck & Co, (1995) US 5 413 999; USA-prior. 1991.
66 Abbott, (1996) US 5 541 206, USA-prior. 1989.
67 Abbott, (1996) US 5 567 823, USA-prior. 1995.
68 Abbott, (1997) US 5 635 523, USA-prior. 1989.
69 Abbott, (2005) US 6 894 171, USA-prior. 1998.
70 Kaldar, S.W., et al. (1997) *J. Med. Chem.* **40**, 3979.
71 Agouron, (1996) US 5 484 926, USA-prior. 1993.
72 Vertex, (1996) US 5 585 397, USA-prior. 1992.
73 Stoner, E.J., et al. (1999) *Org. Process Res. Dev.* **3**, 145. **4**, 264 (2000).
74 Abbott, (1999) US 5 914 332, USA-prior. 1995.
75 Abbott, (2006) US 7 141 593, USA-prior. 1999.
76 Vertex, (2002) US 6 436 989, USA-prior. 1997.
77 Smith Kline Beecham, (2003) US 6 514 953, GB-prior. 1998.
78 Bold, G., et al. (1998) *J. Med. Chem.* **41**, 3387.
79 Novartis, (1998) US 5 849 911, CH-prior. 1996.
80 Bristol-Myers Squibb, (2000) US 6 087 383, USA-prior. 1998.
81 Turner, S.R., et al. (1998) *J. Med. Chem.* **41**, 3467.
82 Fors, K.S., et al. (1998) *J. Org. Chem.* **63**, 7348.
83 Pharmacia & Upjohn, (1998) US 5 852 195, USA-prior. 1994.

84 Upjohn Appl., (1995) EP 758 327, USA-prior. 1994.
85 Gosh, A.K., et al. (2004) *J. Org. Chem.* **69**, 7822.
86 Gosh, A.K., et al. (2006) *J. Med. Chem.* **48**, 5252.
87 Tibotec, (2009) EP 1 725 566, EP-prior. 2003.
88 Vertex, (2010) US 7 820 671, USA-prior. 2000.
89 Venkatraman, S., et al. (2006) *J. Med. Chem.* **49**, 6074.
90 Schering-Plough, (2003) US 7 012 066, USA-prior. 2001.
91 Schering-Plough, (2012) US RE 43298, USA-prior. 2000.
92 Rosenquist, A., et al. (2014) *J. Med. Chem.* **57**, 1673.
93 Tibotec/Medivir, (2011) US 8 012 939, EP-prior. 2005.
94 Glinski, R.P., et al. (1973) *J. Org. Chem.* **38**, 4299.
95 Horwitz, J.P., et al. (1964) *J. Org. Chem.* **29**, 2076.
96 Imazawa, M., et al. (1978) *J. Org. Chem.* **43**, 3044.
97 Burroughs Wellcome, (1988) US 4 724 232, GB-prior. 1985.
98 McLaren, C., et al. (1991) *Antiviral Chem. Chemother.* **2**, 321.
99 Ajinomoto, (1990) US 4 970 148, JP-prior. 1987.
100 Bristol-Myers Squibb, (1991) US 5 011 774, USA-prior. 1987.
101 US Dept. of Health, (1989) US 4 861 759, USA-prior. 1985.
102 Lin, T.-S., et al. (1987) *J. Med. Chem.* **30**, 440.
103 US DHSS, (1989) US 4 879 277, USA-prior. 1985.
104 Horwitz, J.P., et al. (1966) *J. Org. Chem.* **31**, 205.
105 Bristol-Myers, (1992) US 5 130 421, USA-prior. 1988.
106 Beach, J.W., et al. (1992) *J. Org. Chem.* **57**, 2217.
107 IAF, (1991) US 5 047 407, USA-prior. 1989.
108 Crimmins, M.T., et al. (1996) *J. Org. Chem.* **61**, 4192.
109 Burroughs Wellcome, (1991) US 5 034 394, GB prior. 1988.
110 Glaxo SmithKline, (2002) US 6 417 191, GB-prior. 1995.
111 Gilead, (1998) US 5 733 788, USA-prior. 1996.
112 Gilead, (1999) US 5 922 695, USA-prior. 1996.
113 Holy, A., et al. (1987) *Collect, Czech. Chem. Commun.* **52**, 2801.
114 Ceskoslov. akad. vet., (1989) US 4 808 716, CS-prior. 1985.
115 Starrett, J.E., et al. (1994) *J. Med. Chem.* **37**, 1857.
116 Academy of Sci. of Czech Republic and Rega Stichting v.z.w., (1997) US 5 663 159, USA-prior. 1990.
117 Jeong, L.S., et al. (1993) *J. Med. Chem.* **36**, 181.
118 Gaede, B.J., et al. (2005) *Org. Proc. Res. & Dev.* **9**, 23.
119 Emory Univ, (1993) US 5 210 085, USA-prior. 1990.
120 BioChem Pharma, (1996) US 5 538 975, GB-prior. 1991.
121 Emory Univ, (1998) US 5 814 639, USA-prior. 1990.
122 Emory Univ, (1999) US 5 914 331, USA-prior. 1990.
123 Bisacchi, G.S., et al. (1997) *Bioorg, Med. Chem. Lett.* **7**, 127.
124 Opio, C.K., et al. (2005) *Nat. Rev./Drug Discov.* **4**, 535.
125 BMS, (1993) US 5 206 244, USA-prior. 1990.
126 Velasco, J., et al. (2013) *J. Org. Chem.* **78**, 5482.
127 Holy, A. (1972) *Collect. Czech. Chem. Commun.* **37**, 4072.
128 Idenix, (2003) US 6 596 700, USA-prior. 2000.
129 Idenix, (2004) US 6 787 526, USA-prior. 2000.
130 Gilead, (2010) US 7 803 788, USA-prior. 2000.
131 Gilead, (2016) US 9 296 769, USA-prior. 2014.
132 Derstine, B.P., et al. (2020) *Org. Process Res. Dev.* **24**, 1420–1427.
133 Hargrave, K.D., et al. (1991) *J. Med. Chem.* **34**, 2231.
134 Boehringer Ing, (1994) US 5 366 972, USA-prior. 1989.
135 Romero, D.L., et al. (1993) *J. Med. Chem.* **36**, 1505.
136 Upjohn, (1996) US 5 563 142, USA-prior. 1989.
137 Young, S.D., et al. (1995) *Antimicrob. Agents Chemother.* **39**, 2602.
138 Thompson, A.S., et al. (1995) *Tetrahedron Lett.* **36**, 8937.
139 Choudhury, A., et al. (2003) *Org. Process Res. Dev.* **7**, 324.
140 Merck & Co, (1996) US 5 519 021, USA-prior. 1992.
141 Merck & Co, (1997) US 5 698 741, USA-prior. 1995.
142 De Corte, B.L. (2005) *J. Med. Chem.* **48**, 1689.
143 Janssen, (2006) US 7 037 917, EP-prior. 1999

144 Janssen, P.A.J., et al. (2005) *J. Med. Chem.* **48**, 1901.
145 Guillemont, J., et al. (2005) *J. Med. Chem.* **48**, 2072.
146 Janssen, (2006) US 7 125 879, EP-prior. 2001.
147 MSD, (2013) US 8 486 975, USA-prior. 2010.
148 Gauthier, D.R., et al. (2015) *Org. Lett.* **17**, 1353–1356.
149 Campeau, L.-C., et al. (2016) *Org. Process Res. Dev.* **20**, 1476–1481.
150 McMullen, J.P., et al. (2018) *Org. Process Res. Dev.* **22**, 1208–1213.
151 Enserink, M. (2013) Dueling Reviews for Controversial Flu Drug, *Science* **340**, 508–509.
152 von Itzstein, M. (2007) The War Against Influenza: Discovery and Development of Sialidase Inhibitors, *Nat. Rev. Drug Discovery* **6**, 967–974.
153 Biota, (1994) US 5 360 817, AUS-prior. 1990.
154 Chandler, M., et al. (1995) *J. Chem. Soc. Perkin Trans.* **1**, 1173.
155 Gilead, (1998) US 5 763 483, USA-prior. 1995.
156 Chand, P., et al. (2001) *J. Med. Chem.* **44**, 4379–4392.
157 Biocryst, (2003) US 6 562 861, USA-prior. 1997.
158 Biocryst, (2014) US 8 778 997, USA-prior. 2006.
159 Belema, M., et al. (2014) *J. Med. Chem.* **57**, 2013.
160 Belema, M., et al. (2014) *J. Med. Chem.* **57**, 5057.
161 BMS, (2012) US 8 303 944, USA-prior. 2006.
162 Wang, P., et al. (2009) *J. Org. Chem.* **74**, 6819–6824.
163 Ross, B.S., et al. (2011) *J. Org. Chem.* **76**, 8311–8319.
164 Sofia, M., et al. (2010) *J. Med. Chem.* **53**, 7202–7218.
165 Pharmasset, (2010), US 7,964,580, US-prior. 2007.
166 Abbott, (2012) US 8 188 104, USA-prior. 2007.
167 AbbVie, (2013) US 8 501 238, USA-prior. 2007.
168 Mangion, I.K., et al. (2014) *Org. Lett.* **16**, 2310.
169 Hughes, D.L. (2016) *Org. Process Res. Dev.* **20**, 1404–1415.
170 MSD, (2011) US 7 973 040, USA-prior. 2008.
171 MSD, (2013) US 8 871 759, USA-prior. 2009.
172 Harper, S., et al. (2012) *ACS Med. Chem. Lett.* **3**, 332–336.
173 Kuethe, J., et al. (2013) *Org. Lett.* **15**, 4174–4177.
174 Williams, M.J., et al. (2016) *Org. Lett.* **18**, 1852–1955.
175 Link, J.O., et al. (2014) *J. Med. Chem.* **57**, 2033.
176 Gilead, (2012) US 8 088 368, USA-prior. 2009.
177 Gilead, (2014) US 8 822 430, USA-prior. 2009.
178 Gilead Pharmasset, (2014) WO 2014 120981, USA-prior. 2013.
179 DeGoey, D.A., et al. (2014) *J. Med. Chem.* **57**, 2047.
180 AbbVie, (2010) US 8 691 938, USA-prior. 2009.
181 AbbVie, (2014) US 8 680 106, USA-prior. 2013.
182 Abbott & Enanta, (2013) US 8 420 596, USA-prior. 2008.
183 AbbVie, (2014) US 8 691 938, USA-prior. 2009.
184 Gilead, (2013) US 8 575 135, USA-prior. 2011.
185 Gilead, (2014) US 8 921 341, USA-prior. 2011.
186 Gilead, (2017) US 9 655 944, USA-prior. 2012.
187 Gilead, (2016), US 9 296 782, USA-prior. 2013.
188 Enanta, (2015) US 9 220 748, USA-prior. 2011.
189 Kallemeyn, J.M., et al. (2020) Development of a Large-Scale Route to Glecaprevir: Synthesis of the Macrocycle via Intramolecular Etherification, *Org, Process Res. Dev.* **24**, 1373–1392.
190 Hill, D.R., et al. (2020) Development of a Large-Scale Route to Glecaprevir: Synthesis of the Side Chain and Final Assembly, *Org, Process Res. Dev.* **24**, 1393–1404.
191 AbbVie, (2015) US 8 937 150, USA-prior. 2011.
192 AbbVie, (2017) US 9 586 978, USA-prior. 2011.
193 Di Santo, R. (2014) Inhibiting the HIV Integration Process: Past, Present, and the Future, *J. Med. Chem.* **57**, 539–566.

194 Palani, A. and Tagat, J.R. (2006) Discovery and Development of Small-Molecule Chemokine Coreceptor CCR5 Antagonists, *J. Med. Chem.* **49**, 2851–2857.
195 LaBonte, J., et al. (2003) *Nat. Rev. Drug Discov.* **2**, 345.
196 Matthews, T., et al. (2004) *Nat. Rev. Drug Discov.* **3**, 215.
197 Duke Univ, (1995) US 5 464 933, USA-prior. 1993.
198 Trimeris, (2000) US 6 015 881, USA-prior. 1998.
199 Duke Univ, (2000) US 6 133 418, USA-prior. 1995.
200 Trimeris, (2002) US 6 475 491, USA-prior. 1995.
201 Summa, V., et al. (2008) *J. Med. Chem.* **51**, 5843.
202 Humphrey, G.R., et al. (2011) *Org. Process Res. Dev.* **15**, 73.
203 IRBM, (2007) US 7 169 780, USA-prior. 2001.
204 Merck & Co, (2010) US 7 754 731, USA-prior. 2004.
205 Pfizer, (2003) US 6 667 314, GB-prior. 2000.
206 Pfizer, (2009) US 7 576 097, GB-prior. 2000.
207 Satoh, M., et al. (2006) *J. Med. Chem.* **49**, 1506.
208 Japan Tobacco, (2007) US 7 176 220, JP-prior. 2002.
209 Japan Tobacco, (2009) US 7 531 554, JP-prior. 2004.
210 Johns, B.A., et al. (2013) *J. Med. Chem.* **56**, 5901.
211 Shionogi & GSK, (2012) US 8 129 385, JP-prior. 2005.
212 Shionogi & ViiV, (2014) US 8 778 943, JP-prior. 2005.
213 Trofimov, F.A. et al., inventors, (1993) (US 5 198 552), PCT/SU-prior. 1989.
214 Shenyang Pharmaceut. University, (2011) US "7 960 427", CN-prior. 2004.
215 Shenyang Pharmaceut. University, (2006) EP appl. 1 731 506 A1, CN-prior. 2004.
216 Lischka, P., et al. (2010) *Antimicrob. Agents Chemother.* **54**, 1290.
217 Humphrey, G.R., et al. (2016) *Org. Process Res. Dev.* **20**, 1097–1103.
218 Bayer, (2007) US 7 196 086, DE-prior. 2003.
219 Aicuris, (2011) US 8 084 604, DE-prior. 2005.
220 Aicuris, (2013) US 8 372 972, DE-prior. 2005.
221 Siga Technologies, (2010) US 7 737 168, USA-prior. 2003.
222 Siga Technologies, (2011) US 8 039 504, USA-prior. 2003.
223 Siga Technologies, (2013) US 8 530 509, USA-prior. 2003.
224 Siga Technologies, (2016) US 9 339 466, USA-prior. 2010.
225 Hughes, D.L. (2019) *Org. Process Res. Dev.* **23**, 1298–1307.
226 Shionogi, (2015) US 8 927 710, JP-prior. 2009.
227 Shionogi, (2015) US 8 987 441, JP-prior. 2010
228 Shionogi, (2017) US 9 815 835, JP-prior. 2009.
229 Shionogi, (2019) US 10 392 406, JP-prior. 2015.
230 Shionogi, (2020) US 10 759 814, JP-prior. 2009.
231 Toyama Chemical, (2004) US 6 787 544, JP-prior. 1998, 1999.
232 Toyama Chemical, (2004) US 6 800 629, JP-prior. 2000.
233 Toyama Chemical, (2013), US 8 513 261, JP-prior. 2009.
234 Fujifilm/Toyama Chemical, (2015) US 9 181 203, Jp-prior. 2010, 2011.
235 MYR GmbH, (2018) EP appl. 3 392 267, EP-prior. 2017.
236 MYR GmbH, (2021) US 10 925 925, EP-prior. 2017.
237 MYR GmbH, (2018) US 2018/0228804 A1, EP-prior. 2014.
238 MYR GmbH, (2018) US 2018/0296634 A1, EP-prior. 2017.
239 MYR GmbH, (2016) WO 2016/055534 A2, EP-prior. 2014.
240 Meanwell, N.A., et al. (2018) *J. Med. Chem.* **61**, 62–80.
241 Chen, K., et al. (2014) *J. Org. Chem.* **79**, 8757–8767.
242 BMS-Team (2017) Comprehensive Description of Synthesis of BMS-663068 in 9 Parts, *Org. Process Res. Dev.* **21**, 1091–1185.
243 BMS, (2010) US 7 745 625, USA-prior. 2004.
244 BMS, (2012) US 8 168 615, USA-prior. 2004.
245 BMS, (2013) US 8 461 333, USA-prior. 2010.